TIME SERIES ANALYSIS AND APPLICATIONS

Enders A. Robinson
Visiting Professor of Geophysics
Cornell University

Goose Pond Press
11600 Southwest Freeway, Suite 179
Houston, Texas 77031

Time Series Analysis and Applications
First printing 1981

Library of Congress Catalog Card Number 81-81825
Robinson, Enders Anthony
    Time Series Analysis and Applications.
Houston, Goose Pond Press
628 p.
8105   810420

CONTENTS

PREFACE

Many aspects of science have been influenced by the applications of time series analysis. This book is concerned with the properties of the underlying physical structure of the process which generates the time series data and with methods of fitting mathematical models to this structure. Thus the emphasis is on the structural nature of the process and how the statistical variables interact with this structure. The understanding of the interrelationship between the structural and statistical features brings forth digital data processing methods that can be used to analyze the time series data. The results are estimates of the structural and statistical parameters from which decisions about control and communication can be made. Throughout the book the reader will experience the interplay between structural concepts such as minimum-delay and all-pass and statistical concepts such as white noise. The main applications of this structural and statistical approach to time series analysis are the data processing methods known as deconvolution and the spectral estimation methods derived from deconvolution. Reference is made to pages 443–456 of this volume for the deconvolution of human speech (also called linear predictive coding or LPC) and to pages 457–493 for the deconvolution of geophysical signals. Spectral estimation methods based on deconvolution are the autoregressive (AR) method given on page 341 and various autoregressive-moving-average (ARMA) methods given on pages 513–559.

For the past twenty-five years and more I have sent thousands of reprints of my papers to people who have requested them all over the world. Often a request is made not only for a single paper, but also for all other papers dealing with the same subject. As the supply of the various papers became exhausted, I would make xerox copies and send them. The purpose of the present book is to bring together the most requested time series reprints in one volume in order to ease the burden of meeting individual requests. It is hoped that the collection given here will prove helpful to the many kind people I know only through correspondence.

Enders A. Robinson

v

CHAPTER 1

MODEL BUILDING FOR THE HUMAN SCIENCES
by
Enders A. Robinson

INTRODUCTION

Any body of theory deals with models. *A model is a simplified and idealized abstraction whose purpose is to approximate the behavior of a system.* A model of necessity must always be a compromise between simplicity and reality. Any theory can be broken down into three phases :

(1) Development of suitable models for the underlying processes that determine the externally observed behavior.

(2) Development of methods of analysis by use of the models.

(3) Accumulation of a store of ideas from working with the models in order to provide specific criteria and to obtain general intuition.

Adaptive control systems In the human sciences it is now generally recognized that meaningful models must be *adaptive control systems* : that is, systems in which automatic and continual measurement of the process is used as a basis for the automatic and continuing self-design of the system. The above block diagram

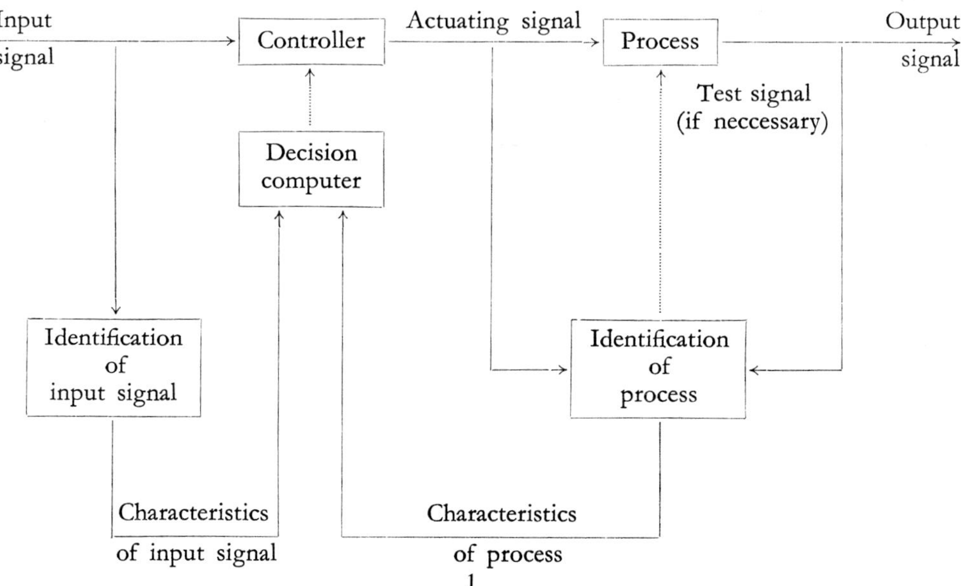

illustrates the essential features of an adaptive control system; practical systems will generally be much much more complex, involve many inputs and outputs, and incorporate simultaneously various combinations of features indicated here.

(1) *Identification of input signal.* The properties of the input signals are estimated in order to provide a basis for the selection of performance or optimization criteria. The usual realization of adaptativity requires the inclusion of nonlinear elements within the system, and so even relatively simple adaptive systems behave nonlinearly with respect to the input signal. Thus for optimum performance a frequent or continuing evaluation of the properties of the input signals is required in order to keep the design criteria properly adjusted.

(2) *Identification of the process.* The identification problem refers to the measurement of the dynamic characteristics of the process to be controlled. By its very nature, adaptivity requires rapid, frequent, and automatic identification of the process. The most general adaptive systems would require a complete process identification in terms of a mathematical model; in practical situations this goal is usually not attained because of limitations on the measuring equipment or constraints on the time permitted for measurement. Because of such limitations, the selection of which quantities are to be measured becomes an important issue.

In addition, identification may require the insertion of a test signal on a regular basis or intermittently.

(3) *The decision computer.* This element utilizes the results of both input signal identification and process identification to determine the required characteristics for the controller. The criteria for system evaluation are stored in the decision computer in order to provide a quantitive basis upon which to make the decisions.

(4) *The controller.* This device acts upon the input signal so as to yield the required actuating signal. The characteristics of the controller are determined by the decision computer.

(5) *The process.* The actuating signal drives the process which yields the output signal of the system.

An example of an adaptive control system is a human being driving a car, which is illustrated in the following block diagram :

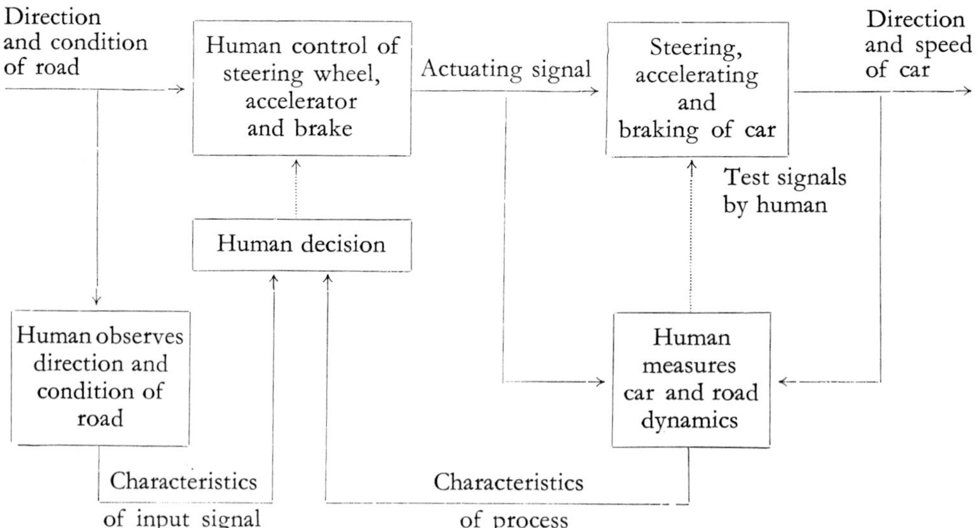

The process identification is made both by (1) the driver continually injecting small variational " test signals " on the steering, acceleration, and braking of the car in order to maintain " the feel of the road and car " and (2) constant measurement of the effect of the control he exerts on the direction and speed of the car under actual driving conditions, for example, when he sees a curve in the road or an icy patch.

Learning in adaptive systems.   The *decision computer* is the most important element in an adaptive situation.   It is for this element that the essential difference between a human and an physical device (such as an electronic computer) becomes apparent.   It has always been hoped that adaptive control systems could be built that would be capable of replacing the human being by an electronic device, and thereby free the human from fatigue, strain, and boredom, and at the same time free the system from various human limitations such as limited response speed and reliability.

Nevertheless, by and large, physical devices fail to implement the learning characteristic of the human being.   For example, let us consider the case of person who has just moved to a northern locality, and must drive on icy and snowy roads.   At first he will try various speeds and measure the effect of the road conditions on the car.   If a slight skid occurs on a slippery patch, he will note the various road conditions and car dynamics that caused this effect.   In time, his knowledge will accumulate, and this experience becomes part of his " decision computer " thereby making him a competent driver under bad road conditions. The rate of this learning process depends on the amount of his experience as well as upon his learning mechanism.

The possibility of introducing learning mechanisms within non-human adaptive systems is a major area for research.   When physical systems can be built that combine learning with adaptivity, then they can be used to take over some of the tasks that require the capabilities and flexibilities of human operators.

Automatic pattern recognition.   Insight into machine learning can be obtained by the study of automatic pattern recognition.   *Automatic pattern recognition is the identification and assignment to classes of patterns by machines.*   Such patterns may be visual, oral, or electromagnetic.   An essential problem is the organization of the use of typical patterns to determine the decision procedure to be employed by the machine.

The theory of pattern recognition involves the use of statistical methods.   Besides exact matching techniques, the most simple method is the based on correlation of the pattern to be identified with various stored reference patterns.   More advanced statistical techniques involve treating a pattern as a vector in multidimensional space.   Thus the pattern recognition problem breaks down into the two component problems :

(1) Selection of the measurement space, the so-called " receptor " problem
(2) Finding a way to partition the space, the so-called " categorizer " problem.

The general selection of the measurement space, that, is, of the receptors, is still an unsolved problem in pattern recognition theory.   One approach would be to find statistics that describe the measure space according to certain desirable properties.   The problem is analagous to statistical problems in the design of experiments.   Measurements for character recognition should be invariant generally with respect to such things as scale change, translations, and rotations and subject to comparison by statistical experiments.

The determination of methods of partitioning the measurement space, that is, of finding categorizers, in a problem that has been more amenable to attack.   Much research has been done in finding optimum categorizers in the case of Gaussian measurements.   As would be expected, the Gaussian hypothesis requires only linear and quadratic computations, and in special cases the quadratic terms can be left out, so that a lincar model os obtained.   The method of partitioning the measure space by the use of hyperplanes has been subjected to considerable research, and in particular discriminant methods, correlation methods, and matched filter methods have been investigated.   Although such linear methods have received the most attention, nonlinear methods such as hyperspheres, polynomial discriminants, and quadratic methods have been tried.

Stochastic learning models.   The stochastic learning models that have been the most developed are essentially descriptive processes in the sense that a certain learning mechanism is assumed instead of decision processes in the sense that the mechanism is derived on the basis of optimization.

For example, let us formulate a Markov model, that, is, one in which only the current state of the system, and not the past history of the system, influences the future course of the system.   At any stage of the process, we suppose that a number of alternative actions are possible.   At the initial stage, the probability that the process chooses any particular alternative is given by some probability distribution.   We now want to specify how the process gradually adapts its behavior to the information that it receives in an environment that possesses features unknown to the process.   A choice of some particular alternative by the process would lead to a particular consequence dependent upon the alternative selected, and moreover this consequence would result in a change of the preference field of the process for different alternatives.   This change in preferences means that the original probability distribution is transformed into some new distribution.   Now the process continues in this fashion, and what is of the most interest is the ultimate behavior of the system.   There are many interesting mathematical as well as practical problems in setting up such models.   In order to compare the results of experiments with animals and with humans with the results obtained mathematically, the nature of the various transformations in the model must be determined from experimental data.   Only in this way can it be decided whether or not such stochastic learning models actually describe real-life learning in animals and humans.

Here the emphasis has been on a descriptive theory, specifically designed so as to help understand how animals and people act in uncertain situations and learn new things.   Beyond this descriptive theory would be a learning theory devoted to the control aspects.

Machine learning.   Much attention in the past few years has been directed at the problem of *machine learning*.   One important aspect is the use of machines in simulating learning processes for pattern recognition.   In fact, machine learning is the single most important aspect of pattern recognition.   There are two types of learning, namely

(1) learning with a teacher, or supervised learning, and
(2) learning without a teacher, or unsupervised learning.

Supervised learning means that the outside source gives the learning machine a sequence of patterns together with labels specifying the category to which each pattern should be assigned. Unsupervised learning means that each pattern is presented without such a label, so the machine never knows for certain to which category each pattern belongs. Both in supervised and unsupervised learning it is required that the machines continually improve their performance. Needless to say, most research to date has been in the area of supervised learning. In the area of unsupervised learning one approach would be to set up certain hypotheses and restrictions about the problem; another approach would be to assume that the unsupervised learning period is preceded by a period of supervised learning during which some characteristics of the categories are established. In any case, unsupervised learning progresses at a slower rate than supervised learning. In machine learning problems it must be recognized that it is a long way before there will be any complete theory, so that practical judgements are at a premium in the solution of problems today.

Feedback *A feedback loop representents a closed chain of dependency.* The type of relationships that is the most useful in thinking about feedback systems is the *cause-and-effect pattern : a thing regarded as a cause when applied to some situation produces an effect.* We have the choice of regarding either the cause acting upon the situation as producing the effect or the situation acting upon the cause as producing the effect. For a sequence, or cascade, of cause-and-effect relationships that are non-interacting, the effect of one becomes the cause of the next, and so we have the concept of a progression of cause-and-effect actions propagating through the cascade : the so-called chain principle. Thus a signal can be visualized as flowing through the cascade. The concept of signal flow is extremely versatile in describing the interdependencies among the signals of a complex system.

The concept of feedback plays an important role in many types of systems. The recognition of feedback in a situation involves the recognition of a pattern of relationships; any situation in which it is perceived that a closed sequence of cause-and-effect relationships exists may be said to be a feedback situation. To perceive feedback the principal emphasis should be placed on the interrelations of signals; the interconnections of parts should be subordinated. That is, the layout of the actual physical or organic parts should be secondary to the cause-and-effect relationships. It is not surprising then, that once a certain feedback pattern is known, a perceptive person can find the same pattern recurring in other situations and experiences.

The relationships existing in any situation or experience may be represented by different models, some containing feedback loops and some not. Thus it might be argued that the presence or absence of feedback in any given situation is more a matter of viewpoint than reality. Nevertheless the different models for a given situation portray different information : a diagram of the actual physical layout represents a model that is useful in describing the situation, whereas a diagram representing the flows of signals is a more abstract model that gives an interpretation of the physical or organic behavior of the situation. Thus at first sight, say upon study of the model of the physical layout, a system may not suggest feedback, whereas further analysis would lead to a signal flow diagram with feedback loops. In other words the signal flow diagram is a way of portraying the cause-and-effect relationships in such a way that brings to the surface the causal dependencies. In this way feedback can be visualized in what would ordinarily be thought of as non-feedback structures and *the emergence of the actual feedback pattern depends upon the particular viewpoint adopted in regard to the interpretation of causal dependencies.*

Let us now contrast the simplest possible feedback configuration with an equally simple cascade, or open-loop, configuration. The signal flow in a simple feedback system is depicted by the block diagram :

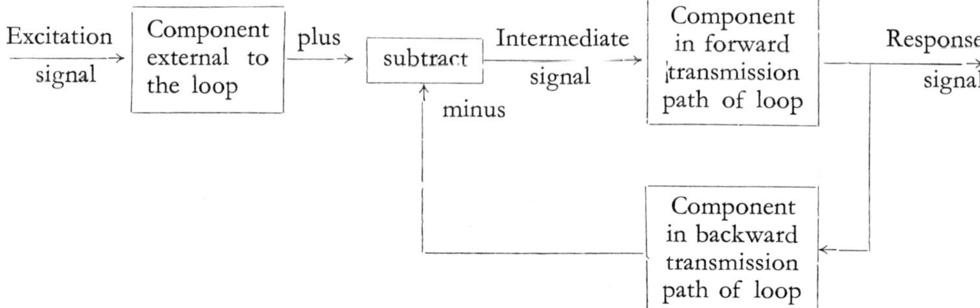

In this feedback system, we see that the intermediate signal is dependent upon the response signal as well as the excitation signal. The signal flow in a simple cascade system is depicted by the block diagram :

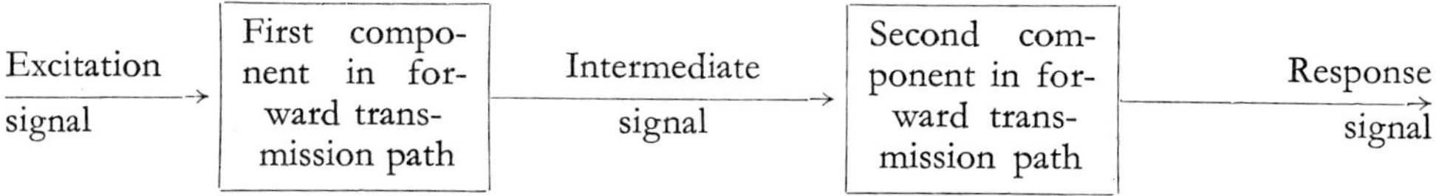

In this cascade system, we see that the intermediate signal depends only on the excitation signal. This difference in dependency has led to the development of a body of theory whose usefulness depends upon the proper exploitation of the feedback pattern.

In order to compare the cascade system and the feedback system we can use a criterion that relates the percentage variation of the system output/input transfer function to the percentage variation of one of the parameters of the system. The ratio of these two percentage variations may be called the *sensitivity*. The usual objective is to have a system that is insensitive to parameter variation; in this way certain desirable signal transmission characteristics can be realized even if there is large variations in a component, and so system performance is less dependent on variations of device properties due to manufacturing tolerances, use of less reliable or cheaper materials, and aging.

When the sensitivity criterion is applied to the cascade system, it is evident that the sensitivity to either component is unity; that is, there is a one-to-one correspondence between percentage variation in system performance and percentage variation in each component.

For the feedback system we have to apply the sensitivity criterion to each of the three components : (1) the component external to the loop, (2) the component in the forward transmission path of the loop, and (3) the component in the backward transmission path of the loop. As we would anticipate, the sensitivity to the component external to the loop is unity, so it is desirable that this component is free of variation, for the feedback cannot reduce the sensitivity to its variation. The sensitivity to the component in the forward transmission path of the loop is given by

$$\begin{pmatrix} \text{Sensitivity to} \\ \text{forward component} \end{pmatrix} = \frac{1}{1 + (\text{forward component parameter})(\text{backward component parameter})}$$

From this equation we see that the sensitivity to the forward component can be made arbitrarily small, provided that the product

(forward component parameter) (backward component parameter)

is made indefinitely large. Therefore *in order to reduce the sensitivity with respect to the variation of some given process or organism, the parameter representing this process or organism should be located within the forward transmission path of the loop.* The sensitivity to the component in the backward transmission path of the loop is

$$\left(\begin{array}{c}\text{Sensitivity to}\\\text{backward component}\end{array}\right) = \frac{\text{(forward component parameter) (backward component parameter)}}{1 + \text{(forward component parameter) (backward component parameter)}}$$

From this equation we see that there is no possibility of reducing the sensitivity of the feedback system to the variations in the backward component. *Thus the component in the return path of the loop, like the component external to the loop, should desirably be free of variation,* because the feedback does not reduce the sensitivity of the system to their variation. In the event that it is necessary ot reduce the sensitivity of a component either in the return path or external to the loop, then we are forced to use additional feedback loops, thereby obtaining a multiple-loop configuration.

EXAMPLES OF SCIENTIFIC MODELS

Probability models. As we have seen, one of the purposes of scientific investigation is the construction of theoretical *models* of real-life (or real-world) situations. A model is conceptual in that it represents a mental image of the underlying situation. All scientific models must be ultimately derived from observations. Every model is intrinsically " open-ended "; that is, it is not perfect but is subject to future improvement.

To attain good models both the scientist and the mathematician must work closely together. The science in which mathematical models have been the most fruitful is physics. A well-known example of a mathematical model is *Newton's second law*. It states that force F, mass m, and acceleration a are related by the formula

$$F = ma.$$

This formula represents the mathematical model of the law of nature known as Newton's second law.

Any real-life (or real-world) situation represents a relative combination of circumstances. A large part of mathematics is concerned with the building of mathematical models of sure situations. A *sure situation* is one in which there is no degree of uncertainty; it has all the precision and predictability of a mathematical equation.

An example of a sure situation is represented by Newton's second law, just dissussed above.　Whenever any 2 of the 3 quantities : force F, mass m, or acceleration a, are given the third quantity may be found by solving the mathematical equation

$$F = ma.$$

A sure situation is identified by
(1) its circumstances
(2) its result.
Its circumstances are the antecedent conditions.　Whenever the circumstances are fulfilled, the result occurs.　For example, from Newton's second law we know that whenever a force is applied to a mass (the circumstances), the mass accelerates (the result).

The branch of science known as *probability theory* is concerned with the building of *mathematical models of unsure situations.*　These models are called *probability models.*

An unsure situation is one in which there is a degree of uncertainty; it does not have the precision and predictability of a mathematical equation.　An unsure situation is identified by
(1) its circumstances
(2) its possible results.
Its circumstances are the antecedent conditions.　We recall that a sure situation has only one result.　In contrast, an unsure situation has many (that is, more than one) possible results.　Whenever the circumstances of an unsure situation are fulfilled, one and only one of the possible results *happens.*

Unsure situations are very commom in human experience.　Here are a few simple examples.

Example	*Circumstances*	*Possible result*
1.	A coin is tossed	Either heads or tails
2.	A die is tossed	Any one of 1, 2, 3, 4, 5, 6.
3.	A card is drawn from the top of a pack of 52 cards	A card of any suite and any value.
4.	A rain drop falls on a piece of paper	Any one of the points on the paper.
5.	A seed falls from a tree and lands in a meadow.	Any one of the points in the meadow.
6.	An arrow is short at at target.	Any one of the points the on target.

For the unsure situation in question, we must agree on what is meant by the possible results.　Thus in Example 1 above, the coin could stand on its edge and thereby not fall heads or tails.　Nevertheless it is expedient in most applications to regrad heads and tails as the only possible results.　Such idealizations are standard practice in scietific model building, and indeed must be made in order to make progress with most problems.

Another idealization is illustrated by Example 4.　A rain drop certainly is not a point, but it is expedient to consider it as such in this problem.　In Example 6, we have idealized the problem so that the target is large enough or the marksman's aim is good enough so that the arrow always hits the target.

In reference to Example 3, an unstated assumption is that the pack of cards has been well shuffled.　On the other hand, if the pack of cards is ordered so that the ace of spades is on top, and if this fact is known, then the drawing of a card from the top of the pack would not be an unsure situation, but would be a sure situation.　Such unstated assumptions as the proper shuffling of cards are usually made in problems connected with probability theory.

In subjects that are more or less directly concerned with specific types of objects, a clear indication of the subject matter presents no serious difficulty. A reference to some of the objetcs can convey a certain amount of preliminary information and thereby will give a somewhat accurate idea of what is to follow. Such is the case in most natural sciences.

In contrast the state of affairs in probability theory is not so easy. As we have seen probability theory is not directly concerned with definite objects, but with a certain underlying characteristic common to some real-life (or real-world) situations. Situations with this characteristic are found in all branches of science. Because often such situations have little else in common, a person unaquainted with probability theory must rearrange in his mind the usual scientific classification of these situations, so as to be able to look at those possessing this unifying characteristic. This unifying characteristic is uncertainty.

One point should be emphasized : the role played by chance in an unsure situation is not required to be complete but may be to any degree.

For example, when a coin or a die is tossed, we usually regard the role played by chance as being entire, and the role played by the tosser as having no effect on the ultimate result. Nevertheless there may be individuals with extraordinary skill who can control some of the conditions in tossing the coin, so that they have some influence on the way it falls.

On the other hand, consider the act of shooting an arrow at a target. Because the man takes aim, the ultimate result to some extent depends upon his skill. Nevertheless this act is an unsure situation because the exact point where the arrow strikes is uncertain. The role played by the man has some effect on the ultimate result, and the role played by chance has some effect. Thus chance may be regarded only as a co-agent in the final result.

In fact most uncertain situations are of the type in which chance may be regarded only as a co-agent in the ultimate result.

With regard to mathematical science versus empirical science. the state of affairs in probability theory is analogous to that in other scientific disciplines, for example geometry. As geometry is employed in everyday life, the terms *point, straight line, plane, angle*, etc. designate certain physical configurations. Consequently the propositions of geometry, such as the sum of the angles of a triangle is equal to 180°, formulate relations between physical configurations that can be measured.

Contrasted to the geometry used in everyday life, there is the geometry as put forth by Euclid. This geometry which consists of axioms and theorems is the geometry studied in high school mathematics. In Euclidean geometry, the terms *point, straight line, plane, angle*, etc. appear as undefined things, and the geometry is concerned solely with relationships among them. The axioms state certain rules which these undefined things must obey, such as two points determine a straight line. From these axioms various theorems are derived, such as the sum of the angles of a triangle is equal to two right angles. But the derivations of geometric theorems from the axioms does not depend upon the likeness that might be established between on the one hand the undefined terms like *point* and *straight line* which appear in the axioms and on the other hand the terms like *point* and *straight line* which appear in everyday life as material physical configurations.

Thus Euclidean geometry is not a branch of physics but instead is a part of conceptual mathematics. The undefined terms in Euclidean geometry such as *point* and *straight line* are not interpreted as physical entities within its framework. Nevertheless it is possible to introduce so-called *coordinating definitions* which make each undefined term, such as *straight line* in Euclidean geometry, correspond to a term employed to designate empirical subject matter, such as *straight line* in everyday geometry.

This distinction between Euclidean geometry and the geometry of everyday life, and the possibility of coordinating the two, is characteristic of scientific method.

Thus on one hand Euclidean geometry is a *mathematical science*. It has certain axioms, and the theorems are derived from these axioms by logical deductions. The axioms are nothing more than a set of rules about undefined things, such as *point* and *straight line*. It is meaningless to talk about the definition or true nature of such undefined things, as the mathematical structure does not care about what a *point* and a *straight line* really are.

On the other hand the geometry of everyday life is an *empirical science*. It has certain findings that have been arrived at as the result of physical observations and measurement. The terms that are employed, such as *point* and *straight line*, represent material physical configurations.

There are advantages to distinguishing between mathematical structure and empirical science. One advantage is that we separate questions about mathematical validity from questions of empirical fact. Another advantage is that we increase the *applicability* of mathematics.

Thus on one hand, the same formal mathematical structure may be applied to one empirical science by means of one set of coordinating definitions, and to another empirical science by means of another set of coordinating definitions. For example, the mathematical structure of the algebra of sets (or Boolean algebra) may be coordinated with many empirical sciences.

On the other hand, different formal mathematical structures may be applied to the same empirical science. For example, Euclidean geometry as well as several different non-Euclidean geometries may be applied to astronomy. One benefit is that it often turns out that one mathematical structure is a more effective means than another for organizing the materials of an empirical science.

In fact there are always many benefits in the applications of mathematics to empirical sciences. There is a constant interplay between mathematical theory and empirical applications. Empirical results open up new areas for mathematical research, whereas mathematical advances open new fields for empirical investigation. Thus there is a synthesis of mathematical and practical thought. The mathematician must learn about the physical essence of a scientific problem and then find some suitable mathematical model in which to express it. The scientist must learn about the mathematical model and then look for new empirical evidence which will lead to further improvements. The end result is that both mathematics and scientific practice gain.

Probability theory, like other branches of mathematics, has evolved out of the needs of practical application.

Today the practical applications range over a tremendous number of areas in all branches of science. The tie that exists between probability theory and practical requirements has been the basic cause of the vigorous development of probability theory in recent years.

Because probability theory is applied in so many diverse fields of science, the theory must be general enough so as to provide appropriate tools for the great variety of needs. Therefore modern probability theory is not connected to any particular fields of interest, but instead represents a mathematical structure in the same way as for example Euclidean geometry does.

Thus the terms used in modern probability theory, like the terms *point*, *straight line*, etc. used in Euclidean geometry are undefined.

The central term is of course the word *probability*. For its mathematical usage, we must divest it of all its many connotations in the English language, and treat it instead as an undefined primitive concept.

It is unfortunate that language has no one single word for *probability in its mathematical sense*, but only the popular terms like *chance*, *probability* and *likelihood* with their manifold conscious and unconscious connotations. However, the term *mathematical probability* or the term *probability measure* can be used for probability in its mathematical sense. Nevertheless these two terms are long, so we will consistently use the word *probability* by itself for probability in its mathematical sense.

Because *probability* is an undefined primitive concept, modern probability theory does not attempt to explain the so-called true meaning of probability even as Euclidean geometry does not discuss the true meaning of a straight line. Instead modern probability theory starts with a set of axoims which specify the relationship among the undefined primitive concepts. From these axioms various theorems are proved, and these theorems constitute the mathematical structure of probability theory.

This mathematical structure is applied in the form of models to practical situations. These abstract mathematical models serve as tools. Of course, the same model can be used to describe different empirical situations, and different models can be used to describe the same empirical situation. The interplay between mathematical theory and empirical practice is always present. The manner in which mathematical theories are applied to practical situations depends upon experience. As our experience increases old methods of application are extended and new methods are found. The result is that both theory and practice are the benefactors.

Causal chain model. Let $x_1(t)$, $x_2(t)$, ..., $x_N(t)$ (t an integer) be a multiple stationary stochastic process in discrete time, and let $x(t)$ be the column vector of the $x_j(t)$, $j = 1, 2, ..., N$. Denote the closed linear manifold spanned by $x_1(s)$, ..., $x_N(s)$ for $s \leqslant t$ by $X(t)$.

The innovation $w_j(t)$ is defined by the equation

$$x_j(t) = x_j(t)\, P_{X(t-1)} + w_j(t)$$

where $P_{X(t-1)}$ denotes the projection operator on the manifold $X(t-1)$. The rank of the matrix

$$[E\,\{\, w_j(t)\, \overline{w_k(t)}\, \}], \qquad j,\, k = 1, 2, ..., N$$

is equal to the maximum number of linearly independent variates among $w_1(t)$, ..., $w_N(t)$. We shall assume that the rank is equal to N. We may orthogonalize $w_1(t)$, ..., $w_N(t)$ to obtain a set of orthogonal variates $\theta_1(t)$, ..., $\theta_N(t)$ that span the same manifold as $w_1(t)$, ..., $w_N(t)$.

For the $x(t)$ process there is a uniquely determined decomposition (the Woldian decomposition)

$$x(t) = u(t) + v(t)$$

where $u(t)$ and $v(t)$ are orthogonal, and $u(t)$ is purely non-deterministic and $v(t)$ is deterministic. The $u(t)$ process can be expressed as a linear combination of those innovations that have occurred at or prior to time t :

$$u(t) = \sum_{s=0}^{\infty} c(s)\, w(t-s)$$

where

$$u(t) = [u_j(t)], \qquad j = 1, 2, ..., N \text{ (column vector)},$$
$$c(s) = [c_{jk}(s)], \qquad j, k = 1, 2, ..., N \text{ (matrix)},$$
$$w(t) = [w_k(t)], \qquad k = 1, 2, ..., N \text{ (column vector)}$$

such that

$$u_j(t) = \sum_{s=0}^{\infty} \sum_{k=1}^{N} c_{jk}(s) w_k(t-s)$$

has finite second absolute moment. That is, letting

$$\alpha_j(s) = \left\| \sum_{k=1}^{N} c_{jk}(s) w_k(t-s) \right\|$$

we have

$$\sum_{s=0}^{\infty} [\alpha_j(s)]^2 < \infty$$

for $j = 1, 2, ..., N$. The $c_{ij}(s)$ and $\alpha_j(s)$ are uniquely determined. We note that the Fourier expansion of $u_j(t)$ has the form

$$u_j(t) = \sum_{s=0}^{\infty} \sum_{k=1}^{N} b_{jk}(s) \theta_k(t-s)$$

with

$$\sum_{s=0}^{\infty} \sum_{k=1}^{N} \left| b_{jk}(s) \right|^2 \sigma^2(\theta_k) < \infty \quad \text{where} \quad \sigma^2(\theta_k) = \text{var } \theta_k(t).$$

The orthogonal variates $\theta_1(t), ..., \theta_N(t)$ are certain (uniquely determined) linear combinations of the innovation variates $w_1(t), ..., w_N(t)$. Hence in the above development of $u_j(t)$, the $\theta_k(t)$ may be replaced by their expressions in terms of the $w_k(t)$, which gives the development of $u_j(t)$ in the Woldian decomposition above. We see that

$$[\alpha_j(s)]^2 = \sum_{k=1}^{N} \left| b_{jk}(s) \right|^2 \sigma^2(\theta_k)$$

We shall now choose the following method of orthogonalizing the innovations $w_k(t)$ to yield the $\theta_k(t)$ so as to form a *causal chain*. We shall assume that the deterministic component is absent; a similiar development can be carried through for the general case. Let

$$\theta_1(t) = w_1(t)$$
$$\theta_2(t) = w_2(t) - w_2(t) P_{\theta_1(t)}$$
$$\theta_3(t) = w_3(t) - w_3(t) P_{\theta_1(t)} - w_3(t) P_{\theta_2(t)}$$
$$\theta_N(t) = w_N(t) - w_N(t) P_{\theta_1(t)} - ... - w_N(t) P_{\theta_{N-1}(t)}$$

Then we have the (moving-average) causal chain representation

$$x(t) = \sum_{s=0}^{\infty} b(s) \theta(t-s)$$

where

$$b(s) = [b_{jk}(s)], \qquad j, k = 1, 2, ..., N$$

with $b_{jk}(0) = 0$ when $k > j$. This representation may be inverted to yield the (autoregressive) causal chain representation

$$\theta(t) = \sum_{r=0}^{\infty} a(r)x(t-r)$$

where

$$a(r) = [a_{jk}(r)], \qquad j, k = 1, 2, \ldots, N$$

with $a_{jk}(0) = 0$ when $k > j$. Moreover the $b_{jk}(r)$ can be uniquely determined from the $a_{jk}(s)$ by recursive deductions. For

$$x(t) = \sum_{s=0}^{\infty} b(s)\theta(t-s) = \sum_{s=0}^{\infty} b(s) \sum_{r=0}^{\infty} a(r)x(t-s-r)$$

$$= \sum_{n=0}^{\infty} \left[\sum_{s=0}^{n} b(s)a(n-s) \right] x(t-n)$$

Hence for $n = 0$ we have

$$\sum_{s=0}^{n} b(s)\,a(n-s) = I$$

(where I denotes the identity matrix)
which is

$$b(0)\,a(0) = I$$

which yields the equations

$$\begin{aligned}
&b_{11}(0)\,a_{11}(0) = I && \text{(gives } b_{11}(0)) \\
&b_{22}(0)\,a_{22}(0) = I && \text{(gives } b_{22}(0)) \\
&b_{21}(0)\,a_{11}(0) + b_{22}(0)\,a_{21}(0) = 0 && \text{(gives } b_{21}(0)) \text{ etc.}
\end{aligned}$$

For $n = 1, 2, 3, \ldots$, we have

$$\sum_{s=0}^{n} b(s)\,a(n-s) = 0$$

which is

$$b(n)\,a(0) = -\sum_{s=0}^{n-1} b(s)\,a(n-s)$$

which yields the equations

$$b_{1N}(n)a_{NN}(0) = -\sum_{s=0}^{n-1} \sum_{j=1}^{N} b_{1j}(s)a_{jN}(n-s) \qquad \text{(gives } b_{1N}(n))$$

$$b_{12}(n)a_{22}(0) + \ldots + b_{1N}(n)a_{N2}(0) = -\sum_{s=0}^{n-1} \sum_{j=1}^{N} b_{1j}(s)a_{j2}(n-s) \qquad \text{(gives } b_{12}(n))$$

$$b_{11}(n)a_{11}(0) + b_{12}(n)a_{21}(0) + \ldots + b_{1N}(n)a_{N1}(0) = -\sum_{s=0}^{n-1} \sum_{j=1}^{N} b_{1j}(s)a_{j1}(n-s) \qquad \text{(gives } b_{11}(n))$$

etc.

Similarly, given the $b_{jk}(r)$, the $a_{jk}(s)$ can be uniquely determined by recursive deductions.

Model for spectral analysis.   In principle, all problems of measurement and detection are problems in the domain of statistical communication theory.   Each requires the extraction of " signal " from a background of " noise ". The detection and identification of weak signal phenomena hidden in noise backgrounds have extensive applications in the electronic communications and data-handling sciences, but applications in other fields are constantly expanding.

Communications and data-processing systems are usually required to handle a large assortment of signals in the presence of noise. The type of noise present must be specified for each problem; noise can be the wrong signal, or the right signal out of place, and what may be noise to one observer may be signal to another observer. In any case, the design of systems depends to a large extent upon the statistical properties of both the signals and the noise.

All communication circuits or channels are imperfect to some extent. In telephone and radio, the desired signal is heard against a background of noise, which may be strong or weak, random or harmonic. In television, an ever-changing granular snow overlays the picture to varying degrees.

In many cases, the noises can be represented, at least approximately, as stationary stochastic time functions. The most relevant parameter of such a time function is its autocorrelation function, or alternatively its power spectrum. Since the autocorrelation function and the power spectrum are Fourier transform pairs, each contains the same information, the autocorrelation in the time domain, and the power spectrum in the frequency domain.

In some cases, the signals can be represented as pulse signals, whereas in other cases they can be represented as stationary time functions.

Thus the measurement of the autocorrelation function or the power spectrum of a time series is a problem of practical as well as theoretical interest.

In some applications it is possible to obtain the autocorrelation or power spectrum by theoretical methods. In most applications, however, the problem becomes one of measurement and computation. Exact determination in these cases, of course, is not theoretically possible, as it would require an infinitely-long recording of the process under consideration. Approximate determination of the autocorrelation or spectrum raises many practical questions : to what accuracy should the time function be measured, how long a stretch of the time function should be used, should digital or analogue equipment be utilized in the computations, and what reliance should be placed upon the final results?

The measurement of power spectra is not intrinsically difficult. It may be done by a good set of band-pass filters. The important thing to remember is that a power spectrum represents the density of power as a function of frequency. It is well-know that density can never be measured at any given point; it is always necessary to estimate the mass over a small interval or band, and then the estimate of the density is the mass estimate divided by the length (or area, or volume, as the case may be) of the band.

For example, suppose that we have a magnetic wire of varying density. To calculate the magnetic density at a given point on the wire, we would first measure the total intensity in a small band surrounding the point in question. Then our estimate of the density would be the total intensity in the band divided by the length of the band.

Similarly, for any stationary time function we cannot directly measure the spectral power density at any given frequency. Instead we must estimate the total power within a band of fequencies centered at the frequency in question. Then our estimate of the spectral density is the band power divided by the width of the band.

The steps necessary for spectral estimation are *filtering*, *rectifying*, and *integrating*, as depicted in the figure.

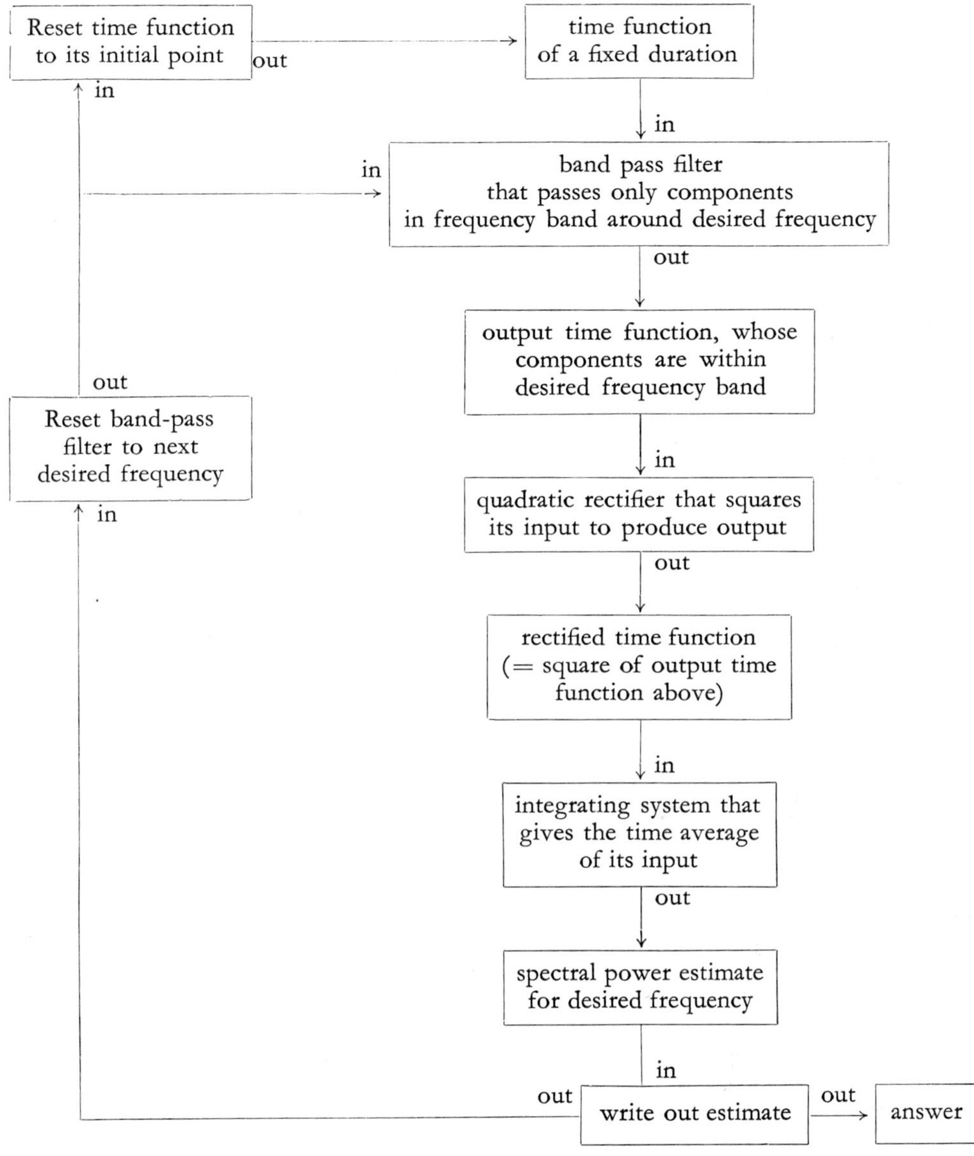

Decisions under uncertainty.   The relevant consequences of any decision always lie in the future : nothing done in the present can change the past.   Thus an essential element in the making of any decision is a forecast.   Because it is a rare occasion when events happen exactly as planned, forecasts are likely to be in error to a greater or less extent.   It is therefore expedient to make an analysis of a decision problem on the basis of several alternative estimates of the unknown future events. Sometimes such an analysis leads to a single definite decision; more often, it will turn out that one act would be preferable if some events happen, while some other acts would be better if other events happen. Since the decision-maker beforehand does not know what event will happen, he must balance one act off against another in order to reach an unequivocal decision. Therefore a decision-maker must look for reasonable criteria for choosing among acts when their consequences cannot be predicted with certainty. If he decides not to make an immediate choice, then he should have ways to obtain additional information to improve his forecasts of the consequences of the various acts open to him.

The increasing emphasis on decision-making has resulted in a considerable body of case studies. Nevertheless, the accumulated body of organized knowledge about decision-making processes is still small. It is difficult to organize and compare different decision processes, for at present there is a lack of basic concepts that can be used to identify common elements in disparate situations. Guidelines are needed to determine what information is relevant and to show how various pieces of data can be fitted together. The advent of the general purpose digital computer has given a tremendous impetus to the resolution of many of these difficulties. Many of the concepts that relate to the handing of information on computers can be used for the study of decision processes. Detailed models of decision-making processes can be represented in sequential form on the computer which then can be used to simulate the operation of the processes, thereby generating a hypothetical stream of behavior that can be compared to the behavior of the original processes.

Once a general model of a decision process has been tried and tested, it is necessary to translate the knowledge gained to a form that the operating people can understand and use. As an example the writer worked upon a general mathematical model of the storage of petroleum and petroleum products for a major international oil company that has affiliated oil companies operating in many of the countries of the world. The model is extremely complex involving not only refined mathematical and probabilistic concepts, but also detailed information about the corporate structure of the company, its affiliates, and competitive oil companies, the general political and economic environments of the countries in which operations are carried out, and much practical knowledge of the oil business. This entire model is set up on several large digital computers, and new data that accumulates day by day in continually fed into the machines. This model gives the management of the company an accurate and meaningful picture of the current status of its oil storage problems as well as of future expectations. Nevertheless the model is much too complex and intricate to be conveyed as it stands to the operating personnel of the affiliated companies. As a result it was found necessary to issue short reports from time to time whose purpose is to give to the operating personnel the pertinent information required to maintain the overall operations of the entire network of companies at the highest possible level. One such memorandum that we prepared is given, in the following section, as an illustration of how practical knowledge gained from an elaborate computer model can be effectively disseminated to the operational decision-makers.

Operational model : Seasonal storage of oil and oil products. The problems of estimating suitable amounts of seasonal storage and preparing an economic justification for such storage may be quite complex. With this in mind, this paper has been prepared as a rough guide to some of the factors entering into an economic evaluation of seasonal storage of oil and oil products. It is not intended as a simplified method of justifying seasonal storage but rather as an outline to stimulate thinking toward the optimum installation of seasonal tankage. In general, it has been found that a corporate approach to seasonal storage justification must be taken; in other words the advantages accruing to the various functional departments (refining, marketing, marine) must be grouped and the project presented as an overall package.

Seasonal tankage should be distinguished from regulatory and working tankage :

1. *Regulatory Tankage* is tankage proided to meet government inventory regulations. Laws concerning regulatory tankage vary from country to country. Although under certain circumstances this tankage may wholly or partially serve as working capacity, it generally does not contribute to company operations.

2. *Working Tankage* permits short term inventory fluctuations due to imbalances between oil receipts and shipments. Such variations, usually of a few days' or weeks' duration, may be assignable to such factors as fluctuations in cargo receipts and in delivery schedules. In addition, at a refinery, variations occur as a result of refinery operational problems such as planned or emergency downtimes, process and blending requirements, or quality accumulation.

3. *Seasonal storage* is a storage over and above working and regulatory requirements used to provide space for the accumulation of inventory during seasons of low consumption to meet demand during seasons of high consumption. Seasonal tankage may often serve two contraseasonal products, such as gasoline and asphalt with peak summer demand and heating oil and fuel oil with peak winter demand. The possibilities for contraseasonal use of seasonal tankage are greatest at refineries.

The following factors affect the need for seasonal storage :

1. *Weather*. The greatest factor affecting seasonal requirements is, of course, the seasonal weather pattern. In addition to the effect which seasonal variations in weather has on oil demand, it must be recognized that individual heating seasons show wide fluctuations from year to year. In the United States, for example, seasonal degree-day totals during the past fifty years have varied over a range of plus or minus 15 per cent of the average.

The extent to which seasonality affects petroleum consumption varies not only from country to country but from area to area within a country. The effect of weather on distillate and fuel oil sales should be carefully analyzed. To this end, maximim use should be made of statistical weather data for key cities within the marketing area. In weight-averaging the weather data, recognition should be taken of the number of burner installations within a given area. The effect that a variation of 5 per cent from the normal seasonal degree-day total will have on demand should always be borne in mind.

Weather affects not only oil demand but supply conditions as well. For example, seasonal tankage must be installed at marine terminals that cannot be supplied by water shipments during the winter because of freezing of harbors.

2. *Minimizing seasonal swings in requirements*. Although there is little possibility of developing sufficient summer outlet to eliminate seasonal swings in distillate demand, without discounts of such magnitude as to be uneconomical, every effort should be made to develop summer outlet for products with high winter demand. Customers (consumers and resellers) should be encouraged to make full use of their own tankage by filling such tankage during the summer months.

3. *Location of seasonal storage.* In general, the closer seasonal storage is to the market the greater will be the transportation savings resulting from such storage. Thus locating seasonal tankage at secondary terminals brings the supply close to points of consumption and thereby tends to even out secondary type transportation.

On the other hand, the closer seasonal storage is to a primary supply source, the greater will be the flexibility to meet variations in demand in the individual markets. Because the volumes handled at refineries are usually larger than at secondary terminals, the location of seasonal storage at refineries usually permits the construction of larger tanks which are normally lower in unit investment. Moreover, by locating seasonal tankage at refineries, storage becomes available for other products with a contraseasonal pattern.

The optimum location, therefore, generally represents a compromise between shipping requirements and operating flexibility.

4. *Operation of seasonal tankage.* Careful utilization of inventory control cannot be stressed too strongly in connection with seasonal tankage operation. Tankage can easily be used for purposes for which it was not intended, such as continuing to run excessive amounts of crude or not changing yield patterns resulting in the buildup of products for which there is not sufficient demand. Consequently, a refinery with seasonal tankage must set definite inventory targets and make every effort to follow them as closely as possible. For example, monthly or semi-monthly inventory checks will reveal whether semi-annual target inventories are being approached too rapidly or too slowly.

Economic consideration should be given to the possibility of installing somewhat larger tanks than initially required for working purposes at marketing terminals. Pending use of the full capacity in the future for working purposes, the surplus space could be temporarily used for seasonal storage.

The economic aspects of seasonal storage involve the following factors :

1. *Transportation savings.* Seasonal storage may effect savings in tanker, barge, rail and/or tank truck transportation costs. In general, the greatest of these would be tanker transportation cost savings. Such savings are generally twofold. First, as the peaks and lows of shipments are smoothed out, winter spot charters may be replaced by less expensive summer spot charters. Second, as the level of summer shipments increase due to the use of seasonal tankage at a location close to the consuming market, spot and short term winter charters may be converted to long term charters (or owned tonnage) at a considerable savings.

Transportation savings should be based on a reasonable estimate of rates which might prevail under normal market conditions over the long range. In a tight tanker market the savings should be greater, whereas in a weak tanker market the savings might be less or even eliminated.

As might be expected, there are diminishing savings per unit of tankage capacity as the volume of tankage is increased. As the quantity of seasonal tankage is increased and the peaks and lows of shipments are smoothed out, an increasing volume of oil would be moved in the border months of the summer season to replace movements in the border months of the winter season. Since the differential spot rates between these border months is less than the differential between mid-summer and mid-winter, the savings are correspondingly lower. Furthermore, each additional increment of tankage increases by a lesser amount the potential conversion of short term to long term charters.

2. *Refinery savings.* Seasonal tankage may be considered as an alternate for spare refining capacity needed solely to meet variations in seasonal demand patterns. Since most

seasonal products have their peak demand in winter (motor gasoline and asphalt being exceptions) the use of seasonal storage permits a higher level of runs in summer and lower in winter than otherwise would be possible. In this way, the amount of capacity required to handle the same volume of business is reduced. Such an evening out in refinery thruput should also afford greater inherent flexibility to meet unusual variations in product demand.

It is quite likely that operating cost economies would result from operating a refinery at a more nearly uniform year-round rate. In addition to the refinery throughput considerations, seasonal storage might well result in considerable savings due to refinery operations at a more nearly uniform yield pattern. It might be possible, for example, to minimize degradation of distillates to fuel oil in the winter months, and uneconomical summer operations designed to lessen the production of fuel oils.

In evaluating the economics of seasonal storage, working capital required to cover the necessary inventory should not be overlooked. On the average oil would be in storage about six months of the year and account should be taken of the necessary capital being tied up for that period of time.

3. *Producing savings.* Seasonal storage permits a direct savings in producing investment and operating costs which would otherwise be required to handle peak winter demands. Such savings, although quite real, are usually not readily evaluated by refiners and marketers who purchase the major portion of their crude and product requirements.

4. *Other factors.* The per barrel investment cost of tankage varies over rather wide ranges and therefore often has a substantial bearing on the economic justification. The load-bearing characteristics of available land, the need for auxiliary facilities such as connecting pipelines, power facilities, roadways, fencing, spacing and size regulations, etc., may so affect the cost of new tankage that sites which would otherwise be optimum cannot be economically justified.

A refiner or marketer who depends on increasing product imports for covering seasonal demand should of course recognize that such action shifts the burden to his supplier. It is felt that the refiner or marketer should develop the justification that exists for his installing tankage which would even out seasonal import swings. The importing affiliate would realize not only transportation savings but also product price appreciation which usually exists for heating fuels. Then it will be up to a coordinating group to determine whether the tankage should preferentially be installed by the supplying affiliate.

In summary, the economic evaluation of seasonal storage involves a detailed study of costs and investments associated with the alternates of building seasonal storage versus meeting a market demand without such storage. Obviously many of the important factors such as weather, tanker rates, refinery operations, etc., may vary considerably from year to year. It is therfore necessary to make a good many assumptions—as realistically as possible—and to understand and appreciate the effect that variations in these assumptions might have on the economics of the project.

Incomplete information and dynamic programming. Many decision processes involve the use of channels whose statistical properties are not completely known. Such problems are fundamental in prediction theory. Many of the specific scientific results that heve been obtained require exact knowledge of the statistical properties of the process; such assumptions of perfect knowledge represent a fundamental weakness in many existing theories. Nevertheless methods are being developed to handle these formidable problems arising from incomplete information.

A general model of a communication system would consist of :

(1) a source that emits signals,
(2) a communication channel that transmits these signals to an observer,
(3) a terminus that is influenced by the decisions of the observer.

The following factors must be built into the information pattern of the model :

(1) the types of signals emitted at the source,
(2) the transformations of the signals by the channel,
(3) the resolving power of the observer and his knowledge about the effects of his decisions.

The system may be depicted as :

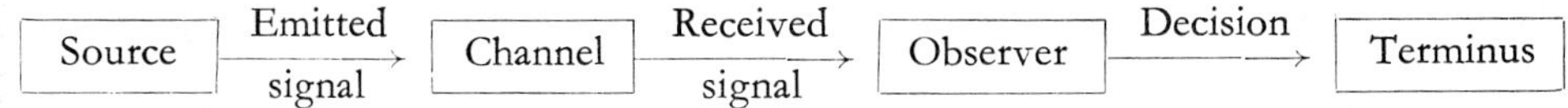

The fundamental problem is one of designing and utilizing such a system so that it will perform various tasks as efficiently as possible.

Let us now look at the problems that confront the observer. Upon receiving a signal, he is required to make a decision which affects the terminus. He will base his decision upon his knowledge of the probabilities with which the various signals are emitted, the channel probabilities which due to various imperfections and external influences distort and change the emitted signals as they are being transmitted by the channel, and the present state of the system.

We can consider the entire system as a unit that represents a multistage decision process. A stage of the process consists of a signal being emitted from the source, transformed by the channel, received by the observer, and made the basis for a decision. The result of the decision is to transform the state of the system into another state. The process continues for a certain number of stages; at the end of which, the process is evaluated in terms of some function of the final state of the system.

We may use the criterion that the decisions are made so as to maximize the expected value of the agreed-upon function of the final state of the system. In the language of dynamic programming, any sequence of allowable decisions is called a *policy* and a policy that maximizes the expected value of the prescribed function is called an *optimal policy*.

A determination of the optimal policy can be made by the use of Bellman's *principle of optimality*, which states that an optimal policy has the property that, whatever the initial state and initial decision are, the remaining decisions must contribute an optimal policy with regard to the state resulting from the first decision.

Let us now consider this communication process for the case in which we do not have information about the properties of the channel. There are various ways to approach this problem. For example, we can assume that the channel probabilities are chosen, either at the beginning of the process or at each stage, by an opponent who wants to minimize our maximum expected return. This may be called the game theory approach. Another approach would be to assume that we are given an initial probability distribution for the channel probabilities and then proceed to determine narrower and narrower limits on the basis of observational experience. A third approach would be to assume that we are given that the channel probabilities are fixed, but unknown, parameters, and proceed as done in the sequential design of experiments.   It is plausible that a method can be designed for this case that will " zero in " on the true channel probabilities, and hence asymptotically the expected gain will approach that corresponding to the case where the channel probabilities are known.

Discontinuance of underground nuclear weapon testing[1].   The possibility of the discontinuance of underground nuclear weapon testing presents a challenge to the human sciences.   The technical objectives are to produce a system that would be able to detect nuclear explosions of one kiloton (that is, the equivalent of 1000 tons of TNT) fired underground.   Although the radioactive debris is permanently trapped underground, present political considerations prevent the possibility of on-site inspection that would open this radioactivity to discovery and subsequent study.   As a result a detection method must be used that can identify a underground explosion at a long distance, and the only possible approach is one based upon observations of the seismic signals caused by the blast.   Unfortunately the seismic signals received at a distance are largely governed by the nature of the transmission path instead of by the source itself, and as a result the signatures from explosions and those from naturally occurring earthquakes are very similar and hard to discriminate.

A major problem then is the setting up of *research and evaluation centers* that can evolve better and better methods of underground nuclear detection.   In this context it is necessary to include the major subgroupings of personnel at such a center, their aims, activities, and interactions as well as the principal hardware facilities.   Any *a priori* attempt to define these entities in detail would no doubt be premature as they will be in fact shaped and reshaped by the developing staff.   Nevertheless it is necessary to present various concepts as to the initial state would serve as the starting point in the center's development.

Two primary considerations that would shape the technical approach to the center are :

(1) the technical requirements of its mission of research and evaluation in terms of required facilities and goals,

(2) the international character of the center in the context of the present state of science and engineering.

One thing is clear : the research and evaluation goals of such a center in a sense are in constant conflict; research sails with the creative drives of individuals, whereas evaluation is anchored to the broad affairs of man in a disciplined fashion.   Nevertheless, there is a common scientific ground that can serve both purposes.   For example, an empirical discovery about the spectral character of microseisms could on the one hand demand further research into the transmission mechanisms of the oceans and the earth's crust and on the other hand suggest certain engineering techniques that would improve the detection system. As another example, we can visualize that better seismic data collection and interpretation facilities can simultaneously provide increased research capabilities in seismology and underlie improvement in the detection system evaluation and performance.

The strong linkage between research and evaluation means that a mixture of these activities at a single center strongly enhances each function.   However, a strong argument

---

1.   This section is written with the help of
        Stephen M. Simpson, Jr.
        Massachussetts Institute of Technology.

can be made that the mixture should consist as the juxtaposition of the research group and the evaluation group as separate entities instead of a smooth blend with undefined boundaries. The two groups should not be set up so as to discourage individuals from alternatively serving in both capacities, but rather they should provide an atmosphere of relatively homogeneous motivation for each of the two functions. Moreover in this way each group can have a well-defined responsibility paralleling its own particular yardstick of achievement.

It is always difficult to define the function of a *research group* in any detail because of the very nature of research; it represents a multistage decision process with unknown channel probabilities. The center should provide excellent opportunities to merge theory and experiments. The tremendous data collecting network that is envisioned opens up new dimensions for the study of the earth, and the research group's responsibility would be to help maximize the scientific potential of the information obtained, and in turn this will lead to better possibilities for improving the detection aspects of the systems. The knowledge gained from research studies would feed-back into the overal evaluation program, and similarly there would be a feedback from the evaluation aspects to the research program. In this way both the research and evaluation functions would be enhanced. To help stimulate the research atmosphere, other earth studies less directly concerned with the detection problem should also be encouraged. And, of course, it is expected that parallel efforts going on at other institutions would contribute to the cross-fertilization of all these ideas. Moreover the applied research, which involves to a large extent problems of analyzing and interpreting data, would have wide applications to other sciences handling large volumes of data.

Let us now turn to the activities to be performed by the *evaluations group* so as to bring out the important problems to be faced and to determine some areas of primary emphasis. The evaluation group is directly concerned with what we may call the *systems development program*, namely, that complex of men, equipment, and information making up the international effort on the detection of underground nuclear explosions. The evaluation group may be thought of as playing two main roles, namely :

(1) *Direct participation* in certain aspects of systems development program.

(2) *Evaluation* of the systems development program and helping to give it *guidance*.

A foremost responsibility of the evaluation group in the *area of direct participation* would be the critical examination of ideas for system improvement and the design and performance of definitive experiments. Items involved would be the preliminary selection of the most promising ideas, definition of critical experiments for testing these ideas, comparison of the experimental requirements against the data and the computational facilities available, with corresponding modifications to these, detailed specification of the data processing facility necessary, performance and evaluation of experiments run on the data processing facility, and general encouragement of work in the systems development program in areas considered most promising.

In regard to the responsibility of the evaluation group in the *area of evaluation and guidance* the first problem is to determine precisely where the matter now stands, and to this end it would be necessary to cooperate closely with those parts of the systems development program which have been engaged in evaluation studies. Historically, the initial system structuring and evaluation was performed by the conference of experts at the 1958 Geneva meetings. The measure of system value chosen in their evaluation procedure was the number of " undefined " events at a given energy level, where " undefined " was determined on the basis of first motion and geographic criteria derived from seismic theory and other considerations.   A more recent evaluation summary has dropped the first motion criteria and added associated-shock criteria.   One important initial problem for the evaluation group, then, would be the establishment of evaluation groundrules to specify what type of measures should be used for what type of purposes.   Undoubtedly several points of view would have to be taken, the two main groupings being internal scientific and subsystem measures and overall measures meaningful in a military-political sense.   The maintenance of up-to-date system evaluations for both presently practical and projected system configurations will be a continuing responsibility.   Primary weaknesses of present evaluation can be traced to uncertainties in our knowledge of critical seismic environmental parameters and of system threat parameters.   Environmental parameters that require closer understanding include the distribution of earthquakes and microseisms in energy, geographic location, depth, time, associated events, spectral content, source radiation patterns, etc.   The threat para-meters uncertainly known at present include many details of weapon-to-earth coupling, —for example, coupling variations with strata and depth, muffling and big-hole decoupling,— as well as the possibility of various specialized counter measures open to a moratorium violater.

Biological models : Automatic microscopy[1].   Images from a microscope, often of life signi-ficance to patients, may be detected photo-electrically as well as by the human eye.   Quantitative photoelectric data can then be trans-mitted to automatic electronic processing equipment for analysis.   The speed, accuracy and capacity of machines which perform this automatic microscopy is a promising avenue of research in diagnostic microscopy in the service of medicine and mankind in general.

As Cleveland, Ohio, the Cancer Detection Center labors in the analysis of 30 000 cases of suspected cancer a year.   This figure is not unusual for area service of this sort in large population centers.   Here is a frontier, where the most elegant available medical diagnostic information is studied to make decisions which may spell the life or death of the person involved.

Let us look for a moment at how such procedures are carried out today.

In the great task of studying 100 000 slides per year from the 30 000 cases, in order to make these vital decisions, four human microscopists repeat the monotonous scan of the test slides under personal work loads of 80-190 cases per day, regardless of human factors such as fatigue, strain and illness.   These four persons, at the possible cost of a life, observe unerringly the most subtle microscopic anatomic changes.

This typical American Cancer Center must and does look toward the day when auto-matic microscopy will reduce the burden on their staff and at the same time increase their efficacy and accuracy.

The need for automatic microscopy is not only felt in cancer centers, but every place where medical work is being carried out.   The medical tasks of a daily nature, for example, the red blood cell count, the white blood cell count, the " differential white blood cell stain "

---

1.   This section is written with the help of
     Charles N. Loeser, M. D.
     Western Reserve University School of Medicine
     and
     Leopold Rovner
     Optionic Research, Inc., Cambridge, Massachusetts.

(often diagnostic of infection types (viral or bacterial) and indicator of leukemia) and other routine tests cry for accurate automatic processing.

This is not to say that active interest is not present in the industrial world in these matters. The following are some current developments.

(1) *micrometron.* A development of Optionic Research, Inc., under direction of Leopold ROVNER. A device for the automatic high speed scanning and counting of red and white blood cells in numbers of up to 100,000 per slide, with a accuracy of one to two percent.

(2) *cytoanalyser.* Reported by Walter TOLLES of Airborne Instruments. Outgrowth of R. E. MELLORS and Sloan-Kettering's interest in automatic cancer diagnosis. A preliminarily encouraging high speed aid to mass screening for cancer.

(3) *iconumberator.* A Dumont development under Carl BERKLEY for general counting : Fluorescent particles, bacterial colonies, red blood cells, etc.

(4) *flying spot microscope.* A number of optical and electronic companies, Philco and others, are interested. Device uses cathode ray oscillograph beam as light source. Can accurately scan, resolve, and display microscopic particles at rates of hundreds of times a second. Used on biological specimens at the Southwestern Medical School, University of Texas by P. O. MONTGOMERY.

(5) *television for study of living cells.* Dumont, RCA, and CBS have all experimented along these lines, even using color. Drs. Geroge WILLIAMS at NIH, Washington, and C. N. LOESER at Western Reserve University, Cleveland, have helped to develop the tool for spectrographic interpretation of the televised image of the cells. Addition of counting and photo-electric discrimination techniques to this television image, as well as to the flying spot microscope image, would permit collection of massive amounts of data, available now only in statistically scant amounts.

This development of tools to ease routine tasks gives an indication of what the future holds.

Today's research microscopic anatomist still looks at differences in color and brightness through his microscope, and with these two elements he can delve far into actual molecular structure by extrapolation of the meaning of the images he sees or records photographically. By utilizing the quantitative determination of the amount of light that a chemical structure within a cell absorbs, or the fluorescent light that it can give off, the microscopist can extend the limits of the study of mere color or brightness. The greater the number of such color and intensity changes discovered and understood, the greater the chance that physicians' diagnostic range may be extended. Biochemical demonstrations of delicate enzymatic changes in cancer,—for example the recent findings of Dr. WROBLEWSKI—should they be made visibly detectable by delicate stain technology, would open a whole world of diagnostic possibilities.

The biochemist crushes tissue, and studies extractives thereof, chemically, and in so doing he may study millions of cells together. But there is a growing world of microscopists who believe that data from individual living cells,—data collected by careful microscopy in pure wave lengths of light,—may yet reveal innermost cell secrets of division, respiration, nutrition or death, not observable in the cells studied in test tubes.

Of the astronomical number (about 10^{14}) of living cells which form the basic structure of a single human being, very small physical-size samples can thus reveal a clarifying and detailed story of the body's condition of health or disease. In such samples the number-count of cells, which are individually studied, by quantitative means, must tend to match the numbers of individual observations that represent good statistical merit of observation.

The rapid search procedures of electronic data processing (EDP) systems, when applied to research and diagnostic microscopy, make it practical to study numbers of cells as great as 10,000; or 100,000; or even 1,000,000 cells, if necessary. These numbers of cells are entirely too large and impractical for search and detailed analytic study by the eye of a human observer.

In good routine medical hematology practice, a blood sample is diluted in a small pipet and then placed by capillarity into a calibrated chamber. The number of cells which occupy a field of demarcated squares is then counted by a microscopist. Usually 80 squares are counted, which under a dilution of 1 : 200 would lead to a count of 500 cells on the average for representative human blood (5,000,000 cells per cubic millimeter). The random settling of cells on the slide, however, leads to a deviation from this expected number of 500 cells. This deviation is the so-called field error.

The number of cells falling within the 80 squares in repeated counts from the blood specimen follows the normal distribution of error, with mean value of 500 and standard deviation $\sqrt{500}$ or 22.4. Since about 95 per cent of observations taken at random falls within plus or minus two standard deviations from the mean, 95 % of the counts will fall within a range of 500 plus or minus 44.8, i.e., within a range of 455 to 545. As a percentage of the mean of 500, this error range is from -9 % to $+9$ %.

With the micrometron, however, many more than 80 squares can be counted automatically. For example, suppose 20 times 80 or 1600 squares were counted. Then in a single high speed observation run, an average of 20 times 500 or 10,000 cells would be counted. The standard deviation would be $\sqrt{10,000} = 100$, so that the 95 per cent range would be between 9,800 and 10,200 or from -2 per cent to $+2$ per cent. Thus the field error would be reduced from the plus or minus 9 per cent range of good routine practice today to a plus or minus 2 per cent range by increasing the number of squares counted by twenty times. in observations done by the machine at rates of from 500 to 1000 cells observed per second.

Today, in general, the intrinsic quality of actual *information* derived in any one day from tissue or blood samples serially studied by the human microscopist, in the research and clinical medicine laboratory—however well he may be trained in his basic medical science —corresponds to the plus or minus 9 per cent error range corresponding to a count of only 80 squares. On the other hand, the intrinsic quality of information derived from the more extensive samples made possible by automatic microscopy corresponds to the plus or minus 2 per cent error range corresponding to a count of 1600 squares for the same blood sample

While it is evident that some of the major aspects of nature's picture of body and tissue conditions are clearly revealed within what is now considered to be a practical period of time for a human observation procedure, a great many hitherto unknown values will be obtainable if Electronic Data Processing of the microscope image is used.

As we have seen, a problem of importance in hematology is the quantitatively accurate and rapid observation of the average number of blood cells per unit volume of blood in the human body. Statistical aspects of the problem are treated by Joseph BERKSON (1944). Berkson's article deals with the present state of cell counting as it is done by human observers using good routine practice in medical laboratories across the country.

We shall discuss the counting of large quantities of blood cells in numbers and accuracies and with speeds beyond the practical reach of a human observer. This work was done for the development of an automatic blood cell counter machine.

Each individual count produced in an automatic apparatus for counting single cells makes up a complete single observation in the statistical framework. Thus a machine for making complete single cell counts at high speeds is a device which automatictally combines great numbers of single observations into measuring data of high precision, in a short time.

We shall consider both the count of red cells and white cells. We shall present the theory as initially applied to counting large numbers of red cells. The mathematical derivations given for red cell counting are equally valid for white cell counting, provided that proper sample preparation and machine adjustements are made so that the leukocytes being counted satisfy approximately the same mathematical conditions as assumed in the derivations for erythrocytes. After the presentation of the theory for erythrocytes, we shall treat problems peculiar to leukocytes.

First of all, we wish to treat the problem of determining the average number of cells per unit volume in a cell counting chamber. The usual model uses the following two assumptions about the position of cells in a chamber :

(1) the position of a cell is equilikely anywhere in the chamber. In more precise terms, the position of a cell is a random variable with a uniform (equiprobable) distribution.

(2) the position of one cell does not influence the position of any other cell. In more precise terms, the position of each cell is statistically independent of the position of every other cell.

It is not practical to count every cell in the chamber. Instead the number of cells in a subvolume of the chamber is counted. Such a subvolume is called a field in cell counting terminology. Now it is readily seen the number of cells which lie in any field is a random variable. The problem, then, is to determine the average or mean value of the cell count. Then the mean divided by the volume of the field gives the average number of cells per unit volume in the counting chamber.

In order to estimate the mean it is convenient to find the distribution of the cell count. It can be shown that the count follows the binomial distribution law. If the sub-volume which we consider is very small compared to the total chamber volume, the binomial distribution is approximated by a Poisson distribution, which has the same mean and which has a standard deviation equal to the square root of the mean. When the mean is a relatively large number, for example greater than 100, the Poisson distribution is approximated by a normal distribution with the same mean and same standard deviation.

Thus to a high degree of precision the number of cells in a given field is normally distributed with standard deviation equal to the square root of the mean, under the stated conditions. Let the mean count which we wish to estimate be m, and the number of cells actually counted be x. Now it is well known that for a normal distribution the following inequality

$$- 2 \leqslant \frac{m - x}{\sqrt{m}} \leqslant 2 \tag{1}$$

holds about 95 percent of the time. Dividing through by $\sqrt{m}$ the inequality becomes

$$- \frac{2}{\sqrt{m}} \geqslant \frac{m - x}{m} \leqslant \frac{2}{\sqrt{m}} \tag{2}$$

The expression $\frac{m - x}{m}$ represents the ratio of the error m — x to the true count m, and hence represents the relative error of the field when one field is compared with the true count.

Instead of speaking of " errors " the more modern statistical concept of " confidence limits " may be used. As is common practice we estimate the standard deviation $\sigma = \sqrt{m}$ by $\sqrt{x}$. Then our inequality becomes

$$-2 \leqslant \frac{m - x}{\sqrt{x}} \leqslant 2 \tag{3}$$

which gives the confidence interval

$$x - 2\sqrt{x} \leqslant m \leqslant x + 2\sqrt{x} \tag{4}$$

at a 95 percent confidence level. Rewriting this expression we have

$$x\left(1 - \frac{2}{\sqrt{x}}\right) \leqslant m \leqslant x\left(1 + \frac{2}{\sqrt{x}}\right) \tag{5}$$

For example, suppose a count of $x = 20{,}000$ has been made. At this count the value of $\frac{2}{\sqrt{x}}(100)$ is 1.4 percent, or 0.014. Thus we write the confidence interval

$$20{,}000(1 - .014) \leqslant m \leqslant 20{,}000(1 + .014) \tag{6}$$

which is $19{,}720 \leqslant m \leqslant 20{,}280$. We then state that the unknown true count m is situated between 19,720 and 20,280. If we constantly apply such a rule of behavior, then in the long run our statements will be correct for about 95 percent of the cases, and incorrect for only about 5 percent of the cases.

Let us now consider the field accuracy when two distinct fields are compared, which gives limits for the relative variation of two counts as a percentage of the true count, where the two counts x_1 and x_2 are taken over two different volumes v_1 and v_2 which are of equal magnitude. Since both x_1 and x_2 are statistically independent, normally-distributed random variables each with mean m and standard deviation $\sqrt{m}$, it follows that their difference $x_1 - x_2$ is a normally distributed random variable with mean zero and standard deviation $\sqrt{2\,m}$. Then the inequality

$$-2 \leqslant \frac{x_1 - x_2}{\sqrt{2\,m}} \leqslant 2 \tag{7}$$

holds about 95 percent of the time. Rewriting this expression we have the equivalent inequality

$$-\frac{2}{\sqrt{m}} \cdot \sqrt{2} \leqslant \frac{x_1 - x_2}{m} \leqslant \frac{2}{\sqrt{m}} \cdot \sqrt{2} \tag{8}$$

Hence $\frac{2}{\sqrt{m}} \cdot \sqrt{2}$ gives limits which hold 95 percent of the time for the relative variation $\frac{x_1 - x_2}{m}$ of the counts x_1 and x_2 in two distincts fields.

Because the curves for field accuracy start to level off in the range of 10,000 to 50,000 cells counted, it is desirable to have red cell counts of this magnitude.

The analysis up to this point has been concerned with field accuracy limitations. We now wish to include the volumetric accuracy limitations of the chamber and the pipette. If deviations of chamber and pipette accuracy are independent and are approximately normally distributed, then the distribution of the count x of a blood sample measured by them will also be approximately normally distributed. The mean of x is the true count of the blood sample. The standard deviation σ is given by

$$\sigma = \sqrt{\sigma_f^2 + \sigma_c^2 + \sigma_p^2} \tag{9}$$

where $\sigma_f = \sqrt{m}$ is the standard deviation of the cells counted within the chamber where field accuracy limitations apply, σ_c is the standard deviation of chamber accuracy deviations, and σ_p is the standard deviation of pipette accuracy deviations. For a 95 percent confidence level, the x confidence limits, as a percentage of the true count m, are given by the positive and negative values of

$$\frac{200\,\sigma}{m} = \sqrt{\frac{(200)^2}{m} + \left(\frac{200\,\sigma_c}{m}\right)^2 + \left(\frac{200\,\sigma_p}{m}\right)^2} \tag{10}$$

By plotting curves that combine volumetric accuracy limitations with the field accuracy limitations in the comparison of one field with the true count, it is seen that red cell counts should be of the magnitude of 10,000 to 50,000 cells per count.

Berkson experimental verified, and we therefore assume, that for counts of different magnitude both σ_c and σ_p are proportional to the mean count m. In good routine practice Berkson observed that the standard deviation σ_c of the chamber accuracy deviations is given by $\sigma_c = .046\ m$; that is, chamber accuracy is good to within $\pm$ 4.6 % about 68 percent of the time, or at a 68 percent confidence level. At a 95 percent confidence level, that is about 95 percent of the time, the chamber accuracy deviations are within $\pm$ 9.2 percent, i.e., twice the standard deviation of chamber accuracy data. In an analogous manner Berkson observed that the standard deviation σ_p of the pipette accuracy deviations is given by $\sigma_p = .047\ m$; that is, pipette accuracy is good to within $\pm$ 4.7 percent about 68 percent of the time. At a 95 percent confidence level, the pipette accuracy deviations are within $\pm$ 9.4 percent, i.e., twice the standard deviation of pipette accuracy data.

In view of Berkson's observed values for accuracy of routine clinical red cell counting and considering the action of the especially careful laboratory worker, there is reason to believe that accuracies to within $\pm$ 2 percent can be achieved, with pipette and chamber respectively. This belief assumes careful control of pipette and hemacytometer volumetric accuracy obtained with calibrated pipettes of the conventional 1 : 200 type now in use and with conventional hemacytometer chambers of 0.1 mm depth used for the visual counting of cells.

This limit of $\pm$ 2 percent corresponds to a confidence level of 68 percent, i.e., the volumetric accuracies just cited lie within $\pm$ 2 percent about 68 percent of the time. This circumstance corresponds to a standard deviation σ_p of the pipette equal to .02 m and the standard deviation σ_c of the chamber also equal to .02 m. To continue reasoning as previously stated, then at a 95 percent confidence level, where we like to operate, the accuracies with these carefully calibrated pipettes and hemacytometers intended for visual count will be good to within $\pm$ 4 percent about 95 percent of the time.

The *Micrometron* is a digital computer type optical electronic blood cell counter manufactured by Optionic Research, Inc., Cambrigde 38, Massachusetts. It was developed by

Optionic Research, Inc., under the direction of Leopold Rovner, for the United States Air Force School of Aviation Medicine.

The hemacytometer chamber of the Micrometron is built upon a circular quartz optical flat of approximately three inches diameter.   Spiral scan is used.   The depth of the hemacytometer chamber is 0.1 mm.   Dilutions for red cell counts are usually of the order of 1 : 200 to 1 : 1,000, and actual red cell counts usually are of the order of 5,000 to 50,000, as representative of 5,000,000 cells per cubic millimeter.   The Micrometron, however, can scan 500,000 red cells, or more, if required for accurate size-distribution studies.   A central slit-width of 12 microns is used for red cells.

For a red cell count of 5,000 per cu. mm. and dilution of 1 : 400, the cells are spread out at about 1,250 cells per square mm.   At a 12 micron slit width about 83.3 mm. of path are traversed to observe 1,250 cells, so that cells are encountered at an average rate of 15 per mm. of path traversed.   With an annular chamber of 2.5 inches center-line diameter (radius = 31.7 mm.) about 3,000 cells are counted per turn of the hemacytometer chamber.

The following table (Table I) gives data for cell-count and time-for-cell-count based on rotational speeds from 0.1 rps to 10. rps for a rotatable table on a machanical stage :

TABLE I.

Total cells counted vs. time to count cells.

Total Cells Counted	Time to Count Cells			
	0.1 *rps*	1. *rps*	5. *rps*	10. *rps*
5,000	16.7 sec.	1.67 sec.	.334 sec.	.167 sec.
50,000	167.	16.7	3.34	1.67
500,000	1,670	167.	33.4	16.7

(Trace-speeds are based on center-line radius of counting chamber = 31.7 mm.; or single turn = 200 mm.)

The following table (Table II) gives the microseconds time-lapse to traverse central diameter of an 8 micron red cell under the stated rotational speeds of the counting chamber :

TABLE II

Rotational speed vs. microsecond pulse width

Rotational Speed	Microsecond Pulse Width
0.1 rps (20. mm./sec.)	400. microseconds
1. rps (200. mm./sec.)	40. microseconds
5. rps (1 000. mm/sec.)	8. microseconds
10. rps (2 000. mm./sec.)	4. microseconds

The selection of chamber-diameter, rotational speeds, and depth of chamber was made for :

A. Best chamber area design for uniform distribution of cells from sample.

B. Minimal acceleration effects on displacement of blood cells and streaming of diluting fluids.

C. Maximal effectiveness of coincidence and anti-coincidence electrical circuits.

D. Completion of significant count within a reasonable time interval.

The Micrometron has such features as : visual inspection of each sample, uniform accuracy of 2 %, electronic true-focus control, automatic precision pipetting, automatic chamber wash, direct printed cell count available. It saves the work of 6 to 8 technicians, and provides economy of time, space and costs. Scanning 10,000 to 100,000 cells per blood count, it can provide blood counts at an operational rate faster than 1 blood count per minute.

The following table shows representative savings available in a central laboratory operation with six hematology technicians making routine blood cell counts.

TABLE III

Cost per month for human observers vs. micrometron system.

Human Observers		Micrometron System	
Wages, Six persons at $ 250.	$ 1,500	60 % of Time of one Hematology	
Overhead at 50 %	750.	Technician at $ 250.	$ 150.
		60 % Time, Assistant at $ 150.	90.
		Wages	$ 240.
Supplies and Services	250	Overhead at 50 %	120.
		Supplies	100.
		Lease (5-Year Term)	685.
		Service Contract	125.
Per Month Total	$ 2,500.	Per Month Total	$ 1,270.
Per Month Savings.	$ 1,230.		

Savings Per Year with the Micrometron : $ 14,760.

An ideal counting apparatus would be one that would respond to any portion whatever of a cell impinging on the receiving slit or aperture. Let us now deal with the problems of physical counting systems which require a certain detectable part of an element to pass through an aperture in order for that aperture to give a signal. The signal-to-noise ratio and amplifier response will then determine the observable limits of signal perception.

A *two beam system* such as is used in the Micrometron is composed of two adjacent apertures oriented perpendicular to the scanning direction. We assume that the apertures are of rectangular shape, and that both have the same width in the scanning direction. Let us designate the upper aperture by A and the lower aperture by B. The logical operation of a two beam system is as follows : If the photomultiplier tubes of apertures A and B are driven simultaneously, then no count is recorded because of the anti-coincidence circuitry of the counting system. If the photomultiplier tube of A alone is driven, then no count is recorded. On the other hand, if the photomultiplier tube of B alone is driven, then one count is recorded. Because of the symmetrical properties of the two beam system, it can be shown that cell counts made by it are independent of cell size parameter, within limits.

To this point we have considered only the effects of counting cells which lie sufficiently far apart so that the counting apertures encounter but one cell at a time. If two cells modulate an aperture simultaneously, then there is a good chance that only one count would result, and consequently the count of one cell would be missed. Because of the random distribution of cells in the chamber, some cells will lie close together.   Hence a certain number of cells will not be counted, thereby giving rise to a counting loss.   We shall examine the factors which cause this counting loss in order to determine the statistical corrections which should be made.

As we have seen, in a physical system at least a certain critical minimum percent of an aperture must be modulated in order for that aperture to drive its photomultiplier tube.   Let the aperture scan two cells which lie close together.   If the cells are sufficiently close together, the percent of the aperture modulated will never fall below the minimum, and hence the cells will be recorded as one unit.   Thus for any given orientation of the cells with respect to the aperture, and for any given cell sizes, one may define a critical minimum distance, which we shall denote by q, such that two cells separated by a distance less than q will not be recorded as separate units by the aperture.   For definiteness, we shall let q be the projection of the center-to-center distance between the cells into the direction of scan. We shall call q the *resolving distance* of the aperture.

For a given cell size distribution, the effects due to the orientation and sizes of the cells may be averaged out.   We may think of q as depending upon the dimensions of the aperture and the sensitivity of the recording system.   More precisely the resolving distance q depends upon the width and height of the aperture, and upon the minimum critical percent of aperture area which must be modulated in order to drive its photomultiplier tube.   In turn, the critical percent of aperture area which must be modulated depends upon the signal-to-noise ratio amplifier response.

For any specific counting apparatus the value of the resolving distance q would have to be evaluated.   One way to evaluate q experimentally is by counts on a known blood sample at various dilution ratios.

Once the value of q has been estimated for a specific counting apparatus, it is possible to make statistical corrections for the elements which are not counted because of minimum resolving distance.   Suppose that n elements lie in the field scanned by the counting apparatus, but only n′ elements are recorded because of the counting loss due to the resolving distance q of the counter.   That is, we suppose that there are n — n′ elements whose counts are not recorded for any given count.   If we let $\overline{n}'$ denote the average value of n′, then n — $\overline{n}'$ represents the average number of cells whose counts are not recorded.   Then the average number of cells missed as a percent of total cells present is given by

$$\frac{n - \overline{n}'}{n} . \quad (100) \qquad\qquad (11)$$

and for brevity we shall call this expression the *average counting loss* in percent.

Let us now determine the average counting loss as a function of the resolving distance q, and of the number of cells per unit area in the field.   As we have seen, when the distance in the scanning direction between two consecutive cells is less than the resolving distance applicable to the given situation only one count is recorded and one cell is therefore missed. The resolving distance for this situation depends upon the orientation and sizes of the cells, the dimensions of the aperture, and the sensitivity of the recording system.   Since we are treating the statistical aspects of large cell counts, it is reasonable to average the effects of the orientation and sizes of the cells.

The average counting-loss may be shown to be a function of the mean resolving distance q, and the number of cells per unit area in the field, as follows.   Since on the average n — 1 out of n cells are not counted if these n cells have mutual distances in the scanning direction less than the mean resolving distance q, we see that a cell is counted on the average only if the interval to the left of the cell is greater than q. (The direction of scan is to the right). As a result the probability of counting n' cells out of a total of n cells in the field scanned by the counting apparatus is equal to the probability of encountering n' intervals (between consecutive cells) which are greater than the mean resolving distance q out of the total of n intervals on the scanning path.

In order to compute this probability, let us consider a scanning path of lenght L, onto which we project the centers of n cells which lie in the field scanned by the counting apparatus. The average number of cell centers per unit length on the scanning path is then equal to u = n/L. By the equiprobable assumption of fall-out of cells on the counting chamber, the probability that a cell falls on a given segment of length q is q/L. If we let n cells fall at random on the scanning path, the probability that exactly x of them fall on the given segment is given by the binomial distribution.

$$B(x) = \frac{n!}{x!(n-x)!} \left(\frac{q}{L}\right)^x \left(1 - \frac{q}{L}\right)^{n-x} \tag{12}$$

In counting large numbers of cells the length of the scanning path L and the total number of cells n on the scanning path are large, but the average number of cells per unit length (equal to u = n/L) remains constant. Under these conditions the binominal distribution in expression (12) may be approximated by the Poisson distribution

$$B(x) \approx P(x) = \frac{e^{-uq}(uq)^x}{x!} \tag{13}$$

This follows since the binominal distribution becomes the Poisson distribution in the limit as n and L increase indifinitely in such a way that the average number $\left(u = \frac{n}{L}\right)$ of cells per unit length remains constant.

Now the probability that the distance between the centers of two consecutive cells is greater than q is equal to the probability that no cells fall within a given segment of length q, which, by equation (13) is

$$P(x = 0) = e^{-uq}. \tag{14}$$

As a result the mutually exclusive and exhaustive probability that the distance between the centers of two consecutive cells is less than (or equal to) q is

$$1 - P(x = 0) = 1 - e^{-uq}. \tag{15}$$

In the case we are considering, there are n cells on the counting path; i.e., there are n consecutive intervals. Because of the random independent fall-out of cells in the chamber, and because the scanning path L is large compared to the average distance between cells on the scanning path, we may consider the lengths of the intervals between consecutive cells to be independent of each other.

Since only cells separated by intervals of length greater than the resolving distance q are recorded, the probability of a cell being recorded is equal to the probability that an interval is greater than q. As we have seen in expression (14), this probability which we denote by h is equal to

$$h = e^{-uq}. \tag{16}$$

Consequently, the probability that n' cells will be recorded out of a total of n cells on the scanning path is given by the binomial distribution

$$B(n') = \frac{n!}{n'! \, (n-n')!} h^{n'} (1-h)^{n-n'} \qquad (17)$$

Therefore the mean number $\bar{n}'$ of cells recorded is given by

$$n' = nh = ne^{-uq}, \qquad (18)$$

and the standard deviation $\sigma(n')$ of the number of cells recorded by

$$\sigma(n') = \sqrt{nh(1-h)} = \sqrt{ne^{-uq}(1-e^{-uq})}. \qquad (19)$$

The average counting loss is therefore

$$\frac{n-\bar{n}'}{n} = 1 - \frac{\bar{n}'}{n} = 1 - e^{-uq} \qquad (20)$$

where u is the average number of cells per unit length on the scanning path and q is the resolving distance of the counter.

Using equation (20), we obtain the following table for average-counting loss where we have used the parametric values :

Average Red Blood Cell Count = 5,000,000 per cu. mm.
Effective Counting Height of Slit = 12.0 μ
Slit Width = 3.0 μ
Average Cell Radius = 4.0 μ
(Note : μ is the symbol for micron = 0.001 millimeter).

TABLE IV

Average counting loss for various resolving distances.

Dilution Ratio	Average Counting Loss		
	For resolving distance : $q = 11\,\mu$	$q = 8\,\mu$	$q = 4\,\mu$
1 : 100	.483	.381	.213
1 : 200	.281	.213	.113
1 : 400	.152	.113	.058
1 : 800	.080	.058	.030
1 : 1 000	.064	.048	.024
1 : 1 600	.041	.030	.015
1 : 2 000	.033	.024	.012

In regard to field and volumetric accuracy, the confidence limit curves previously derived for erythrocytes are also valid for leukocytes, provided sufficiently large dilution ratios are used. Leukocyte populations, with their wide variations in cell sizes and shapes, however, require more detailed statistical treatment than the nearly homogeneous erythrocyte populations.

In counting leukocytes with a blood cell counter machine, using an optical slit system which scans the cells, a leukocyte may be detected photoelectrically by its cytoplasm, or by its nucleus, depending upon the technique desired to receive signals from the cells.

The basic mathematical derivations presented here apply equally as well, within close quantitative limits, to counts based on the apparent cytoplasm shapes, or to counts based on the apparent shapes of the nucleii of white cells as they are scanned by the slit system.

There is a characteristic analytical difference between aspects of erythrocyte and leukocyte counting with an optical scanning machine. This difference resides in the appropriate evaluation of counting loss correction factors for the red, and the white cells.

Corrections for counting loss of red cells, which are required because of the large numbers of erythrocytes in the optical field, are made quite simply because the cells have a nearly homogeneous size distribution and, in actual practice, quite small variations of size range.

Corrections for counting loss of white cells, which have wide variations in size range, at first appeared difficult but have been found quite feasible and practical because, in the first instance, white cells are small in number in the blood volume and are relatively dispersed in the optical focal plane on which they are distributed while being scanned by the slit system.

In either case large errors in assumption of true average cell size, in the specimen being observed, have relatively small effect on the accuracy of the final count. This is due to the fact that size factor (in correction of counting loss for the optical slit scanning procedure) is a second-order correction factor, and does not change the end count by much more than about 1. or 2. percent.

The data of this report deal primarily with the count of leukocytes by an optical scanning machine which receives a signal from the cytoplasm of the cell. The significant parameters in this case are the dimensions of the whole cell as viewed by the optical slit of the counting system. Where counting is performed by optical signals (light pulse modulations) from nucleii of white cells, a valid count of these leukocytes can be made largely independent of the effect of variations in sizes and shapes of their nucleii as these variations affect counting loss correction factors.

The field and volumetric accuracy derivations given previously apply equally as well for white cell count as for red cell count.

In regard to *field accuracy*, the basic assumptions concerning the distribution of cells in the chamber hold, with good reason, for the count of white cells as well as for the count of red cells. When sufficiently large dilution ratios exist, we are able to assume that each white cell has a uniform random distribution in the chamber, and each leukocyte acts independently of other white cells in fall out to the base plane of the couting chamber. These assumptions are the " equiprobable " and " independence " assumptions with regard to distribution of cells upon the base plane of the counting chamber. With such uniformly random fall out, the confidence limits for field accuracy given previously are applicable for the number of cells counted, whether red cells or white cells.

It should be noted that, in using the confidence limits for the comparison of the count of one field with the true count, the red or white cell count observed is compared to an average count, called the true count, which it is in a practical sense as derived from an effective infinity of cells of the respective categories in the human body. We may expect, when suitable samples are taken, that 95 percent of the time the absolute magnitude of the deviations of the observed and true counts will be under the values given by the confidence limit curve for the comparison of the count of one field with the true count.

In regard to *volumetric accuracy* the same reasoning holds for the derivation of combined field and volume accuracy for white cells as well as for red cells. The confidence limits for field and volumetric accuracy given previously apply with equal validity to the red cell count and to the white cell count.

The intrinsic difference beetween analytical aspects of red cell and white cell counting with an optical-electronic counting machine, having a slit system, rests in the available knowledge of cell-size distribution of red cells, and of white cells, and in the consequent evaluation of suitable *counting loss correction factors* applicable to the appropriate types of cell.

For red cells, the range of size distribution is comparatively small, usually from 4 to 11 microns, and size distribution of these cells is sharply peaked in the neighborhood of 7 to 9 microns. On the other hand, for white cells the range of sizes is larger, usually from 8 to 24 microns. Whereas the red cell population is relatively homogeneous, the white cell population is relatively non-homogeneous. Each variety of white cell has its own size distribution; in other words, the overall white cell size distributions have groups of peaks.

A typical white cell population in health is described by the table below. The values of this table are derived from values of the variety size ranges and the variety frequency ranges of leukocytes, as given by Wintrobe (1952, Table 6, p. 192; Table 8, p. 206).

TABLE V

Typical leukocyte population in health

Leukocyte Variety	Average radius in microns	Frequency of the variety
Segmented Neutrophils	6.25	.59
Lymphocytes	6.25	.30
Monocytes	8	.05
Juvenile Neutrophils	7	.04
Eosinophils	6.25	.02
Basophils	6.25	.0
Myelocytes	7.5	.0

As discussed previously counting loss results from a group of two or more cells modulating the counting aperture simultaneously, so that only the count of one cell is recorded, and the counts of the other cells in this group are missed. This effect is due to the random distribution of cells in the chamber which allows some cells to lie close together by chance. One may control this effect by varying the dilution ratio. That is, as the blood sample becomes more dilute, the likelihood that cells lie close enough together to produce counting losses becomes lessened.

As we have seen in order to evaluate the counting loss, the resolving distance q of the aperture must be evaluated. As it was pointed out, the resolving distance q of the aperture in the scanning of two adjacent cells is a function of the following variables : the orientation and sizes of the cells, the dimensions of the aperture, and the sensitivity of the recording system.

In our previous discussion on the resolving distance, we assumed that the cell population was effectively homogeneous, which is the case for red cells. This discussion can be extended to the case of non-homogeneous cell populations, that is, the case of white cell populations.   Thus we can estimate a resolving distance q which is characteristic of a physical apparatus scanning a white cell population.

Once the value of q has been estimated for a specific counting apparatus, it is possible to make statistical corrections for the elements which are not counted because of minimum resolving distance.   Suppose that n elements lie in the field scanned by the counting apparatus, but only n′ elements are recorded because of the counting loss due to the resolving distance q of the counter.   That is, we suppose that there are n — n′ elements whose counts are not recorded for any given count.   If we let $\bar{n}'$ denote the average value of n′, then n — $\bar{n}'$ represents the average number of cells whose counts are not recorded.   Then the ratio of the average number of cells missed to the total number of cells present is given by

$$y = \frac{n - \bar{n}'}{n} \qquad (21)$$

For brevity we call y the average counting loss, and 100 y the percentage (average) counting loss.

As we have shown, the average counting loss is a function of the resolving distance q and of the number of cells per unit area in the field.   Thus we may determine counting loss correction factors to apply to the number of cells recorded by the blood cell counter machine to give the true (average) count.   We now want to show that despite the wide variations in size of the different leukocyte varieties it is possible to obtain the true count within certain predeterminable limits.   As we shall see, these limits are sufficiently small so that it is possible to count leukocytes with the same order of magnitude of statistical accuracy as erythrocytes.

From the above derivations we can determine the average error in the percent counting loss which would occur by assuming an incorrect average leukocyte radius.   For example, by assuming the average leukocyte radius to be 6.5 microns, whereas it was actually 4 or 10 microns respectively, Table VI gives the magnitude of the average error in the counting loss for various leukocyte counts :

TABLE VI

Error in counting loss
for assuming that average radius = 6.5 microns

Leukocyte Count (cells per cubic mm)	Percent Error if actual average radius = 4 microns	Percent Error if actual average radius = 10 microns
2,500	0.15	0.21
5,000	.30	.42
7,500	.44	.62
10,000	.60	.81
15,000	.89	1.25
25,000	1.40	2.00

We see from Table VI that the estimate of average cell size, when made, in the use of counting loss correction procedures, has relatively small effect, of the order of 1. or 2. percent in the final value of observed leukocyte count.

In general, it would be quite safe, within the accuracies of the order of 2 percent, to use an average radius for leukocytes of the order of 6.5 microns. In this case one would be able to expect to have corrections valid to within 2 percent, as shown.

In physical counting systems one may expect an electronic deadtime following the response of the apparatus to the transit of an impulse producing element. On basis of known characteristics of the electronic cell counting system the electronic dead-time may be as low as six or seven microseconds and generally will not be greater than ten microseconds. For example, if the scanning velocity is .167 microns per microsecond, a distance of 1.67 microns will be scanned by the apparatus during a dead time of ten microseconds. This distance may be called the dead-time distance and we shall denote it by d.

Any new element encountered by the scanning apparatus in the dead-time distance will ordinarily not be recorded as a separate count. As a result, there is a counting loss due to this electronic dead-time.

The counting loss due to dead-time is analogous to the counting loss due to the resolving distance of a physical counter. As a result, both effects may be included in a resolving factor f for the physical counter. More precisely, the resolving factor f, expressed as a distance, will be the sum of the resolving distance q and the dead-time distance d.

Thus we have seen that correction factors for counting loss of white cells can be derived for an optical-electronic machine, such as the *Micrometron*, and that larger variations in true average cell size from specimen to specimen have relatively small effect on the correction factors, so that a valid count is reached largely independent of variations in cell sizes and shapes.

CONCLUDING REMARKS

In this paper we have emphasized a number of specific models as illustrations of general principles and methods of model building in the human sciences. Our approach has been on the pluralism of model building in line with the general concept of Wold (1961) that science does not aim at constructing an all embracing model for the entire body of scientific knowledge, but that each science can be seen as a collection of models. Throughout we have tried to emphasize the dualism between the theoretical model and its empirical aspects and to illustrate that the actual product of scientific research is a flow of models at different aspiration levels.

ACKNOWLEDGEMENT

I would like to express my sincere thanks to Professor Herman Wold and Mr. George Stojkovic, Uppsala, for helpful comments and discussions.

REFERENCES

ABRAMSON, N., BRAVERMANN, D. and SEBESTYEN, G. (1963). Pattern recognition and machine learning. *I.E.E.E. Transactions on Information Theory*, vol. **9**, pp. 257-261.
BELLMAN, Richard (1961). *Adaptive Control Processes*, Princeton University Press, Princeton, N. J.
BERKSON, Joseph (1944). Blood cell count : Error, *Medical Physics*, Year Book Publishers, Chicago.
BLACKMAN, R. B, and TUKEY, J. W. (1959). *The measurerment of power spectra*, Dover, N. Y
BODE, H. W. (1945). *Network analysis and feedback amplifier design*, Van Nostrand, Princeton, N. J.

BROWNING, RUFUS P. (1962).   Computer programs as theories of political processes.   *The Journal of Politics*, vol. **24**, pp. 562-582.
DAVENPORT, W. B. and ROOT, W. L. (1958).   *Theory of random signals and noice*, McGraw-Hill, N. Y.
HANNAN, E. J. (1960).   *Time series analysis*, Methuen, London.
KAC, M. (1962).   A note on learning signal detection, *IRE Transactions on Information Theory*, vol. **8**. pp. 126-128.
KENDALL, M. G. and STUART, A. (1958).   *The Advanced Theory of Statistics* (3 volumes), Charles Griffin and Co., London and Stechert-Hafner, N. Y.
LANING, J. H. and BATTIN, R. H. (1956).   *Random processes in automatic control*, McGraw-Hill, N. Y.
LYNCH, William A. (1961).   Linear control systems, a signal-flow-graph viewpoint.   *Adaptive Control Systems*; McGraw-Hill, N. Y. pp. 23-49.
PARZEN, Emanuel (1962).   *Stochastic processes*, Holden-Day, SanFrancisco.
PETERSON, E. L. (1961).   *Statistical analysis and optimization of systems*, John Wiley, N. Y.
ROBINSON, ENDERS A. (1959).   *An introduction to infinitely many variates*.   Charles Griffin and Company, London and Stechert-Hafner, New York.
ROBINSON, ENDERS A. (1962).   *Random wavelets and cybernetic systems*.   Charles Griffin and Company, London and Stechert-Hafner, New York.
SHANNON, C. E. (1948).   *The mathematical theory of communication*, University of Illinois Press, Urbana.
SIMPSON, Stephen M. (1961-1964).   Reports to the Advanced Research Projects Agency, Washington, D. C.
TRUXAL, John G. (1961).   The concept of adaptive control.   *Adaptive control systems*, McGraw-Hill, N. Y., pp. 1-19.
TSIEN, H. S. (1954).   *Engineering cybernetics*, McGraw-Hill, N. Y.
WHITTLE, P. (1963).   *Prediction and Regulation*, English Universities Press.
WIENER, Norbert (1948).   *Cybernetics, or control and communication in the animal and the machine*, John Wiley, New York.
WINTROBE, M. M. (1952).   *Clinical hematology*, Third Edition, Philadelphia.
WOLD, Herman (1938).   *A study in the analysis of stationary time series*, Almqvist and Wicksells, Uppsala.
WOLD, Herman (1952).   *Demand analysis*, Almqvist and Wicksells, Uppsala and Wiley, N. Y.
WOLD, Herman (1961).   The approach of model building and the possibilities of its utilization in the human sciences.   *Monoco Entretiens of* 1960.

CHAPTER 2

Econometrica, Vol. 27, 4 (October 1959)

A STOCHASTIC DIFFUSION THEORY OF PRICE

By Enders A. Robinson[1]

1. STOCHASTIC MARSHALLIAN THEORY

LET $\bar{x}(p)$ DENOTE the classical excess demand as a function of p, where p is the deviation of price from its equilibrium level (i.e., $\bar{x}(0) = 0$). Let the price unit be denoted by π, and the time unit by τ. We may write price $p = i\pi$ and time $t = j\tau$, where i and j are integers. We assume that excess demand x_t at time t is given by

$$(1) \qquad x_t = \bar{x}(p_t) + u_t$$

where p_t is price at time t and u_t is a variate with zero mean occurring at time t. Furthermore, we assume that u_t and u_s are statistically independent for $t \neq s$. Our dynamic assumption about the market is the following:

$$(2) \qquad p_{t+\tau} = \begin{cases} p_t + \pi \text{ if } x_t \geqslant 0, \\ p_t - \pi \text{ if } x_t < 0. \end{cases}$$

That is, price rises by one price unit π if excess demand is nonnegative, and falls by $-\pi$ if excess demand is negative. Hence the dynamic assumption (2) represents a market in which abrupt price changes can not take place in one time interval; that is, a market in which price is "sticky." As we shall see, this assumption that in small times price can change only by small amounts leads to the so-called stochastic diffusion processes.

A classical stochastic diffusion process is the Ehrenfest model of diffusion, which received wide attention in connection with Boltzmann's statistical explanation of irreversibility and recurrence in thermodynamics. To reduce our economic model of supply and demand to the Ehrenfest model, we assume:

(a) The excess demand curve $\bar{x}(p)$ is a negatively sloping straight line given by

$$(3) \qquad \bar{x}(p) = -ap, \, a > 0;$$

(b) The variates u_t have identical rectangular distributions, with range $-aR \leqslant u_t \leqslant aR$ (with R being taken to be an integer for simplicity).

[1] This paper is a section of a thesis submitted in January 1952 to the Massachusetts Institute of Technology in partial fulfillment of the requirements for the degree of Master of Science. My sincere thanks go to my thesis advisor Professor Robert M. Solow for his help and encouragement, and to Professor William M. Gorman whose helpful suggestions made this paper more readable. The preparation of this paper for publication was sponsored by the United States Army under Contract No. DA-11-022-ORD-2059 and was done at the Mathematics Research Center, United States Army, Madison, Wisconsin.

Under these assumptions, price p_t can take only the values

(4) $-R\pi, (-R+1)\pi, \ldots, -2\pi, -\pi, 0, \pi, 2\pi, 3\pi, \ldots, (R-1)\pi, R\pi.$

If these possible prices are considered as the states of a Markov process, then the dynamic assumption (2) shows that price p_t is undergoing a random walk in which transitions are possible only to the right neighbor $p_t + \pi$ or to the left neighbor $p_t - \pi$. Because the variate u_t has a rectangular distribution, we see that the one-step transition probabilities, defined by

(5) $q_{ij} \equiv \Pr\{p_{t+\tau} = j\pi \mid p_t = i\pi\}$

(where the vertical bar stands for "given that"), are

$$q_{i,i+1} = \Pr\{x_t \geqslant 0 \mid p_t = i\pi\} = \frac{1}{2} - \frac{i}{2R},$$

(6) $q_{i,i-1} = \Pr\{x_t < 0 \mid p_t = i\pi\} = \frac{1}{2} + \frac{i}{2R}, \; q_{ij} = 0 \text{ for } j \neq i+1, \, i-1.$

Here i ranges through integer values from $-R$ to R. Such a random walk describes the Ehrenfest model of diffusion, and is the random walk of an elastically bound particle (P. and T. Ehrenfest [2], E. Schroedinger and F. Kohlrausch [6], Ming and Uhlenbeck [5], Kac [3]).

   Let

(7) $q_{ij}(k) \equiv \Pr\{p_{t+k\tau} = j\pi \mid p_t = i\pi\}$

denote the higher transition probabilities. Since the one-step transition probabilities do not depend upon the initial time t, the higher transition probabilities also do not depend upon t; that is, the transition probabilites are stationary. The Kolmogorov [4] equation for Markov processes reduces to the difference equation (for $k > 0$)

(8) $q_{ij}(k) = q_{i,j+1}(k-1) \left[\frac{1}{2} + \frac{j+1}{2R}\right] + q_{i,j-1}(k-1) \left[\frac{1}{2} - \frac{j-1}{2R}\right]$

with the initial condition

(9) $q_{ij}(0) = \begin{cases} 1 \text{ for } i = j, \\ 0 \text{ for } i \neq j. \end{cases}$

From equations (8) and (9) we have respectively

(10) $\sum_j j\pi \, q_{ij}(k) = \left(1 - \frac{1}{R}\right) \sum_j j\pi q_{ij}(k-1),$

(11) $\sum_j j\pi q_{ij}(0) = i\pi$

so that the conditional mean of $p_t = j\pi$, given $p_0 = i\pi$, is

(12) $E\{p_t \mid p_0\} = p_0 \left(1 - \frac{1}{R}\right)^k,$

where $t = k\tau$. Hence the conditional mean price does not oscillate as $k \to \infty$, but instead damps out exponentially from the initial price p_0 to the equilibrium price 0. Professor Gorman points out that this is as one might have expected from the slow manner (2) in which the process gropes its way toward equilibrium. Similarly the conditional variance is found to be

$$(13) \quad \mathrm{var}\{p_t \mid p_0\} = \frac{\pi^2 R}{2}\left[1 - \left(1 - \frac{2}{R}\right)^k\right] + p_0^2\left(1 - \frac{2}{R}\right)^k - p_0^2\left(1 - \frac{1}{R}\right)^{2k}$$

where $t = k\tau$. As $k \to \infty$ the conditional variance tends to $\pi^2 R/2$. The stationary distribution of p_t then is given by the binomial distribution

$$(14) \qquad\qquad \Pr\{p_t = j\pi\} = \binom{2R}{j} 2^{-2R}.$$

2. ALTERNATIVE REPRESENTATION FOR DISCRETE TIME

Equations (2) and (6) may be written

$$(15) \qquad p_{t+\tau} - p_t = \varepsilon_{t+\tau} \equiv \begin{cases} \pi \text{ with probability} = \left(\dfrac{1}{2} - \dfrac{p_t}{2R\pi}\right), \\[2ex] -\pi \text{ with probability} = \left(\dfrac{1}{2} + \dfrac{p_t}{2R\pi}\right). \end{cases}$$

The conditional mean of $\varepsilon_{t+\tau}$ given p_t is seen to be

$$(16) \qquad E\{\varepsilon_{t+\tau} \mid p_t\} = \pi\left(\frac{1}{2} - \frac{p_t}{2R\pi}\right) - \pi\left(\frac{1}{2} + \frac{p_t}{2R\pi}\right) = -\frac{p_t}{R}.$$

Define the deviate $\delta_{t+\tau}$ by

$$(17) \qquad\qquad \delta_{t+\tau} \equiv \varepsilon_{t+\tau} - E\{\varepsilon_{t+\tau} \mid p_t\} = \varepsilon_{t+\tau} + \frac{p_t}{R}.$$

Then we have

$$E\{\delta_{t+\tau} \mid p_t\} = 0$$

and

$$(18) \quad E\{\delta_{t+\tau}^2 \mid p_t\} = \left(\pi + \frac{p_t}{R}\right)^2\left(\frac{1}{2} - \frac{p_t}{2R\pi}\right) + \left(-\pi + \frac{p_t}{R}\right)^2$$
$$\left(\frac{1}{2} + \frac{p_t}{2R\pi}\right) = \pi^2 - \frac{p_t^2}{R^2},$$

and equation (15) becomes

$$(19) \qquad\qquad p_{t+\tau} - \left(1 - \frac{1}{R}\right)p_t = \delta_{t+\tau}.$$

(It is interesting to compare equation (19) with the discrete time-parameter Markov process of the autoregressive type (Wold [7]) given by

$$(20) \qquad\qquad p_{t+\tau} - \varrho p_t = \xi_{t+\tau}, \varrho = 1 - \frac{1}{R} < 1$$

where $\xi_{t+\tau}$ and p_t are statistically independent.)

3. PASSAGE TO CONTINUOUS TIME

Let us now consider the limiting case

$$(21) \qquad \tau \to 0,\ \pi \to 0,\ R \to \infty$$

such that:

(a) The conditional mean $E\{\varepsilon_{t+\tau} \mid p_t\}$ becomes proportional to τ, as $\tau \to 0$; i.e.,

$$(22) \qquad E\{\varepsilon_{t+\tau} \mid p_t\} = -\frac{p_t}{R} \to -p_t\,\alpha\tau$$

where $\alpha \equiv \lim 1/R\tau$ is a constant.

(b) The conditional variance $E\{\delta_{t+\tau}^2 \mid p_t\}$ becomes proportional to τ, as $\tau \to 0$; i.e.,

$$(23) \qquad E\{\delta_{t+\tau}^2 \mid p_t\} = \pi^2 - \frac{p_t^2}{R^2} \to \sigma^2\tau$$

where $\sigma^2 \equiv \lim \pi^2/\tau$ is a constant.

Under these limiting conditions the conditional moments $E\{\delta_{t+\tau}^n \mid p_t\}$, $n > 2$, and the conditional covariance $E\{\delta_{t+\tau}\delta_{t+2\tau} \mid p_t\}$ tend to zero in a higher order. The limiting conditions insure that in small time intervals price can only change in small amounts.

Equation (19) may be written

$$p_{t+\tau} - p_t = -\frac{p_t}{R} + \delta_{t+\tau}$$

$$(24)$$

which under the limiting conditions becomes (with $\tau = dt$)

$$(25) \qquad dp_t = -\alpha p_t\,dt + dy_t$$

where y_t is a stochastic process with orthogonal (uncorrelated) increments dy_t, i.e.,

$$(26) \quad E\{dy_t\} = 0, \quad E\{(dy_t)^2\} = \sigma^2 dt, \quad E\{(y_{t4} - y_{t3})(y_{t2} - y_{t1})\} = 0,$$
$$t_1 < t_2 \leqslant t_3 < t_4.$$

That is, the increments dy_t are white noise with variance $\sigma^2 dt$. Integrating the stochastic differential equation (25) as a first order linear equation we obtain

$$(27) \qquad p_t = \int_{-\infty}^{t} e^{-\alpha(t-s)} dy_s = p_0 e^{-\alpha t} + \int_{0}^{t} e^{-\alpha(t-s)}\,dy_s$$

which exhibits price p_t as a linear combination of the past increments dy_s, $s \leqslant t$, weighted by $e^{-\alpha(t-s)}$. Equation (27) is thus the Wold [7] decomposition of price p_t (for continuous time-parameter t). We regard the p_t process as being started up in the remote past.

Under the same limiting conditions, equations (12) and (13) become

$$(28) \qquad E\{p_t \mid p_0\} = p_0 e^{-\alpha t}, \quad \mathrm{var}\{p_t \mid p_0\} = \frac{\sigma^2}{2\alpha}\left[1 - e^{-2\alpha t}\right]$$

and the Kolmogorov difference equation (8) becomes the parabolic partial differential equation of the Fokker-Planck (or diffusion) type

$$(29) \qquad \frac{\partial f}{\partial t} = \alpha \frac{\partial (p_t f)}{\partial p_t} + \frac{\sigma^2}{2} \frac{\partial^2 f}{\partial p_t^2}.$$

Here $f \equiv f(p_t \mid p_0)$ is the conditional density function of p_t given p_0, where $t > t_0 = 0$. The limiting conditions (21), (22), (23) would indicate that the p_t process is normal, in which case f would be a normal density function with mean and variance given by (28), i.e.,

$$(30) \qquad f(p_t \mid p_0) = \frac{\sqrt{2\alpha}}{\sigma \sqrt{2\pi(1 - e^{-2\alpha t})}} \exp \left[-\tfrac{1}{2} \frac{2\alpha(p_t - p_0 e^{-\alpha t})^2}{\sigma^2(1 - e^{-2\alpha t})} \right].$$

That this normality supposition (30) is correct can be immediately verified by substituting (30) into (29). By letting $t \to \infty$ in (30), or by taking the limit of (14), we find that the stationary distribution of price p_t is normal with mean zero and variance $\sigma^2/2\alpha$. Hence price p_t $(-\infty < t < \infty)$ is a normal stationary time series with autocovariance

$$(31) \qquad \mathrm{cov}\,\{p_{t+s} p_t\} = \frac{\sigma^2}{2\alpha} e^{-\alpha|s|}$$

and spectral density

$$(32) \qquad F'(\omega) = \frac{\sigma^2}{2\alpha} \int_{-\infty}^{\infty} e^{-\alpha|s|} \cos \omega s \; ds = \frac{\sigma^2}{\alpha^2 + \omega^2}, \qquad -\infty < \omega < \infty.$$

Since the p_t process is normal, the increments dy_t are also normally distributed, and so are not only orthogonal but statistically independent. The least squares prediction $E\{p_t \mid p_s, s \leqslant 0\}$ of the price p_t from the complete past p_s, $s \leqslant 0$, is thus equal to the least squares *linear* prediction $p_0 e^{-\alpha t}$, as given by the Wold decomposition (27). That price p_t is a Markov process is clearly seen from

$$(33) \qquad E\{p_t \mid p_s, s \leqslant 0\} = E\{p_t \mid p_0\} = p_0 e^{-\alpha t}.$$

As in the discrete case the conditional mean price $p_0 e^{-\alpha t}$ damps exponentially from the initial price p_0 to the equilibrium price 0 as $t \to \infty$.

Denote the conditional mean price $E\{p_t \mid p_0\}$ by $\bar{p}_t$. Then by equations (28) and (3) we have

$$(34) \qquad \frac{d\bar{p}_t}{dt} = -\alpha \, \bar{p}_t = \frac{\alpha}{a} \, \bar{x}(\bar{p}_t).$$

Professor Gorman points out that the similarity to Samuelson's system is remarkable when one considers how different is the basic assumption (2), and explains it by saying that what happens, of course, is that the proportion of adjustments which go in the "correct direction" is a linear function of $\bar{x}$.

In this paper we have restricted ourselves to Markov processes of the diffusion type: i.e., Markov processes in which price can only change by small amounts in small times. Markov processes in general have the local structure of additive processes, and hence can contain two possible components: the *diffusion* (normal and continuous) term and the *transition* (Poisson and discontinuous) term (Bartlett [1]). The transition component would have arisen if price were allowed to change abruptly (discontinuously) in small time intervals. In such a case the Kolmogorov equation would not become in the limit a diffusion equation such as (29) but instead would become in the limit an integro-differential equation of the Boltzmann type in the kinetic theory of gases.

The University of Wisconsin and Mathematics Research Center, U. S. Army

REFERENCES

[1] BARLETT, M. S.: *An Introduction to Stochastic Processes with Special Reference to Methods and Applications*, Cambridge: Cambridge University Press, 1955.

[2] EHRENFEST, P. AND T.: "Über zwei bekannte Einwände gegen das Boltzmannsche H-Theorem," *Physikalische Zeitschrift*, Vol. 8 (1907), pp. 311-314.

[3] KAC, M.: "Random Walk and the Theory of Brownian Motion," *American Mathematical Monthly*, Vol. 54 (1947), pp. 369-391.

[4] KOLMOGOROV, A.: "Über die analytischen Methoden in der Wahrscheinlichkeitsrechnung," *Mathematische Annalen*, Vol. 104 (1931), pp. 415-458.

[5] MING CHEN WANG AND G. E. UHLENBECK: "On the Theory of the Brownian Motion II," *Reviews of Modern Physics*, Vol. 17 (1945), pp. 323-342.

[6] SCHRÖDINGER, E. AND F. KOHLRAUSCH: "Das Ehrenfestsche Model der H-Kurve," *Physikalische Zeitschrift*, Vol. 27 (1926), pp. 306-313.

[7] WOLD, H.: *A Study in the Analysis of Stationary Time Series*, Stockholm: Almqvist and Wiksell, 1938; Second Edition, 1954.

CHAPTER 3

PREDICTION AND FORECASTING

Enders A. Robinson

Lincoln, Massachusetts

and

Herman Wold

Uppsala, Sweden

20.1. THE NOTION OF PROBABILITY

THE concept of probability is very deeply embedded in human experience. Let us look at its manifestation in language.* An English word whose origin is the Old Norse word *happ* is *hap*. *Hap* (n.) means *luck* or *chance*, whereas *to hap* (v.i.) means *to befall*. From this basic word the following English words are derived:

  (1) *to happen* (v.i.): to occur by chance.
  (2) *happening* (n.): occurrence, that which happens.
  (3) *haply* (adv.): by chance.
  (4) *haphazard* (adj.): random, determined by chance.
  (5) *perhaps* (adv.): by some chance, maybe.
  (6) *happy* (adj.): favored by luck or fortune.
  (7) *happily* (adv.): by good fortune.
  (8) *happiness* (n.): good luck, a state of well-being.
  (9) *happy-go-lucky* (adj.): trusting to luck, easygoing.
(10) *hapless* (adj.): unlucky.
(11) *mishap* (n.): ill luck, an injurious accident.

The words (n.)

hap	*luck*	*chance*
fortune	*lot*	*hazard*

---

* The following abbreviations are used: adj. = adjective; adv. = adverb; n. = noun; v.i. = verb intransitive; v.t. = verb transitive.

cannot be properly defined. They are more or less synonymous. They all agree in designating that which *happens*, that which befalls either partially or entirely as the result of unknown, unconsidered, or unpredictable forces. Synonymous words such as these continually grope their way through language by connotation, that is by their suggestive significance. There is nothing less precise than connotation which changes from person to person and from time to time. It is based on associations, many of which are below the conscious level. Nevertheless, without connotation we could not have language as we know it.

The word *luck* (n.) usually refers to that which happens to one personally or individually. The word *luck* is associated etymologically and by continual use with gambling, as:

It was a wonderful run of luck.
Don't abuse luck by playing all over the roulette table layout.
Playing the roulette wheel is strictly luck and nothing else.
It would be a kick in the teeth to the luck which had been given him.
Luck only enters the game at sporadic intervals.
Luck has no influence in the long run.

The word *luck* is used constantly in colloquial speech, as:

His winning was just dumb luck.
It was just his luck to fall off the boat.
He had bad luck at his job today.
The hunter had bad luck.
The man tried his luck at the old mine.

The word *luck* unqualified can imply a happy outcome, as:

Wish me luck.
He had luck.

The derived word *lucky* (adj.) means favored by luck, fortunate, as:

My lucky number is 3.
I have no lucky numbers.
He was lucky to win again.
You were lucky to find your lost ring.
Lucky in cards, unlucky in love.

An American would say *good luck*, a Frenchman for the same occasion would say *bonne chance*. The English word *chance* (n.) comes from the Old French *cheance*, which in turn comes from the Late Latin *cadentia*, which means a falling, especially of dice or of fortune. The Late Latin word *cadentia* itself is derived from the Latin word *cadere* meaning to fall, to happen. Thus we have found an important link: the word *chance* has its roots with the falling of dice.

The English word *chance* (n.) serves often as the general name for the incalculable and fortuitous element in human existence and in nature as:

Most phenomena are influenced by chance.

In common usage, *chance* seldom loses implications derived from its original association with the throwing of dice and the selection of one outcome out of many possible outcomes by this means. Consequently, it may mean determination by irrational, uncontrollable forces, as:

He left things to chance.

It may mean any one of the contingencies on which a player takes a risk in a game of chance, as:

Number 7 is my chance.

It may mean any risk or gamble, as:

This farm is my chance.
Take a chance and accept the position.
It is my only chance.

It may mean an opportunity that comes seemingly by luck or accident, as:

When I get a chance, I'll leave.
It was my last chance to escape.
He has long hoped for a chance of promotion.

It may mean a possibility of something happening, as:

There is little chance of rain.
The chances are even.

It may mean a possibility of success among many possibilities of failure, as:

He is always willing to take a chance.
What are his chances?
The chances are against him.
He has one chance out of a million.

It may mean a measure that an event happens, as:

What is the chance of the bullet hitting the target?
Its chance of happening is 2 out of 10, or 0.2.
What is the chance of drawing two kings in a poker hand?
What chance is there of rain this weekend?
There is little chance.

The English word *chance* (v.i.) means to happen, come or arrive without design or expectation, as:

He chanced upon this new concept.

Chance (v.t.) means to take the chances of, to risk, as:

To chance a swim in such cold water is foolish.

A word etymologically related to *chance* is *case*. *Case* (n.) comes from the Old French *cas*, which is from the Latin *casus*, which in turn is from the Latin *cadere* meaning to fall, to happen.

The English word *chance* (adj.) means happening by chance. Synonymous are *haply, accidental, fortuitous*, and *aleatory*.

A word related to *chance* is *accident*. The word *accident* came into English through French from the Latin word *accidere*, which means to happen and which is a compound of *ad* plus *cadere* (to fall). Thus an accident is a happening or an event which takes place without one's foresight or expectation, and now often means one of an afflictive, injurious, or unfavorable character, as:

He was in an automobile accident.

The derived word *accidental* (adj.) means happening by chance or unexpectedly. Although the usage of the word *accidental* usually stresses chance, it now sometimes may stress nonessentiality.

The word *fortune* (n.) came from the Old French *fortune*, which in turn came from Latin *fortuna*. In addition to its meaning of luck, chance, it also means (sometimes capitalized) the personified power of chance, the personification of the cause of that which befalls in a sudden or unexpected manner. Thus we hear about *Dame Fortune* who is the same mysterious woman as *Lady Luck*. Sometimes she is pictured as having a large wheel, which she spins to determine the various happenings of life.

The derived word *fortunate* (adj.) means coming by good luck, auspicious, lucky. The synonyms *lucky, fortunate*, and *happy* all connote having a favorable issue. *Lucky* implies success by chance rather than as the result of merit. *Fortunate* is less suggestive of a favorable accident and may carry the connotation of being watched over by a higher power or of being favored beyond one's desserts. *Happy* combines the implications of *lucky* and *fortunate* to express gratification in the sense of well-being and of complete satisfaction. Thus *happy* becomes synonymous with *glad* (from Anglo-Saxon *glaed* meaning bright, glad) to mean characterized by joy or pleasure, pleased.

The English word *fortuitous* (adj.) comes from the Latin word *fortuitus*, which is from the Latin *forte* (adv.) meaning by chance, which in turn is from the Latin *fors, fortis* meaning chance. *Fortuitous* means happening by chance or accident and is synonymous with *accidental*. Whereas both *accidental* and *fortuitous* mean not expected, outside of the regular course of things, *fortuitous* more strongly suggests chance than accidental and often connotes absence of a cause.

The word *lot* comes from the Anglo-Saxon word *hlot*, which was an object used as a counter or check in determining a question by chance. *To choose by lot* is the use of such a counter as a means of deciding something. *Lot* has come to mean that which befalls one from a choice of lot, and hence a share or allotment. *Lot* has also come to mean hazard, fortune, especially the fate which falls to one by the will of an overruling power, as:

He is a man content with his lot.

Always in the usage of *lot* to mean the state or end predetermined for one, there is the suggestion of the operation of blind chance. Like the word *luck*, the word *lot* usually refers to that which happens to a person as an individual.

The derived word *lottery* (n.) means a scheme for the distribution of prizes by lot, especially such a scheme in which lots, or chances, are sold, and has come to mean figuratively an affair of chance.

The word *hazard* (n.) comes to English through the Old French word *hasard*, which is from Arabic *al-zahr*, which means the die. Originally *hazard* was a dice game of which the present-day craps is a simplified form. Thus *hazard* came to mean chance. In addition *hazard* also has come to mean risk, danger, peril.

The derived word *hazardous* (adj.) thus means both depending upon chance or luck and dangerous, risky. In fact hazardous implies so many chances of evil and/or harm that the thing so described is exceedingly dangerous.

The Latin word *alea* meaning die, chance, did not come directly into the English language. Nevertheless, this word is known from the famous expression of Caesar when he crossed the Rubicon River: *alea jacta est* (which means *the die is cast*).

The derived Latin word *aleatorius* has come into English as *aleatory* (adj.). It means pertaining to or resulting from luck. In legal terminology it means depending upon an uncertain event or contingency as to both profit and loss. Thus *aleatory contracts* include lottery agreements, wagering contracts, and insurance contracts.

Let us now look at the word *random*. Its earliest known meaning is that of furious action, such as a charge of cavalry. It seems to be connected with the Teutonic word *rand*, which means *brim*, and implies the furious and irregular action of a river full to the brim. The Swedish word *rand* means verge, brink, brim. The English word *random* is related to the Old French *randon*, which means violence, rapidity. *Random* (n.) means an aimless course or progress. It is used chiefly, in the expression *at random* which means without definite aim, direction, rule, or method. *At random* and *at haphazard* are synonymous in the sense now described.

Random (adj.) and *haphazard* (adj.) are synonyms. Both mean having a cause or a character that is determined by accident rather than by design or by method.

Random signifies that there is no fixed or clearly defined aim, purpose, or evidence of method or system or direction, and hence implies no or little guidance by a governing mind, eye, objective, or the like, as:

> The man aimed the bow at random.
> The knight wandered at random.

Here *random* indicates the aimless character of the performance as contrasted to the definite intention to hit a certain mark. Some examples of this usage are:

> He only has a random collection of books on science.
> It was only a random allusion to the subject.
> It was a random shot into the woods.

Haphazard, on the other hand, signifies an aimlessness, or randomness, that is more or less at the mercy of chance, as:

> The garden was a haphazard arrangement of shrubs and plants.
> The rain fell in a haphazard pattern on the lawn.
> The foreign policy is haphazard.

In mathematics the word *random* is used as the word haphazard is used in ordinary language, that is, with a definite connotation of chance.

Now we want to look at the English words *probable* (adj.) and *likely* (adj.), and the corresponding words *probability* (h.) and *likelihood* (n.).

First let us look at the correspondence:

> Latin: probabilis
> English: probable
> French: probable
> Italian: probabile

The Latin word *probabilis* (adj.) comes from the Latin word *probare* meaning to try, to test, to prove. The English word *probable* comes through French from the Latin *probabilis*. Thus etymologically *probable* is related to *provable*.

The Latin word *verisimilis* (adj.) comes from the Latin *verus* true (genitive: *veri*) + *similis* like. The following correspondence exists among modern languages:

> Latin: verisimilis   (*veri* true + *similis* like)
> English: verisimilar   (*veri* true + *similar* like)
> German: wahrscheinlich   (*wahr* true + *scheinlich* appearing)

Dutch:	waarschijnlijk	(*waar* true + *schijnlijk* appearing)
Swedish:	sannolik	(*sanno* true + *lik* like)
	trolig	(*tro* true + *lig* like)
Danish:	sandsynlig	(*sand* true + *synlig* appearing)
	trolig	(*tro* true + *lig* like)
Russian:	veroyatnii	(*vernii* true)
French:	vraisemblable	(*vrai* true + *semblable* like)
Italian:	verisimile	(*veri* true + *simile* like)

All the above words were built up from the basic stems shown in the parentheses to express the meaning: true-like, true-seeming, true-appearing, true-resembling, true-similar. The English word *verisimilar* is not commonly used, but instead the corresponding word *likely* is used. The word *likely* is made up of Anglo-Saxon roots as opposed to the Latin derivation of its counterpart *verisimilar*.

The English word *likely* (adj.) is made up of *like* (adj.) (which means having the same, or nearly the same, appearance, qualities, or characteristics as another or others referred to as a basis of comparison) + *ly* (which is a suffix that forms adjectives and mean like in appearance, manner, or nature.) *Like* comes from Anglo-Saxon *lic* body, and it originally means having the same body or shape, and hence, like; whereas *-ly* also comes from Anglo-Saxon *lic* body.

The English words *probable* and *likely* are synonyms. They signify uncertainty that may be, or become, true, real, or actual. Something is *probable* if it has so much evidence in its support or seems so reasonable that it commends itself to the mind as worthy of belief, although not to be accepted as a certainty. Thus the *probable* conclusion from evidence at hand is the one which the weight of evidence supports even though it does not provide proof, as:

He is the probable author of the medieval manuscript.
The probable cause of the fire was faulty wiring.
What are the probable expenses for a trip to Europe?
What is the probable origin of the folk song?

The word *likely* is very close to the word *probable,* and so they can often be used interchangeably. In contrast to *probable, likely* does not as often or as invariably suggest grounds sufficient to warrant a presumption of truth, as:

It is not the probable place where they will meet, but it is still likely.

Also *likely* is sometimes used in the sense of *promising* because of appearances, ability to win favor, etc., as:

She is a likely candidate.

Using these words we may establish a scale of expressions, as:

It is *certain* to snow.
It is *very likely* to snow.
It is *likely* to snow.
It is *as likely* to snow *as not*.
It is *unlikely* to snow.
It is *very unlikely* to snow.
It is *certain not* to snow.

Or as:

He is the *certain* winner.
He is the *probable* winner.
He is a *likely* winner.
He is an *unlikely* winner.

The word *likelihood* (n.) is derived from *likely*. We have the correspondence:

English:	verisimilitude	
	likelihood	
German:	wahrscheinlichkeit	(*keit* hood)
Dutch:	waarschijnlijkheid	(*heid* hood)
Swedish:	sannolikhet	(*het* hood)
Danish:	sandsynlighed	(*hed* hood)
Russian:	veroyatnost	
French:	vraisamblance	
Italian:	verisimiltudine	

The English word *probability* (n.) is derived from *probable*. The corresponding word in French is *probabilité*, and in Italian *probabilità*.

Likelihood and *probability* are synonyms. They mean quality or state of being likely or probable, as

The likelihood of winning is small.
His probability of being chosen is great.
The probability of hitting the target is small.

As we have seen, the word *chance* (n.) etymoiogically is related to the falling of a die. On the other hand, the word *probability* (n.) etymologically is related to trying or testing a thing against some truth. In mathematical usage, the words *probability* and *chance* are overlapped so that they both signify

a measure that an event happens.

The words that we have considered so far have many connotations. Nevertheless, we may be comforted by the word *stochastic*, introduced by

James Bernoulli (1654–1705) explicitly for the purpose of probability theory. *Stochastic* (adj.) is taken from the Greek word *stochazesthai* (v.), which means to shoot at a mark, and which was used by Plato in the sense of to aim at an ideal. Although mathematicians have been using the word *stochastic* since the time Bernoulli introduced it, it has not yet entered common language, and so lacks such connotations as are associated with the other words examined in this chapter.

The manyfold connotations of words used in everyday life can cause trouble when these same words are used in a precise scientific sense. Lichtenberg has said:

"All our philosophy is a correction of the common usage of words."

Here we do not wish to undertake a correction of common word usage. Instead we wish to develop certain concepts on a mathematical basis with a view toward their intuitive background and their applications. The content of the concepts are not derived from the meanings popularly given to the words, and is therefore independent of current word usage. The correctness of these concepts can only be judged from their fruitfulness, both as creative ideas and as useful instruments for advancing scientific investigation.

20.2. INTRODUCTION TO PREDICTION THEORY

A stochastic phenomenon well known to everyone is:

A baby is to be born. What is its sex?

Science up to this point has not found a way to *predict* the sex of a newborn baby. The biological mechanism that determines the sex is not understood, and research shows that this problem is exceedingly complex. Nevertheless, the possible results of the phenomenon are very simple, namely the result must be one of two alternatives, namely *boy* or *girl*. (Note: we treat the birth of twins as two separate births, the birth of triplets as three separate births, etc.).

For this stochastic phenomenon, and for other stochastic phenomenon as well, mankind has been lead to the study of *statistical data* in order to gain *empirical experience*.

To illustrate what we mean let us look at the birth of a baby. Of course, for each individual birth we are unable to *predict* whether the baby will be a boy or a girl. But the accumulated *empirical experience* of mankind tells us that out of a large number of newborn babies, approximately 50% will be boys.

Thus in ancient times it was observed that the ratio of male births to the total number of births in entire countries remained almost unchanged from year to year. This ratio was found to be half by the Chinese over 2200 years ago. How did the ancient Chinese attain the empirical knowledge that the sex ratio is 50%? They arrived at it by means of statistical data collected by censuses of their population.

In modern times, of course, statistics are kept much more accurately, and so today we know the sex ratio more precisely. Thus out of a large number births, approximately 51% will be boys and 49% will be girls.

It is important to bear in mind that over the years the sex ratio has been determined from the available statistical data of actual births and not by theoretical reasoning. Thus out of many births the number of boys and the number of girls are counted. Such numbers form the pertinent statistical data. For example, the statistics for the United States in the year 1950 show that there were

1,823,555 boys born

and

1,730,594 girls born

out of a total of

3,554,149 births.

From this data, the ratio of the number of boys to the total number of births is

$$\frac{1823555}{3554149} = 0.513.$$

That is, 51.3% of the babies born in 1950 were boys.

The lesson we draw is this. Statistical data such as the number of boys and girls born in the United States in 1950 establishes and supports our empirical knowledge. Thus, although the sex of any particular birth is unknown in advance and so represents a stochastic phenomenon, we are comforted by our accumulated empirical experience which tells us that out of 100 births approximately 51 will be boys and 49 will be girls.

From statistical evidence, the empirical *experience* of mankind grows, thereby making the study of stochastic phenomena more amenable, and the uncertainties of anticipation more bearable. Our empirical experience is that 51% of births are boys. A remarkable feature of this statistical ratio is its stability, that is, it does not noticeably change from place to place and from year to year. Of course there is some fluctuation, but this fluctuation is relatively small.

The stability of the statistical ratio of sex is well understood and appreciated by mankind. We might say that our entire social structure is based on this stability. Think of the repercussions that would result if for some generation or for some countries there were approximately twice as many girls born as boys.

Despite the fact that there are many babies being born all the time, we cannot get away from the fact that any particular birth is a stochastic phenomenon *by itself*. Its stochastic nature does not depend upon the fact that other births are taking place. We can always look at any particular birth *by itself*, isolated from other births. The fact remains that the particular birth is stochastic, that is, either a boy or girl may happen.

Let us consider an event associated with a stochastic phenomenon. The fundamental notion is that this event has a *probability*. This probability is a number, between 0 and 1. This number is called:

> *the probability that the event happens*

or, when the context is clear,

> *the probability of the event,*

or

> *the probability of happening,*

or simply

> *the probability.*

More specifically, consider the event "a boy" for the stochastic phenomenon "birth". This event has a probability, which we call

> *the probability the event "a boy" happens*

or simply

> *the probability of a boy.*

To illustrate the usage of mathematical notation, we may let B designate the *event* " a boy". Then we may let $P(B)$ (which is read "P of B") designate the *probability of a boy*. The probability $P(B)$ is a number. This number must lie between 0 and 1, that is, it must not be less than 0 nor greater than 1.

But what is this number? No one knows exactly. But the notion of this probability is natural to each of us, and through experience and reflection we may obtain what we consider a reasonable evaluation of its numerical value.

As we have seen, mankind has not yet been successful in the use of theoretical methods to evaluate this probability. Nevertheless, through experience we do know that the 51 % of all births are boys, that is, the ratio is 0.51. On this basis we have reason to place the probability of a boy to be 0.51. In mathematical notation we have

$$P(B) = 0.51.$$

There are two *means* of evaluating probabilities of events:

(1) by accumulated experience;

(2) by theoretical reflection and argument.

In the case of the probability of a boy, means (1) was used. Nevertheless, a breakthrough in biological research might allow us also to use means (2).

Let us now illustrate the use of means (2), namely, the use of theoretical reflection and argument, to evaluate a probability.

Let the stochastic phenomenon be the tossing of a homogeneous and symmetrical twenty-sided die. There is no accumulated empirical experience on this phenomenon. Nevertheless, we do have much accumulated experience on the tossing of a six-sided die which very accurately meets the ideal homogeneous and symmetrical properties. Our experience there is that the probability of each face is 1/6.

By theoretical reflection and argument we may therefore evaluate the probability of each face of homogeneous, symmetrical twenty-sided die to be 1/20.

Science is concerned with the study of phenomena. A *phenomenon* may be described as anything of scientific interest susceptible to theoretical description and explanation.

Natural sciences deal with the theory of various natural phenomena. Chemistry studies chemical phenomena. There are divisions, such as organic chemistry, inorganic chemistry, and physical chemistry, depending upon the type of phenomena. Physics is divided into fields such as mechanics, optics, thermodynamics, electromagnetism, and nuclear physics. Biology is concerned with the phenomena associated with living organisms. It is divided into botany and zoology, that is the study of plant and animal phenomena respectively. There is the science of genetics which studies hereditary phenomena. Geology, oceanography, and meteorology are concerned with phenomena associated with the earth, the sea, and the air. Astronomy studies planetary and stella phenomena.

Behavioral sciences are concerned with the theory of phenomena associated with relationship among men and also among higher animals. Psychology studies behavioral phenomena of humans and animals. Sociology studies phenomena associated with human social structures and economics studies economic phenomena.

One of the purposes of mathematics is the construction of *mathematical models* of phenomena. A model is conceptual in that it represents a mental image of the underlying phenomenon. No model is perfect in the sense that further improvements are not possible. To attain good models both the scientist and the mathematician must work closely together.

The science in which mathematical models have been the most fruitful is physics. A well-known example of a mathematical model is Newton's second law. It states that force F, mass m, and acceleration a are related by the equation

$$F = ma.$$

This equation is the mathematical model. This equation could be given to a mathematician, who would then study its properties. The mathematician might hypothesize that force F may be various mathematical functions, from which he would arrive at the corresponding solutions of this equation. Carried to an extreme, the mathematician might conclude that there is no physical phenomenon at all, for all that he is left with are certain mathematical problems associated with the solution of equations, which in turn can be considered as belonging to a well-established mathematical domain. From this point of view this part of the science of physics appears to lose its individual existence and becomes part of the branch of mathematics.

This identification of two things belonging to different categories is not allowable. A mathematical model can never be identical to a physical theory. Physics remains the natural science concerned with the description and explanation of physical phenomena. Mathematics remains the conceptual science concerned with the construction of mathematical models of such phenomena. The role of the *task* physics and the *tool* mathematics must never be identified as the same thing.

A large part of mathematics is concerned with the building of mathematical models of sure phenomena. A sure phenomenon is one in which there is no degree of uncertainty. A sure phenomenon is identified by:

(1) its circumstances;

(2) its result.

Its circumstances are the antecedent conditions. Whenever the circumstances are fulfilled, the result occurs. Such phenomena are very common, so let us list a few examples:

Sure Phenomena

Example	*Circumstances*	*Result*
1.	Water is cooled to a temperature of 0°C at normal atmospheric pressure	The water freezes
2.	Water is heated to a temperature of 100°C at normal atmospheric pressure	The water boils

3.	Bar of iron is put into water	The bar sinks
4.	An acid is mixed with a base	A salt and water is formed until chemical equilibrium is reached
5.	A force is applied to a mass	The mass accelerates
6.	Two parents, each with a pair of dominant genes, mate. The offspring gets one gene from each parent	The offspring has a dominant pair of genes

The prediction of events related to a sure phenomenon belongs to the subject matter of the discipline itself and is not the type of prediction with which we are concerned in this essay.

The branch of mathematics known as *probability theory* is concerned with the building of *mathematical models of stochastic phenomena*. These models are called *probability models*.

A stochastic phenomenon is one in which there is a degree of uncertainty. A stochastic phenomenon is identified by:

(1) its circumstances;

(2) its possible results.

Its circumstances are the antecedent conditions. We recall that a sure phenomenon has only one result. In contrast, a stochastic phenomenon has many (that is, more than one) *possible results*. Whenever the circumstances of a stochastic phenomenon are fulfilled, one and only one of the possible results *happens*.

Here and always we use the word *to happen* with its English language meaning:

to happen (v.i.):   to occur by chance.

A stochastic phenomenon is therefore a *happening*.

Stochastic phenomena are very common in human experience. Here are a few examples:

Stochastic Phenomena

Example	*Circumstances*	*Possible Result*
1.	A coin is tossed	Either heads or tails
2.	A die is tossed	Any one of 1, 2, 3, 4, 5, 6

3.	A card is drawn from the top of a pack of 52 cards	A card of any suite and any value
4.	A rain drop falls on a piece of paper	Any one of the points on the paper
5.	A seed falls from a tree and lands in a meadow	Any one of the points in the meadow
6.	An arrow is shot at a target	Any of the points on the target

For the stochastic phenomenon in question, we must agree on what is meant by the possible results. Thus in Example 1 above the coin could stand on its edge and thereby not fall heads or tails. Nevertheless, it is expedient in most applications to regard heads and tails as the only possible results. Such idealizations are standard practice in scientific model building, and indeed must be made in order to make progress with most problems.

Another idealization is illustrated by Example 4. A rain drop certainly is not a point, but it is expedient to consider it as such in this problem. In Example 6 we have idealized the problem so that the target is large enough so that the arrow always hits the target.

In reference to Example 3, an unstated assumption is that the pack of cards has been well shuffled. On the other hand, if the pack of cards is ordered so that the ace of spades is on top, and this fact is known, then the drawing of the card from the top of the pack would not be a stochastic phenomenon, but would be a sure phenomenon. Such unstated assumptions as the proper shuffling of cards are usually made in problems connected with probability theory.

First of all we need a shorter, and easier to say, expression for *possible result*. The word that we will use instead of the expression *possible result* is *case* (n.). The use of the word *case* for this purpose is traditional, and is traceable to the origin of the word in the Latin *cadere* meaning to fall, to happen.

Thus, when a coin falls, there are two cases, namely heads and tails. When a die falls, there are six cases, namely 1, 2, 3, 4, 5, 6.

Let us now look more closely into what we mean by a possible result, or case.

We recall that whenever the circumstances of a stochastic phenomenon are fulfilled, one and only one case happens. The phrase *one and only one case happens* means the following three things:

(1) Less than one case is not admitted, that is, the cases are *elemental.*

(2) One case must happen, that is, the cases are *exhaustive.*

(3) More than one case cannot happen, that is, the cases are *exclusive.*

Fix these three things in your mind. The possible results, or cases, of a stochastic phenomenon are EEE:

(1) E for *Elemental.*

(2) E for *Exhaustive.*

(3) E for *Exclusive.*

For example, the *cases* in tossing a die are the *faces*

  1, 2, 3, 4, 5, 6.

The faces are EEE:

(1) E for Elemental: we do not allow the faces of the die to be subdivided, so that only part of the face shows. That is, *a part of one face is not admitted.*

(2) E for Exhaustive: we do not allow the die to stand on an edge, so that no face shows. That is, *one face must show.*

(3) E for Exclusive: we do not allow the die to show two faces, say face 4 and face 5, at the same time. That is, *only one face must show.*

In saying that the cases are *elemental,* we mean that *the cases are elements* in that they are indecomposable. The individual cases cannot be subdivided or cut up in any way.

In saying that the cases are *exhaustive,* we mean that *the cases exhaust all possibilities.* No cases have been left out.

In saying that the cases are *exclusive,* we mean that *the happening of one case excludes the happening of the other cases.* Two cases cannot happen in any given instance of the phenomenon.

When we are confronted with a stochastic phenomenon, our first task is to make an analysis of its possible results, or cases. Generally the analysis can be done in several different ways, and the particular way chosen depends upon the problem at hand.

For example, suppose two coins are tossed, namely coin 1 with sides heads H and tails T and coin 2 with sides heads h and tails t.

Let us ask what are the possible results, the cases? Suppose there are three problems that we are interested in:

  Problem 1:   Which side appears for each coin?

  Problem 2:   How many heads appear?

  Problem 3:   Whether the two coins match (that is, both heads or both tails) or not match (that is, heads on one and tails on the other)?

As a graphic aid, we may consider the three diagram shown in Fig. 20.1.

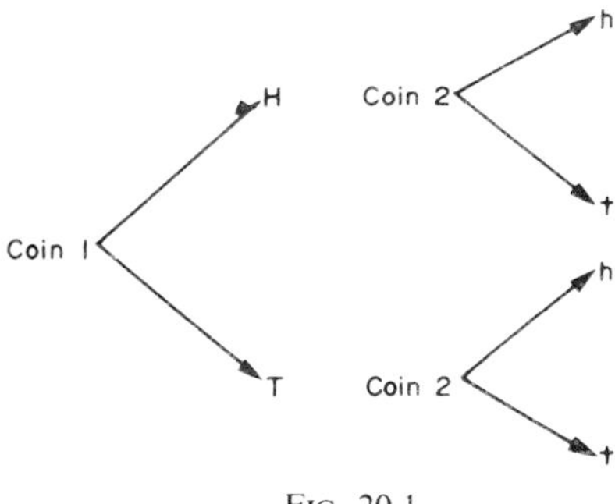

FIG. 20.1

Thus for the three problems we might propose the following three sets of cases:

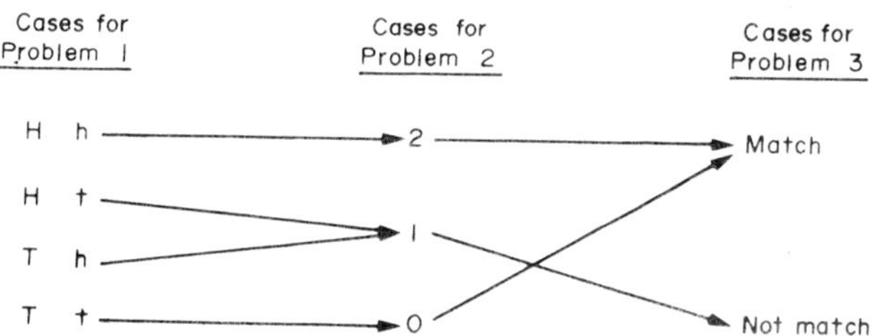

Each set of cases is an admissible set, that is, one and only one case happens. Nevertheless it is seen that the cases for Problem 1 represent a finer analysis than the cases for Problem 2. In other words, if we know which case in the listing for Problem 1 happens, then we may tell which case happens in the listing for Problem 2 (as shown by the arrows). Nevertheless, the reverse is not true. Also we see that the analysis for Problem 2 is finer than the analysis for Problem 3.

For some purposes the finer analysis of cases may be necessary, whereas for other purposes the grosser analysis of cases may be adequate. It is important to realize that the cases in a given problem may be analyzed in many different ways, from a very rough analysis to a highly refined one. The two conditions that must be met for any analysis of cases are:

Condition 1: under the circumstances of the stochastic phenomenon, one and only one of the cases happens, that is, the cases are EEE:

> E for elemental
>
> E for exhaustive
>
> E for exclusive

Condition 2: the analysis is fine enough to meet the needs of the problem.

One thing is clear: *if a problem makes sense for a rough analysis of the cases, it will still make sense for any finer analysis.* Nevertheless, the converse is not always true. That is, some problems which make sense for a fine analysis may not make sense for a rougher analysis.

In subjects that are more or less directly concerned with specific types of objects, a clear indication of the subject matter presents no serious difficulty at the outset of a book. A reference to some of the objects can convey a certain amount of preliminary information and thereby will give the student a somewhat accurate idea of what is to follow. Such is the case in most natural sciences.

In contrast the state of affairs in prediction theory is not so easy. Prediction theory is not directly concerned with definite objects, but with a certain underlying characteristic common to some phenomena. Phenomena with this characteristic are found in all branches of science. Because often such phenomena have little else in common, a student unacquainted with prediction theory must rearrange in his mind the usual scientific classification of these phenomena, so as to be able to look at those possessing this unifying characteristic.

For this purpose we have classified phenomena into two categories, namely sure phenomena and stochastic phenomena.

The stochastic phenomena are the ones that possess the unifying characteristic that we are interested in from the point of view of prediction.

A stochastic phenomenon has this characteristic: *for each instance that the circumstances are fulfilled, one and only one case happens.* This is the unifying characteristic of the phenomena that are studied in prediction theory.

Therefore in order to obtain a clear understanding of prediction theory one must first have a clear grasp of what constitutes stochastic phenomena. A stochastic phenomenon is a *happening.* It is not an occurrence that can be exactly fixed in advance, perfectly predicted, etc., but instead it is an occurrence that depends in some way and to some extent upon chance, luck, hap, etc. All that we mean is that there is something about a stochastic phenomenon that is uncertain, unsure. This uncertainty as a rule is not entirely unlimited, but only prevails in certain directions and up to certain points.

For example, the statement

A man attains some age before death

represents a stochastic phenomena. The age to which a man lives cannot be exactly fixed in advance. No person can calculate what may be the length of any particular life. The hour of a man's death is not predetermined within man's knowledge. When death comes, it *happens.* This happening is a stochastic phenomenon. The universe that represents his age consists of all the numbers from 0 as the lower limit to say 150 years as the upper limit, as we can feel perfectly certain that his life will not stretch out more than 150 years. Thus the uncertainty about a man's final age is not entirely unlimited, but can be said to lie within bounds extending from 0 to 150 years.

We toss a penny into the air. It happens to fall heads or it happens to fall tails. This falling is a stochastic phenomenon. Its universe is *heads, tails,* so that the uncertainty here is limited to two cases.

Suppose that we measure the height to which a (normal) man grows. His height will of course lie between certain extremes, say between 1 meter and 3 meters. The height to which he grows represents a stochastic phenomenon. His height is a happening, that is an occurrence dependent to some extent upon unknown and unobtainable factors.

As another example, consider the sex of a newborn baby. The baby may happen to be a boy or happen to be a girl. Its sex represents a stochastic phenomenon. The universe consists of two cases, namely *male, female.*

One point should be made very clear. Some of the previous examples should have brought this point out, but anyhow it is worth stressing again. The point is this: the role playing by chance, luck, hap, etc., in a stochastic phenomenon is not required to be complete but may be to any degree.

For example, when a coin or a die is tossed we usually regard the role played by chance as being entire and the role played by the tosser as having no effect on the ultimate result. Nevertheless there may be individuals with extraordinary skill who can control some of the conditions in tossing the coin, so that they have some influence on the way it falls.

On the other hand, consider the act of shooting an arrow at a target. Because the man takes aim, the ultimate result to some extent depends upon his skill. Nevertheless this act is a stochastic phenomenon because the exact point where the arrow strikes is uncertain. The role played by the man has some effect on the ultimate result and the role played by chance has some effect. Thus chance may be regarded only as a co-agent in the final result.

In fact most stochastic phenomena are of the type in which chance may be regarded only as a co-agent in the ultimate result, and these are the phenomena which properly constitute the subject matter of prediction theory.

We have classified phenomena into two categories, namely:

(1) sure phenomena;

(2) stochastic phenomena.

There are other classifications. In particular, we may classify phenomena into the following two categories:

(1) experimental phenomena;

(2) non-experimental phenomena.

An *experimental phenomenon* is one which can be repeated for an indefinite number of instances. In each instance the same circumstances are fulfilled and one result is produced. Thus if the experimental phenomenon is repeated five times, a series of five results is produced. All other phenomena fall into the categories of non-experimental phenomena.

Phenomena that are both stochastic and experimental are very useful in gaining insight to the methods of prediction theory. For brevity we shall replace the long expression *experimental, stochastic phenomenon* by the single word *experiment*. Thus an *experiment* is a stochastic phenomenon that can be repeated indefinitely, where each repetition is made under the same circumstances and produces one result. Twenty repetitions of the experiment would produce a series of 20 results. Each result, of course, must be one and only one of the possible results, or cases, of the stochastic phenomenon.

An example of an experiment is the tossing of a coin. This experiment can be repeated an indefinite number of times. For each repetition, either H or T happens. Thus a series produced by 20 repetitions might be

HHHTTHHTHTHHTTTHTHTT

The intuitive concept of experiments *repeated under identical conditions* leads to the notion of *stochastic independence*.

When a scientist says that two experiments are performed under identical conditions, he implies independence, that is, he implies that the result of either experiment has no influence on the result of the other experiment.

For example, if we toss a coin twice *under identical conditions*, then the two tosses are *independent*. The appearance of heads or tails on toss 1 does not influence which side appears on toss 2.

The notion of stochastic independence applies not only to repetitions of the same experiment under identical conditions, such as successive tossings of a coin, but also applies to different experiments that have *nothing to do with each other*. Such experiments can either be done simultaneously (i.e. at the same time) or sequentially (i.e. in a time succession).

For example, suppose we toss a coin and throw a die, either simultaneously or sequentially. It seems evident that the fall of the coin has nothing to do with the fall of the die, and accordingly we say that these two experiments are *independent*.

20.3. ARITHMETIC AND GEOMETRIC SMOOTHING

A major aim of the studies given in this chapter is to discover those factors that will enable us to predict stochastic events. Stochastic events which occur in a time sequence can often be represented as a time series. A time series is a sequence of observations on a variable of interest. The variable is observed at discrete time points, usually equally spaced in time. Time series analysis is involved with the description of the process or phenomena that generate the sequence. Thus, in order to forecast a time series it is necessary to represent the behavior of the process by a mathematical model that can be extended into the future. For good forecasts it is required that the model be a good representation of the observations in any segment of time close to the present time. Once a valid model for the process has been determined, an appropriate forecasting technique can usually be developed.

As a preliminary example let us consider a time series generated by a constant process plus an independent random error. We can write the time series x_t as generated by this constant model as

$$x_t = b + e_t$$

where t is an integer representing time, b is an unknown parameter, and e_t is an independent random variable with mean 0 and variance σ^2. It is possible that in different widely separated parts of the time series the value of b will be different, but in any local segment of the data we assume that b is a constant. To forecast future values of the time series we must estimate the value of the unknown parameter b. We suppose that all observations from $t = 1$ to $t = N$ are available. If we consider all these observations to be equally important in estimating b, then they would be weighted equally, and the least-squares criterion requires that b is chosen so as to minimize

$$\sum_{t=1}^{N} (x_t - b)^2.$$

By setting the derivative of this expression with respect to b equal to zero, we obtain the estimated value $\hat{b}$ as

$$\hat{b} = \frac{1}{N} \sum_{t=1}^{N} x_t.$$

Thus the estimated value of b is the arithmetic mean, or sample mean, of the N observations.

The arithmetic mean includes all the past observations of the time series. If the value of the unknown parameter b changes slowly with time, it would be reasonable to place more weight on the most current observations than those observed a long time ago. Let us decide to include only the most recent n observations and to weight these observations equally. That is, we assign weight $1/n$ to $x_N, x_{N-1}, \ldots, x_{N-n+1}$ and weight zero to observations $x_{N-n}, x_{N-n-1}, \ldots, x_1$. The least-squares criterion now requires minimizing

$$\sum_{t=N-n+1}^{N} (x_t - b)^2,$$

which yields the estimate M_N for b given by

$$M_N = \frac{1}{n} \sum_{t=N-n+1}^{N} x_t.$$

That is, M_N is simply the average of the most recent n observations:

$$M_N = \frac{x_N + x_{N-1} + \cdots + x_{N-n+1}}{n}.$$

At each time point the oldest observation is discarded and the newest one is added to the set making up the estimate. For this reason M_N is called an n-period simple moving arithmetic average. At time instant N, the forecast for any future time $N + p$ is just

$$\hat{x}_{N+p} = M_N.$$

We may call this the arithmetic smoothing method for forecasting. Let us now give an alternative equation for computing the simple moving average, namely

$$M_N = M_{N-1} + \frac{x_N - x_{N-n}}{n}.$$

From this formula it is possible to obtain M_N directly from the previous value M_{N-1} and the two end observations x_N and x_{N-n} of the time period in question.

Let us now give a description of geometric smoothing methods for forecasting. Geometric smoothing is one of the most widely used methods for smoothing time series in order to forecast the immediate future. This popularity can be attributed to its computational efficiency, its simplicity and accuracy, and the ease of adjusting its responsiveness to any changes in the time series being forecast.

Again let us consider a time series generated by a constant process plus an independent random error:

$$x_t = b + e_t.$$

Let us assume that at time N we have available the estimate S_{N-1} of b made at the previous time $N - 1$. Using S_{N-1} and x_N we want to calculate an updated estimate S_N. We will obtain this new estimate by modifying the old estimate S_{N-1} by some fraction of the forecast error resulting from using the old estimate to forecast the present value x_N. This forecast error is

$$x_N - S_{N-1}$$

so that if α is the desired fraction, the new estimate of b is

$$S_N = S_{N-1} + \alpha(x_N - S_{N-1}).$$

If we rearrange terms we have

$$S_N = \alpha x_N + (1 - \alpha)S_{N-1}.$$

This operation is called simple (or first-order) geometric smoothing, and S_N is called the smoothed value of the time series x_N. The fraction α is called the smoothing constant. The smoothing constant α is a weighted average of all past observations. To see this, let us first substitute for S_{N-1} in the above equation to obtain

$$\begin{aligned} S_N &= \alpha x_N + (1 - \alpha)[\alpha x_{N-1} + (1 - \alpha)S_{N-2}] \\ &= \alpha x_N + \alpha(1 - \alpha)x_{N-1} + (1 - \alpha)^2 S_{N-2}. \end{aligned}$$

If we continue to substitute recursively for $S_{N-2}, S_{N-3}, \ldots$, we finally obtain

$$S_N = \alpha \sum_{i=0}^{N-1} (1 - \alpha)^i x_{t-i} + (1 - \alpha)^N S_0$$

where S_0 is the initial estimate of b used to start the geometric smoothing process. It can be shown that the weights sum to unity, since

$$\alpha \sum_{i=0}^{N-1} (1 - \alpha)^i = 1 - (1 - \alpha)^N.$$

Because the weights given to past observations are non-negative and add to one, we may interpret S_N as a weighted average. With the exception of the coefficient of S_0, the weights decrease geometrically with the age of the observations, so the name geometric smoothing has been attached to this procedure. Because these weights decline exponentially when connected by a smooth curve, the name exponential smoothing has also been used to describe this method.

The weights are

$$\alpha, \ \alpha(1 - \alpha), \ \alpha(1 - \alpha)^2, \ \alpha(1 - \alpha)^3, \ \ldots, \ \alpha(1 - \alpha)^{N-1}, \ (1 - \alpha)^N.$$

If the smoothing constant is 0.25, then the weights are 0.25, 0.1875, 0.1406, 0.1055, 0.0791, . . .
In comparison the weights of a four-point moving arithmetic average are 0.25, 0.25, 0.25, 0.25, 0, 0 . . .

The value S_N of the geometric smoothing can be used as an estimator of the unknown parameter b, so the forecast of the time series at any future time $N + p$ would be

$$\hat{x}_{N+p} = S_N.$$

The choice of the smoothing constant α is important in determining the characteristics of the geometric smoothing. The response of the forecast to changes of the parameter b is a function of the magnitude of α. The larger the value of α, the faster the response. That is, larger values of α cause the smoothed value to react quickly, not only to real changes but also to random fluctuations. We can judge the effect of α on responsiveness by comparing geometric and arithmetic smoothing. The average age of the data in an n-time moving average is

$$\frac{1}{n} \sum_{i=0}^{n-1} i = \frac{n-1}{2}.$$

In geometric smoothing the weight given to data i time units ago is $\alpha(1 - \alpha)^i$ so the average age is

$$\alpha \sum_{i=0}^{\infty} (1 - \alpha)^i i = \frac{1 - \alpha}{\alpha}.$$

As a result, if we want to define a geometric smoothing system that is equivalent to a n-time moving average, we set

$$\frac{1 - \alpha}{\alpha} = \frac{n-1}{2}, \quad \text{which is} \quad \alpha = \frac{2}{n+1}.$$

The smoothing constant α controls the number of past values of the time series that influence the forecast. Small values of α give significant weight to many prior observations and thus result in a slow response of the forecasting system to changes in the parameter b of the time series model. Larger values of α give weight to only the more recent historical data and thus cause the forecasting system to respond more rapidly to shifts in the parameter b. However a large value of α may cause the system to be oversensitive, i.e. a large smoothing constant may cause the system to respond to random variations e_t, where actually the model parameter b has not changed. Such oversensitivity is generally not desirable. A widely used technique to determine α is to carry out a sequence of trials on a set of actual historical data using several different values of α, and then to select that value of α that minimizes the sum of squared errors or alternatively that optimizes some other measure of effectiveness. As a general rule the smoothing constant α for the constant model discussed above usually turns out to be between 0.01 and 0.30.

20.4. PREDICTION OF STATIONARY TIME SERIES

The final aim of science is prediction. To predict is to tell or declare beforehand, to forecast. In science it is necessary to base predictions on observable phenomena, which means that our predictions must be based on present and past events but not on future events. In order to develop a mathematical theory of prediction we must represent events by quantitative variables.

Associated with each of these variables is an instant of time, and the entire past, present, and future of such a variable constitutes a time series.

Time series may be either discrete or continuous. A discrete time series is one in which the quantitative variable can take on values only at discrete instants of time. An example of a single discrete time series would be the day-to-day sequence of the closing prices of General Motors common stock on the New York Stock Exchange, for here the quantitative variable, closing price, is defined only at time 4.00 p.m. on each trading day. An example of multiple discrete time series would be the sequence of closing prices of all stocks on the New York Stock Exchange.

A continuous time series is one in which the quantitative variable is defined at all instants of time. For example, the altitude of an airplane as a function of time would represent a single continuous time series. The altitude, azimuth, and velocity of the airplane as a function of time would then be an example of a multiple continuous time series.

Generally speaking, in the social sciences, such as economics, one more often encounters discrete time series than continuous ones, whereas in the physical sciences and in engineering one more often encounters continuous time series. For analysis on a digital computer, however, it is necessary to convert continuous time series to discrete time series by sampling the continuous time series at equal increments of time.

The science of prediction deals with the forecasting of future values of time series from their observable past and present values. The first successful solution to the prediction problem for a general class of time series was given by Herman Wold in his book, *A Study in the Analysis of Stationary Time Series*, published in Uppsala, Sweden, in 1938. In this book Wold gave the general solution to the linear least-squares prediction problem for discrete stationary time series. Wold's solution represented the breaking of a barrier, and in the few years following his work all the important mathematical problems concerning the prediction of stationary time series were solved, since their solutions depend upon or can be derived from Wold's original solution.

The more important of these mathematical problems were the following three. The first is the setting of Wold's solution into analytic form from its original statistical form, and this result was given by the Russian mathematician Andre Kolmogorov in 1941. The second important problem was the solution of the prediction problem for continuous stationary time series. This difficult result was given by the American mathematician Norbert Wiener in his book, *Extrapolation, Interpolation, and Smoothing of Stationary Time Series with Engineering Applications*, published in Cambridge, Massachusetts, in 1942. The third important problem was the solution of the prediction problem for multiple stationary time series, the first results being given by the young Russian mathematician V. Zasuhin in 1941 just before he was killed at the front, and also by Norbert Weiner in his 1942 book. All these results had an important influence on the development of radar and gunnery during the war.

In this section we shall heuristically review Wold's solution to the prediction problem in such a way as to make clear what the mathematics is doing. Any discrete time series may be represented by a sequence of observations x_t. We assume that the observations are equally spaced in time, and for convenience take the spacing between each successive observation to be one unit of time. Thus the time index t may take any integer value from minus infinity to infinity.

Any observation x_t of the time series is a random variable, and thereby has a distribution function which generally depends upon time t. Any sequence of observations of the time series has a joint distribution function which depends upon the absolute times of these observations. A time series is called stationary if the distribution function of any set of observations does not depend upon the absolute times of these observations, but only upon the relative times between observations. In other words, if the joint distribution of any sequence of observations is the same no matter where in time we start the sequence, then the time series is stationary. In this sense we may say that a stationary time series is one which is not tied down to an absolute origin in time. Thus for a stationary time series the mean and standard deviation of any observation does not depend on t since none of the distributions depend on absolute t. That is, the mean and standard deviation are each constant for all values of t. It is customary to center the units of measurement so the mean is zero.

Let us think of the time value t as being the present time. As a result the observation x_t is the present value of the time series. The observations x_{t-1}, x_{t-2}, $\ldots$, preceding x_t in time are the past values, and the observations x_{t+1}, x_{t+2}, $\ldots$, following x_t in time are the future values.

The prediction problem may be formulated in the following way. We wish to find a best estimate of the future value x_{t+p}, which will occur p units of time from the present time t. This prediction will be designated as $\hat{x}_{t+p}$, and will be optimum in the sense that the mean square prediction error is a minimum, i.e. that the expected value of the squared discrepancy between $\hat{x}_{t+p}$ and the true value x_{t+p} will be a minimum. At the time we wish to make the prediction, namely, the present time t, we have only the present and past observations of the time series available. That is, in computing our predicted value x_{t+p} we shall be able only to utilize the values x_t, x_{t-1}, x_{t-2}, $\ldots$, of the time series which have occurred by the present time t, but shall not be able to utilize any of the future observations which are unobtainable at the present time. Our prediction will be a linear prediction in that the predicted value will be a linear combination of the present and past observations. The solution of the prediction problem will consist of finding the coefficients of the linear combination in terms of statistics, which may be computed from the past history of the time series.

A time series is called deterministic if the mean square prediction error is zero, in which case the future is completely determined from the past. A time series is called nondeterministic if the mean square prediction error is not zero, in which case the future cannot be completely determined by a linear operation on the past. In this section we shall consider the prediction of nondeterministic time series from which any deterministic component has been removed. We shall therefore be faced with the major problem of prediction theory, for the prediction of deterministic time series may be determined readily by straight-forward mathematical methods.

The first step in the solution of the prediction problem is to represent the present variable x_t as a linear regression on past variables plus an error term. As the number of past variable becomes indefinitely large, we have

$$x_t = (-a_1 x_{t-1} - a_2 x_{t-2} - a_3 x_{t-3} - \cdots) + e_t$$

where the regression coefficients are $-a_1, -a_2, -a_3, \ldots$, and e_t is the error term. The reason we have used minus signs in the regression coefficients is so that we can bring all the x terms to one side in the above equation and have all positive signs:

$$x_t + a_1 x_{t-1} + a_2 x_{t-2} + a_3 x_{t-3} + \cdots = e_t.$$

This equation is known as the autoregressive form of the time series x_t, where e_t may be thought of as the innovation in the x time series at time t. By the correlation properties of regression residuals, it follows that the innovations e_t and e_s are two different times t and s are uncorrelated. Thus the innovations e_t for all values of t form an orthogonal set. This situation is in contrast with the time series x_t, where generally values for two different time points are correlated, i.e. the self-correlation or autocorrelation of the time series.

As a result of the uncorrelated property of the innovations we can write the time series x_t as a Fourier series in the innovations e_t, i.e.

$$x_t = e_t + b_1 e_{t-1} + b_2 e_{t-2} + \cdots$$

where the Fourier coefficients are given by

$$b_s = E(x_t e_{t-s}) \quad (s = 1, 2, 3, \ldots).$$

This Fourier series representation of the time series in terms of present and past innovations represents the Predictive Decomposition Theorem of

Herman Wold (1938). Given the set of either the a or b coefficients, we can find the other by means of the formula

$$b_t + a_1 b_{t-1} + \cdots + a_{t-1} b_1 + a_t = 0 \quad (t = 1, 2, 3, \ldots).$$

From this decomposition Wold (1938) gave the explicit solution to the prediction problem, which may be derived as follows. The future value x_{t+p} will occur p units of time from the present time t. The future value is given by the predictive decomposition.

$$x_{t+p} =$$
$$(e_{t+p} + b_1 e_{t+p-1} + \cdots + b_{p-1} e_{t+1}) + (b_p e_t + b_{p+1} e_{t-1} + b_{p+2} e_{t-2} + \cdots)$$

where the first term in parentheses involves future (and hence unknown) innovations, and the second term in parentheses involves present and past (and hence known) innovations. Hence the first term in parentheses represents the unpredictable part of x_{t+p}, and the second term in parentheses represents the predictable part of x_{t+p}. The first term in parentheses therefore is the prediction error. We have just said that the present and past innovations are known. They are known because we can obtain them from present and past values of the time series by use of the autoregressive representation, i.e.

$$e_t = x_t + a_1 x_{t-1} + a_2 x_{t-2} + \cdots$$
$$e_{t-1} = x_{t-1} + a_1 x_{t-2} + a_2 x_{t-3} + \cdots$$
$$e_{t-2} = x_{t-2} + a_1 x_{t-3} + a_2 x_{t-4} + \cdots$$

As a result the second term in parentheses may be computed at time t and is the optimum least-squares prediction $\hat{x}_{t+p}$, i.e.

$$\hat{x}_{t+p} = b_p e_t + b_{p+1} e_{t-1} + b_{p+2} e_{t-2} + \cdots$$

If we substitute the above expression for the innovations into this expression for the optimum prediction, we obtain the prediction equation

$$\hat{x}_{t+p} = c_0 \hat{x}_t + c_1 x_{t-1} + c_2 x_{t-2} + \cdots$$

where the predictor coefficients are given by

$$c_0 = b_p,$$
$$c_s = b_p a_s + b_{p+1} a_{s-1} + \cdots + b_{p+s-1} a_1 + b_{p+s}, \quad (s = 1, 2, 3 \ldots).$$

This last equation, then, is the equation for the calculation of the coefficients of the linear operator to be used in making an optimum prediction with a prediction lead of p units. This equation was first given by Wold (1938), and represents the explicit solution of the least-squares prediction problem for stationary time series.

A problem of practical as well as theoretical interest is the prediction of share prices listed on the various stock exchanges. In a study of stock prices, a least-squares prediction operator

$$\hat{x}_{t+1} = c_o x_t + c_1 x_{t-1} + c_2 x_{t-2} + \cdots$$

was constructed from a sequence of 364 consecutive nonoverlapping 4-week averages of Standard and Poor's index covering a period from 1918 to 1945 from which the general trend had been removed. It was found that the prediction operator, which had nonzero coefficients up to c_{80}, gave substantially the same forecast as the one with only one coefficient

$$\hat{x}_{t+1} = 0.986 x_t$$

because all the coefficients except the first were so small.

20.5. PREDICTION OF HUMAN SPEECH

An important application of prediction theory is the analysis and synthesis of human speech. Speech represents the spoken word, whereas text represents the written word. When one reads, he converts text to speech; when one writes, he converts speech to text. However, there is an intermediate point between speech and text; this intermediate point is characterized by numbers, which we call the (numerical) parameters. For the purposes of this chapter, we make the following definitions. We define speech analysis as the conversion of speech to the numerical parameters; we define speech synthesis as the conversion of the numerical parameters to speech. In this chapter we shall examine the predictive mechanism that makes speech analysis and speech synthesis possible with the digital computer technology available today.

Let us further define text analysis as the conversion of text to the numerical parameters as required for speech synthesis. Also let us define text synthesis as the conversion of the numerical parameters as generated by speech analysis to text. Text analysis and text synthesis are not yet perfected but may be within the next few years. At that time an automatic reading machine could be constructed for which the input is printed text and the output is speech. Even more useful would be the automatic typewriter for which the input is speech and the output is typewritten text. The automatic reading machine would be made up of a text analysis machine followed by a speech synthesis machine. The automatic typewriter would be made up of a speech analysis machine followed by a text synthesis machine.

At the present time both speech analysis and speech synthesis can be implemented by use of a digital computer. There are three main types of applications: the first involving speech analysis only; the second, speech synthesis only; and the third, both analysis and synthesis. An application of speech analysis is an automatic speech recognition system, such as one in which an airline pilot would speak into the system and the desired result is some action based on the recognition of his speech. Another application of speech analysis would be one of speaker identification or verification. An application of both speech analysis and synthesis would be secure voice transmission; that is, the sender's speech would be analyzed into its parameters, the parameters would be transmitted, and the receiver would synthesize the speech from the parameters. For even more secure transmission, the parameters could be coded before being sent. In this way an alien receiver would first have to break the code and then synthesize the speech, a difficult task to say the least. Another application of speech analysis and synthesis is the data rate compression of speech. If speech is transmitted simply by sampling and digitizing, the data rate required is 100,000 bits per second of speech. The numerical parameters, however, can represent a reduction of about 50 to 1, so that by transmitting the numerical parameters the data rate required would only be 2000 bits per second of speech. A third application of speech analysis and synthesis would be speech retrieval systems. In this application the analysis of speech can be carried out at leisure, and the results stored in a computer. For example, the speech might be medical records in a spoken form. Then a physician would be able to telephone the computer, and the computer would synthesize the stored parameters into speech, which the physician would hear on the telephone in answer to his questions.

A speech waveform is transmitted from the lips. Sounds are generated by forcing air through the human vocal tract. The vocal tract runs from the glottis in the throat to the lips. Thus the speech mechanism consists of the vocal tract excited by an appropriate source in the throat. Different sounds are produced by changing the shape of the vocal tract as by moving the lips and tongue and by changing the type of excitation. Voiced sounds are uttered by means of vibration of the vocal chords, as are the consonants d and b. Thus in the case of voiced sounds the excitation is a periodic pulse train produced by the air flow through the vibrating vocal chords. Fricative sounds are produced by the forcing of breath through a constricted passage, as are such consonantal sounds as f and v, s and z, sh and zh. Because the constriction produces turbulence, the excitation for fricative sounds may be represented as a white noise source.

A speech waveform is made up of a time sequence of the various voiced, fricative, and other sounds making up the words. Ideally one would like to cut the speech waveform into adjacent segments, each segment representing a unique sound. Each such segment would then be represented by a given shape of the vocal tract and a given excitation. Prediction methods would be used to decompose each segment into its two components; namely, vocal-tract shape and excitation type. These two components are represented by numbers that make up the parameters for the sound corresponding to that segment. The determination of these parameters represent the analysis of that sound segment. The analysis of the entire speech waveform would therefore result in a time sequence of such parameter sets, each parameter set representing one sound.

Thus the basic problem of speech analysis is one of the decomposition, or deconvolution, of each sound segment to separate its excitation function from its vocal tract shape. In order to solve this problem, it is first necessary to obtain a good working model of the vocal tract. The model is based on an approximation to the vocal tract made up of a set of interconnected sections. Such a model is called an acoustic tube model. If we assume that the width of each section is the same, then the acoustic tube model is specified by the cross-sectional area of each section. At each interface between two adjacent sections, a travelling wave will be partially reflected and partially transmitted, the division of energy between the reflected and transmitted waves being governed by the reflection coefficient associated with that interface. The acoustic tube can support traveling wave-motion from the glottis to the lips, which we call upgoing waves, and also traveling wave motion from the lips to the glottis, which we call downgoing waves. The upgoing wave and the downgoing wave in any section will be generally different from the respective upgoing wave and downgoing wave in any other section, because of the effects of the reflections and transmissions at each interface. In other words, the vocal tract represents a reverberating system with the full array of multiple reflections and transmissions.

In summary, the acoustic properties of the human vocal tract can be determined by considering it as an acoustic tube of variable cross-sectional area. More specifically, we consider a model formed by cascading N cylindrical sections, each of the same width, and each with a certain cross-sectional area. This cross-sectional area will vary from sound to sound uttered by the person, so in our mathematical analysis we consider one sound at a time. For a given sound let the cross-sectional area of the nth section be denoted by S_n, where n runs from 1 to N. Let section 1 be the one closest to the lips and let section N be the one closest to the glottis. For the purpose of description, we speak of the lips as being at the top and the glottis as being at the bottom of a vertically standing tube made up of the N sections. Sound waves traveling in the direction from the lips to the glottis are called downgoing waves and are denoted by the symbol x. Sound waves traveling in the other direction, that is, from the glottis to the lips, are called upgoing waves and are denoted by the symbol y. The velocity of the sound waves is a constant, and it is convenient to choose the width of each section to be such that it takes exactly one-half time-unit for a wave to travel from one side of the section to the other. In other words, the finer we choose our model the smaller the time-unit, and the grosser we choose our model the larger the time-unit.

Let $x_1(t)$ and $y_1(t)$ be the downgoing and upgoing waves respectively at the top of section 1 and let $x_2(t)$ and $y_2(t)$ be the downgoing and upgoing waves respectively at the top of section 2. The basic idea in speech analysis, as well as in the analysis of many other physical and biological systems, is one of cross-prediction. More specifically, we want to consider the lagged value of the downgoing wave, namely $x_1(t-1)$, and the value of the upgoing wave, namely $y_1(t)$, of section 1, and we want to predict each from the other. In order to simplify the mathematics, we require all our variables to have zero mean and unit variance. We assume this requirement is satisfied for x_1 and y_1, and in order to satisfy this requirement for the other variables we will introduce suitable normalization factors as needed. Our problem therefore is to predict $x_1(t-1)$ from $y_1(t)$; we shall call the prediction error $\lambda_1 x_2 (t-0.5)$. Our problem also is to predict $y_1(t)$ from $x_1(t-1)$; we shall call the prediction error $\lambda_1 y_2 (t-0.5)$. These two prediction error equations can be written as:

$$\lambda_1 x_2(t - 0.5) = x_1(t - 1) - c_1 y_1(t),$$
$$\lambda_1 y_2(t - 0.5) = y_1(t) - c_1 x_1(t - 1).$$

The constant c_1, which appears in each of these two equations, is the regression coefficient. The constant c_1 is the same for both equations because in both equations c_1 is equal to the covariance of $x_1 (t-1)$ and $y_1(t)$. The constant λ, which appears in each of these two equations, is a normalization factor chosen so that each of $x_2 (t-0.5)$ and $y_2 (t-0.5)$ has unit variance; hence λ_1 is not an independent constant but is related to c_1 by the equation $\lambda_1 = \sqrt{1 - c_1^2}$.

The above two equations comprise the basic predictive deconvolution algorithm. They show how to go from x_1 and y_1 to x_2 and y_2. Instead of considering the subscripts 1 and 2 we may consider the subscripts n and $n + 1$, and thus obtain the algorithm to go from x_n and y_n to x_{n+1} and y_{n+1}, namely

$$\lambda_n x_{n+1}(t - 0.5) = x_n(t - 1) - c_n y_n(t),$$
$$\lambda_n y_{n+1}(t - 0.5) = y_n(t) - c_n x_n(t - 1),$$

where c_n is computed as the covariance of $x_n (t-1)$ and $y_n(t)$ and λ_n is computed as $\sqrt{1 - c_n^2}$.

We now must relate this mathematical algorithm to the physical model of the acoustic tube approximation to the human vocal tract. Speech sounds are produced as a result of acoustical excitation of the vocal tract. The vocal tract is regarded as an acoustic tube with a varying cross-sectional area. For any given sound, the tube is divided into N cylindrical sections, each of the same width. The following assumptions are made for this model. We assume that the waves propagated through the acoustic tube are plane waves. We assume that the tube is rigid, and that the losses in the sound wave due to viscosity and heat conduction are negligible. As the boundary conditions, the top of the tube (which corresponds to the lips) is assumed to be connected to another section (which corresponds to the open air) with an infinite area, which means there is perfect reflection at the lips. The radiation effects from the lips into the air must therefore be handled separately. It is assumed that the bottom of the tube (which corresponds to the glottis) is excited by a source that is an upgoing wave. A resulting downgoing wave will be returned to the glottis.

By using the boundary condition at the top of the tube (i.e. perfect reflection), the downgoing wave in the first section is equal to the negative of the upgoing wave in the first section, i.e. $x_1 = -y_1$. Thus by recording the speech at the lips we know both the upgoing and downgoing waves in the first section. We then apply the predictive deconvolution algorithm to obtain c_1, λ_1 and the upgoing wave y_2 and the downgoing wave x_2 in the second section. It turns out that the computed regression coefficient c_1 is equal to the physical reflection coefficient of the interface between sections 1 and 2, and the computed scale factor λ_1 is equal to the physical (root-energy) transmission coefficient between sections 1 and 2. Hence our deconvolution computations result in physical measures of the acoustical tube. Given x_2 and y_2 just computed, we can apply the deconvolution algorithm again and obtain c_2, λ_2, x_3 and y_3. The computed coefficient c_2 gives the physical reflection coefficient of the interface between sections 2 and 3, so we have obtained the next physical measure of the acoustical tube. We can then repeat the algorithm again, and keep repeating it in a stepwise fashion until we obtain all the reflection coefficients up to and including c_N, the reflection coefficient between the last section and the glottis. We also obtain x_{N+1}, which is the upgoing wave in the glottis, or in other words the excitation source to the vocal tract.

If S_n is the cross-sectional area of section n, and S_{n+1} is the cross-sectional area of section $n + 1$,then from physical considerations it can be shown that the reflection coefficient at the interface between sections n and $n + 1$ is given by

$$c_n = \frac{S_n - S_{n+1}}{S_n + S_{n+1}}.$$

Solving this equation for S_{n+1}, we have

$$S_{n+1} = \frac{1 - c_n}{1 + c_n} S_n.$$

Thus, given the reflection coefficients $c_1, c_2, \ldots, c_N$ just obtained, and given the cross-sectional area S_1 at the lips, we can compute, in a stepwise fashion, the cross-sectional areas of all the sections from the lips to the glottis S_1, $S_2, \ldots, S_N$ .Thus the deconvolution process has given us the shape of the vocal tract and the excitation source at the glottis from knowledge of the wave motion y_1 of a particular sound at the lips.

For example, suppose we have a male American speaker with a vocal tract 17.5 cm long from lips to glottis. Suppose that we divide this length into seven sections of width 2.5 cm each, and number the sections from 1 at the lips to 7 at the glottis. Then the normalized cross-sectional areas for the seven sections from lips to glottis for the American vowel i would be 2.2, 0.5, 0.5, 1.1, 4.0, 4.3, 1.9. For the American vowel a, these areas would be 1.4, 4.0, 3.7, 0.6, 0.2, 0.3, 0.3. For the American vowel u, these areas would be 0.2, 4.0, 2.5, 1.0, 1.2, 1.8, 2.8. In practice, a finer division of the vocal tract would be made, but these examples illustrate the principle of obtaining the cross-sectional areas for a given sound.

The reflection coefficients, or equivalently the cross-sectional areas, make up the numerical parameters necessary to characterize the vocal tract for a given sound. The deconvolution procedure also produced the excitation waveform at the glottis for the given sound. This excitation waveform must also be reduced to a few numerical parameters. These excitation parameters result from a determination of whether the sound is voiced or unvoiced, and for a voiced speech a determination of the fundamental frequency. Also the gain (or volume) of the speech sound must be determined. Thus the excitation parameters can be reduced to as few as three numbers: the first parameter representing the gain, the second parameter representing whether the sound is voiced or not voiced, and the third parameter the fundamental frequency in case the speech sound is voiced. In more complex systems, more parameters would be used, but still the number of parameters would be relatively few. The determination of the above vocal-tract and excitation parameters are the end result of the deconvolution of the particular speech sound.

The analysis of actual speech first consists of breaking the speech up into segments, where each segment represents a single pure sound. Then each segment is deconvolved into its set of parameters. The collection of these parameter sets is equivalent to the original speech, but represents a great reduction in the volume of numbers required to characterize the given speech. An automatic typewriter would then have to substitute a single phonetic symbol for each set of numerical parameters so that a text of the speech can be typed out automatically.

The synthesis of speech consists of taking the sets of speech parameters, and for each set first constructing the equivalent of the excitation from the excitation parameters, and then reconvolve this excitation with the vocal tract parameters to produce each sound making up the speech. In case the appropriate excitation parameter indicating the speech sound is voiced, then an artificial excitation function is mathematically constructed, made up of a periodic pulse sequence, with gain and period given by the excitation parameters. In case the appropriate excitation parameter indicating the speech sound is unvoiced, then another artificial excitation function is mathematically constructed of a noise-like sequence with gain given by the excitation parameter. In any case the artificial excitation waveform so generated is not the same waveform as the excitation waveform produced by the deconvolution, but the two are similar only in that certain gross features are the same in both, like gain, and periodic or noiselike character. If this difference between the two waveforms was not permissible, and as a result if we had to preserve the entire excitation waveform and not simply its gross features, then there would be essentially no compression resulting from the deconvolution process, and its effectiveness would be minimal. Happily, this unfortunate situation is not the case, and hence the deconvolution of speech is, indeed, a powerful process. The shape of the vocal tract, however, is very important; in our simple examples we used seven parameters to represent the vocal tract but twice that number, or 14, is usually required to produce excellent results.

Let us now take the speech of a human and analyze it by deconvolving it into parameter sets. Let us now take these parameter sets and synthesize speech by reconvolving it. Both the deconvolution and reconvolution take place within a digital computer at high speed. The quality of the synthetic speech is so close to that of the original speech that it is difficult or impossible to tell the two apart. This is true for a wide range of speakers and spoken material including both male and female speakers. The essence of the entire processes of speech analysis and synthesis is based on deconvolution and reconvolution methods in which the concept of prediction error is the key. The essence of the physical process of speech can be unraveled by a separation of the predictable and unpredictable components. In fact, in the analysis of almost any physical process, prediction theory plays an essential role.

20.6. PREDICTION OF PRINTED ENGLISH

An interesting application of the statistical theory of prediction is the problem of the prediction of printed English. A simple experiment to demonstrate to what extent the English language is predictable would be the following. One person selects a passage from a book. The other person who is to do the predicting tries to guess the first letter of the passage. After his guess he is told what is the correct letter, and then he proceeds to try to guess the next. Thus the person is able to write down the correct text up to the current point to use in making the next prediction.

Dr. C. E. Shannon (1951), of the Bell Telephone Laboratories, had this experiment carried out on a text with 129 letters, in which a space is counted as an additional letter, thereby making a 27-letter alphabet. Of the total of 129 letters in the text, 89 letters or 60% were guessed correctly. In an experiment of this type he found that the errors occur most frequently at the beginning of words and syllables, as expected, because the line of thought has more possibility of branching out at such points.

An extension of this experiment would be one in which if the person fails to guess the next letter, he is asked to guess again, and must continue guessing until he is correct. If he guesses the first time, he is scored 1; if he guesses incorrectly the first time, but correctly the second, he is scored 2; if he guesses incorrectly the first time and the second time, but correctly the third, he is scored 3. In this way, with a 27-letter alphabet, the lowest score that any letter in the text may receive is 1, and the highest score possible would be 27. If in this experiment the person makes his guess on the basis of knowing the preceding N letters, then Shannon calls this scheme N-gram prediction. If we allow N to approach infinity, then the person guesses with knowledge of the entire past history of the text.

In the following example the first line is the original text and the number in the second line indicates the score the person received for that letter. The test was taken at random from Dana's *Two Years Before the Mast*.

WHEN ALL SAIL HAD BEEN SET AND THE DECKS
3 1 1 2 1 2 3 1 1 4 2 9 1 1 2 1 1 1 2 1 1 1 19 1 1 1 3 1 1 11 1 1 1 3 2 2 1 1

CLEARED UP THEN CALIFORNIA WAS A SPECK ON
1 6 3 1 1 2 1 1 1 15 2 1 1 1 1 1 8 2 4 2 1 1 1 1 1 12 1 1 1 1 14 3 1 3 1 1 4 1

THE HORIZON AND TH
1 1 1 1 14 2 2 2 1 1 1 11 1 1 11 1

Out of the 100 letters and spaces, the person guessed correctly the first time for 68 of them, the second time for 15 of them, the third time for 7 of them, the fourth time for 5 of them, and only 5 of them required five or more guesses.

The sequence of numbers indicating the number of guesses constitutes what is called the reduced text. This experiment then translated the original text, made out of the alphabet with the 27 symbols, A, B, C, D, E, . . . , X, Y, Z, space, to the reduced text in a new language with the symbols 1, 2, 3, . . . , 25, 26, 27. This new language has this interesting property. The symbol 1 occurs most frequently, and the symbols 2, 3, 4, 5, . . . have successively smaller frequencies, the final symbols 25, 26, 27 occurring very rarely.

Now the original text exhibited the constraints imposed upon it by the nature of the English language. That is, the statistical structure of the language in the original text is based on such peculiarities of English as the strong tendency of H to follow T, or of U to follow Q. The translation of the original English text to the reduced text has simplified to a large extent the statistical structure. That is, the complicated constraints among groups of letters in the English language has been reduced to the fact that in the reduced language the symbol 1 occurs most frequently, the symbol 2 occurs next most frequently, until finally the symbol 27 occurs least frequently.

Shannon (1951) has developed the properties of the ideal N-gram predictor. This predictor would be a machine which would give the best possible prediction of a letter, when the preceding N-gram (i.e., the preceding N letters) is known. In order to code this ideal predictor for a digital computer, one would need to know the conditional probabilities that a certain letter will occur, given that a particular N-gram has occurred. Then once a particular N-gram is observed by the machine, the machine will predict that letter which has the highest conditional probability of occurring. If this prediction is correct, that letter in the original text will be replaced by a 1 in the reduced text. If this prediction is incorrect, the machine will then predict as a second guess the letter with the second highest conditional probability. If this second guess is right, then the letter in the original text will be replaced by a 2 in the reduced text. If the second guess is wrong, the process will be repeated until a correct guess is made.

Once the original text has been translated into the reduced text by the machine, the reduced text may be transmitted from one location to another by means of a communication system. Because in the reduced text the symbol 1 has the highest frequency, the symbol 2 the next highest, the symbols 3, 4, 5, ... with successively lower frequencies, the reduced text is an easier message to communicate than the original text in English where the frequencies of the symbols are more nearly equal.

In telegraphy a dot is the easiest coded symbol to transmit. As a result the letter E is represented by a dot in the Morse code. In our example from Dana the letter E occurred 11 times, and as a result we would benefit in this way 11 times in the transmission of Dana's text. On the other hand, if we coded the symbol 1 in the reduced language by a dot, we would benefit 68 times because there are 68 occurrences of the symbol 1 in the reduced text.

Another advantage of transmitting the reduced text instead of the English text would be the greater accuracy obtained in filling in any part of the message that was lost in the transmission. For example, if a letter was lost in Dana's text and the receiver substituted an E for the lost letter, he would be right 11 times out of 100. If a symbol was lost in the reduced text and the receiver substituted the symbol 1 for it, he would be right 68 times out of 100.

Once the message has been received, the reduced text must be translated back into the original text. This can be done with an identical prediction device at the receiving end. That is, in the recovering process the inverse operation operates on the preceding symbols of the reduced text with the same conditional probabilities. For example, suppose that the machine having recovered all the preceding letters observes the symbol 2 in the reduced text. Then the machine would pick the second most probable letter for the preceding N-gram, and that would be the letter in the original text. In this way the original message can be completely recovered.

Now let us compare Shannon's method of prediction of printed English with the mathematical theory of prediction for stationary time series developed by Wold, Kolmogorov, and Wiener which we presented earlier. There we started with a time series with autocorrelation and constructed a time series of innovations which was equivalent to the original time series, each innovation being a linear function of the present and past of the original time series. The essential property of the innovation time series is that the individual innovations are uncorrelated, whereas the individual observations in the original time series are correlated. In other words, prediction theory is based on the reduction of dependent observations to independent observations. This concept parallels Shannon's method of building the reduced text from the original English text. That is, although the successive symbols in the reduced text are not perfectly uncorrelated, they are much closer to noncorrelation than the original text.

REFERENCES

KOLMOGOROV, A. (1941) Stationary sequences in Hilbert space. *Bull. State Univ. Moscow, Math.*, **2**: 40.
ROBINSON, E. A. (1967) *Statistical Communication and Detection*, Charles Griffin, London.
ROBINSON, E. A. (1967) Multichannel Time Series Analysis with Digital Computer Programs, *Holden-Day, San Francisco.*
SHANNON, C. E. (1951) Prediction and entropy of printed English, *Bell System tech. J.*, **30**: 50.
WIENER, N. (1942) The Extrapolation, Interpolation and Smoothing of Stationary Time Series with Engineering Applications, MIT Press, Cambridge, Mass.
WOLD, H. (1938) *A Study in the Analysis of Stationary Time Series*, Almquist & Wiksells, Uppsala.
WOLD, H. (1953) *Demand Analysis*, Wiley, New York.
ZASUHIN, V. (1941) On the theory of multidimensional stationary random processes, *C.R. (Doklady) Acad. Sci. USSR*, N.S., **33**: 435–437.

CHAPTER 4

WAVELET COMPOSITION OF TIME-SERIES

ENDERS A. ROBINSON

University of Wisconsin and University of Uppsala

1. Wavelet Composition of Time-Series

Time-series occur in nearly every branch of science, and provide basic data by which models are constructed to describe and explain dynamic behavior. Accordingly, the analysis of time-series is a central problem in model building.

Figure 1 illustrates a portion of a typical time-series. The horizontal axis represents time, and the vertical axis gives the value of the time-series at the specified time. In figure 1 the time unit is one day, and in general the time unit will depend upon the particular application under consideration and the data that is available. Time-series such as the one shown are not perfectly predictable, but instead exhibit disturbances and fluctuations which lend themselves more readily to a probabilistic treatment than to a deterministic one.

The general approach which we wish to develop in this paper is the follow-

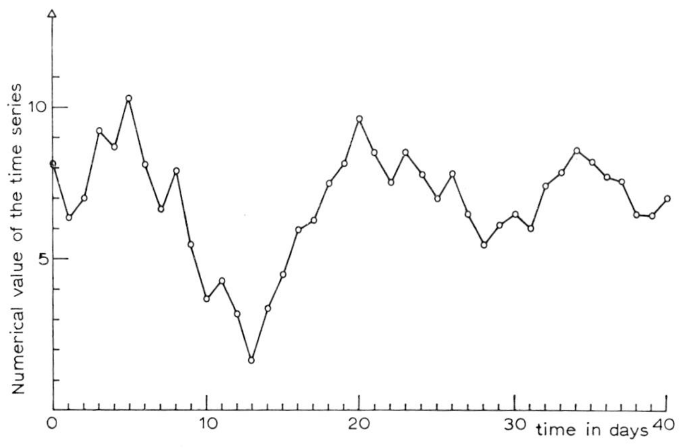

Figure 1.

ing. A time-series such as the one shown in figure 1 is additively composed of *wavelets*. The wavelets arrive in succession, and each wavelet eventually dies out. The wavelets all have the same basic form or shape, but the strength or impetus of each wavelet is random and uncorrelated with the strength of the other wavelets.

For example, we might hypothesize that the wavelet represents the reaction, or response, to some news. Each day brings forth fresh news, which generates a new wavelet. This time of origin, or birth date, of the wavelet is called the wavelet's arrival time.

Thus we see that, on any given day, the news of the day generates a wavelet. This wavelet persists for a time interval spanning several days or longer. Nevertheless, as the days pass, this particular wavelet will eventually die out. In other words, the news which generated this wavelet becomes so remote, so far in the past, that it no longer has any significant influence on the present value of the time-series. The result is that, after a certain number of days following its birth date, the wavelet damps out (or becomes infinitesimally small).

Despite the foreordained death of any individual wavelet, the time-series does not die. The reason is that a new wavelet is born each day to take the place of the one that does die. On any given day, the time-series is composed of many living wavelets, all of a different age,—some young, others old.

More precisely, on a certain day, say day t_0, the time-series is made up of:

(1) The wavelet born on day t_0
(2) the wavelet born the preceding day, that is $t_0 - 1$
(3) the wavelet born on day $t_0 - 2$
(4) the wavelet born on day $t_0 - 3$

    . . .

As we go down this list, the wavelets become older and older, until finally their effect on day t_0 is so feeble that we may neglect them from that point on. That is, the wavelets beyond that point are dead.

As we know, each day brings fresh news. The news of some days are more important than that of other days. There is no reason to expect that each day's news will have the same strength. Moreover, in the framework the time-series under consideration, this news is not predictable because it comes from outside, or exogeneous, factors. That is, the news of an outside event is exogeneous to the time-series under consideration, but this news may generate a wavelet which persists for several days in the time-series.

Thus, some wavelets are born strong, and other weak, depending upon the strength of the news of the day. That is, every wavelet has a different strength, or impetus. These strengths, being dependent on the outside news are random with respect to the time-series under consideration. Moreover, we count only fresh news. That is, news that is a direct consequence of some previous news is lumped with the previous day, and so does not give an impetus as fresh news would. In this way, new impetuses are uncorrelated with the old.

In this section we have outlined our method of attack. A time-series is additively composed of many overlapping wavelets. All these wavelets have the same shape or form, and they arrive in succession with strengths which are random and uncorrelated with each other. Each wavelet in time dies out, giving place to the new wavelets that are continually being generated.

2. Wavelets

In this section we want to introduce some ways to represent wavelets. Because wavelets are our basic building blocks we shall spend several sections to describe their characteristics.

When we plot a wavelet, we see its entire life history, from the time it is born until it damps out. Its origin time, or birth date, is called time 0, which serves as the reference date for the wavelet. At time 0 we plot the amplitude of the wavelet at its birth. At time 1 we plot the amplitude of the wavelet when it is 1 unit old. At time 2 we plot the amplitude of the wavelet when it is 2 units old, and so on.

For example, suppose the wavelet has amplitude 4 at time 0, amplitude 2 at time 1, amplitude 1 at time 2, and zero amplitude for all succeeding times. Table 1 summarizes this wavelet:

TABLE 1

Time	0	1	2	3	4	...
Wavelet	4	2	1	0	0	...

More concisely, we may summarize this wavelet by the row vector

$$(4, 2, 1)$$

where it is understood that 4 is the initial amplitude (i.e. at time 0) (so that all preceding amplitudes would be zero), 2 is the amplitude at time 1, and

1 is the amplitude at time 2 or the final amplitude (so that all following amplitudes would be zero).

There are various ways to plot wavelets. Figure 2 shows several alternative ways, each giving a picture of the wavelet (4, 2, 1). Still other ways are possible.

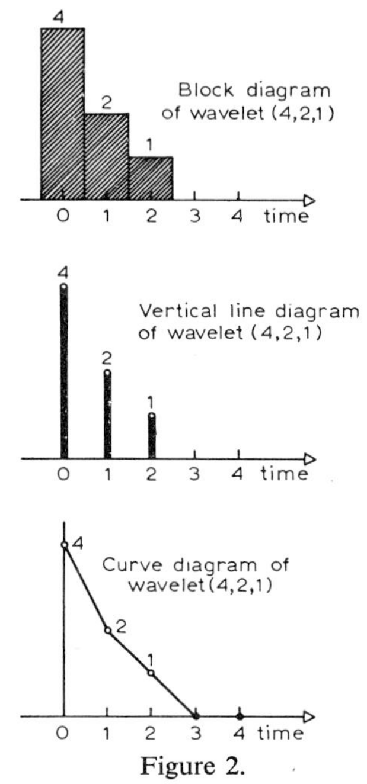

Figure 2.

Also, wavelets may be represented algebraically. For example, we may write the wavelet

$$b = (b_0, b_1, b_2) \, .$$

This notation means that wavelet b has amplitude b_0 at time 0, amplitude b_1 at time 1, and amplitude b_2 at time 2. For example, the equation

$$(b_0, b_1, b_2) = (4, 2, 1) \, ,$$

means that $b_0 = 4$, $b_1 = 2$, $b_2 = 1$.

It is possible to consider wavelets with complex amplitudes, that is, amplitudes of the form $u + iv$ where u and v are real numbers and $i = \sqrt{-1}$. An example is the wavelet $(i, 0.5)$ where i is the amplitude at time 0 and 0.5 is the amplitude at time 1. Another example is the wavelet $(2 + i, 1 - i, i)$, where

$2+i$ is the amplitude at time 0, $1-i$ is the amplitude at time 1, and i is the amplitude at time 2.

The amplitudes of a wavelet are sometimes called the coefficients of the wavelet. For example, for the wavelet (b_0, b_1, b_2), we may call b_0, b_1 and b_2 either the amplitudes or the coefficients of the wavelet. That is, b_0 would be the (amplitude) coefficient at time 0, etc. Because the wavelet (b_0, b_1, b_2) has three coefficients, we say that it has *length* (or *time duration*) equal to 3, or it is a 3-length wavelet.

Up to now we have considered only finite wavelets, that is, wavelets with a finite number of coefficients, or, in other words, wavelets with finite length. These wavelets die out completely (i.e. become zero after a certain age). For example, wavelet (4, 2, 1) dies out (i.e. becomes zero) at time 3.

It is possible to have infinite wavelets, but for stability we must require that they have *finite energy*. Thus an infinite-length wavelet may be represented by

$$(b_0, b_1, b_2, \ldots)$$

where b_0 is the coefficient for time 0, b_1 is the coefficient for time 1, b_2 is the coefficient for time 2, and the three dots indicate that the coefficients extend on for all positive times to form an infinite sequence. For stability it is required that the wavelet's energy given by

$$b_0^2 + b_1^2 + b_2^2 + \ldots$$

be finite. Note: In case the coefficients of the wavelet are complex, then the energy is given by

$$b_0 \bar{b}_0 + b_1 \bar{b}_1 + b_2 \bar{b}_2 + \ldots$$

where the horizontal bar indicates the complex conjugate of the quantity beneath it. (Thus, for example, if $b_0 = u + iv$, then $\bar{b}_0 = u - iv$.) An example of an infinite length wavelet is

$$b = (1, \tfrac{1}{2}, \tfrac{1}{4}, \tfrac{1}{8}, \tfrac{1}{16}, \tfrac{1}{32}, \tfrac{1}{64}, \ldots)$$

where the first coefficient 1 is the coefficient for time 0, the next coefficient $\tfrac{1}{2}$ is the coefficient for time 1, etc. An alternate way of representing this wavelet would be

$$b = (b_0, b_1, b_2, \ldots) \quad \text{where} \quad b_t = (\tfrac{1}{2})^t \quad \text{for} \quad t = 0, 1, 2, \ldots.$$

Because of the stability property, we see the magnitude of the coefficients asymptotically approach zero as time increases.

3. Convolution of Two Wavelets

The *convolution* of the two wavelets

$$a = (a_0, a_1, a_2 \ldots)$$

and

$$b = (b_0, b_1, b_2, \ldots)$$

is defined to be the wavelet

$$c = a * b = (c_0, c_1, c_2, \ldots)$$

where the coefficients of c are given by the formula

$$c_t = \sum_{s=0}^{t} a_s b_{t-s} .$$

The asterisk $*$ denotes convolution.

To illustrate the use of this formula, suppose

$$a = (a_0, a_1) = (2, 1)$$

and

$$b = (b_0, b_1) = (3, 4) .$$

Each of these wavelets has two coefficients or has length two. Using the convolution formula, we see that the convolution of a and b is

$$c = a * b = (c_0, c_1, c_2)$$

where

$$
\begin{aligned}
c_0 &= a_0 b_0 &&= 2 \cdot 3 = 6 \\
c_1 &= a_0 b_1 + a_1 b_0 &&= 2 \cdot 4 + 1 \cdot 3 = 11 \\
c_2 &= a_1 b_1 &&= 1 \cdot 4 = 4 .
\end{aligned}
$$

Hence

$$c = a * b = (6, 11, 4) .$$

We shall now interpret what is meant by convolution in three different ways: (1) as a folding operation, (2) as a sliding strip operation, and (3) as obtainable by means of polynomial multiplication.

Convolution is a *folding operation*. To see this, construct the table

	b_0	b_1
a_0	$a_0 b_0$	$a_0 b_1$
a_1	$a_1 b_0$	$a_1 b_1$

or

	3	4
2	6	8
1	3	4

whose entries are the products of the wavelets *a* and *b* (in the margin).

Thus we have the table of entries (without the margins):

which we are going to fold successively on the dotted lines. The element in the upper left hand corner, that is $a_0 b_0$, or 6, is c_0. Now fold this entry over, thus obtaining the table

The sum of the two entries before the next fold is

$$a_0 b_1 + a_1 b_0 \quad \text{or} \quad 8 + 3 = 11$$

which is c_1. Now fold these entries over, thus obtaining

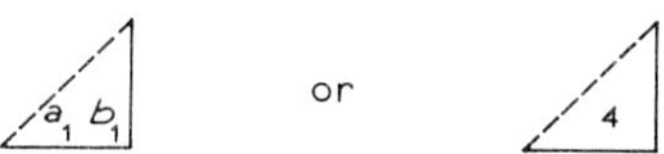

This last entry is c_2. Thus we have found the convolution series

$$(c_0, c_1, c_2) = (a_0 b_0, a_0 b_1 + a_1 b_0, a_1 b_1) = (6, 11, 4).$$

Convolution may also be regarded as operation on series written on *movable strips*. Thus we write

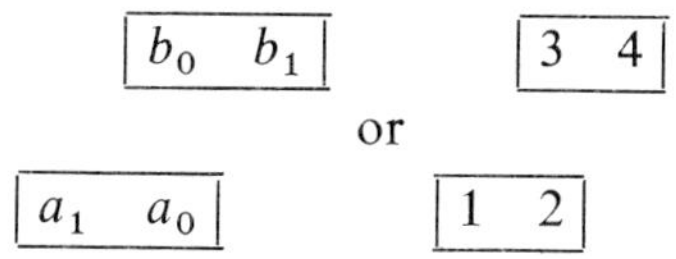

where we have reversed the direction of the *a* series. (In other words, the *a* series is written in reverse order from right to left on its strip, whereas the *b* series is written in natural order from left to right on its strip.)

To obtain c_0, we slide the *a* strip to the position:

$$\boxed{b_0 \quad b_1} \qquad \boxed{3 \quad 4}$$

or

$$\boxed{a_1 \quad a_0} \qquad \boxed{1 \quad 2}$$

and then multiply adjacent entries. Thus

$$c_0 = a_0 b_0 \quad = \quad 2 \cdot 3 = 6$$

To obtain c_1, we slide the **a** strip one more notch

| b_0 | b_1 | | 3 | 4 |
|-------|-------| |---|---|

or

| a_1 | a_0 | | 1 | 2 |

and multiply adjacent entries. Thus

$$c_1 = a_0 b_1 + a_1 b_0 = 2 \cdot 4 + 1 \cdot 3 = 11 \ .$$

Finally we slide the **a** strip one more notch

| b_0 | b_1 | | 3 | 4 |

or

| a_1 | a_0 | | 1 | 2 |

thus obtaining

$$c_2 = a_1 b_1 \quad = \quad 1 \cdot 4 = 4 \ .$$

Convolution may be obtained by *multiplication of polynomials*. Thus we write the polynomials

$$A(z) = a_0 + a_1 z \qquad\qquad A(z) = 2 + z$$

or

$$B(z) = b_0 + b_1 z \qquad\qquad B(z) = 3 + 4z$$

and then multiply them, obtaining

$$
\begin{array}{ll}
\begin{array}{r}
b_0 + b_1 z \\
a_0 + a_1 z \\
\hline
a_0 b_0 + a_0 b_1 z \\
a_1 b_0 z + a_1 b_1 z^2 \\
\hline
a_0 b_0 + (a_0 b_1 + a_1 b_0)z + a_1 b_1 z^2
\end{array}
& \quad\text{or}\quad
\begin{array}{r}
3 + 4z \\
2 + z \\
\hline
6 + 8z \\
3z + 4z^2 \\
\hline
6 + 11z + 4z^2
\end{array}
\end{array}
$$

The resulting polynomial

$$C(z) = a_0 b_0 + (a_0 b_1 + a_1 b_0)z + a_1 b_1 z^2 = 6 + 11z + 4z^2$$

has coefficients which are equal to the convolution

$$c_0 = a_0 b_0 = 6 \ , \quad c_1 = a_0 b_1 + a_1 b_0 = 11 \ , \quad c_2 = a_1 b_1 = 4 \ .$$

Thus multiplication of polynomials corresponds to convolving their coefficients.

One implication is that convolutions can be taken in anv order (this is, convolution is *commutative*) since polynomial products can be taken in any order. For example,

$$a * b = b * a$$

since

$$A(z)B(z) = B(z)A(z) .$$

By the same reasoning, we see that convolution is *associative*, that is

$$(a * b) * c = a * (b * c) :$$

and convolution is *distributive* with respect to addition, that is

$$a * (b + c) = (a * b) + (a * c) .$$

At this point we would like to define the *z-transform* of a wavelet. The z-transform of a finite wavelet is the polynomial in z associated with the wavelet. In other words, the z-transform of the wavelet

$$b = (b_0, b_1, b_2, \ldots, b_n)$$

is

$$B(z) = b_0 + b_1 z + b_2 z^2 + \ldots + b_n z^n .$$

The z-transform of an infinite wavelet is the power series in z associated with the wavelet. In other words, the z-transform of the wavelet

$$b = (b_0, b_1, b_2, \ldots)$$

is

$$B(z) = b_0 + b_1 z + b_2 z^2 + \ldots .$$

In this section we have defined convolution of two wavelets by a formula, and then have interpreted this formula
   (1) as a folding operation
   (2) as a sliding strip operation, with one wavelet reversed
   (3) as obtainable by means of polynomial multiplication.
Finally we defined the z-transform of a wavelet.

4. Convolution of a Wavelet with a Time-Series

In the preceding section we defined the convolution of two wavelets. There is no reason, however, why this definition cannot be extended to the convolution of a time-series with a wavelet. Let the wavelet be

$$a = (a_0, a_1, a_2, \ldots)$$

and let the time-series be

$$x = (\ldots, x_{-2}, x_{-1}, x_0, x_1, x_2, x_3, \ldots),$$

which is infinitely long in both the positive and negative directions. Then their convolution is the time-series

$$y = (\ldots, y_{-2}, y_{-1}, y_0, y_1, y_2, y_3, \ldots)$$

where y_t is given by the formula

$$y_t = \sum_{s=0}^{\infty} a_s x_{t-s}.$$

Like the x time-series, the y time-series is also infinitely long in both directions.

Now let us illustrate this convolution process schematically. We suppose that the numbers x_t are going into a *box*; that is,

$$
\begin{aligned}
& \cdot \ \cdot \ \cdot \\
x_{-1} \ & \text{enters at time } t = -1 , \\
x_0 \ & \text{enters at time } t = \ \ \ 0 , \\
x_1 \ & \text{enters at time } t = \ \ \ 1 , \\
x_2 \ & \text{enters at time } t = \ \ \ 2 , \\
x_3 \ & \text{enters at time } t = \ \ \ 3 , \\
& \text{etc.}
\end{aligned}
$$

so that the time-series x is the *input* to the box. Coming out of the box are the numbers y_t; that is,

$$
\begin{aligned}
& \cdot \ \cdot \ \cdot \\
y_{-1} \ & \text{emerges at time } t = -1 , \\
y_0 \ & \text{emerges at time } t = \ \ \ 0 , \\
y_1 \ & \text{emerges at time } t = \ \ \ 1 , \\
y_2 \ & \text{emerges at time } t = \ \ \ 2 , \\
& \text{etc.}
\end{aligned}
$$

so that the time-series y is the *output* from the box. The wavelet a is inside the box and is called the *memory function* of the box, or the *impulse response* of the box, or the *operator*. Thus we have figure 3.

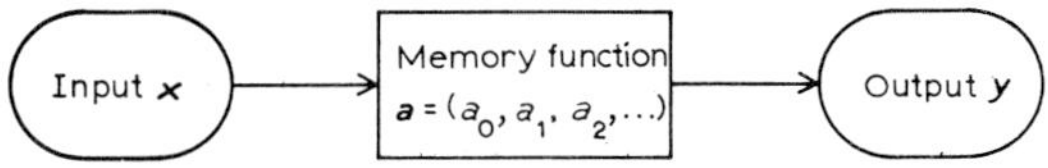

Figure 3. Illustration of input, box, and output.

Whenever we illustrate an input/output relationship as shown in figure 4.1, we mean the following: The output y is equal to the convolution of the input x and the memory function a, or

$$y = a * x \, .$$

A box that generates its output in this way from its input is called a *linear system*, or a *linear filter*.

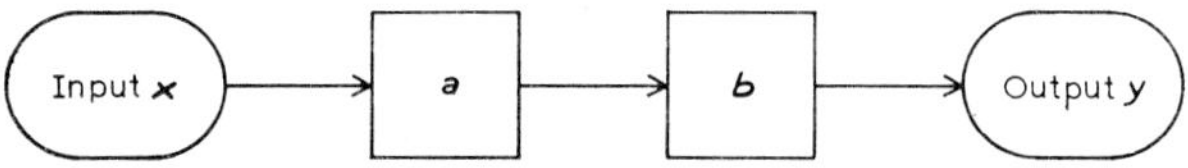

Figure 4. Illustration of two boxes in tandem.

Consider the situation shown in figure 4 where two boxes are hitched in tandem.

The output of the first box is $a * x$, which is the input to the second box. Hence the output of the second box is

$$y = b * (a * x) \, .$$

Hence the two boxes may be replaced with one box whose memory function c is the convolution of the memory functions of the two boxes, that is

$$c = b * a \, .$$

Thus figure 4 is equivalent to figure 5 below.

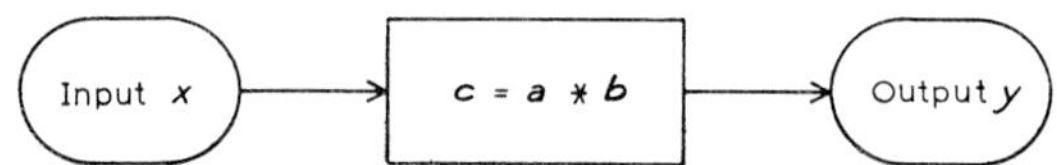

Figure 5. Single box equivalent to the two boxes of figure 4.

5. Autocorrelation, Minimum-Delay, and Maximum-Delay

The autocorrelation of a wavelet $(b_0, b_1, b_2, \ldots)$ is defined to be

$$r_j = \sum_{i=0}^{\infty} b_{i+j} \bar{b}_i \qquad \text{(for all } j\text{)} \, .$$

For a finite wavelet $(b_0, b_1, \ldots, b_n)$, this equation becomes

$$r_j = \begin{cases} \sum_{i=0}^{n-j} b_{i+j}\bar{b}_i & \text{for} \quad j=0, 1, 2, \ldots, n-1, n \\ \bar{r}_{-j} & \text{for} \quad j= -n, -n+1, \ldots, -2, -1 \,. \\ 0 & \text{for} \quad j< -n \text{ and } j>n \,. \end{cases}$$

The horizontal bar indicates the complex conjugate, applicable to those cases when the wavelet coefficients are complex.

Let us now define the reverse wavelet. If $(b_0, b_1, \ldots, b_n)$ is a finite wavelet, then $(\bar{b}_n, \bar{b}_{n-1}, \ldots, \bar{b}_1, \bar{b}_0)$ is called the *reverse wavelet*. For example, $(3, 0, 1)$ is the reverse of the wavelet $(1, 0, 3)$. Another example is $(i, 0.5)$ which is the reverse wavelet of $(0.5, -i)$, since $-i=i$. (Note: $i=\sqrt{-1}$). Also we may say that the wavelets $(1, 0, 3)$ and $(3, 0, 1)$ are the reverses of each other. So also are $(0.5, -i)$ and $(i, 0.5)$.

To autocorrelate a finite wavelet, we merely convolve the wavelet with its reverse. For example, let us find the autocorrelation of the wavelet (b_0, b_1, b_2). Its reverse wavelet is $(\bar{b}_2, \bar{b}_1, \bar{b}_0)$. The folding table is

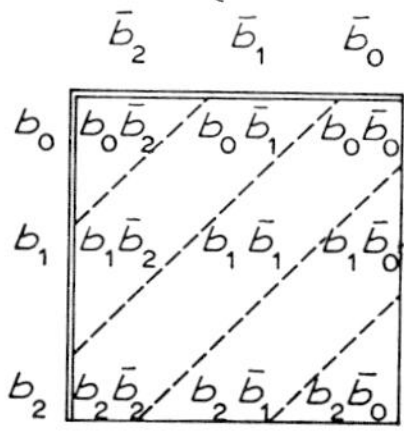

The diagonal term is always r_0, that is

$$r_0 = b_0\bar{b}_0 + b_1\bar{b}_1 + b_2\bar{b}_2 \,.$$

(For each summand, the first index minus the second index is 0.)

Working up from the diagonal we always get the r_j's for the negative j's, that is,

$$r_{-1} = b_0\bar{b}_1 + b_1\bar{b}_2$$

(For each summand, the first index minus the second index is -1.) and

$$r_{-2} = b_0\bar{b}_2 \,.$$

(The first index minus the second index is -2.)

Working down from the diagonal we always get the r_j's for positive j's, that is,

$$r_1 = b_2\bar{b}_1 + b_1\bar{b}_0$$

(For each summand, the first index minus the second is 1.)

and

$$r_2 = b_2 \bar{b}_0 \; .$$

(The first index minus the second index is 2.)

Let us now consider the wavelet $(2, 1)$. Its reverse is $(1, 2)$, so its autocorrelation table is

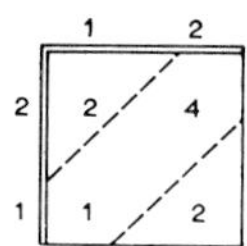

and so its autocorrelation is

$$r_{-1} = 2$$
$$r_0 \;\; = 4 + 1 = 5$$
$$r_1 \;\; = 2 \; .$$

On the other hand, consider the wavelet $(1, 2)$. Its reverse is $(2, 1)$, so its autocorrelation table is

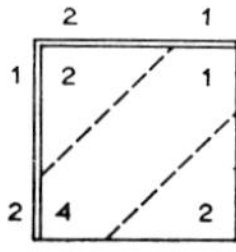

and so its autocorrelation is

$$r_{-1} = 2$$
$$r_0 \;\; = 1 + 4 = 5$$
$$r_1 \;\; = 2 \; .$$

Thus we see that the wavelet $(2, 1)$ has the *same* autocorrelation as its reverse $(1, 2)$. This is always so. That is, any wavelet has the same autocorrelation as its reverse.

But the above examples illustrate something further. It is: Let (b_0, b_1) be a 2-length wavelet. Then the *only* other 2-length wavelet with the *same* autocorrelation as (b_0, b_1) is its reverse $(\bar{b}_1, \bar{b}_0)$. Thus, the only other 2-length wavelet with the same autocorrelation as $(2, 1)$ is $(1, 2)$. Of course, there are wavelets of greater length with the same autocorrelation, but there is no other wavelet of length 2 with the same autocorrelation.

Consequently, we may pair every 2-length wavelet with another 2-length wavelet, namely its reverse. Such a pair is called a *dipole* and is of the form

$$(b_0, \, b_1) \quad \text{and} \quad (\bar{b}_1, \bar{b}_0) \; .$$

An example of a dipole is

$$(2, 1) \quad \text{and} \quad (1, 2) \,.$$

Another example is

$$(0.5, -i) \quad \text{and} \quad (i, 0.5) \,.$$

It is convenient to name each member wavelet of the dipole. Hence one 2-length wavelet of the dipole is called the minimum-delay wavelet, and the other is called the maximum-delay wavelet. The *minimum-delay wavelet* is the one in which the largest coefficient (in magnitude) is at the front, whereas the *maximum-delay wavelet* is the one in which the largest coefficient (in magnitude) is at the end. Thus (2, 1) is minimum-delay, and (1, 2) is maximum-delay. Also $(i, 0.5)$ is minimum-delay (because $|i| = 1 > 0.5$), whereas $(0.5, -i)$ is maximum-delay (because $|-i| = 1 > 0.5$).

The concepts of minimum-delay and maximum-delay have been defined for 2-length wavelets. Now let us extend these definitions to wavelets of any length. To do so, take a *group of dipoles*, each with the minimum-delay wavelet on the left. For example

	Min-delay		*Max-delay*
Dipole 1:	$(2, 1)$	and	$(1, 2)$
Dipole 2:	$(i, 0.5)$	and	$(0.5, -i)$
Dipole 3:	$(-i, 0.5)$	and	$(0.5, i)$

The convolution of the minimum-delay wavelets in the group (that is, the convolution of the wavelets on the left) yields a minimum-delay wavelet. Thus

$$(2, 1) * (i, 0.5) * (-i, 0.5) = (2, 1, 0.5, 0.25)$$

is a minimum-delay wavelet. (Note: We have evaluated its coefficients as follows. First, we construct the table

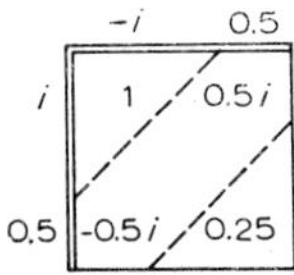

so

$$(i, 0.5) * (-i, 0.5) = (1, 0, 0.25) \,.$$

Next, construct the table

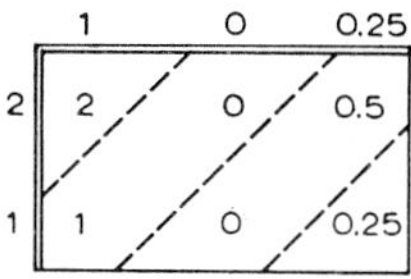

so

$$(2, 1) * (1, 0, 0.25) = (2, 1, 0.5, 0.25)$$

which is the minimum-delay wavelet given above.)

On the other hand, the convolution of the maximum-delay wavelets in the group of dipoles (that is, the convolution of the wavelets on the right) yields a maximum-delay wavelet. Thus

$$(1, 2) * (0.5, -i) * (0.5, i) = (0.25, 0.5, 1, 2) \, .$$

(It is seen that the maximum-delay wavelet is the reverse of the minimum-delay wavelet $(2, 1, 0.5, 0.25)$ found above, as we would expect.)

Furthermore, the convolution of one wavelet from each dipole in the group, but a mixture of minimum-delay and maximum-delay wavelets, gives a so called mixed-delay wavelet. Thus

$$(1, 2) * (i, 0.5) * (-i, 0.5) = (1, 2) * (1, 0, 0.25) = (1, 2, 0.25, 0.5)$$

is mixed-delay. (Here we have used the table

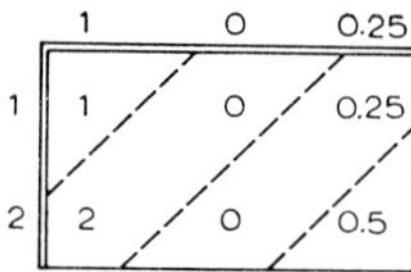

to perform the convolution.)

If every 2-length wavelet in a group of n dipoles is distinct, then there are 2^n possible $(n+1)$-length wavelets that can be generated, by taking one 2-length wavelet from each dipole. If, of course, some of the 2-length wavelets are not distinct, then some of the $(n+1)$-length wavelets will not be distinct either. In any case, all the $(n+1)$-length wavelets generated in this way from one group of dipoles will have the *same* autocorrelation. These $(n+1)$-length wavelets, so generated, are said to form a *suite of wavelets* (all with the same autocorrelation).

Another way of describing minimum-delay is by means of *partial energy*. Consider the (real) wavelet

$$(b_0, b_1, b_2, b_3) \, .$$

The partial energy for time 0, denoted by p_0, is b_0^2. Since partial energy is cumulative, the partial energy p_1 at time 1 is $b_0^2 + b_1^2$, and so on. That is

$$p_0 = b_0^2$$
$$p_1 = b_0^2 + b_1^2$$
$$p_2 = b_0^2 + b_1^2 + b_2^2$$
$$p_3 = b_0^2 + b_1^2 + b_2^2 + b_3^2 \,.$$

The last value of partial energy, in this case p_3, is in the total energy of the wavelet and is equal to r_0, the autocorrelation at index 0. (If the wavelet is complex, then we should use $b_0 \bar{b}_0$ instead of b_0^2, $b_1 \bar{b}_1$ instead of b_1^2, etc.) For example, the partial energy of the minimum-delay wavelet (2, 1, 0.5, 0.25) is

$$p_0 = 4$$
$$p_1 = 4 + 1 = 5$$
$$p_2 = 4 + 1 + 0.25 = 5.25$$
$$p_3 = 4 + 1 + 0.25 + 0.0625 = 5.3125 \,.$$

On the other hand, the partial energy of the maximum-delay wavelet (0.25, 0.5, 1, 2) is

$$p_0 = 0.0625$$
$$p_1 = 0.0625 + 0.25 = 0.3125$$
$$p_2 = 0.0625 + 0.25 + 1 = 1.3125$$
$$p_3 = 0.0625 + 0.25 + 1 + 4 = 5.3125 \,.$$

By comparing the two partial energy curves (see figure 6) we see that the partial energy of the maximum-delay wavelet never exceeds that of the minimum-delay wavelet. We would expect just this behavior from the way we constructed the two wavelets: The minimum-delay wavelet being the one with the energy concentrated at the front, and the maximum-delay wavelet being the one with the energy concentrated at the end.

The partial energy curves of the mixed-delay wavelets in the same suite lie between the partial energy curves of the minimum-delay and maximum-delay wavelets of the suite. That is, the mixed-delay wavelets have their energy concentrated between the two extremes. Thus the mixed-delay wavelet (1, 2, 0.25, 0.5) has partial energy

$$p_0 = 1$$
$$p_1 = 1 + 4 = 5$$
$$p_2 = 1 + 4 + 0.0625 = 5.0625$$
$$p_3 = 1 + 4 + 0.0625 + 0.25 = 5.3125$$

and this curve in figure 6 lies between the partial energy curves of the minimum-delay wavelet and the maximum-delay wavelet of the suite.

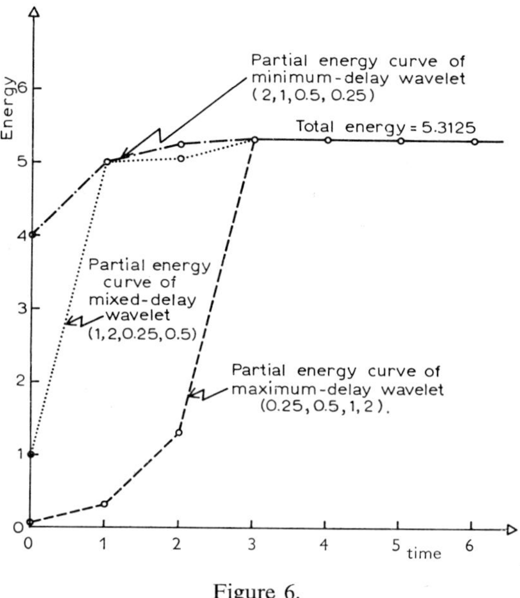

Figure 6.

The development in this section may be summarized as follows. A dipole consists of a 2-length wavelet and its reverse, that is

$$(a, b) \quad \text{and} \quad (\bar{b}, \bar{a}).$$

One of these two wavelets is called minimum-delay, and the other maximum-delay. If $|a| > |b|$, then (a, b) is the minimum-delay one, and $(\bar{b}, \bar{a})$ is the maximum-delay one. On the other hand, if $|b| > |a|$, then (a, b) is the maximum-delay one, and $(\bar{b}, \bar{a})$ is the minimum-delay one. Next, we consider a group of n dipoles, say

	Minimum-delay Wavelet	Maximum-delay Wavelet
Dipole 1	(α_0, α_1)	$(\bar{a}_1, \bar{a}_0)$
Dipole 2	(β_0, β_1)	$(\bar{\beta}_1, \bar{\beta}_0)$
$\vdots$	$\cdots$	$\cdots$
Dipole n	(ω_0, ω_1)	$(\bar{\omega}_1, \bar{\omega}_0)$

where

$$|\alpha_0| > |\alpha_1|$$
$$|\beta_0| > |\beta_1|$$
$$\cdots$$
$$|\omega_0| > |\omega_1|$$

(so that the 2-length wavelets on the left hand side are minimum-delay, and on the right hand side are maximum-delay). We now wish to form a $(n+1)$-length wavelet by convolving n of the 2-length wavelets, one from each dipole. Since there are n dipoles, there are 2^n possible $(n+1)$-length wavelets. These 2^n possible wavelets, it turns out, all have the same autocorrelation, and so they are said to form a suite of $(n+1)$-length wavelets, all with the same autocorrelation. (Of course, all of the 2^n wavelets in the suite do not need to be distinct). The wavelet in the suite formed by the convolution of all the minimum-delay 2-length wavelets is given by

$$(\alpha_0, \alpha_1) * (\beta_0, \beta_1) * \ldots * (\omega_0, \omega_1)$$

and is called the minimum-delay wavelet of the suite. The wavelet in the suite formed by the convolution of all the maximum-delay 2-length wavelets is given by

$$(\bar{\alpha}_1, \bar{\alpha}_0) * (\bar{\beta}_1, \bar{\beta}_0) * \ldots * (\bar{\omega}_1, \bar{\omega}_0)$$

and is called the maximum-delay wavelet of the suite. Clearly the maximum-delay wavelet of a suite is the reverse of the minimum-delay wavelet of the suite. Any other wavelet of the suite, that is any wavelet formed by a mixture of 2-length minimum-delay and maximum-delay wavelets, is called mixed-delay. Let us note that the coefficient for time index 0 of the minimum-delay wavelet of the suite is

$$\alpha_0 \beta_0 \ldots \omega_0 \, ,$$

whereas the coefficient for time index 0 of the maximum-delay wavelet is

$$\bar{\alpha}_1 \bar{\beta}_1 \ldots \bar{\omega}_1 \, .$$

Because

$$|\alpha_0| > |\alpha_1|$$
$$|\beta_0| > |\beta_1|$$
$$\cdots$$
$$|\omega_0| > |\omega_1|$$

it follows that

$$|\alpha_0 \beta_0 \ldots \omega_0| > |\bar{\alpha}_1 \bar{\beta}_1 \ldots \bar{\omega}_1| \, ,$$

which says that the coefficient for time index 0 of the minimum-delay wavelet is greater in magnitude than the coefficient for time index 0 of the maximum-delay wavelet of the same suite. Consider the mixed-delay wavelet of the suite given by

$$(\bar{\alpha}_1, \bar{\alpha}_0) * (\beta_0, \beta_1) * \ldots * (\omega_0, \omega_1) .$$

It coefficient for time index 0 is

$$\bar{\alpha}_1 \beta_0 \ldots \omega_0 .$$

Since

$$|\alpha_0| > |\bar{\alpha}_1|$$

it follows that

$$|\alpha_0 \beta_0 \ldots \omega_0| > |\bar{\alpha}_1 \beta_0 \ldots \omega_0|$$

which says that the coefficient for time index 0 of the minimum-delay wavelet is greater in magnitude than the coefficient for time index 0 of this mixed-delay wavelet of the suite. By the same argument, it is clear that the coefficient for time index 0 of the minimum-delay wavelet of a suite is greater in magnitude than the coefficient for time index 0 of any other wavelet in the same suite.

6. Inverse to Memory Function of Length 2

Suppose we have a box whose memory function is the 2-length wavelet

$$b = (b_0, b_1) .$$

Without loss of generality, we may normalize this wavelet so $b_0 = 1$ and $b_1 = k$. That is, we consider the wavelet

$$b = (b_0, b_1) = (1, k) .$$

(Note: k may be complex.)

Let the time-series x be the input to the box, and let the time-series y be the output. See figure 7.

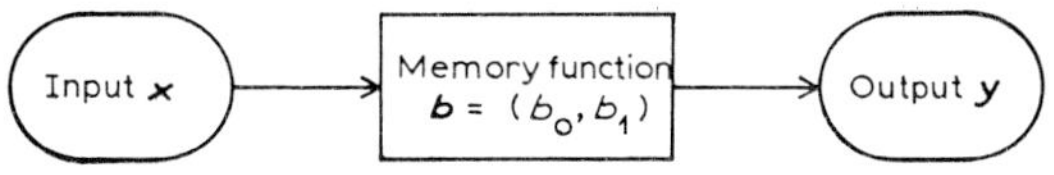

Figure 7. Box with 2-length memory.

Given the input x and the memory function b, we know how to calculate the output y. The output y is the convolution of input x and memory function

b, that is

$$y = b * x$$

or

$$y_t = b_0 x_t + b_1 x_{t-1} = x_t + k x_{t-1} \, .$$

On the other hand, suppose we are not given the input **x** but instead we are given the output **y**. In other words, given the output **y** and the memory function **b**, we wish to calculate the input **x**. This is the *inverse problem*, a type of problem that continually crops up in science.

Let us then try to find another system, one with memory function denoted by $a = b^{-1}$, and with the property that

$$a * b = \delta$$

where δ is the *unit impulse wavelet*, that is δ is defined by

$$\delta = (1, 0, 0, 0, \ldots) \, .$$

In other words, δ is a *unit spike* at time 0. It is easy to see that δ convolved with any **x** is equal to the same **x**, that is

$$\delta * x = x \, ,$$

so δ plays the role of the identity wavelet. Because $a * b = \delta$, we see the *output* from box **a** hitched in tandem with box **b** is the same as the input to box **b**. In other words **a** and **b** are inverses to each other. See figure 8.

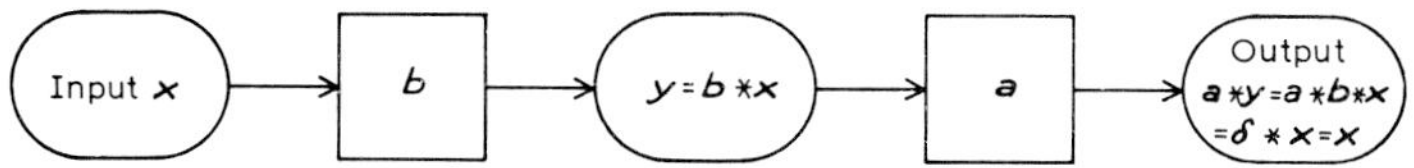

Figure 8. Mutually inverse boxes.

In brief, then, we wish to find the wavelet

$$a = (a_0, a_1, a_2, \ldots)$$

such that the convolution of **a** and **b** is a spike at time 0; that is,

$$a * b = (a_0, a_1, a_2, \ldots) * (1, k) = (1, 0, 0, 0, \ldots) \, .$$

At this point let us use the *z*-transform representation, so the above equation becomes

$$(a_0 + a_1 z + a_2 z^2 + \ldots)(1 + kz) = 1 \, .$$

Now divide each side by $1 + kz$, thus obtaining

$$a_0+a_1z+a_2z^2+\ldots=\frac{1}{1+kz}.$$

But the right hand side can be expressed in powers of z by polynomial long division. That is,

$$
\begin{array}{r}
1-kz+k^2z^2-k^3z^3+\ldots \\[2pt]
\hline
1+kz\,\overline{)\,1} \\
\underline{1+kz} \\
-kz \\
\underline{-kz-k^2z^2} \\
k^2z^2 \\
\underline{k^2z^2+k^3z^3} \\
-k^3z^3\ldots
\end{array}
$$

so

$$\frac{1}{1+kz}=1-kz+k^2z^2-k^3z^3+k^4z^4-\ldots\ .$$

Identifying coefficients of like powers of z, we have

$$\boldsymbol{b}^{-1}=\boldsymbol{a}=(a_0,\,a_1,\,a_2,\,a_3,\,\ldots)=(1,\,-k,\,k^2,\,-k^3,\,\ldots)$$

or

$$a_t=\begin{cases}0 & t=-1,\,-2,\,-3,\,\ldots \\ (-k)^t, & t=0,\,1,\,2,\,\ldots\end{cases}$$

which shows that the $\boldsymbol{a}$ box has an infinite memory function.

For example, suppose $k=0.5$, so $\boldsymbol{b}=(1,\,0.5)$ is a minimum-delay wavelet. Let us check the property that $\boldsymbol{a}\circledast\boldsymbol{b}=\boldsymbol{\delta}$ by folding:

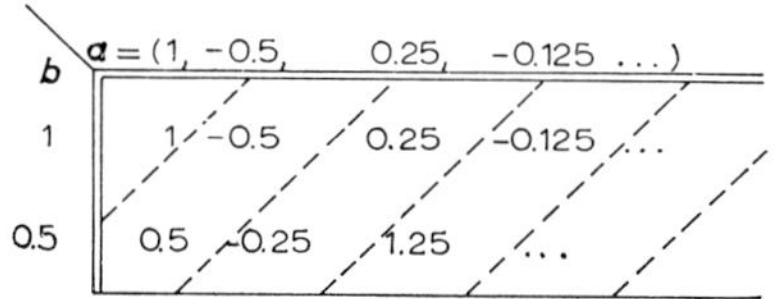

The first fold gives $\delta_0=1$, the second fold gives

$$\delta_1=-0.5+0.5=0\,,$$

the third fold gives

$$\delta_2=0.25-0.25=0\,,$$

and it is easy to see that all the remaining folds give 0 also.

As another example, suppose k is 2. Then $\boldsymbol{b}=(1,\,2)$ is a maximum-delay wavelet. Using our formula for $\boldsymbol{a}=\boldsymbol{b}^{-1}$ we have

$$\boldsymbol{a}=(1,\,-2,\,4,\,-8,\,16,\,-32,\,64,\,-128,\,\ldots)\,.$$

The property $a * b = \delta$ is satisfied, but we note that the coefficients of a are growing in magnitude. Hence, if we truncate a after a finite number of coefficients, then a tremendous truncation error will result. Because the magnitudes of the coefficients of this a increase without limit as time increases, we see that a does not have finite energy and hence is not stable. That is, a is *unstable*.

In fact we see that a will be a (stable) wavelet whenever $|k| < 1$, and will be unstable whenever $|k| > 1$. In other words, if $b = (1, k)$ is a minimum-delay, then its inverse a is a stable wavelet. If $b = (1, k)$ is a maximum-delay, then its inverse a is unstable.

Is there any way that we can avoid this unfortunate condition of instability? The answer is no, if we require the inverse a to have the form

$$a = (a_0, a_1, a_2, \ldots)$$

where a_0 is the coefficient at time 0, a_1 is the coefficient at time 1, etc., and where it is understood that all coefficients *before* time 0 are automatically zero (that is, $a_t = 0$ for time $t = -1, -2, -3, \ldots$).

Let us now introduce the notion of antiwavelets. An *antiwavelet* has the form

$$a = (\ldots, a_{-3}, a_{-2}, a_{-1})$$

where a_{-1} is the coefficient at time -1, a_{-2} is the coefficient at time -2, a_{-3} is the coefficient at time -3, etc. and where it is understood that the coefficient at time 0, and all following coefficients, are automatically zero (that is, $a_t = 0$ for time $t = 0, 1, 2, 3, \ldots$). For stability, an antiwavelet

$$a = (\ldots, a_{-3}, a_{-2}, a_{-1})$$

is required to have finite energy, that is,

$$\ldots + a_{-3}^2 + a_{-2}^2 + a_{-1}^2 < \infty \, .$$

It follows then that the magnitude of its coefficients asymptotically approach zero as time increases in the *negative* direction, that is $|a_t| \to 0$ as $t \to -\infty$.

Now let us ask the same question. Given the maximum-delay wavelet

$$b = (1, k) \quad \text{with} \quad |k| > 1 \, ,$$

can we find an inverse $a = b^{-1}$ which is stable? The answer is yes, provided we allow the inverse a to be an antiwavelet.

Hence we wish to find the antiwavelet

$$a = (\ldots, a_{-3}, a_{-2}, a_{-1})$$

such that the convolution of a and b is a spike at time 0; that is

$$a \divideontimes b = (\ldots, a_{-3}, a_{-2}, a_{-1}) \divideontimes (1, k) = (1, 0, 0, 0, \ldots) .$$

Using the z-transform representation (where antiwavelets require negative powers of z), the above equation becomes

$$(\ldots + a_{-3} z^{-3} + a_{-2} z^{-2} + a_{-1} z^{-1})(1 + kz) = 1 .$$

Now divide each side by $1 + kz$, so as to obtain

$$(\ldots + a_{-3} z^{-3} + a_{-2} z^{-2} + a_{-1} z^{-1}) = \frac{1}{1 + kz} .$$

But the right hand side can be expressed in powers of z^{-1} by polynomial long division. That is,

$$
\begin{array}{r}
(kz)^{-1} - (kz)^{-2} + (kz)^{-3} - \ldots \\[2pt]
kz + 1 \overline{)\ 1 } \\
\underline{1 \quad + (kz)^{-1}} \\
- (kz)^{-1} \\
\underline{- (kz)^{-1} - (kz)^{-2}} \\
(kz)^{-2}
\end{array}
$$

so

$$\frac{1}{kz + 1} = k^{-1} z^{-1} - k^{-2} z^{-2} + k^{-3} z^{-3} - k^{-4} z^{-4} + k^{-5} z^{-5} - \ldots .$$

Identifying coefficients in like powers of z we obtain the inverse

$$
\begin{aligned}
a &= (\ldots, a_{-5}, a_{-4}, a_{-3}, a_{-2}, a_{-1}) \\
  &= (\ldots, k^{-5}, -k^{-4}, k^{-3}, -k^{-2}, k^{-1})
\end{aligned}
$$

or

$$a_t = \begin{cases} -(-k)^t, & t = -1, -2, -3, \ldots \\ 0, & t = 0, 1, 2, \ldots . \end{cases}$$

For example, suppose $k = 2$, so $b = (1, 2)$ is maximum-delay. Let us check the property that

$$a \divideontimes b = (1, 0, 0, 0, \ldots) = \delta$$

by folding. We have the table:

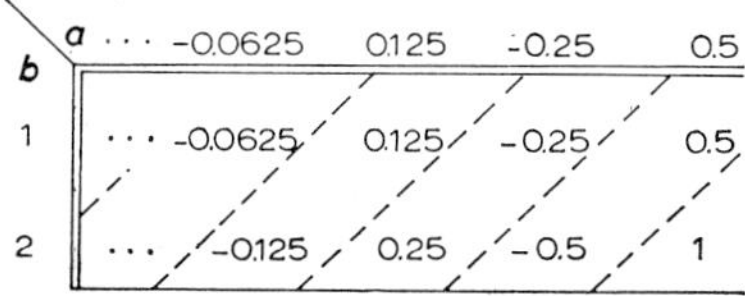

which gives:

$$
\begin{array}{ll}
\cdots \qquad\qquad\qquad \cdots & \\
0.125-0.125=0 & \text{at time} \;-3 \\
-0.25+0.25=0 & \text{at time} \;-2 \\
0.5-0.5=0 & \text{at time} \;-1 \\
1 & \text{at time} \quad 0
\end{array}
$$

and gives 0 for all positive times. Thus we have obtained the spike at time 0, as desired.

In this section, we considered the problem of finding the inverse of the 2-length wavelet

$$b=(1,\,k)\,.$$

(where k is allowed to be a complex number). We found that a stable inverse can always be found, but that its form depends upon whether $|k|<1$ or $|k|>1$, that is, upon whether b is minimum-delay or maximum-delay. In both cases, the stable inverse has infinite length. In the first case (i.e. $|k|<1$), the stable inverse is an infinite wavelet (and so its coefficients are automatie cally zero for negative times.) In the second case (i.e. $|k|<1$), the stably inverse is an infinite antiwavelet (and so its coefficients are automaticall- zero for non-negative times). To summarize:

If $b=(1,\,k)$ is a *minimum-delay wavelet* (that is, if $|k|<1$), then its *stable inverse* is the infinite-length *wavelet*

$$a=(a_0,\,a_1,\,a_2,\,a_3,\,\ldots)=(1,\,-k,\,k^2,\,-k^3,\,\ldots)$$

If $b=(1,\,k)$ is a *maximum-delay wavelet* (that is, if $|k|>1$), then its *stable inverse* is the infinite-length *antiwavelet*

$$a=(\ldots,\,a_{-4},\,a_{-3},\,a_{-2},\,a_{-1})=(\ldots,\,-k^{-4},\,k^{-3},\,-k^{-2},\,k^{-1})\,.$$

7. Approximate Inverse to 2-Length Minimum-Delay Memory Function

The problem we wish to consider now is the following. The inverses given in the preceding section are exact inverses, and they are infinitely long. For practical reasons, we would like an inverse which is finite. Such a finite inverse could not be exact, so it would be only an approximate inverse.

In this section, we shall consider the case when $b=(1,\,k)$ is minimum-delay, i.e. when $|k|<1$. For simplicity we shall carry through the following analysis for the case when k is a real number, as it may be easily modified to cover the case when k is a complex number. The first thing that comes to mind is to form approximate inverses by truncating the true exact inverse. Such ones are called *truncated approximate inverses.*

From the preceding section, we recall that the *exact inverse* of the minimum-delay wavelet b is

$$b^{-1} = a = (a_0, a_1, a_2, \ldots) = (1, -k, k^2, -k^3, \ldots) .$$

Truncating all the coefficients except the first, we obtain the 1-length truncated approximate inverse which is

$$a = (1) .$$

Convolving this approximate inverse with the *input b*, we obtain the (*actual*) *output c* given by

$$c = b * a = (1, k) * (1) = (1, k) .$$

The *desired output* is the spike at time 0 represented by

$$\delta = (1, 0)$$

so the error (between desired output and actual output) is

$$e = \delta - c = (1, 0) - (1, k) = (0, -k) .$$

The sum of squares of the coefficients of the error wavelet, or squared norm of the error, is

$$0^2 + (-k)^2 = k^2 .$$

Thus we have found the error produced by the 1-length truncated approximate inverse.

Let us now turn to the 2-length approximation. Truncating all the coefficients of the exact inverse a except the first two, we obtain the 2-length truncated approximate inverse

$$a = (1, -k) .$$

Convolving this approximate inverse with b, we obtain the output c given by

$$c = b * a = (1, k) * (1, -k) = (1, 0, -k^2) .$$

The desired output is the spike at time 0 given by

$$\delta = (1, 0, 0)$$

so the error is

$$e = \delta - c = (1, 0, 0) - (1, 0, -k^2) = (0, 0, k^2)$$

with squared norm

$$0^2 + 0^2 + (k^2)^2 = k^4 .$$

Thus we have found the error produced by the 2-length truncated approximate inverse. This error is smaller than that for the 1-length truncated approximate inverse.

Continuing in this way, we obtain table 2.

TABLE 2

Approximate inverses by truncation ($|k| < 1$)

Length	Approximate inverse	Error	Squared Norm of error
1	(1)	$(0, -k)$	k^2
2	$(1, -k)$	$(0, 0, k^2)$	k^4
3	$(1, -k, k^2)$	$(0, 0, 0, -k^3)$	k^6
4	$(1, -k, k^2, -k^3)$	$(0, 0, 0, 0, k^4)$	k^8
...	...	...	...

We now ask the question: Is there another way to find an approximate inverse of a given length which has error with smaller squared norm than that of the same-length truncated approximate inverse. The answer is yes. In fact, we shall now develop a method to find approximate inverses whose error has *minimum* squared norm. Such ones are called *least-squares approximate inverses*.

Let

$$a = (a_0)$$

be the desired approximate inverse (of length 1) for the minimum-delay wavelet

$$b = (1, k) \quad \text{where} \quad |k| < 1 .$$

We recall that the autocorrelation of this wavelet is

$$r_{-1} = k, \quad r_0 = 1 + k^2, \quad r_1 = k \quad \text{and all other} \quad r_j = 0 .$$

The output is

$$c = a * b = (a_0) * (1, k) = (a_0, a_0 k)$$

whereas the desired output is the spike at time 0 given by

$$\delta = (1, 0) .$$

Hence the error is

$$e = \delta - c = (1, 0) - (a_0, a_0 k) = (1 - a_0, -a_0 k)$$

with squared norm

$$E = (1 - a_0)^2 + (-a_0 k)^2 = 1 - 2a_0 + a_0^2 (1 + k^2)$$
$$= 1 - 2a_0 + a_0^2 r_0 .$$

We wish to minimize E with respect to a_0. We may do so by equating the derivative of E with respect to a_0 equal to zero; that is,

$$\frac{\partial E}{\partial a_0} = -2 + 2a_0 r_0 = 0 .$$

Now let us solve for a_0. We have

$$a_0 = \frac{1}{r_0} .$$

This is the coefficient of the desired approximate inverse $a = (a_0)$ of length 1 which produces the error with minimum squared norm.

Let us now consider the case for length 2.

Let

$$a = (a_0, a_1)$$

be the desired approximate inverse. The output is

$$c = a * b = (a_0, a_1) * (1, k) = (a_0, a_0 k + a_1, a_1 k)$$

whereas the desired output is the spike at time 0 given by

$$\delta = (1, 0, 0) .$$

It follows that the error is

$$e = (1, 0, 0) - (a_0, a_0 k + a_1, a_1 k)$$
$$= (1 - a_0, -a_0 k - a_1, -a_1 k)$$

with squared norm

$$E = (1 - a_0)^2 + (-a_0 k - a_1)^2 + (-a_1 k)^2$$
$$= 1 - 2a_0 + a_0^2 + a_0^2 k^2 + 2a_0 a_1 k + a_1^2 + a_1^2 k^2$$
$$= 1 - 2a_0 + a_0^2 (1 + k^2) + 2a_0 a_1 k + a_1^2 (1 + k^2)$$
$$= 1 - 2a_0 + a_0^2 r_0 + 2a_0 a_1 r_1 + a_1^2 r_0 .$$

To find the a_0, a_1 which minimize E, we solve the simultaneous equations

$$\frac{\partial E}{\partial a_0} = -2 + 2a_0 r_0 + 2a_1 r_1 = 0$$
$$\frac{\partial E}{\partial a_1} = \qquad 2a_0 r_1 + 2a_1 r_0 = 0$$

or

$$r_0 a_0 + r_1 a_1 = 1$$
$$r_1 a_0 + r_0 a_1 = 0$$

The solution is

$$a_0 = \frac{r_0}{\Delta},$$

$$a_1 = -\frac{r_1}{\Delta}.$$

(where $\Delta = r_0^2 - r_1^2$.) These are the coefficients for the desired approximate inverse $a = (a_0, a_1)$ of length 2 which produces the error with minimum squared norm. Expressed in terms of k, these coefficients are

$$a_0 = \frac{1+k^2}{1+k^2+k^4},$$

$$a_1 = -\frac{k}{1+k^2+k^4}.$$

It follows that $|a_0|$ is always greater than $|a_1|$ regardless of k. Hence the least-squares approximate inverse (a_0, a_1) is always minimum-delay regardless of the delay properties of the wavelet $b = (1, k)$.

Approximate inverses of greater length may be found by the same least-squares procedure. As we shall see in section 9, these longer least-squares approximate inverses are also always minimum-delay, regardless of the delay properties of the wavelet $b = (1, k)$.

In this section we have considered the minimum-delay wavelet $(1, k)$ where $|k| < 1$. We have presented two ways to approximate its exact inverse (which has infinite length) by an approximate inverse with finite length. The first way is to obtain the approximate inverse by merely truncating the exact inverse. The second way is to obtain the approximate inverse by minimizing the squared norm of the error between the desired output (i.e. the delta function) and the actual output (i.e. the convolution of the wavelet and the approximate inverse).

8. Inverse to Memory Function of Arbitrary Length

Let the wavelet

$$b = (b_0, b_1, b_2, \ldots, b_n)$$

be the memory function of a linear system. Its associated polynomial (or z-transform) is

$$B(z) = b_0 + b_1 z + b_2 z^2 + \ldots + b_n z^n.$$

This polynomial can be factored, as done in high-school algebra, and so written as a product of n-factors,

$$B(z) = (\alpha_0 + \alpha_1 z)(\beta_0 + \beta_1 z)(\gamma_0 + \gamma_1 z) \ldots (\omega_0 + \omega_1 z)$$

each factor being the z-transform of a 2-length wavelet. These 2-length wavelets are denoted by

$$(\alpha_0, \alpha_1), \quad (\beta_0, \beta_1), \quad (\gamma_0, \gamma_1), \ldots, (\omega_0, \omega_1)$$

where alpha indicates the first and omega the last. Since multiplying polynomials is equivalent to convolving their coefficients, we have

$$b = (\alpha_0, \alpha_1) * (\beta_0, \beta_1) * (\gamma_0, \gamma_1) * \ldots * (\omega_0, \omega_1) .$$

This tandem group of convolutions has physical meaning in terms of our boxes (i.e. linear systems). It means that a box with a long memory function **b** can be looked upon as a sequence of boxes each with a memory of length 2. See figure 9.

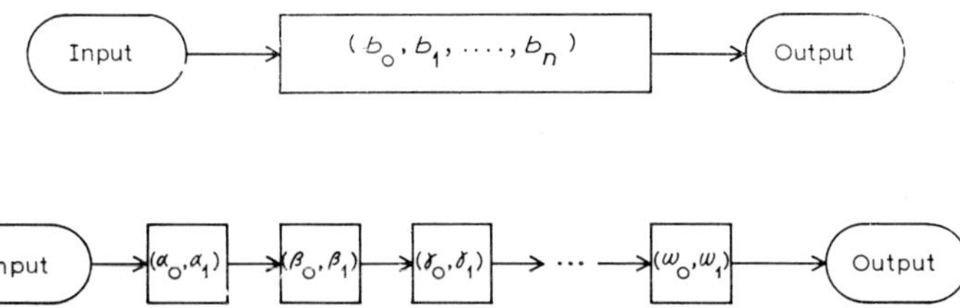

Figure 9. Equivalent systems.

For example, the memory function (6, 11, 4) has z-transform

$$6 + 11z + 4z^2$$

which may be factored as

$$(2 + z)(3 + 4z) .$$

Hence the box

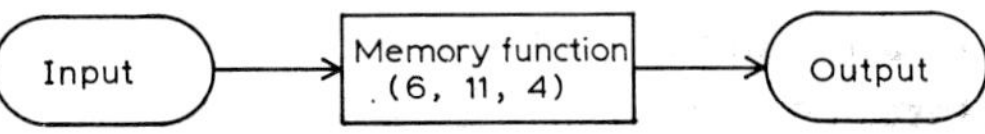

is equivalent to the two tandem boxes

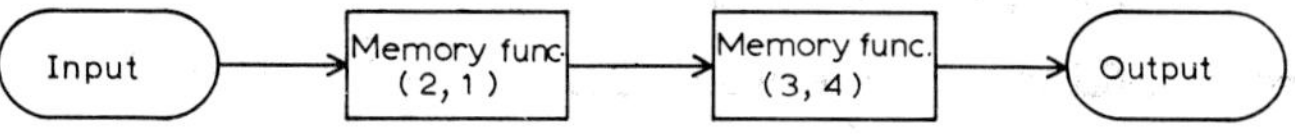

In factoring polynomials, we must be prepared to admit factors which have complex coefficients, as it is known from high-school algebra that the roots of a polynomial may be complex. Nevertheless, if the original polynomial has real coefficients, it will turn out that complex factors occur in pairs, so that the coefficients of one of the pair are the complex-conjugates of their counterparts in the other of the pair. For example,

$$2+z+0.5z^2+0.25z^3$$

may be factored as

$$(2+z)(i+0.5z)(-i+0.5z) .$$

There is one pair of complex factors, namely

$$(i+0.5z) \quad \text{and} \quad (-i+0.5z) .$$

The coefficients $(i, 0.5)$ of the first are seen to be the complex-conjugates of the coefficients $(-i, 0.5)$ of the other, i.e. i is the complex-conjugate of its counterpart $-i$, and 0.5 is the complex-conjugate of its counterpart 0.5.

Let us now investigate the inverse a of a many-coefficient memory function

$$b=(b_0, b_1, \ldots, b_n) .$$

We need only utilize results we have already developed.

First, as above, represent the b box as a tandem combination of systems with 2-length memory functions. That is, b is the convolution of n 2-length wavelets:

$$(b_0, b_1 \ldots, b_n)=(\alpha_0, \alpha_1)*(\beta_0, \beta_1)* \ldots *(\omega_0, \omega_1) .$$

In terms of polynomials (i.e. z-transforms) we have

$$(b_0+b_1z+ \ldots +b_nz^n)=(\alpha_0+\alpha_1 z)(\beta_0+\beta_1 z) \ldots (\omega_0+\omega_1 z) .$$

The inverse a of the b box has transfer function

$$\frac{1}{b_0+b_1z+ \ldots +b_nz^n} = \left(\frac{1}{\alpha_0+\alpha_1 z}\right)\left(\frac{1}{\beta_0+\beta_1 z}\right) \ldots \left(\frac{1}{\omega_0+\omega_1 z}\right) .$$

Now, if each component 2-length wavelet is minimum-delay, then the composite wavelet b is minimum-delay (and vice versa). In this minimum-delay case, we may perform the polynomial division

$$\frac{1}{b_0+b_1z+ \ldots +b_nz} = a_0+a_1 z+a_2 z^2+ \ldots$$

to find the coefficients of the wavelet a which is the stable inverse of b. Alternatively, we may find the inverse wavelet of each 2-length wavelet, and then convolve these n inverses to yield the wavelet a.

On the other hand, if each component 2-length wavelet is maximum-delay then the composite wavelet b is maximum-delay (and vice versa). In this maximum-delay case, we may perform the polynomial division

$$\frac{1}{b_0 + b_1 z + \ldots + b_n z^n} = a_{-1} z^{-1} + a_{-2} z^{-2} + a_{-3} z^{-3} + \ldots$$

to find the coefficients of the antiwavelet a which is the stable inverse of b. Alternatively, we may find the inverse antiwavelet of each 2-length wavelet, and then convolve these n inverses to find the antiwavelet a.

Finally, if the component 2-length wavelets are a mixture of minimum-delay ones and maximum-delay ones, then we must find the inverse of each one. Each of these component wavelet inverses will either be a wavelet or an antiwavelet, depending upon whether the component wavelet was minimum-delay or maximum-delay. Then we must convolve these n inverses together to find a which is the stable inverse of b. Now a has both a wavelet and an antiwavelet component, so it may be written

$$a = (\ldots, a_{-2}, a_{-1}, a_0 a_1, a_2, \ldots) .$$

Its z-transform therefore contains both positive and negative powers of z and so has the form of a Laurent series:

$$\ldots + a_{-2} z^{-2} + a_{-1} z^{-1} + a_0 + a_1 z + a_2 z^2 + \ldots .$$

9. Approximate Inverse to Arbitrary-Length Minimum-Delay Memory Function

We proceed the same way as we did in the case of a 2-length minimum-delay wavelet (see Section 7).

Let the (real valued) memory function

$$b = (b_0, b_1, \ldots, b_n)$$

be a $(n+1)$-length minimum-delay wavelet. Let the $(m+1)$-length wavelet

$$a = (a_0, a_1, \ldots, a_m)$$

be the desired approximate inverse of b. We wish to choose a so that the convolution of a with b will produce as nearly as possible the desired output which is the spike at time 0 given by

$$\delta = (1, 0, 0, \ldots, 0) .$$

We recall that when input b goes into the box a, an output wavelet c is produced according to the convolution formula. Figure 10 sums up the situation, showing the desired output versus the actual output.

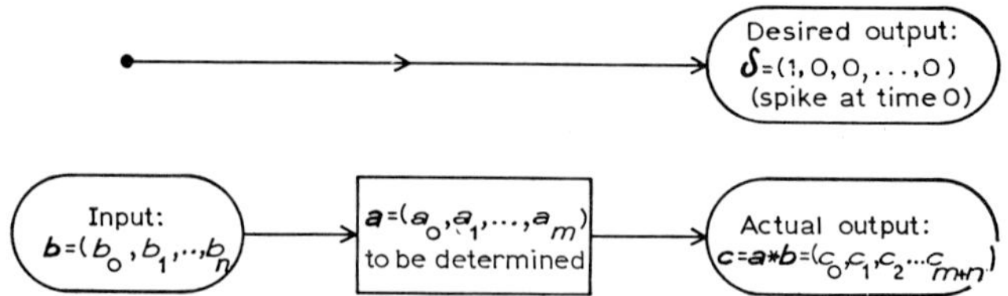

Figure 10. Desired output vs. actual output.

In other words, we want to build a box $(a_0, a_1, \ldots, a_m)$ which will condense the input $(b_0, b_1, \ldots, b_n)$ as well as possible into the spike $(1, 0, 0, \ldots, 0)$.

Thus we shall choose the approximate inverse a such that the difference vector or error

$$(1, 0, 0, \ldots, 0) - (c_0, c_1, c_2, \ldots, c_{m+n}) = (1 - c_0, -c_1, -c_2 \ldots, -c_{m+n})$$

has minimum squared norm. Hence a will be the *least-squares approximate inverse* to b. That is, we wish to determine a by minimizing the squared norm

$$E = (1 - c_0)^2 + (-c_1)^2 + (-c_2)^2 + \ldots + (-c_{m+n})^2$$

which is

$$E = (1 - c_0)^2 + c_1^2 + c_2^2 + \ldots + c_{m+n}^2 ,$$

or

$$E = 1 - 2c_0 + (c_0^2 + c_1^2 + c_2^2 + \ldots + c_{m+n}^2) = 1 - 2c_0 + \sum_{t=0}^{m+n} c_t^2 .$$

Now we must substitute into E the expression for c in terms of a and b, that is

$$c_t = \sum_{s=0}^{m} a_s b_{t-s} .$$

Hence

$$E = 1 - 2a_0 b_0 + \sum_{t=0}^{m+n} \left(\sum_{s=0}^{m} a_s b_{t-s} \right)^2 .$$

We minimize E by setting its partial derivatives with respect to each of $a_0, a_1, \ldots a_m$ equal to zero, thus obtaining the $(m+1)$ equations in the $(m+1)$ unknowns $a_0, a_1, \ldots, a_m$:

$$\frac{\partial E}{\partial a_0} = -2b_0 + \sum_{t=0}^{m+n} 2\left(\sum_{s=0}^{m} a_s b_{t-s}\right) b_t = 0$$

$$\frac{\partial E}{\partial a_1} = \sum_{t=0}^{m+n} 2\left(\sum_{s=0}^{m} a_s b_{t-s}\right) b_{t-1} = 0$$

$$\cdots \qquad \cdots$$

$$\frac{\partial E}{\partial a_m} = \sum_{t=0}^{m+n} 2\left(\sum_{s=0}^{m} a_s b_{t-s}\right) b_{t-m} = 0 .$$

We may cancel out the 2's and interchange the summation signs, thus obtaining:

$$-b_0 + \sum_{s=0}^{m} a_s \sum_{t=0}^{m+n} b_{t-s} b_t = 0$$

$$\sum_{s=0}^{m} a_s \sum_{t=0}^{m+n} b_{t-s} b_{t-1} = 0$$

$$\cdots$$

$$\sum_{s=0}^{m} a_s \sum_{t=0}^{m+n} b_{t-s} b_{t-m} = 0 .$$

Next we note that the summations on t give the autocorrelation r_j of b, that is

$$r_{-s} = \sum_{t=0}^{m+n} b_{t-s} b_t$$

$$r_{-s+1} = \sum_{t=0}^{m+n} b_{t-s} b_{t-1}$$

$$\cdots$$

$$r_{-s+m} = \sum_{t=0}^{m+n} b_{t-s} b_{t-m} .$$

Inserting these r_j into our $(m+1)$ simultaneous equations we obtain

$$-b_0 + \sum_{s=0}^{m} a_s r_{-s} = 0$$

$$\sum_{s=0}^{m} a_s r_{-s+1} = 0$$

$$\cdots$$

$$\sum_{s=0}^{m} a_s r_{-s+m} = 0 .$$

Moving the $-b_0$ to the right hand side, we finally obtain

$$\begin{bmatrix} a_0 r_0 + a_1 r_{-1} + a_2 r_{-2} + \ldots + a_m r_{-m} = b_0 \\ a_0 r_1 + a_1 r_0 + a_2 r_{-1} + \ldots + a_m r_{-m+1} = 0 \\ \ldots \qquad\qquad \ldots \\ a_0 r_m + a_1 r_{-1+m} + a_2 r_{-2+m} + \ldots + a_m r_0 = 0 . \end{bmatrix}$$

These $(m+1)$ equations in the $(m+1)$ unknowns $a_0, a_1, \ldots, a_m$ are called the *normal equations*. Using the fact that the autocorrelation is an even function, that is

$$r_{-1} = r_1, \quad r_{-2} = r_2, \quad r_{-3} = r_3, \ldots, r_{-m} = r_m$$

the normal equation may be written

$$\begin{bmatrix} a_0 r_0 + a_1 r_1 + a_2 r_2 + \ldots + a_m r_m = b_0 \\ a_0 r_1 + a_1 r_0 + a_2 r_1 + \ldots + a_m r_{m-1} = 0 \\ \ldots \\ a_0 r_m + a_1 r_{m-1} + a_2 r_{m-2} + \ldots + a_m r_0 = 0 . \end{bmatrix}$$

At first glance, one might expect that the computation involved in the solution of this set of $(m+1)$ simultaneous equations would be approximately the same as any other set with a symmetrical positive definite matrix. Nevertheless, this is not so because of the much greater symmetry involved: namely, all the elements on any super-diagonal or sub-diagonal are equal. Hence the left hand side of each equation involves, at most, only the $(m+1)$ autocorrelation coefficients

$$r_0, r_1, r_2, \ldots, r_m .$$

These known coefficients are merely shifted to the right one notch as we go from one equation to the next, a fact which we can use for tremendous computational saving.

In other words, although it appears that we are faced with familiar computational work of solving $(m+1)$ equations in $(m+1)$ unknowns, we are in fact only faced with solving a specialized type of vector operation. This operation involves the determination of the $(m+1)$ unknowns

$$a_0, a_1, \ldots, a_m$$

from the known $(m+1)$ autocorrelation coefficients

$$r_0, r_1, \ldots, r_m$$

and from the known b_0.

Let us now talk about the squared norm

$$E = 1 - 2c_0 + c_0^2 + c_1^2 + c_2^2 + \ldots + c_{m+n}^2 .$$

First let

$$\rho_{-m}, \rho_{-m+1}, \ldots, \rho_{-1}, \rho_0, \rho_1, \ldots, \rho_{m-1}, \rho_m$$

denote the autocorrelation of the wavelet

$$a = (a_0, a_1, \ldots, a_m) .$$

Then it may be shown that

$$c_0^2 + c_1^2 + c_2^2 + \ldots + c_{m+n}^2 = \sum_{s=-m}^{m} \rho_s \, r_s .$$

Hence it follows that

$$E = 1 - 2c_0 + \sum_{s=-m}^{m} \rho_s r_s .$$

Second, it is not difficult to show that minimum value of E (obtained by using $a_0, a_1, \ldots, a_m$ determined from the solution of the normal equations) is

$$E_{\min} = 1 - c_0 = 1 - a_0 b_0 .$$

Hence

$$a_0 b_0 = c_0 = c_0^2 + c_1^2 + \ldots + c_{m+n}^2$$

so c_0 is non-negative, and a_0 and b_0 both have the same algebraic sign.

The question of the delay property of the least-squares approximate inverse now comes up. Because $c_0 = a_0 b_0$, the minimum squared norm of the error is

$$E_{\min} = 1 - 2a_0 b_0 + \sum_{s=-m}^{m} \rho_s r_s$$

where $(a_0, a_1, \ldots, a_m)$ is the wavelet determined from the solution of the normal equations. *Suppose* that $(a_0, a_1, \ldots, a_m)$ is *not* minimum-delay. (We shall show that this supposition leads to a contradiction). There is a suite of wavelets each of which has the same autocorrelation ρ_s as $(a_0, a_1, \ldots, a_m)$. Pick from this suite the minimum-delay wavelet $(\alpha_0, \alpha_1, \ldots, \alpha_m)$ whose leading coefficient α_0 has the same algebraic sign as b_0. We recall that in any suite of wavelets with the same autocorrelation, the leading coefficient of a minimum-delay wavelet is greater (in magnitude) than the leading coefficient of any non-minimum delay wavelet in the suite. Hence

$$|\alpha_0| > |a_0| .$$

Now if the memory function of the box is $(\alpha_0, \alpha_1, \ldots, \alpha_m)$ then the squared norm of the error is

$$E_\alpha = 1 - 2\alpha_0 b_0 + \sum_{s=-m}^{m} \rho_s r_s .$$

Comparing $E_{\min}$ and E_α, we see

$$E_\alpha - E_{\min} = -2\alpha_0 b_0 + 2a_0 b_0 .$$

Because $|\alpha_0| > |a_0|$ and because α_0, b_0, a_0 all have the same algebraic sign, it follows that

$$E_\alpha - E_{\min} < 0$$

or

$$E_\alpha < E_{\min} .$$

But this is a contradiction, since $E_{\min}$ is by definition the minimum value of E. Hence our supposition that $(a_0, a_1, \ldots, a_m)$ is not minimum-delay is contradicted. Thus $a = (a_0, a_1, \ldots, a_m)$ is minimum-delay. (Moreover, an inspection of the proof shows that a is minimum-delay regardless of whether $b = (b_0, b_1, \ldots, b_n)$ is minimum-delay or not, as no reference to the delay properties of b is used in the proof.)

We have therefore proved the following theorem.

Let b be any given wavelet, and let a be the least-squares approximate inverse to b. Then a is minimum-delay.

In this section we have shown how to determine a box with memory function $(a_0, a_1, \ldots, a_m)$ which approximately converts an input wavelet $(b_0, b_1, \ldots, b_n)$ into the spike $(1, 0, 0, \ldots, 0)$ by the principle of least-squares. We have shown that the box a, which is called the least-squares approximate inverse of b, is *minimum-delay*. Moreover a is minimum-delay regardless of whether b is minimum-delay or not.

10. Shaping Filters

In the preceding section, we learned how to build a box $(a_0, a_1, \ldots, a_m)$ which condenses in a least-squares sense the input $(b_0, b_1, \ldots, b_n)$ into the spike $(1, 0, 0, \ldots, 0)$.

More generally, we may use the same approach to build a box $(a_0, a_1, \ldots, a_m)$ which converts in a least-squares sense the input $(b_0, b_1, \ldots, b_n)$ into the wavelet $(d_0, d_1, d_2, \ldots, d_{m+n})$. This wavelet, called the desired output, may be of any shape, and hence the box $(a_0, a_1, \ldots, a_m)$ is called a *least-squares shaping filter*, or *least-squares wavelet shaper*. See figure 11.

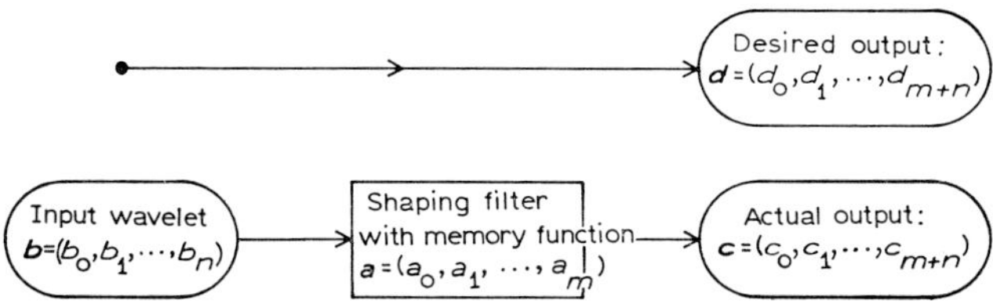

Figure 11. Shaping filter.

The actual output c is the convolution of the input b and the memory function a of the shaping filter, that is

$$c = a \ast b \ .$$

The problem is to determine the memory function a so that the actual output c is as close as possible (in a least-squares sense) to the desired output d. That is, we wish to minimize the squared norm of the error given by

$$E = (d_0 - c_0)^2 + (d_1 - c_1)^2 + (d_2 - c_2)^2 + \ldots + (d_{m+n} - c_{m+n})^2$$

by varying $(a_0, a_1, \ldots, a_m)$. Using summation notation, we have

$$E = \sum_{t=0}^{m+n} (d_t - c_t)^2 \ .$$

Inserting into E the convolution

$$c_t = \sum_{s=0}^{m} a_s b_{t-s}$$

we obtain

$$E = \sum_{t=0}^{m+n} \left(d_t - \sum_{s=0}^{m} a_s b_{t-s} \right)^2 \ .$$

The squared norm E is minimized if its partial derivatives with respect to each of the coefficients $a_0, a_1, \ldots, a_m$ equal zero. Setting the partial derivative of E with respect to a_j equal to zero (where j may equal $0, 1, 2, \ldots, m$), we obtain

$$\frac{\partial E}{\partial a_j} = \sum_{t=0}^{m+n} 2 \left(d_t - \sum_{s=0}^{m} a_s b_{t-s} \right)(-b_{t-j}) = 0$$

which gives

$$-\sum_{t=0}^{m+n} d_t b_{t-j} + \sum_{t=0}^{m+n} \left(\sum_{s=0}^{m} a_s b_{t-s}\right) b_{t-j} = 0$$

or

$$\sum_{s=0}^{m} a_s \sum_{t=0}^{m+n} b_{t-s} b_{t-j} = \sum_{t=0}^{m+n} d_t b_{t-j} . \tag{1}$$

Equation (1) represents the jth equation of a set of $m+1$ equations given by $j=0, 1, 2, \ldots, m$. From section 5, we see that

$$\sum_{t=0}^{m+n} b_{t-s} b_{t-j} = r_{j-s} \tag{2}$$

where r_{j-s} is the autocorrelation (for index $j-s$) of the input wavelet b. At this point, we shall define g_j to be

$$g_j = \sum_{t=0}^{m+n} d_t b_{t-j} \qquad (j=0, 1, 2, \ldots m) \tag{3}$$

which is the *cross-product* of the wavelet d with the wavelet b. Hence substituting expressions (2) and (3) into equation (1) we obtain

$$\sum_{s=0}^{m} a_s r_{j-s} = g_j \qquad (j=0, 1, 2, \ldots, m) \tag{4}$$

where, we repeat:

The *unknown* a_s are the memory coefficients of the shaping filter.

The *known* r_{j-s} are the autocorrelation coefficients of the input wavelet b.

The *known* g_j are the cross-product coefficients between desired output d and input b.

Writing out the set of $m+1$ equations, whose j^{th} equation is (4), we obtain

$$\begin{cases} a_0 r_0 + a_1 r_{-1} + a_2 r_{-2} + \ldots + a_m r_{-m} = g_0 \\ a_0 r_1 + a_1 r_0 + a_2 r_{-1} + \ldots + a_m r_{1-m} = g_1 \\ a_0 r_2 + a_1 r_1 + a_2 r_0 + \ldots + a_m r_{2-m} = g_2 \\ \qquad \ldots \\ a_0 r_m + a_1 r_{m-1} + a_2 r_{m-2} + \ldots + a_m r_0 = g_m . \end{cases} \tag{5}$$

This set of equations is called the *normal (or perpendicular) equations.*

Let us now use matrix algebra, and rederive the normal equations. Let the so-called regressor matrix B be the $(m+1) \times (m+n+1)$ matrix defined by

$$B = \begin{bmatrix} b_0 & b_1 & b_2 & \ldots & b_n & 0 & 0 \ldots 0 & 0 \\ 0 & b_0 & b_1 & \ldots & b_{n-1} & b_n & 0 \ldots 0 & 0 \\ & & \ldots & & & & & \\ 0 & 0 & 0 & & \ldots & & b_{n-1} & b_n \end{bmatrix}.$$

We recall that the memory function is the $(m+1) \times 1$ row vector

$$a = (a_0, a_1, \ldots, a_m) ,$$

the actual output is the $(m+n+1) \times 1$ row vector

$$c = (c_0, c_1, \ldots, c_{m+n}) ,$$

and the desired output is the $(m+n+1) \times 1$ row vector

$$d = (d_0, d_1, \ldots, d_{m+n}) .$$

The convolution

$$c = a * b$$

thus may be written as the matrix product

$$c = a B .$$

The error vector is

$$e = d - c$$

with squared norm given by

$$E = e\,e' = (d-c)(d-c)' ,$$

where the prime indicates matrix transpose. From the theory of least-squares, we know that the squared norm of the error is a minimum if and only if the regressor B is *normal* (or *perpendicular*) to the error $e = (d-c)$, that is, if and only if

$$(d-c)B' = 0 .$$

This set of equations is thus the *normal* (or *perpendicular*) *equations*. These equations become

$$c B' = d B' .$$

By substituting

$$c = a B ,$$

the normal equations take the form

$$a B B' = d B' . \tag{6}$$

To see that the normal equations (6) are indeed the same as the normal equations (5) previously derived, we proceed as follows. Multiplying B with

B', we obtain the $(m+1) \times (m+1)$ matrix

$$BB' = \begin{bmatrix} r_0 & r_1 & \cdots & r_m \\ r_{-1} & r_0 & \cdots & r_{m-1} \\ & & \cdots & \\ r_{-m} & r_{1-m} & \cdots & r_0 \end{bmatrix}$$

where

$$r_t = \begin{cases} \sum_{i=0}^{n-t} b_{i+t} b_i & \text{for} \quad t = 0, 1, 2, \ldots, n \\ r_{-t} & \text{for} \quad t = -n, -n+1, \ldots, -1 \\ 0 & \text{for} \quad t > n . \end{cases}$$

That is, BB' is the autocorrelation matrix of the input wavelet b. Multiplying d with B' we obtain

$$dB' = (g_0, g_1, \ldots, g_m)$$

where

$$g_0 = d_0 b_0 + d_1 b_1 + \ldots + d_n b_n$$
$$g_1 = d_1 b_0 + d_2 b_1 + \ldots + d_{n+1} b_n$$
$$g_2 = d_2 b_0 + d_3 b_1 + \ldots + d_{n+2} b_n$$
$$\cdots$$
$$g_m = d_m b_0 + d_{m+1} b_1 + \ldots + d_{m+n} b_n .$$

That is, the right hand side dB' of the normal equations is the cross-product vector $(g_0, g_1, \ldots, g_m)$. Therefore the normal equations (6) are

$$(a_0, a_1, \ldots, a_m) \begin{bmatrix} r_0 & r_1 & \cdots r_m \\ r_{-1} & r_0 & \cdots r_{m-1} \\ & \cdots & \\ r_{-m} & r_{1-m} & \cdots r_0 \end{bmatrix} = (g_0, g_1, \ldots, g_m)$$

which is identical to the normal equations (5), which we wished to confirm.

In the normal equations (6), the autocorrelation matrix $B'B$ and the right hand side dB' are known, and the memory function a is unknown.

Although at first sight it may appear that we have the familiar task of solving a set of $(m+1)$-equations in the $(m+1)$ unknowns $(a_0, a_1, \ldots, a_m)$, it turns out that the computational work involved is much less than usual. This easing of computational work results from the fact that all the elements on any given diagonal (i.e. sub-diagonal, main-diagonal, or super-diagonal) of the autocorrelation matrix are equal.

Also we note that in the case of real-valued, scalar wavelets, which we are considering here, the autocorrelation is an even function, that is

$$r_{-t}=r_t ,$$

which further simplifies computation.

The following two figures (figures 12 and 13) give two examples of shaping filters, with the numerical values (rounded off) of the wavelets filled in.

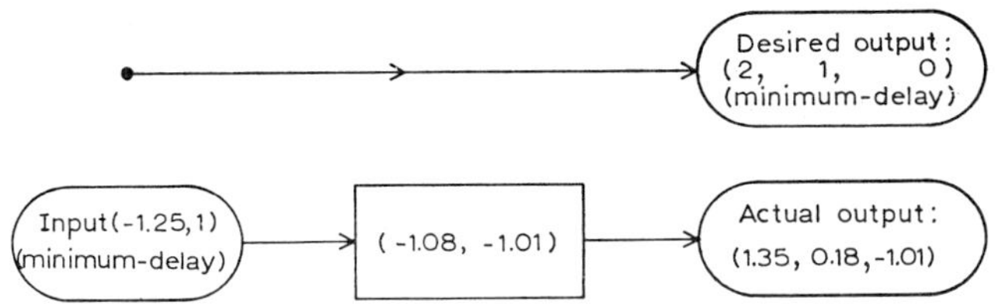

Figure 12. Shaping filter with minimum-delay input.

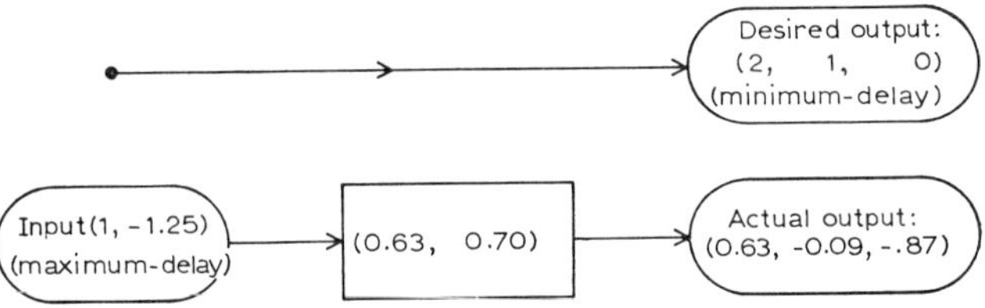

Figure 13. Shaping filter with maximum-delay input.

In both figures 12 and 13, the desired output is the minimum-delay wavelet $(2, 1, 0)$, whereas in figure 12 the input is the minimum-delay wavelet $(-1.25, 1)$ and in figure 13 the input is the maximum-delay wavelet $(1, -1.25)$. In figure 12, where both the input and the desired output have the same delay property (i.e. both minimum-delay), the shaping filter is able to do a better job than in figure 13, where input and desired output have opposite delay properties.

11. Spiking Filters

The least-squares approximate inverse a treated in section 9 is nothing more than the least-squares shaping filter whose desired output is the spike $(1, 0, 0, \ldots, 0, 0)$. The spike 1 occurs at time index 0, so the least-squares approximate inverse a may alternatively be called the *least-squares zero-delay spiking filter*.

Now, more generally, we wish to consider the set of spikes:

$$(1, 0, 0, \ldots, 0, 0) : \quad \text{spike at time index } 0$$
$$(0, 1, 0, \ldots, 0, 0) : \quad \text{spike at time index } 1$$
$$(0, 0, 1, \ldots, 0, 0) : \quad \text{spike at time index } 2$$
$$\cdots$$
$$(0, 0, 0, \ldots, 1, 0) : \quad \text{spike at time index } m+n-1$$
$$(0, 0, 0, \ldots, 0, 1) : \quad \text{spike at time index } m+n.$$

For any given input, we may compute the *least-squares spiking filter* for each of these spikes as desired output. These spiking filters are special cases of shaping filters, and are called respectively the least-squares:

0-delay spiking filter (or approximate inverse)
1-delay spiking filter (or 1-delay approximate inverse)
2-delay spiking filter (or 2-delay approximate inverse)
$\cdots$
$(m+n-1)$-delay spiking filter (or $(m+n-1)$-delay approximate inverse)
$(m+n)$-delay spiking filter (or $(m+n)$-delay approximate inverse).

The following figures show the 2-length least-squares spiking filters (a_0, a_1) for the 2-length input wavelet $(b_0, b_1) = (-1.25, 1)$:

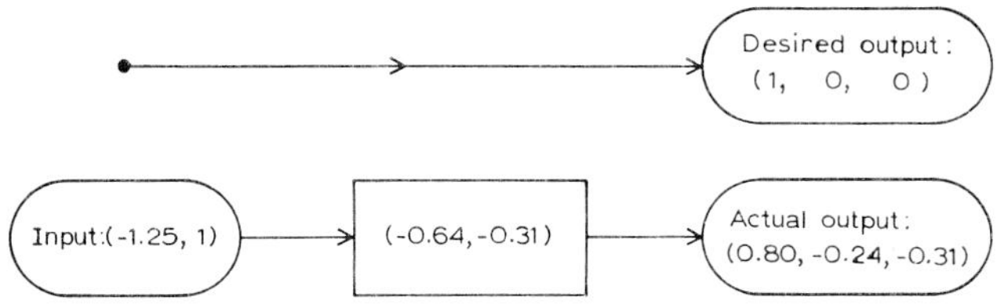

Figure 14. 0-delay spiking filter.

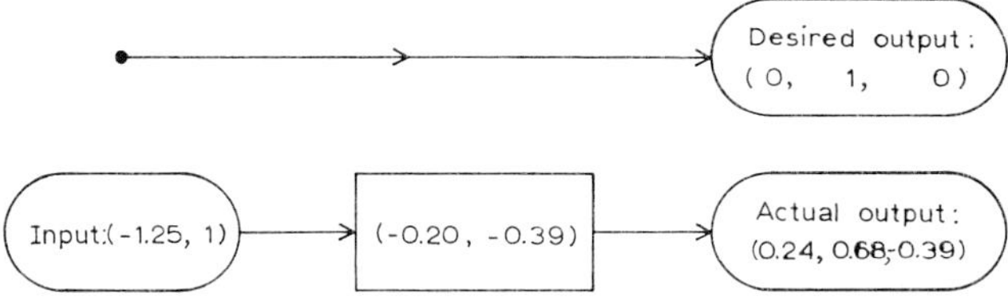

Figure 15. 1-delay spiking filter.

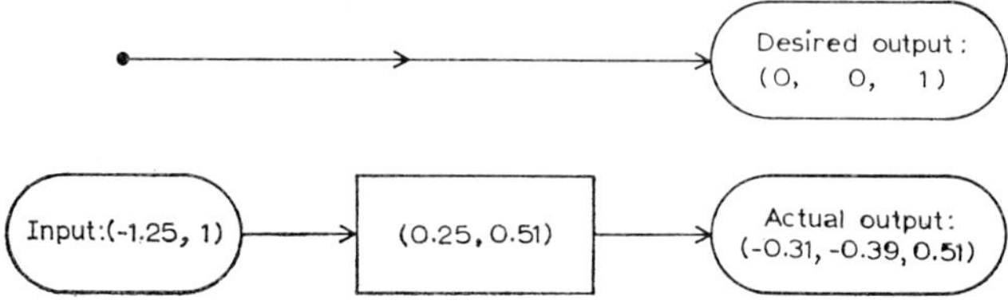

Figure 16. 2-delay spiking filter.

In these figures, the input wavelet is minimum-delay, so as we expect the best job is done by the spiking filter with the minimum-delay desired output $(1, 0, 0)$. Also the worst job is done by the spiking filter with the maximum-delay desired output $(0, 0, 1)$. This is shown very clearly by the squared-norms of the errors, which are shown in the following table:

TABLE 3

Errors for the spiking filters.

Spike at index	*Squared-norm of error (for 2-length spiking filter)*			
0	$(1 - 0.80)^2 + (0.24)^2$	$+ (0.31)^2$	$= 0.19984$	(Smallest error)
1	$(-0.24)^2 + (1 - 0.68)^2$	$+ (0.39)^2$	$= 0.31226$	
2	$(0.31)^2 + (-0.39)^2$	$+ (1 - 0.51)^2$	$= 0.48790$	(Greatest error)
	Total of above		1.00000	

We see that the total of these squared-norms is equal to 1, which is the value of n, where $n+1$ is the length of the input wavelet $(b_0, b_1, \ldots, b_n)$. In fact, this is always true, that is, *the total of the squared-norms is equal to n, regardless of the numerical values of the coefficients $b_0, b_1, b_2, \ldots, b_n$ and regardless of the value of m.*

In the above case we have considered the 2-length spiking filters (a_0, a_1) for the minimum-delay input $(-1.25, 1)$. The following table shows the squared-norms of the errors for spiking filters of greater length, all for the same minimum-delay input $(-1.25, 1)$.

TABLE 4

Errors for minimum-delay input.

Spike at index	*Squared-norm of error for $(m + 1)$-length spiking filter.*			
	$m = 1$	$m = 2$	$m = 3$	$m = 4$
0	0.19984	0.11340	0.06766	0.04151
1	0.31226	0.17718	0.10572	0.06485
2	0.48790	0.27685	0.16519	0.10134
3		0.43257	0.25811	0.15834
4			0.40330	0.24740
5				0.38656
Totals	1.00000	1.00000	1.00000	1.00000

This table shows that the squared norm of the error for the spike at any fixed index decreases as we increase the length of the spiking filter. Also we see that because the input is minimum-delay, the smallest error is always for the minimum-delay spike $(1, 0, 0, \ldots, 0, 0)$.

The corresponding table for the maximum-delay input wavelet $(1, -1.25)$ is:

TABLE 5

Errors for maximum-delay input.

Spike at index	Squared norm of error for $(m + 1)$-length spiking filter.			
	$m = 1$	$m = 2$	$m = 3$	$m = 4$
0	0.48790	0.43257	0.40330	0.38656
1	0.31226	0.27685	0.25811	0.24740
2	0.19984	0.17718	0.16519	0.15834
3		0.11340	0.10572	0.10134
4			0.06766	0.06485
5				0.04151
Totals	1.00000	1.00000	1.00000	1.00000

Again we see that the squared-norm of the error for the spike at any fixed index decreases as we increase the length of the spiking filter. In this case, the smallest error is always for the maximum-delay spike $(0, 0, 0, \ldots, 0, 1)$, because the input is maximum-delay. The reverse symmetry of the above two tables is as we would expect.

Thus we see that for a minimum-delay input the spiking filter should be chosen for which the desired output is the minimum-delay spike $(1, 0, 0, \ldots, 0, 0)$. On the other hand, for a maximum-delay input the spiking filter should be chosen for which the desired output is the maximum-delay spike $(0, 0, 0, \ldots, 0, 1)$. We recall that a mixed-delay wavelet is a convolution of a minimum-delay component and a maximum-delay component. It follows that for a mixed-delay input, the least error would occur for the spiking filter with desired output $(0, 0, \ldots, 0, 1, 0, \ldots, 0, 0)$, that is, a spike at an intermediate time index.

12. White Noise

We recall that a wavelet

$$\mathbf{b} = (b_0, b_1, b_2 \ldots)$$

is characterized by the properties that it has a certain origin time and that it has finite energy, that is,

$$b_0^2 + b_1^2 + b_2^2 + \ldots < \infty .$$

The property of having finite energy means that the wavelet is a transition or dying-out, phenomenon, and not a continuing one.

We now want to introduce a certain type of time series, called *white noise*. White noise is not a transient phenomenon but a continuing one. White noise is represented by a sequence, infinitely long in both directions,

$$\varepsilon = (\ldots, \varepsilon_{-2}, \varepsilon_{-1}, \varepsilon_0, \varepsilon_1, \varepsilon_2, \varepsilon_3, \ldots)$$

showing that the phenomenon has been underway from the remote past $(t = -\infty)$ and will continue to the far-off future $(t = \infty)$.

Before we are able to define white noise, we must first define "time average" of a time-series, such as ε. To do so we must use a limiting process. First we take the average of the value ε_t. This average is

$$\frac{1}{1} (\varepsilon_t) .$$

Then we take the average of this value (that is ε_t) and its two neighbors ε_{t-1} and ε_{t+1}. This new average is

$$\frac{1}{3} (\varepsilon_{t-1} + \varepsilon_t + \varepsilon_{t+1}) .$$

The next step is to take the average of these values (that is $\varepsilon_{t-1}, \varepsilon_t, \varepsilon_{t+1}$) and their two neighbours ε_{t-2} and ε_{t+2}. This new average is

$$\frac{1}{5} (\varepsilon_{t-2} + \varepsilon_{t-1} + \varepsilon_t + \varepsilon_{t+1} + \varepsilon_{t+2}) .$$

We keep continuing this process *ad infinitum*, and the limiting average value thus obtained is called the *time average* or *time mean* of the time series. This time average is denoted by

$$M\{\varepsilon_t\}$$

which we may read as "mean of ε_t time-series".

Let us now give the properties which characterize white noise ε. The first characteristic is that it has zero mean, that is,

$$M\{\varepsilon_t\}=0 .$$

The squared values ε_t^2 form the time-series

$$(. . ., \varepsilon_{-1}^2, \varepsilon_0^2, \varepsilon_1^2, \varepsilon_2^2, . . .)$$

whose mean is given by

$$M\{\varepsilon_t^2\} .$$

This mean is called the *mean squared value* of the ε_t time-series. Because the mean squared value $M\{\varepsilon_t^2\}$ represents the *average energy per unit time* of the ε_t time-series, we may also call $M\{\varepsilon_t^2\}$ the *power* of the ε_t time-series. The second characteristic of white noise ε is that it has finite power, that is

$$M\{\varepsilon_t^2\} < \infty .$$

The products $\varepsilon_{t+j}\varepsilon_t$ (for time index j) form the time-series

$$(. . ., \varepsilon_{-1+j}\varepsilon_{-1}, \varepsilon_j\varepsilon_0, \varepsilon_{1+j}\varepsilon_1, \varepsilon_{2+j}\varepsilon_2, . . .)$$

whose mean is given by

$$M\{\varepsilon_{t+j}\varepsilon_t\} .$$

This mean is called the *autocorrelation (for time index j)* of the time-series ε. (Note that the power of the time-series ε is the autocorrelation for index $j=0$.) The third characteristic of white noise is that its autocorrelation is zero for all indices except index 0, that is

$$M\{\varepsilon_{t+j}\varepsilon_t\}=0 \quad \text{for all } j, \text{ expect } j=0 .$$

The autocorrelation of a time-series may be denoted by the Greek letter phi, so we may write

$$\phi(j)=M\{\varepsilon_{t+j}\varepsilon_t\}$$

or, to explicitly denote that it is the autocorrelation of the time-series ε, we may write

$$\phi_{\varepsilon\varepsilon}(j)=M\{\varepsilon_{t+j}\varepsilon_t\} .$$

To summarize, a time-series ε is called *white noise* provided it satisfies
(1) $M\{\varepsilon_t\}=0$
(2) $\phi_{\varepsilon\varepsilon}(0)$ is finite
(3) $\phi_{\varepsilon\varepsilon}(j)=0 \quad$ for all j, except $j=0$.
An example of white noise is the time-series obtained by successively flipping a coin, writing 1 for heads, and -1 for tails:

$$\ldots, 1, 1, -1, -1, -1, 1, 1, 1, -1, -1, -1, -1, -1, 1, 1, -1, 1, \ldots .$$

Another example may be obtained by tossing a die, respectively writing 1, 2, 3 for the 1, 2, 3 on the die, and writing $-1, -2, -3$ for the 4, 5, 6 on the die:

$$\ldots, 3, 3, 3, 1, -1, 2, 1, -2, -1, 2, -1, -2, \ldots .$$

Still another example would be successive independent observations from a normal distribution with zero mean.

13. Time-Series Model

A white noise series can represent the news, or innovations, which come successively day by day. In other words, the innovation series may be represented by the white noise

$$\varepsilon = (\ldots, \varepsilon_1, \varepsilon_2, \varepsilon_3, \varepsilon_4, \varepsilon_5, \ldots) .$$

That is, the quantity ε_1 gives the strength of the news on day 1, the quantity ε_2 on day 2, the quantity ε_3 on day 3, the quantity ε_4 on day 4, etc.

As we know, the news of the day generates a wavelet. This wavelet is weighted by the strength of the news. Each day a new wavelet is generated, and the time-series is the resultant of all these wavelets.

Let us suppose that the basic wavelet, or wavelet shape, is

$$b = (b_0, b_1, b_2) .$$

On day 1, the wavelet

$$\varepsilon_1(b_0, b_1, b_2) = (\varepsilon_1 b_0, \varepsilon_1 b_1, \varepsilon_1 b_2)$$

is born. In other words $\varepsilon_1 b_0$ is the value of this wavelet on day 1, $\varepsilon_1 b_1$ on day 2, and $\varepsilon_1 b_2$ on day 3. On day 2, the wavelet

$$\varepsilon_2(b_0, b_1, b_2) = (\varepsilon_2 b_0, \varepsilon_2 b_1, \varepsilon_2 b_2)$$

is born. This means that $\varepsilon_2 b_0$ is the value on day 2, $\varepsilon_2 b_1$ on day 3, and $\varepsilon_2 b_2$ on day 4. Hence we may construct the table:

TABLE 6
Wavelet composition of the time-series x.

Time	1	2	3	4	5	6	7	8
...	...	...						
	$\varepsilon_1 b_0$	$\varepsilon_1 b_1$	$\varepsilon_1 b_2$	← Wavelet born on day 1				
		$\varepsilon_2 b_0$	$\varepsilon_2 b_1$	$\varepsilon_2 b_2$	← Wavelet born on day 2			
			$\varepsilon_3 b_0$	$\varepsilon_3 b_1$	$\varepsilon_3 b_2$	← Wavelet born on day 3		
	Wavelet born on day 4 →			$\varepsilon_4 b_0$	$\varepsilon_4 b_1$	$\varepsilon_4 b_2$		
		Wavelet born on day 5 →			$\varepsilon_5 b_0$	$\varepsilon_5 b_1$	$\varepsilon_5 b_2$	
			Wavelet born on day 6 →			$\varepsilon_6 b_0$	$\varepsilon_6 b_1$	$\varepsilon_6 b_2$
						...	...	...
Time series is the sum of the columns	...	...	x_3	x_4	x_5	x_6	...	...	...

As we know, the time-series on a given day, say day 3, is the sum of
(1) the value of the wavelet born on day 3, i.e. $\varepsilon_3 b_0$
(2) the value of the wavelet born on day 2, i.e. $\varepsilon_2 b_1$
(3) the value of the wavelet born on day 1, i.e. $\varepsilon_1 b_2$.

In other words, the value of the time-series at day 3 is

$$x_3 = \varepsilon_3 b_0 + \varepsilon_2 b_1 + \varepsilon_1 b_2 .$$

Similarly, the value of the time-series at day 4 is

$$x_4 = \varepsilon_4 b_0 + \varepsilon_3 b_1 + \varepsilon_2 b_2$$

In general, the value of the time-series on day t is

$$x_t = \varepsilon_t b_0 + \varepsilon_{t-1} b_1 + \varepsilon_{t-2} b_2$$

or

$$x_t = \sum_{s=0}^{2} b_s \varepsilon_{t-s} .$$

This equation states that the time-series

$$\boldsymbol{x} = (\ldots, x_1, x_2, x_3, x_4, x_5, \ldots)$$

is the convolution of the wavelet shape $\boldsymbol{b}$ and the white noise ε. In other words, the above table and the folding table (table 7) are equivalent.

TABLE 7

Folding table for the time-series x.

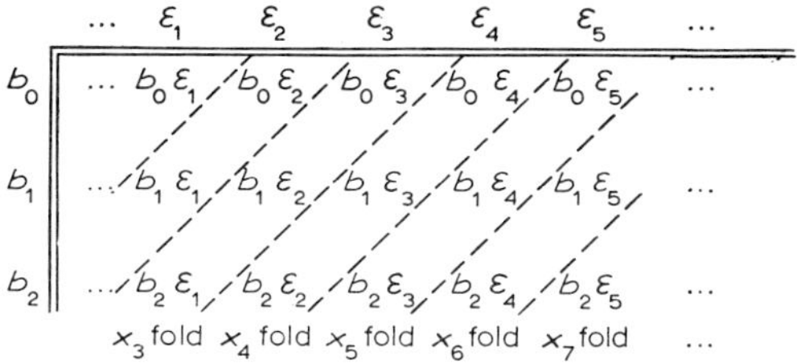

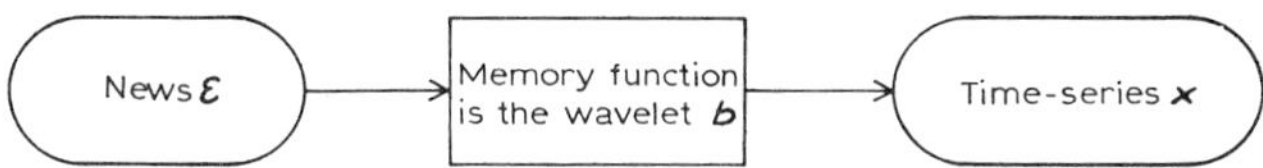

Hence the time-series x is the output of a box b which has input ε, as shown in figure 17.

News ε ⟶ Memory function is the wavelet b ⟶ Time-series x

Figure 17. Model of time-series x.

Thus the time-series x is the convolution of wavelet b and white noise ε, that is,

$$x = b \ast \varepsilon$$

or

$$x_t = \sum_{s=0}^{\infty} b_s \varepsilon_{t-s}.$$

This representation of a time-series as the convolution of a wavelet with white noise is our basic *model*. In this paper we restrict ourselves to only those time-series which have this model. Such time-series are called *regular, stationary time-series*.

In the model of a time-series

$$x = b \ast \varepsilon$$

the wavelet b, of course, need not be minimum-delay. If, however, the wavelet b is minimum-delay, then the time-series model is called a *minimum-delay model*.

Note that white noise ε itself has the (formal) model

$$\varepsilon = b \ast \varepsilon$$

where the wavelet b is the unit spike at time index 0, that is,

$$b = (1, 0, 0, 0, \ldots).$$

In this section we have introduced the basic model for the type of time-series which we consider in this paper. The model states that the time-series is the convolution of a wavelet with white noise. If the wavelet is minimum-delay, then the model is called minimum-delay also.

14. Autocorrelation and Cross-Correlation of Time-Series

The autocorrelation (for time index k) of the time-series

$$x = (\ldots, x_{-1}, x_0, x_1, x_2, \ldots)$$

is defined to be

$$\phi_{xx}(k) = M\{x_{t+k}x_t\} \, .$$

We recall that our model of the time-series is

$$x_t = \sum_{i=0}^{\infty} b_i \varepsilon_{t-i} \, .$$

Replacing t by $t+k$, and the dummy variable i by j, we obtain the equivalent model

$$x_{t+k} = \sum_{j=0}^{\infty} b_j \varepsilon_{t+k-j} \, .$$

Substituting the above two expressions into the expression for the auto-correlation of x, we obtain

$$\phi_{xx}(k) = M\left\{ \sum_{j=0}^{\infty} b_j \varepsilon_{t+k-j} \sum_{i=0}^{\infty} b_i \varepsilon_{t-i} \right\}$$

$$= M\left\{ \sum_{j=0}^{\infty} \sum_{i=0}^{\infty} b_j b_i \varepsilon_{t+k-j} \varepsilon_{t-i} \right\}$$

$$= \sum_{j=0}^{\infty} \sum_{i=0}^{\infty} b_j b_i M\{\varepsilon_{t+k-j} \varepsilon_{t-i}\} \, .$$

We recognize the M term to be the autocorrelation of ε, that is,

$$M\{\varepsilon_{t+k-j} \varepsilon_{t-i}\} = \phi_{\varepsilon\varepsilon}(k-j+i) \, .$$

Because ε is white noise, we see that

$$\phi_{\varepsilon\varepsilon}(k-j+i) = \begin{cases} p & \text{when } k-j+i=0 \\ 0 & \text{otherwise} \end{cases}$$

where p is the power of ε. In other words, the M term is zero unless $j = i+k$.

Hence the autocorrelation of x is

$$\phi_{xx}(k) = \sum_{j=0}^{\infty} \sum_{i=0}^{\infty} b_j b_i \left(\begin{matrix} p & \text{when} & j=i+k \\ 0 & \text{when} & j \neq i+k \end{matrix} \right)$$

$$= \sum_{i=0}^{\infty} b_{i+k} b_i \, p = p \sum_{i=0}^{\infty} b_{i+k} b_i \, .$$

But

$$r_k = \sum_{i=0}^{\infty} b_{i+k} b_i$$

is the autocorrelation of the wavelet b, so

$$\phi_{xx}(k) = p r_k \, .$$

Hence we have the following theorem. *Let x be a time-series whose model is*

$$x = b \ast \varepsilon$$

where b is a wavelet and ε is white noise. Then the autocorrelation of the time-series x is equal to the power p of the white noise times the autocorrelation of the wavelet b, or, in symbols

$$\phi_{xx}(k) = p r_k \, .$$

We may think about this equation as follows. The white noise ε is an uncorrelated series so it adds no correlation by itself to the time-series x. Thus all the correlation within the x series comes as a result of the memory function b. In other words, the time-series x has exactly the same autocorrelation function as the memory function b, except for a scale factor. This scale factor is the power of the white noise.

Of course, in any actual situation, only a finite portion of a time-series can be measured, that is, the portion from some fixed time to a later time. For example, for a given time-series, we may have available only the segment from time $t=1$ to time $t=50$:

$$(x_1, x_2, \ldots, x_{50}) \, .$$

Such a finite segment of a time-series is called a *sample time-series*, as it represents a sample of the infinite-length time-series.

We must always be careful to distinguish between the concepts of *wavelet* on the one hand and *sample time-series* on the other hand. The point of difference is that a wavelet is a self-contained or entire entity, and this entity is only a portion of an entire entity, namely the time-series, and this entity is a continuing phenomenon.

A *wavelet* has finite energy over all time. A *time-series* has infinite energy over all time, but has finite energy per unit of time, or, in other words, finite power. Nevertheless, a *sample time-series*, being measured only over a finite time interval has finite energy. This finite energy divided by the length of the time interval should be thought of as an approximation to the power of the infinite time-series.

Consequently, the autocorrelation of a sample time-series is in effect an average computed from a sample. Thus the autocorrelation of the sample time-series

$$x_1, x_2, \ldots, x_N$$

is

$$r_j = \begin{cases} \dfrac{1}{N} \sum_{i=1}^{N-j} x_{i+j} x_i & \text{for} \quad 0 \le j \le N-1 \\ 0 & \text{for} \quad j \ge N \\ r_{-j} & \text{for} \quad j < 0 . \end{cases}$$

The difference between this definition and that for a wavelet is that we have divided by N, the number of values in the sample time-series. As we have seen, the autocorrelation of the infinite-length time-series is defined to be the limit of the autocorrelation of the sample time-series as N becomes infinite. Of course, there is a discrepancy between the autocorrelation of the sample time-series and the autocorrelation of the infinite-length time-series. This discrepancy is due to the sampling effects arising from the finite length of the sample time-series.

Let us now suppose that we have two time-series, say

$$\boldsymbol{x} = (\ldots, x_{-1}, x_0, x_1, x_2, \ldots)$$

and

$$\boldsymbol{y} = (\ldots, y_{-1}, y_0, y_1, y_2, \ldots) .$$

The products $x_{t+j} y_t$ (for time index j) form the time-series

$$(\ldots, x_{-1+j} y_{-1}, x_j y_0, x_{1+j} y_1, x_{2+j} y_2, \ldots)$$

whose mean is given by

$$M\{x_{t+j} y_t\} .$$

This mean is called the *cross-correlation (for time index j) between the time-series x and the time-series y.* (Note: The order of the time-series, that is *x* first and *y* second, is significant.) The cross-correlation is denoted by the Greek letter phi with subscripts x and y, that is,

$$\phi_{xy}(j)=M\{x_{t+j}\,y_t\}\;.$$

Similarly

$$\phi_{yx}(j)=M\{y_{t+j}\,x_t\}$$

is the *cross-correlation (for time index j) between the time-series y and the time-series x.*

Because

$$\phi_{xy}(-j)=M\{x_{t-j}\,y_t\}$$

and because (letting $t-j=s$)

$$M\{x_{t-j}\,y_t\}=M\{x_s\,y_{s+j}\}=M\{y_{s+j}\,x_s\}=\phi_{yx}(j)$$

it follows that the two cross-correlations are related by

$$\phi_{xy}(-j)=\phi_{yx}(j)\;.$$

Two time-series are said to be *uncorrelated* provided their cross-correlation is zero for all time indices, including time index zero.

15. Smoothing and Prediction

In this section we want to show how to design boxes to perform various tasks. One such task is the smoothing of unwanted random roughness in data. Another task is the prediction of future values of a time-series. A more general task incorporates both of the above tasks, that is, both smoothing and prediction.

The over-all problem of designing boxes to perform smoothing and/or prediction can be readily handled by using the methods developed to this point. Our performance index is the criterion of least-squares. Although the least-squares criterion has serious limitations for some types of applications, it is the criterion which is mathematically the most tractable and is by far the most used one in all branches of science. In principle, our methods permit the use of other criteria, but then one would have to rely more heavily on computing machines, as few non-trivial cases have yet been solved analytically on the bases of other criteria.

Smoothing may be described as the problem of separating as well as possible a signal time-series from an unwanted noise time-series.

Let us now describe the process of smoothing in more detail. The time-series x_t which we have available is a polluted time-series. The polluted time-series x_t consists of a pure component s_t, called the signal, and an impure component n_t, called the noise, that is $x_t = s_t + n_t$. This is an unfortunate situation, because we are only interested in the signal s_t, the noise n_t being unwanted. We therefore want to build a box, with memory function $(f_0, f_1, f_2, \ldots)$, which will filter out, as well as possible, the noise which is contaminating the signal. The box should be designed so that its actual output y_t is close as possible to the desired output s_t, the signal. See figure 18.

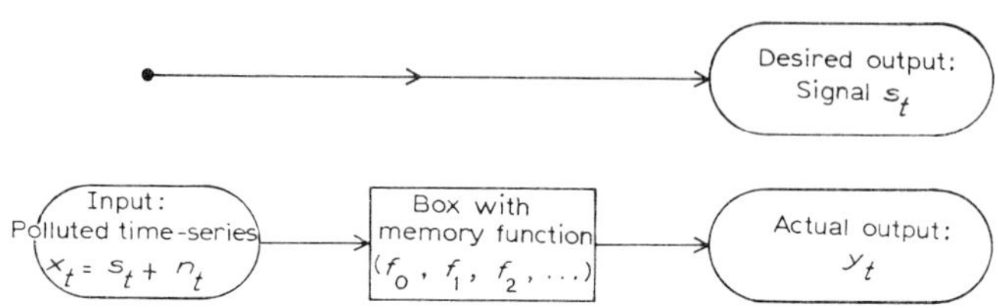

Figure 18. Smoothing or decontamination.

This process is called smoothing because it often happens that the contamination n_t is much rougher than the signal s_t, so that the actual output time-series y_t is much smoother than the polluted time-series x_t. Of course, this is not always so, but nevertheless the term "smoothing" is still used as the generic term for this decontamination operation.

The smoothing, or decontamination, operation may be modified in several ways. For example, the desired output may be the signal delayed by a certain time constant δ. This is the case of *smoothing with delay.* The polluted time-series x_t is fed into the box. The box is designed so that its actual output y_t is the best approximation to the desired output $s_{t-\delta}$, which is the signal delayed by the positive time delay δ. See figure 19.

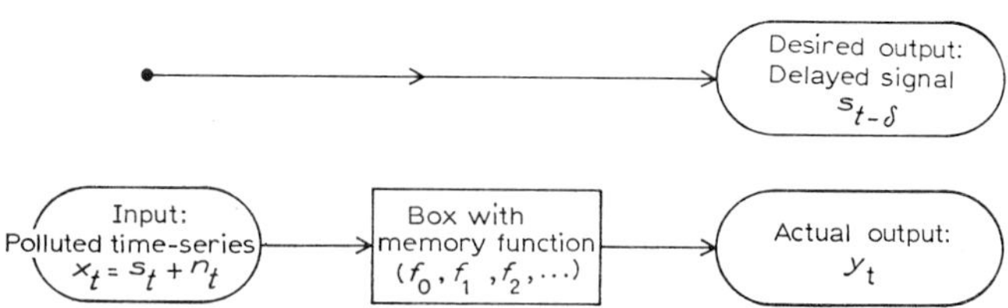

Figure 19. Smoothing with time delay δ.

Another example is the case in which the desired output is the signal advanced by a certain time constant α. The introduction of this time advance

α means that the box is required to predict future values of the signal. This leads us into a discussion of *prediction*.

The simplest case of prediction is that in which the pure signal s_t is processed. That is, we suppose that the signal time-series s_t is available without any contamination. We know the signal s_t for past values of time, and we require a device to yield, as well as possible, the value at a future, or advanced, time. In other words, we wish to feed the signal time-series s_t into the box. The box is designed so that its actual output y_t is the best approximation to the desired output $s_{t+\alpha}$, which is the signal advanced by the positive time advance α. See figure 20.

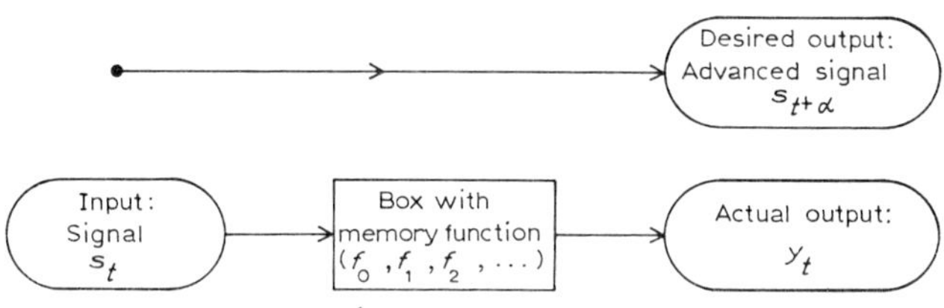

Figure 20. Prediction.

Now let us discuss the case of smoothing, mentioned above, in which the desired output is the signal advanced by a certain time constant. This case which incorporates both smoothing and prediction is called *smoothing with advance*. Here we feed the polluted time-series x_t into the box. The box is designed so that its actual output y_t is the best approximation to the desired output $s_{t+\alpha}$, which is the signal advanced by the positive time advance α. See figure 21.

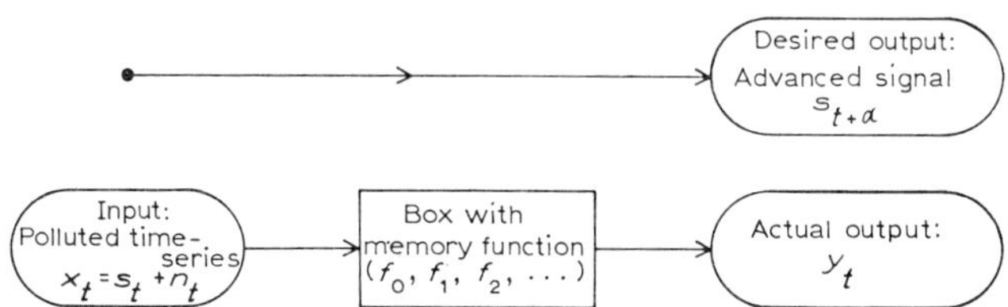

Figure 21. Smoothing with time advance α.

Finally, let us formulate the general case, which includes the above as special cases. The desired output may be any time-series z_t. We wish to feed a time-series x_t into a box. The box is designed so that its actual output y_t is the best approximation to the desired output z_t. See figure 22.

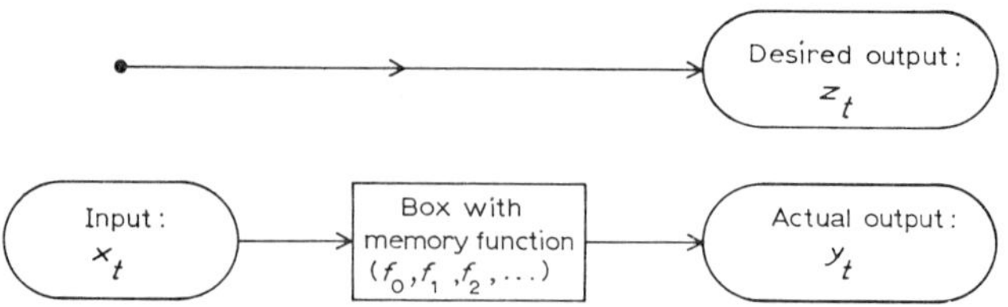

Figure 22. General case.

Now let us turn our attention to the problem of determining the desired memory function. We shall restrict ourselves to the case when the memory function is a finite wavelet $(f_0, f_1, \ldots, f_m)$ of length $(m+1)$. The actual output, which is the convolution of the input with the memory function, is

$$y_t = f_0 x_t + f_1 x_{t-1} + \ldots + f_m x_{t-m} .$$

The error between desired output and actual output is then

$$e_t = z_t - y_t = z_t - (f_0 x_t + f_1 x_{t-1} + \ldots + f_m x_{t-m}) .$$

For simplicity, let us carry out the derivation for $m = 1$, that is, for a 2-length memory function. When we arrive at the normal equations for this case we will easily be able to extend them by eye to the case of arbitrary m.

The error is

$$e_t = z_t - y_t = z_t - (f_0 x_t + f_1 x_{t-1}) .$$

The squared error is

$$\begin{aligned} e_t^2 &= [z_t - (f_0 x_t + f_1 x_{t-1})]^2 \\ &= z_t^2 - 2z_t(f_0 x_t + f_1 x_{t-1}) + (f_0 x_t + f_1 x_{t-1})^2 . \end{aligned}$$

Let us now find the mean of each of the terms in this expression. The mean of the first term is

$$M\{z_t^2\} = \phi_{zz}(0) ,$$

which is the autocorrelation, at index 0, of the desired output. The mean of the second term is

$$\begin{aligned} M\{-2z_t(f_0 x_t + f_1 x_{t-1})\} &= -2f_0 M\{z_t x_t\} - 2f_1 M\{z_t x_{t-1}\} \\ &= -2f_0 \phi_{zx}(0) - 2f_1 \phi_{zx}(1) \end{aligned}$$

where $\phi_{zx}(0)$ is the cross-correlation, at index 0, of the desired output with the input, and $\phi_{zx}(1)$ is the cross-correlation, at index 1, of the desired output with the input. The mean of the third term is

$$M\{(f_0 x_t + f_1 x_{t-1})^2\} = M\{f_0^2 x_t^2 + 2f_0 f_1 x_t x_{t-1} + f_1^2 x_{t-1}^2\}$$
$$= f_0^2 M\{x_t^2\} + 2f_0 f_1 M\{x_t x_{t-1}\} + f_1^2 M\{x_{t-1}^2\}$$
$$= f_0^2 \phi_{xx}(0) + 2f_0 f_1 \phi_{xx}(1) + f_1^2 \phi_{xx}(0)$$
$$= (f_0^2 + f_1^2) \phi_{xx}(0) + 2f_0 f_1 \phi_{xx}(1)$$

where $\phi_{xx}(0)$ is the autocorrelation at index 0, and $\phi_{xx}(1)$ is the autocorrelation at index 1, of the input. Hence the mean squared error is

$$E = M\{e_t^2\} = \phi_{zz}(0) - 2f_0 \phi_{zx}(0) - 2f_1 \phi_{zx}(1) + (f_0^2 + f_1^2)\phi_{xx}(0) + 2f_0 f_1 \phi_{xx}(1) \ .$$

We wish to determine those values of f_0 and f_1 such that the mean squared error E is a minimum. Setting the partial derivation of E with respect to f_0 and f_1 equal to zero, we obtain

$$\begin{cases} \dfrac{\partial E}{\partial f_0} = -2\phi_{zx}(0) + 2f_0 \phi_{xx}(0) + 2f_1 \phi_{xx}(1) = 0 \\ \dfrac{\partial E}{\partial f_1} = -2\phi_{zx}(1) + 2f_1 \phi_{xx}(0) + 2f_0 \phi_{xx}(1) = 0 \ . \end{cases}$$

Because the time-series x_t is real valued, $\phi_{xx}(1) = \phi_{xx}(-1)$, and so these equations may be rearranged into the form

$$\begin{cases} f_0 \phi_{xx}(0) + f_1 \phi_{xx}(-1) = \phi_{zx}(0) \\ f_0 \phi_{xx}(1) + f_1 \phi_{xx}(0) = \phi_{zx}(1) \ . \end{cases}$$

These are the normal equations for the 2-length memory function (f_0, f_1). For the $(m+1)$-length memory function $(f_0, f_1, \ldots, f_m)$, the *normal equations* are seen to be

$$f_0 \phi_{xx}(0) + f_1 \phi_{xx}(-1) + \ldots + f_m \phi_{xx}(-m) = \phi_{zx}(0)$$
$$f_0 \phi_{xx}(0) + f_1 \phi_{xx}(0) + \ldots + f_m \phi_{xx}(1-m) = \phi_{zx}(1)$$
$$\cdots \qquad\qquad \cdots \qquad\qquad \cdots \tag{7}$$
$$f_0 \phi_{xx}(m) + f_1 \phi_{xx}(m-1) + \ldots + f_m \phi_{xx}(0) = \phi_{zx}(m) \ .$$

Now let us consider the following *application*. Let b be the signal wavelet, and let the white noise ε be the signal news, so that the signal time-series s has the model

$$s = b * \varepsilon \ .$$

Also let w be the noise wavelet, and let the white noise θ be the noise news, so that the noise time-series has the model

$$n = w * \theta \ .$$

Furthermore, let us assume that the signal news ε is uncorrelated with the noise news $\boldsymbol{\theta}$. The polluted time-series

$$x = s + n$$

thus has the model shown in figure 23.

We now wish to design a box f into which we will feed the polluted time-series $\boldsymbol{x}$. We want the box to convert (as well as possible) each signal wavelet

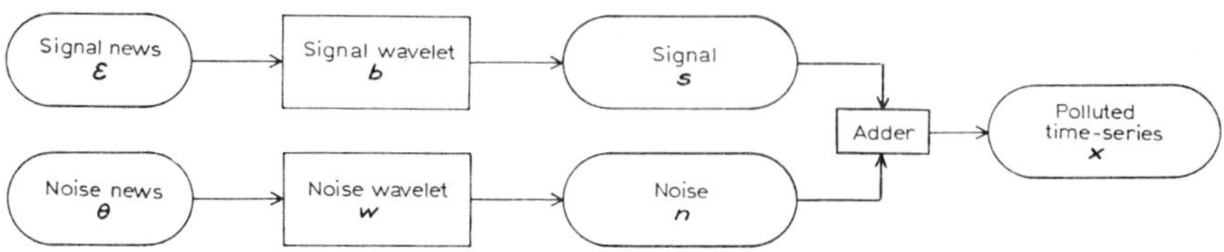

Figure 23. Model of polluted time-series.

$\boldsymbol{b}$ contained in the polluted time-series $\boldsymbol{x}$ into some other wavelet $\boldsymbol{d}$. For example, $\boldsymbol{b}$ may be a long-duration blunt wavelet, and so, for greater resolution, we desire to convert it into a short-duration sharp wavelet $\boldsymbol{d}$, such as a spike. Another example would be the case in which we want to convert the wavelet $\boldsymbol{b}$ with certain delay properties to the wavelet $\boldsymbol{d}$ with other delay properties. While the box f carries out this conversion, it should respond as little as possible to the noise $\boldsymbol{n}$. Thus the desired output is the wavelet $\boldsymbol{d}$ weighted by the signal news ε, and with no weighting given to the noise news $\boldsymbol{\theta}$. That is, the desired output is $\boldsymbol{d} * \varepsilon$. See figure 24.

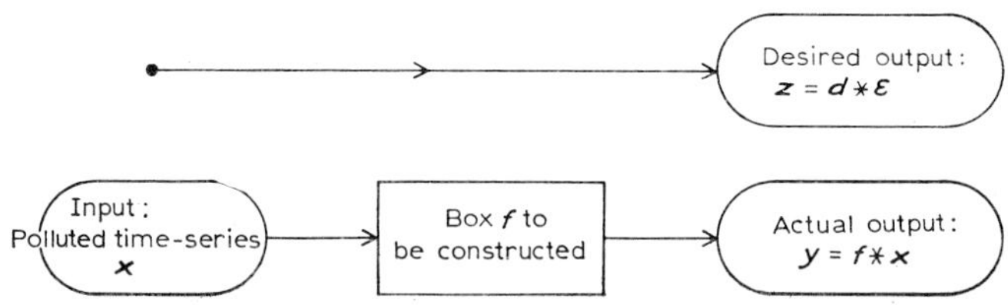

Figure 24. Filtering of polluted time-series.

Let us now set up the normal equations, the solution of which yields the desired memory function

$$f = (f_0, f_1, \ldots, f_m) \, .$$

First let us find an expression for the autocorrelation of the polluted time-series. We shall let p denote the mean squared value of the signal news, or, in other words, the power of the signal news, that is

$$p = M\{\varepsilon_t^2\} .$$

Let q represent the power of the noise news, that is

$$q = M\{\theta_t^2\} .$$

Letting r_j denote the autocorrelation of the signal wavelet b, it follows that the autocorrelation of the signal time-series is

$$\phi_{ss}(j) = pr_j .$$

Letting ρ_j denote the autocorrelation of the noise wavelet w, it follows that the autocorrelation of the noise time-series is

$$\phi_{nn}(j) = q\rho_j .$$

Because the signal news ε and the noise news θ are uncorrelated, it follows that the signal time-series s and the noise time-series n are also uncorrelated. Hence the autocorrelation of the polluted time-series

$$x = s + n$$

is the sum of the autocorrelation of the signal time-series and the auto-correlation of the noise time-series, that is,

$$\begin{aligned}
\phi_{xx}(j) &= \phi_{ss}(j) + \phi_{nn}(j) \\
&= pr_j + q\rho_j .
\end{aligned} \tag{8}$$

This is the desired expression for the autocorrelation of the polluted time-series.

Next, let us find an expression for the cross-correlation of the desired output with the polluted time-series. This cross-correlation, which is on the right hand side of the normal equations (7), is given by

$$\begin{aligned}
\phi_{zx}(j) &= M\{z_{t+j} x_t\} \\
&= M\{z_{t+j}(s_t + n_t)\} = M\{z_{t+j} s_t\} + M\{z_{t+j} n_t\} .
\end{aligned}$$

Because the signal news ε and the noise news θ are uncorrelated, it follows that the desired output time-series $z = d * \varepsilon$ and the noise time-series $n = w * \theta$ are also uncorrelated, that is

$$M\{z_{t+j} n_t\} = 0 \quad \text{for any } j .$$

Hence

$$\phi_{zx}(j) = M\{z_{t+j} s_t\} = p \sum_{=0}^{\infty} d_{t+j} b_t = pg_j \tag{9}$$

where g_j is defined to be

$$g_j = \sum_{t=0}^{\infty} d_{t+j} b_t .$$

Hence, for this application, the autocorrelation function is given by (8) which is

$$\phi_{xx}(j) = pr_j + q\rho_j$$

and the right hand side is given by (9) which is

$$\phi_{zx}(j) = pg_j .$$

Substituting these expressions into the normal equations (7), we thus have the *normal equations* for this application:

$$f_0(pr_0 + q\rho_0) + f_1(pr_{-1} + q\rho_{-1}) + \cdots + f_m(pr_{-m} + q\rho_{-m}) = pg_0$$

$$f_0(pr_1 + q\rho_1) + f_1(pr_0 + q\rho_0) + \cdots + f_m(pr_{1-m} + q\rho_{1-m}) = pg_1$$

$$\cdots \qquad\qquad \cdots \qquad\qquad\qquad\qquad \cdots$$

$$f_0(pr_m + q p_m) + f_1(pr_{m-1} + q\rho_{m-1}) + \cdots f_m(pr_0 + q\rho_0) = pg_m . \qquad (10)$$

Looking at these normal equations, we observe the following:

To set up the left hand side, we must know the numerical values of the autocorrelation

$$\phi_{xx}(j) = pr_j + q\rho_j \quad \text{for} \quad j = -m, -m+1, \ldots, m .$$

of the polluted time-series x_t. Nevertheless we do not need to know the actual coefficients of the signal wavelet b and the noise wavelet w, although such knowledge, together with knowledge of the powers p and q, would allow us directly to compute $\phi_{xx}(j)$. Usually a sample of the polluted time-series x_t is available, from which we may compute the sample autocorrelation and then use it as an approximation to $\phi_{xx}(j)$.

To set up the right hand side, we must know the numerical values of the cross-correlation

$$\phi_{zx}(j) = pg_j \quad \text{for} \quad j = 0, 1, 2, \ldots, m .$$

Obtaining these numerical values can be a difficult problem in practice. Therefore it is instructive to look at the following special cases:

(1) The desired output is the signal s_t (as depicted in figure 18). Hence the wavelet d is the same as the signal wavelet b. The cross-correlation (9) thereby becomes

$$\phi_{zx}(j) = p \sum_{t=0}^{\infty} b_{t+j} b_t = pr_j .$$

Hence the setting-up of the right hand side requires only knowledge of the autocorrelation of the signal time-series

$$\phi_{ss}(j)=pr_j \quad \text{for} \quad j=0, 1, 2, \ldots, m .$$

and does not require knowledge of the actual coefficients of the signal wavelet $\boldsymbol{b}$.

(2) The desired output is the delayed signal $s_{t-\delta}$ (as depicted in figure 19). Hence

$$d_t=b_{t-\delta} ,$$

so the cross-correlation (9) becomes

$$\phi_{zx}(j)=p\sum_{t=0}^{\infty} b_{t-\delta+j}b_t=pr_{j-\delta} .$$

Hence the setting-up of the right hand side requires only knowledge of the autocorrelation of the signal time-series

$$\phi_{ss}(j-\delta)=pr_{j-\delta} \quad \text{for} \quad j=0, 1, 2, \ldots, m .$$

(3) The desired output is the advanced signal $s_{t+\alpha}$ (as depicted in figure 20). Hence

$$d_t=b_{t+\alpha}$$

so that the cross-correlation (9) is

$$\phi_{zx}(j)=p\sum_{t=0}^{\infty} b_{t+\alpha+j}b_t=pr_{j+\alpha} .$$

Hence the setting-up of the right hand side requires only knowledge of the autocorrelation of the signal time-series

$$\phi_{ss}(j+\alpha)=pr_{j+\alpha} \quad \text{for} \quad j=0, 1, 2, \ldots, m .$$

(4) The wavelet $\boldsymbol{d}$ is the unit spike at time index 0, that is,

$$\boldsymbol{d}=(1, 0, 0, \ldots, 0) .$$

The cross-correlation (9) is then

$$\phi_{zx}(j) = \begin{cases} pb_0 & \text{for} \quad j=0 \\ 0 & \text{for} \quad j=1, 2, \ldots, m . \end{cases}$$

Hence the setting-up of the right hand side requires only knowledge of pb_0.

(5) The wavelet $\boldsymbol{d}$ is the unit spike at time index 1, that is

$$\boldsymbol{d}=(0, 1, 0, \ldots, 0) .$$

The cross-correlation (9) is then

$$\phi_{zx}(j) = \begin{cases} pb_1 & \text{for} \quad j=0 \\ pb_0 & \text{for} \quad j=1 \\ 0 & \text{for} \quad j=2, 3, \ldots, m. \end{cases}$$

Hence the setting-up of the right hand side requires knowledge of pb_0 and pb_1. (End of special cases.)

In this section we have considered the problem of designing boxes to perform smoothing and/or prediction. We have shown how to set up the normal equations whose solution gives the coefficients of the memory function of the desired box.

16. Solution of Normal Equations

As we have seen, we are often encountered in time-series analysis with the solution of normal equations which have the general form

$$f_1 r_0 + f_2 r_{-1} + \ldots + f_m r_{1-m} = g_0$$
$$f_1 r_1 + f_2 r_0 + \ldots + f_m r_{2-m} = g_1$$
$$\ldots$$
$$f_1 r_{m-1} + f_2 r_{m-2} + \ldots + f_m r_0 = g_{m-1}$$

where r_j is any arbitrary autocorrelation function. Here the r_j and the g_j are known, and the f_j are unknown *filter coefficients*. At first sight it might appear that we have the familar task of solving m linear equations for m unknowns. Nevertheless, it turns out that the computational work is much lighter than usual because of the fact that on the left hand side all the r_j along any diagonal are the same. In this section we wish to give a method which takes account of this diagonal symmetry. This method works, in addition, for the cases of complex- and/or matrix-valued f_j, r_j, and g_j, as long as the usual complex and/or matrix mathematical conventions are followed.

First of all, it is convenient to introduce a special notation for simultaneous linear equations. In this notation, the above normal equations are written as

f_1	f_2		f_m	
r_0	r_1		r_{1-m}	g_0
r_1	r_0		r_{2-m}	g_1
		$\cdots$		
r_{m-1}	r_{m-2}		r_0	g_{m-1}

Our method is one in which we solve the equations for the case $m=1$, then for case $m=2$, then for case $m=3$, and so on until we arrive at the desired value of m.

The procedure for the case $m=1$ is as follows. We compute α, β, λ, μ by the equations

$$
\begin{array}{cc|c}
1 & 0 & \\
\hline
r_0 & r_{-1} & \alpha \\
r_1 & r_0 & \lambda
\end{array}
\qquad
\begin{array}{cc|c}
0 & 1 & \\
\hline
r_0 & r_{-1} & \mu \\
r_1 & r_0 & \beta
\end{array} .
$$

Next we compute ε by

$$
\begin{array}{c|c}
0 & \\
\hline
r_0 & \varepsilon
\end{array} .
$$

(Hence, for this case, $\varepsilon=0$.) Then we multiply the equation for β, which is,

$$
\begin{array}{c|c}
1 & \\
\hline
r_0 & \beta
\end{array}
$$

by the unknown θ to obtain

$$
\begin{array}{c|c}
\theta & \\
\hline
r_0 & \theta\beta
\end{array} .
$$

Now we add the two equations

$$
\begin{array}{c|c}
0 & \\
\hline
r_0 & \varepsilon
\end{array}
+
\begin{array}{c|c}
\theta & \\
\hline
r_0 & \theta\beta
\end{array}
$$

and hence obtain the equation

$$
\begin{array}{c|c}
\theta & \\
\hline
r_0 & \varepsilon+\theta\beta
\end{array} .
$$

We now require that the identity

$$
\begin{array}{c|c}
\theta & \\
\hline
r_0 & \varepsilon+\theta\beta
\end{array}
=
\begin{array}{c|c}
f_1' & \\
\hline
r_0 & g_0
\end{array}
$$

be true. This identity yields the two equations

$$
\varepsilon+\phi\beta=g_0
$$
$$
\theta=f_1' .
$$

Thus we compute θ by

$$
\varepsilon+\theta\beta=g_0 \quad \text{or} \quad \theta=\frac{g_0-\varepsilon}{\beta}
$$

and then compute f_1' by

$$f_1' = \theta .$$

Let us now drop the prime on f_1', so it becomes the new f_1. Because this new f_1 (i.e. the old f_1') satisfies the equation

$$\begin{array}{c|c} f_1 & \\ \hline r_0 & g_0 \end{array} ,$$

it is the desired filter coefficient for the case $m=1$.

The procedure for the case $m=2$ is as follows. From the procedure for the case $m=1$, we have the two equations

$$\begin{array}{cc|c} 1 & 0 & \\ \hline r_1 & r_0 & \lambda \end{array} \quad \text{and} \quad \begin{array}{cc|c} 0 & 1 & \\ \hline r_1 & r_0 & \beta \end{array} .$$

Multiply the second equation by the unknown u, and add the result to the first equation to obtain:

$$\begin{array}{cc|c} 1 & 0 & \\ \hline r_1 & r_0 & \lambda \end{array} + \begin{array}{cc|c} 0 & u & \\ \hline r_1 & r_0 & u\beta \end{array} = \begin{array}{cc|c} 1 & u & \\ \hline r_1 & r_0 & \lambda+u\beta \end{array} .$$

We set up the identity

$$\begin{array}{cc|c} 1 & u & \\ \hline r_1 & r_0 & \lambda+u\beta \end{array} = \begin{array}{cc|c} 1 & a_1' & \\ \hline r_1 & r_0 & 0 \end{array} .$$

Thus we compute u by

$$\lambda+u\beta=0 \quad \text{or} \quad u=-\lambda/\beta$$

and then compute a_1' by

$$a_1' = u .$$

From the procedure for the case $m=1$, we also have the two equations

$$\begin{array}{cc|c} 0 & 1 & \\ \hline r_0 & r_{-1} & \mu \end{array} \quad \text{and} \quad \begin{array}{cc|c} 1 & 0 & \\ \hline r_0 & r_{-1} & \alpha \end{array} .$$

Let us form

$$\begin{array}{cc|c} 0 & 1 & \\ \hline r_0 & r_{-1} & \mu \end{array} + \begin{array}{cc|c} v & 0 & \\ \hline r_0 & r_{-1} & v\alpha \end{array} = \begin{array}{cc|c} v & 1 & \\ \hline r_0 & r_{-1} & \mu+v\alpha \end{array}$$

$$= \begin{array}{cc|c} b_1' & 1 & \\ \hline r_0 & r_{-1} & 0 \end{array} .$$

Thus we compute v by

$$\mu + v\alpha = 0 \quad \text{or} \quad v = -\mu/\alpha$$

and then compute b'_1 by

$$b'_1 = v \;.$$

Let us now drop the primes on a'_1, b'_1, so they become the new a_1, b_1, respectively. By construction, these new a_1, b_1, (i.e. the old a'_1, b'_1,) satisfy the equations

$$\frac{\begin{array}{cc} 1 & a_1 \end{array}}{\begin{array}{cc} r_1 & r_0 \end{array}}\bigg|\, 0 \qquad \text{and} \qquad \frac{\begin{array}{cc} b_1 & 1 \end{array}}{\begin{array}{cc} r_0 & r_{-1} \end{array}}\bigg|\, 0 \;.$$

Now let us compute the new α, β, λ, μ as follows

$$\frac{\begin{array}{ccc} 1 & a_1 & 0 \end{array}}{\begin{array}{ccc} r_0 & r_{-1} & r_{-2} \\ r_1 & r_0 & r_{-1} \\ r_2 & r_1 & r_0 \end{array}}\left|\begin{array}{c} \alpha \\ 0 \\ \lambda \end{array}\right. \qquad \text{and} \qquad \frac{\begin{array}{ccc} 0 & b_1 & 1 \end{array}}{\begin{array}{ccc} r_0 & r_{-1} & r_{-2} \\ r_1 & r_0 & r_{-1} \\ r_2 & r_1 & r_0 \end{array}}\left|\begin{array}{c} \mu \\ 0 \\ \beta \end{array}\right. \;.$$

Next we compute the new ε by

$$\frac{\begin{array}{cc} f_1 & 0 \end{array}}{\begin{array}{cc} r_1 & r_0 \end{array}}\bigg|\, \varepsilon \;.$$

Let us form

$$\frac{\begin{array}{cc} f_1 & 0 \end{array}}{\begin{array}{cc} r_0 & r_{-1} \\ r_1 & r_0 \end{array}}\left|\begin{array}{c} \\ g_0 \\ \varepsilon \end{array}\right. + \frac{\begin{array}{cc} \theta b_1 & \theta \end{array}}{\begin{array}{cc} r_0 & r_{-1} \\ r_1 & r_0 \end{array}}\left|\begin{array}{c} \\ 0 \\ \theta\beta \end{array}\right.$$

$$= \frac{\begin{array}{cc} f_1 + \theta b_1 & \theta \end{array}}{\begin{array}{cc} r_0 & r_{-1} \\ r_1 & r_0 \end{array}}\left|\begin{array}{c} \\ g_0 \\ \varepsilon + \theta\beta \end{array}\right. = \frac{\begin{array}{cc} f'_1 & f'_2 \end{array}}{\begin{array}{cc} r_0 & r_{-1} \\ r_1 & r_0 \end{array}}\left|\begin{array}{c} \\ g_0 \\ g_1 \end{array}\right. \;.$$

Thus we compute the new θ by

$$\varepsilon + \theta\beta = g_1 \quad \text{or} \quad \theta = \frac{g_1 - \varepsilon}{\beta}$$

and then compute f'_1, f'_2 by

$$f'_1 = f_1 + \theta b_1$$
$$f'_2 = \theta \;.$$

Let us drop the primes on f_1', f_2', so they become the new f_1, f_2 respectively. Because these new f_1, f_2 (i.e. the old f_1', f_2') satisfy the equations

$$
\begin{array}{cc|c}
f_1 & f_2 & \\
\hline
r_0 & r_{-1} & g_0 \\
r_1 & r_0 & g_1
\end{array}
$$

they are the desired filter coefficients for the case $m=2$.

The procedure for the case $m=3$ is as follows. From the procedure for the case $m=2$, we have the equations

$$
\begin{array}{ccc|c}
1 & a_1 & 0 & \\
\hline
r_1 & r_0 & r_{-1} & 0 \\
r_2 & r_1 & r_0 & \lambda
\end{array}
\qquad
\begin{array}{ccc|c}
0 & b_1 & 1 & \\
\hline
r_1 & r_0 & r_{-1} & 0 \\
r_2 & r_1 & r_0 & \beta
\end{array} .
$$

Let us form

$$
\begin{array}{ccc|c}
1 & a_1 & 0 & \\
\hline
r_1 & r_0 & r_{-1} & 0 \\
r_2 & r_1 & r_0 & \lambda
\end{array}
\;+\;
\begin{array}{ccc|c}
0 & ub_1 & u & \\
\hline
r_1 & r_0 & r_{-1} & 0 \\
r_2 & r_1 & r_0 & u\beta
\end{array}
$$

$$
=
\begin{array}{ccc|c}
1 & a_1+ub_1 & 1 & \\
\hline
r_1 & r_0 & r_{-1} & 0 \\
r_2 & r_1 & r_0 & \lambda+u\beta
\end{array}
\;\equiv\;
\begin{array}{ccc|c}
1 & a_1' & a_2' & \\
\hline
r_1 & r_0 & r_{-1} & 0 \\
r_2 & r_1 & r_0 & 0
\end{array}
$$

Thus we compute u by

$$
\lambda+u\beta=0 \quad \text{or} \quad u=-\lambda/\beta
$$

and then compute a_1', a_2' by

$$
a_1'=a_1+ub_1
$$
$$
a_2'=u .
$$

From the procedure for the case $m=2$, we also have the equations

$$
\begin{array}{ccc|c}
0 & b_1 & 1 & \\
\hline
r_0 & r_{-1} & r_{-2} & \mu \\
r_1 & r_0 & r_{-1} & 0
\end{array}
\qquad \text{and} \qquad
\begin{array}{ccc|c}
1 & a_1 & 0 & \\
\hline
r_0 & r_{-1} & r_{-2} & \alpha \\
r_1 & r_0 & r_{-1} & 0
\end{array} .
$$

Let us form

$$
\begin{array}{ccc|c}
0 & b_1 & 1 & \\
\hline
r_0 & r_{-1} & r_{-2} & \mu \\
r_1 & r_0 & r_{-1} & 0
\end{array}
\quad + \quad
\begin{array}{ccc|c}
v & va_1 & 0 & \\
\hline
r_0 & r_{-1} & r_{-2} & v\alpha \\
r_1 & r_0 & r_{-1} & 0
\end{array}
$$

$$
= \quad
\begin{array}{ccc|c}
v & b_1+va_1 & 1 & \\
\hline
r_0 & r_{-1} & r_{-2} & \mu+v\alpha \\
r_1 & r_0 & r_{-1} & 0
\end{array}
\quad \equiv \quad
\begin{array}{ccc|c}
b_2' & b_1' & 1 & \\
\hline
r_0 & r_{-1} & r_{-2} & 0 \\
r_1 & r_0 & r_{-1} & 0
\end{array} .
$$

Thus we compute v by

$$
\mu+v\alpha=0 \quad or \quad v=-\mu/\alpha
$$

and then compute b_1', b_2' by

$$
b_1' = b_1 + va_1
$$
$$
b_2' = v .
$$

Let us now drop the primes on a_1', a_2', b_1', b_2', so they become the new a_1, a_2, b_1, b_2, respectively. By construction these new a_1, a_2, b_1, b_2 (i.e. the old a_1', a_2', b_1', b_2') satisfy the equations

$$
\begin{array}{ccc|c}
1 & a_1 & a_2 & \\
\hline
r_1 & r_0 & r_{-1} & 0 \\
r_2 & r_1 & r_0 & 0
\end{array}
\quad \text{and} \quad
\begin{array}{ccc|c}
b_2 & b_1 & 1 & \\
\hline
r_0 & r_{-1} & r_{-2} & 0 \\
r_1 & r_0 & r_{-1} & 0
\end{array} .
$$

Now let us compute the new α, β, λ, μ as follows.

$$
\begin{array}{cccc|c}
1 & a_1 & a_2 & 0 & \\
\hline
r_0 & r_{-1} & r_{-2} & r_{-3} & \alpha \\
r_1 & r_0 & r_{-1} & r_{-2} & 0 \\
r_2 & r_1 & r_0 & r_{-1} & 0 \\
r_3 & r_2 & r_1 & r_0 & \lambda
\end{array}
\quad \text{and} \quad
\begin{array}{cccc|c}
0 & b_2 & b_1 & 1 & \\
\hline
r_0 & r_{-1} & r_{-2} & r_{-3} & \mu \\
r_1 & r_0 & r_{-1} & r_{-2} & 0 \\
r_2 & r_1 & r_0 & r_{-1} & 0 \\
r_3 & r_2 & r_1 & r_0 & \beta
\end{array} .
$$

Next we compute the new ε by

$$
\begin{array}{ccc|c}
f_1 & f_2 & 0 & \\
\hline
r_2 & r_1 & r_0 & \varepsilon
\end{array} .
$$

150 ENDERS A. ROBINSON

Let us form

$$
\begin{array}{ccc|c}
f_1 & f_2 & 0 & \\
\hline
r_0 & r_{-1} & r_{-2} & g_0 \\
r_1 & r_0 & r_{-1} & g_1 \\
r_2 & r_1 & r_0 & \varepsilon
\end{array}
\;+\;
\begin{array}{ccc|c}
\theta b_2 & \theta b_1 & \theta & \\
\hline
r_0 & r_{-1} & r_{-2} & 0 \\
r_1 & r_0 & r_{-1} & 0 \\
r_2 & r_1 & r_0 & \theta\beta
\end{array}
$$

$$
=
\begin{array}{ccc|c}
f_1+\theta b_2 & f_2+\theta b_1 & \theta & \\
\hline
r_0 & r_{-1} & r_{-2} & g_0 \\
r_1 & r_0 & r_{-1} & g_1 \\
r_2 & r_1 & r_0 & \varepsilon+\theta\beta
\end{array}
\;\equiv\;
\begin{array}{ccc|c}
f_1' & f_2' & f_3' & \\
\hline
r_0 & r_{-1} & r_{-2} & g_0 \\
r_1 & r_0 & r_{-1} & g_1 \\
r_2 & r_1 & r_0 & g_2
\end{array}\; .
$$

Thus we compute the new θ by

$$
\varepsilon+\theta\beta=g_2 \quad \text{or} \quad \theta = \frac{g_2-\varepsilon}{\beta}
$$

and then compute f_1', f_2', f_3' by

$$
\begin{aligned}
f_1' &= f_1+\theta b_2 \\
f_2' &= f_2+\theta b_1 \\
f_3' &= \theta \; .
\end{aligned}
$$

Let us now drop the primes on f_1', f_2', f_3', so they become the new f_1, f_2, f_3 respectively. Because these new f_1, f_2, f_3 (i.e. the old f_1', f_2', f_3') satisfy the equations

$$
\begin{array}{ccc|c}
f_1 & f_2 & f_3 & \\
\hline
r_0 & r_{-1} & r_{-2} & g_0 \\
r_1 & r_0 & r_{-1} & g_1 \\
r_2 & r_1 & r_0 & g_2
\end{array}
$$

they are the desired filter coefficients for the case $m=3$.

It now is apparent that the general procedure may be repeated for the case $m=4$, for the case $m=5$, and so on until the desired value of m is reached.

If we are dealing with real-valued, scalar time-series then $r_{-j}=r_j$. Then, from this symmetry, the coefficient set

$$
(1, a_1, a_2, \ldots, a_{m-1})
$$

is identical to the coefficient set

$$(1, b_1, b_2, \ldots, b_{m-1})$$

for any given value of m. (Also $\alpha = \beta$ and $\lambda = \mu$.) Thus one of these two coefficients sets does not need to be computed in the real-valued scalar case, thereby allowing some gain in efficiency.

In this section we have given a highly efficient method of solving normal equations whose left hand side involves an autocorrelation function.

References

BODE, H. W. (1945). *Network Analysis and Feedback Amplifier Design*. Van Nostrand, Princeton, N.J.

BODE, H. W. and SHANNON, C. E. (1950). A simplified derivation of linear least square smoothing and prediction theory. *Proc. Inst. Rad. Eng.*, **38**, 417–425.

CLAERBOUT, J. (1963). *Digital Filters and Applications to Seismic Detection and Discrimination*. MIT M.S. thesis, Cambridge, Mass.

CRAMÉR, H. (1940). On the theory of stationary random processes. *Ann. Math.*, **41**, 215–230.

DOOB, J. L. (1953). *Stochastic Processes*. John Wiley, New York.

DAVENPORT, W. B. and ROOT, W. L. (1958). *An Introduction to the Theory of Random Signals and Noise*. McGraw-Hill, New York.

HANNAN, E. J. (1960). *Time Series Analysis*. Methuen, London.

KARHUNEN, K. (1949). Über die Struktur stationärer zufälliger Funktionen. *Ark. Mat.*, **1**, 141–160.

KENDALL, M. G. and STUART, A. (1958). *The Advanced Theory of Statistics* (three volumes). Charles Griffin and Co., London.

LANING, J. H. and BATTIN, R. H. (1956). *Random Processes in Automatic Control*. McGraw-Hill, New York.

LEVIN, M. J. (1960). Optimum estimation of impulse response in the presence of noise. *IRE Trans. Circuit Theory*.

RICE, R. B. (1962). Inverse convolution filters. *Geophysics*, **27**, 4–18.

ROBINSON, E. A. (1959). A stochastic diffusion theory of price. *Econometrica*, **27**, 679–684.

— (1960). Sums of stationary random variables. *Proceedings of the American Mathematical Society*, **11**, 77–79.

— (1961). Extremal representation of stationary stochastic processes. *Arkiv för Matematik* (Swedish Academy of Science), **4**, 379–384.

— (1962). *Random Wavelets and Cybernetic Systems*. Charles Griffin and Co., London, and Stechert-Hafner, New York.

— (1963). Extremal properties of the Wold decomposition. *Journal of Mathematical Analysis and Applications*, **6**, 75–85.

— (1963). Structural properties of stationary stochastic processes with applications. *Brown University Symposium on Time Series Analysis*, 170–192, John Wiley, New York.

SIMPSON, S. M. (1956). *Properties, Origin and Treatment of Certain Types of Seismic Noise*. MIT GAG Report **10**, Cambridge, Mass.

— (1957). *The Interrelation of the Deterministic and Probabilistic Approaches to Seismic Problems*. MIT GAG Report **11**, Cambridge, Mass.

— (1961). *Initial Studies on Underground Nuclear Detection with Seismic Data Prepared by a Novel Digitization System*. Scientific Report **1**, Geophysics Research Directorate, Air Force Cambridge Research Center, Office of Aerospace Research, Bedford, Massachusetts.

— (1961). *Time Series Techniques Applied to Underground Nuclear Detection and Further Digitized Seismic Data*. Scientific Report **2** of *ibid*.

— (1962) *Continued Numerical Studies on Underground Nuclear Detection and Further Digitized Seismic Data*. Scientific Report **3** of *ibid*.

WIENER, N. (1942). *The Extrapolation, Interpolation, and Smoothing of Stationary Time Series with Engineering Applications*. MIT DIC Contract 6037, National Defense Research Council, Section D2, Cambridge, Mass. (Reprinted (1949). John Wiley, New York.)

WOLD, H. (1938). *A Study in the Analysis of Stationary Time-series*. Thesis, University of Stockholm. (Second Edition (1954). Almqvist and Wiksell, Uppsala.)

— (1963). Forecasting by the chain principle. *Brown University Symposium in Time Series Analysis*, 471–497, John Wiley, New York.

YULE, G. U. (1927). On a model of investigating periodicities in disturbed series. *Phil. Trans.*, A, **226**, 267–298.

ZADEH, L. A. and RAGAZZINI, J. R. (1952). Optimum filters for the detection of signals in noise. *Proc. IRE*, **40**, 1223–1231.

CHAPTER 5

RECURSIVE DECOMPOSITION OF STOCHASTIC PROCESSES

ENDERS A. ROBINSON

University of Wisconsin and University of Uppsala

1. Predictive Decomposition of Stationary Time Series

In this paper we deal exclusively with discrete time series. *A discrete time series is a sequence of observations x_t which are associated with the discrete time parameter t at equidistant time points.* Without loss of generality we may take the spacing between each successive observation to be one unit of time, so that t takes on integral values. As a result, the minimum period which may be observed is equal to two units, and consequently the maximum frequency which may be observed is equal to $\frac{1}{2}$. Thus we may require all angular frequencies to lie between $-\pi$ and π since no distinction can be made between angular frequencies which differ by an integral multiple of 2π. If, for example, the unit of spacing is taken to be 0.0025 second, then all frequencies of the discrete time series may be required to lie between plus and minus 200 cycles per second.

A linear operator (or linear system) with constant real coefficients a_t acts upon the input x_t to produce the output ε_t according to the *moving summation* or *convolution* formula

$$\varepsilon_t = \sum_{s=-\infty}^{\infty} a_s x_{t-s} = \ldots + a_{-2} x_{t+2} + a_{-1} x_{t+1} + a_0 x_t + a_1 x_{t-1} + \ldots \tag{1}$$

This equation states that *the convolution of the input x_t with the operator coefficients a_t yields the output ε_t.* A linear operator a_t is called *stable* if every bounded input x_t yields a bounded output ε_t. That is, the linear operator (1) is stable if and only if the summation on the right in equation (1) converges for bounded x_t. Thus we see that the output ε_t of a stable linear operator will never become indefinitely large unless the input x_t does so. On the other hand, an *unstable* linear operator a_t is one which gives an unbounded output ε_t for some particular bounded input x_t, though not necessarily for all bounded inputs. This definition of stability is the same as the definition given for the stability of linear filters in electrical engineering. A finite-length linear operator is one for which only a finite number of the operator coefficients a_s do not vanish. In this case the summation on the right of equation (1) contains only a finite number of terms so that the output ε_t will always be bounded for a bounded input x_t. Thus all finite-length linear operators are stable. In general, it may be shown that a necessary and sufficient condition that the summation on the right side of equation (1) converges for bounded x_t is that

$$\sum_{s=-\infty}^{\infty} |a_s| < \infty .\tag{2}$$

Thus *the linear operator* (1) *is stable if and only if condition* (2) *holds*. For a stable linear operator, the infinite summation (1) may always be approximated arbitrarily closely by a partial summation.

A linear operator a_s will be called computationally *realizable* at the time instant t if only values $x_t, x_{t-1}, x_{t-2}, \ldots$ of the input at and prior to time t, and no values $x_{t+1}, x_{t+2}, \ldots$ subsequent to time t, are utilized in the computation at time t of the output ε_t. In other words, the only restriction on a realizable operator is that its output ε_t at time t be affected by inputs which occur at and before time t, and be in no way affected by inputs which occur after time t. That is, a realizable operator is one which does not respond at time t to inputs which have not yet arrived at that time. This definition of realizability is the same as that given for the realizability of filters in electrical engineering.

We shall distinguish between one-sided operators on the one hand and two-sided operators on the other hand. A one-sided operation is obtained from equation (1) by requiring $a_s = 0$ for negative s; that is, the one-sided operator acts upon input x_t to produce output ε_t according to the convolution formula

$$\varepsilon_t = \sum_{s=0}^{\infty} a_s x_{t-s} = a_0 x_t + a_1 x_{t-1} + a_2 x_{t-2} + \ldots .\tag{3}$$

Looking at this formula, we see that the coefficients a_s operate on present and past values, but no future values, of the input. As a result, if time t represents the present calendar time, as is the case for a meteorologist who makes daily weather predictions, then only observations of the input at the present time and at past times and no observations at future times are required to carry out the necessary computation. Thus the one-sided operator (3) is realizable. More precisely, the condition, $a_s = 0$ for $s < 0$, that the linear operator (1) be one-sided is precisely the condition that the linear operator be realizable. In what follows, we shall use the terms realizable operator and one-sided operator interchangeably.

The so called *impulse function* (for discrete time) is defined to be a single unit pulse at time $t=0$. Suppose that we let the input x_t be an impulse function, that is,

$$\begin{cases} x_t = 0 & \text{for} \quad t = -1, -2, -3, \ldots \\ x_0 = 1 \\ x_t = 0 & \text{for} \quad t = 1, 2, 3, \ldots \end{cases}$$

Then the output ε_t of the one-sided operation (3) is zero for negative time and

$$\varepsilon_0 = a_0 , \quad \varepsilon_1 = a_1 , \quad \varepsilon_2 = a_2, \ldots$$

for positive time. Thus the operator coefficients are the output for an impulse-function input. For this reason, we may call the operator coefficients a_0, a_1, a_2, . . . the *impulse response* of the system.

For a two-sided operator, the operator coefficients a_s for negative s do not all vanish in equation (1). Thus a two-sided operator has the property that values of the input . . ., x_{t+2}, x_{t+1} at times subsequent to time t, as well as values x_t, x_{t-1}, x_{t-2}, . . . at time t and prior to time t, are required to compute the output ε_t. Consequently, if time t represents the present calendar time, as in the example of the meteorologist, then ε_t given by a two-sided operator cannot be computed at time t since it involves values of the input at future times which are not observable at the present time t. Similarly, the network which would be equivalent to a two-sided operator would be one which would respond at time t to impulses which have not yet arrived at that time. Consequently, two-sided operators are not realizable. Nevertheless, a very simple trick makes a finite-length two-sided operator computationally realizable. Suppose that a_{-m} is the first non-vanishing operator coefficient in (1), where m is positive. Then in order to compute the finite-length two-sided operator, we must delay our computations until time $t+m$, or later, which is a time delay of m or greater, at which time all the input values needed in the computation will have occurred. This procedure, in effect, transforms the finite-length two-sided operator with respect to time t into a one-sided operator with respect to time $t+m$, or later. Similarly, the engineer may introduce a time delay m, or greater, into his network to transform the non-physically realizable system (at the time instant t) into a physically realizable system (at time $t+m$, or later). Let us make the following observation. The fact that an input time-series is entirely recorded on paper or magnetic tape means that we have waited until all the pertinent information is available. Consequently, the necessary time delay has been introduced to utilize finite-length time-delay two-sided operations, which are computationally realizable. On the other hand, if computations were to be carried out at the same time as the input event is occurring (that is, in real time), then we would not be able to compute such two-sided operations.

As Wiener (1942) points out, the linear operator is the approach from the standpoint of time to a *filter* which is essentially an instrument for the separation of different frequency ranges. The filtering action of the linear operator is brought out by its transfer function. The following interpretation to the transfer function may be made. Let the input $x_t = \exp(i\omega t) = \cos \omega t + i \sin \omega t$ be points from a (complex) sinusoidal wave of angular frequency ω. For a stable linear operation (1) the output ε_t will be a sinusoidal wave of the same frequency but will differ in amplitude and phase; so we may write the output as $\varepsilon_t = A(\omega) \exp(i\omega t)$, where $A(\omega)$ is the transfer function of the linear operator. Using equation (1) we see that the transfer function $A(\omega)$ of a stable operator is given by the Fourier series

$$A(\omega) = \sum_{s=-\infty}^{\infty} a_s e^{-i\omega s} = \sum_{s=-\infty}^{\infty} a_s \cos \omega s - i \sum_{s=-\infty}^{\infty} a_s \sin \omega s \tag{4}$$

in which the operator coefficients a_t are the Fourier coefficients

$$a_t = \frac{1}{2\pi} \int_{-\pi}^{\pi} A(\omega) e^{i\omega t} \, d\omega \,. \tag{5}$$

We note that $A(-\omega) = \overline{A(\omega)}$ where the bar indicates the complex conjugate. The power transfer function $\Psi(\omega)$ is defined to be $A(\omega) A(-\omega)$, which is real and non-negative and contains no phase information. In polar form, the transfer function may be written as

$$A(\omega) = |A(\omega)| e^{iP(\omega)}$$

where the absolute value $|A(\omega)|$ is called the *gain*, and $P(\omega)$ is called the *phase*. The *phase-lag* $\theta(\omega)$ is defined to be the negative of the phase, that is, $\theta(\omega) = -P(\omega)$.

Under the transformation $z = \exp(-i\omega)$, we see that each angular frequency ω in the interval $(-\pi, \pi)$ is represented by a point on the unit circle $|z| = 1$, where ω is the angle. Under this transformation, the transfer function $A(\omega)$, which is a trigonometric series, becomes the function $A^*(z)$, which we shall call the z-transform. For a two-sided operator, we see that $A^*(z)$ is in the form of a Laurent series; that is, $A^*(z)$ contains both positive and negative powers of z. On the other hand, for a one-sided operator, which as we have seen corresponds to a realizable system, the function $A^*(z)$ is the power series

$$A^*(z) = \sum_{s=0}^{\infty} a_s z^s = a_0 + a_1 z + a_2 z^2 + \ldots \qquad (6)$$

Now we know from function theory that the power series (6) converges in a circle with radius equal to R where $1/R$ is the upper limit of the (positive) sequence of numbers $|a_n|^{1/n}$, $n = 1, 2, 3, \ldots$ (Cauchy-Hadamard Theorem), and that $A^*(z)$ may be defined outside of this circle by the process of analytic continuation. Let us now express the stability condition (2) for a one-sided operator or filter in terms of singular points of $A^*(z)$. For a stable filter, condition (2) holds, and hence the power series (6) for $A^*(z)$ converges for $z = 1$. Thus R must be greater than or equal to one. Thus $A^*(z)$ is analytic for $|z| < 1$, and all singularities of $A^*(z)$ are therefore outside the unit circle or on its boundary. Conversely, if the filter is unstable, then condition (2) does not hold so the power series (6) for $A^*(z)$ does not converge absolutely for $z = 1$, and thus $R \leq 1$. Since from function theory we know that there must be at least one singularity on the boundary $|z| = R$ of the circle of convergence, it follows that this point is located either inside or on the boundary of the unit circle. Thus we have the following theorem for one-sided operators. *If all the singular points of $A^*(z)$ are located outside the unit circle, then the one-sided operator (3) is stable. If at least one singular point lies inside the unit circle, the one-sided operator is unstable. If the all singularities are on or exterior to the unit circle, with at least one singularity located on the unit circle, then the one-sided operator may be either stable or unstable.*

Let us consider the complex plane $\lambda = \omega + i\sigma$, where ω denotes the angular frequency. The transformation $z = \exp(-i\lambda)$ maps the exterior $|z| > 1$ of the unit circle in the z-plane into the strip of the upper half λ-plane between $\omega = -\pi$ to $\omega = \pi$. It maps the periphery $|z| = 1$ of the unit circle into the real axis from $\omega = -\pi$ to $\omega = \pi$ of the λ-plane. It maps the interior $|z| < 1$ of the unit circle into the strip of the lower half λ-plane between $\omega = -\pi$ to $\omega = \pi$. Under this transformation $A^*(z)$ becomes the function $A(\lambda)$, the value of which on the real ω-axis is the transfer function $A(\omega)$. Thus the stability condition (2) for a one-sided operator (3) may then be expressed in terms of $A(\lambda)$ as follows. *If all the singular points of $A(\lambda)$ are located in the upper half λ-plane, then the filter is stable; whereas if at least one singular point lies in the lower half-plane, then the filter is unstable.*

In statistical theory our stability requirement that the summation (1) converge for bounded x_t is usually replaced by the weaker requirement that it converge in the mean for an input x_t which is an orthonormal sequence of random variables. In this case we shall call the linear operator stable (in the sense of convergence in the mean) or more briefly stable (in the mean) or, when it is clear from the context, stable. A necessary and sufficient condition that the summation (1) converge in the mean for orthonormal x_t is that

$$\sum_{s=-\infty}^{\infty} \left| a_s \right|^2 < \infty . \tag{7}$$

Thus *the linear operator* (1) *is stable (in the mean) if and only if condition* (7) *holds*. The reason we introduce the concept of stability (in the mean) is that we now wish to present some theorems which are stated in terms of condition (7).

A fundamental theorem of Kolmogorov (1941) is the following. *Let the power transfer function* $\Psi(\omega)$ *be real, non-negative, and integrable over the interval* $(-\pi, \pi)$ *and be positive almost everywhere on this interval. Then a necessary and sufficient condition for us to be able to write*

$$\Psi(\omega) = \left| A(\omega) \right|^2 , \quad A(\omega) = \sum_{s=0}^{\infty} a_s e^{-i\omega s}, \quad \sum_{s=0}^{\infty} \left| a_s \right|^2 < \infty \tag{8}$$

is that

$$\int_{-\pi}^{\pi} \log \Psi(\omega) d\omega > -\infty . \tag{9}$$

We shall call (9) *Kolmogorov's condition* for $\Psi(\omega)$. We see that Kolmogorov's theorem is the discrete time parameter analogue of the Paley-Wiener theorem, whose application is well known to electrical engineers. Kolmogorov's condition (9) then is the analogue of the Paley-Wiener criterion, applied by Wiener (1942, section 1.04) as a criterion for the realizability of filters. As we have seen, the conditions that the linear operator (1) be stable is that condition (7) holds, and be realizable is that $a_s = 0$ for negative s. By application of Kolmogorov's theorem we see that a linear operator a_s is stable and realizable if its power transfer function $\Psi(\omega)$ is real, non-negative, and integrable over $(-\pi, \pi)$ and non-zero almost everywhere for $(-\pi, \pi)$, and if $\Psi(\omega)$ satisfies Kolmogorov's condition (9).

Let $\Psi(\omega)$ be any power transfer function that satisfies the conditions (8) and (9) of Kolmogorov's theorem. Then it is always possible to determine a one sided operator $a_0, a_1, a_2, \ldots$ with squared gain

$$\left| A(\omega) \right|^2 = \Psi(\omega)$$

such that its z-transform satisfies

$$A^*(z) \neq 0 \quad \text{for} \quad |z| < 1 . \tag{10}$$

Condition (10), namely the condition that the z-transform $A^*(z)$ have no zeros within the unit circle (or, equivalently, that $A(\lambda)$ have no zeros in the lower half-plane) is called the minimum phase-lag condition. This condition means that the operator $a_0, a_1, a_2, \ldots$ has the minimum phase-lag possible for its gain $|A(\omega)|$. Any realizable and stable operator with such a minimum phase-lag is called a *minimum-delay operator*. Thus we have the following theorem. *Given any power transfer function that satisfies the conditions of Kolmogorov's theorem, we may always find a minimum-delay operator which has this power transfer function.*

Thus, any one-sided operator that satisfies the conditions (8) and (9) of Kolmogorov's theorem, and in addition satisfies the minimum phase-lag condition (10), is minimum-delay. In other words, a minimum-delay operator satisfies:

(A) $a_s \quad = \quad 0 \quad \text{for} \quad s < 0 \quad$ (the realizability condition)

(B) $\displaystyle\sum_{s=0}^{\infty} a_s^2 < \quad \infty \quad$ (the stability condition)

(C) $\displaystyle A^*(z) = \sum_{s=0}^{\infty} a_s z^s \neq 0 \quad \text{for} \quad |z| < 1 \quad$ (the minimum phase-lag condition)[1]

Because of the minimum phase-lag condition (10), we may take the reciprocal $B^*(z)$ of the z-transform $A^*(z)$ of a minimum-delay operator, the expression for the reciprocal being valid within the unit circle. The reciprocal $B^*(z)$ has no singularities or zeros within the unit circle, and thus has the power series representation

$$B^*(z) = \frac{1}{A^*(z)} = \frac{1}{\displaystyle\sum_{s=0}^{\infty} a_s z^s} = \sum_{s=0}^{\infty} b_s z^s \quad \text{for} \quad |z| < 1 . \tag{11}$$

Concerning this power series with coefficients b_s, two cases may occur: The infinite sum

$$\sum_{s=0}^{\infty} b_s^2$$

---

[1] Here we rule out the presence of the so-called Type 3 all-pass system. For the general case, see Robinson (1962, 1963).

may be either (A) convergent or (B) divergent. If case (A) holds, then we may let $z = e^{-i\omega}$ in equation (11), thus obtaining

$$B(\omega) = \frac{1}{A(\omega)} = \frac{1}{\displaystyle\sum_{s=0}^{\infty} a_s e^{-i\omega s}} = \sum_{s=0}^{\infty} b_s e^{-i\omega s} . \tag{12}$$

The operator $b_0, b_1, b_2, \ldots$ is realizable and stable, and because

$$B^*(z) \neq 0 \quad \text{for} \quad |z| < 1$$

this operator also has minimum phase-lag. That is, the operator $b_0, b_1, b_2 \ldots$ is minimum-delay. If case (B) holds, then it is still possible to speak of the minimum-delay operator $b_0, b_1, b_2, \ldots$ provided we interpret this operator as a *generalized function* as developed in Robinson (1962, 1963).

Both $|A(\omega)|^2$ and $|B(\omega)|^2$ satisfy Kolmogorov's condition (9). Because $B(\omega) = 1/A(\omega)$, it follows that the phase-lag of $B(\omega)$ is the negative of the phase-lag of $A(\omega)$, each of which is the minimum phase-lag associated with its respective gain. That is, $a_0, a_1, a_2, \ldots$ is minimum-delay with gain $|A(\omega)|$, and $b_0, b_1, b_2, \ldots$ is minimum-delay with the reciprocal gain, which is $|B(\omega)| = 1/|A(\omega)|$. Equation (12) yields in the time-domain that

$$a_0 b_0 = 1 , \quad \sum_{s=0}^{n} a_s b_{n-s} = 0 , \quad n = 1, 2, 3, \ldots . \tag{13}$$

From this equation, we see that given the operator coefficients a_s, the operator coefficients b_s may be uniquely determined, and vice versa. The minimum-delay operators a_s and b_s may be represented respectively by the input/output relationships

$$\varepsilon_t = \sum_{s=0}^{\infty} a_s x_{t-s} , \quad x_t = \sum_{s=0}^{\infty} b_s \varepsilon_{t-s} \tag{14}$$

and hence these operators are mutually inverse to each other. Thus the operator a_s operates on the present and past values $x_t, x_{t-1}, x_{t-2}, \ldots$ to yield the present and past values $\varepsilon_t, \varepsilon_{t-1}, \varepsilon_{t-2}, \ldots$, whereas the operator b_s operates on the present and past values $\varepsilon_t, \varepsilon_{t-1}, \varepsilon_{t-2}, \ldots$ to yield the present and past values $x_t, x_{t-1}, x_{t-2}, \ldots$. That is, the present and past of x_t yield the present and past of ε_t, and vice versa. In other words, if we pass the time series x_t through the operator a_s to yield the output ε_t, we may recover x_t by passing ε_t through the inverse operator b_s, where this recovery takes place without loss of time.

Let us now consider the finite-length one-sided operator obtained from (3) by requiring $a_s = 0$ for $s > m$. Thus we have

$$\varepsilon_t = \sum_{s=0}^{m} a_s x_{t-s} = a_0 x_t + a_1 x_{t-1} + \ldots + a_m x_{t-m} .$$

The homogeneous difference equation obtained by setting $\varepsilon_t = 0$ is

$$a_0 x_t + a_1 x_{t-1} + \ldots + a_m x_{t-m} = 0 .$$

We see that the constant coefficients a_s of this difference equation are the operator coefficients. The characteristic equation of this difference equation is

$$a_0 z^m + a_1 z^{m-1} + \ldots + a_m = 0$$

which is

$$z^m A^*(z^{-1}) = 0$$

where $A^*(z)$ is the z-transform of the operator. For an arbitrary set of initial values $x_0, x_1, x_2, \ldots, x_{m-1}$, the series $x_m, x_{m+1}, x_{m+2}, \ldots$ may be generated by recursive deductions from this difference equation, and this series will form a general solution of the difference equation. A necessary and sufficient condition that

$$\sum_{t=0}^{\infty} x_t^2 \quad \text{and} \quad \sum_{t=0}^{\infty} |x_t|$$

converge is that the magnitude of the roots of the characteristic equation be less than one. Equivalently, this condition is that all the zeros of $A^*(z)$ have magnitude greater than one; that is, $A^*(z)$ have no zeros within or on the unit circle. In this case the general solution x_t of the homogeneous difference equation describes a damped oscillation, and the difference equation is said to be stable. If any zeros of $A^*(z)$ have magnitude equal to one, the difference equation is said to be on the borderline of unstability (or semi-stable); whereas if $A^*(z)$ has any zeros with magnitude less than one, the difference equation is unstable. Thus *the coefficients $a_0, a_1, \ldots, a_m$ are the coefficients of a finite-length minimum-delay operator if and only if they are the coefficients of a stable or semi-stable difference equation.* This result is useful since in practice we must restrict ourselves to finite-length operators.

Up to now we have examined the properties of the discrete linear operator, with coefficients a_t, for a single time series. We have seen that the terms "linear operator", "linear system", and "linear filter" may be used interchangeably; "linear operator" usually being associated with the time domain, "linear filter" with the frequency domain.

Any observational time series x_t ($-\infty < t < \infty$) may be considered to be a realization of a so-called *random process*, or *stochastic process*, which is a mathematical abstraction defined with respect to a probability field. For any stochastic process, one may form averages with respect to the statistical population or "ensemble" of realizations x_t for a fixed value of time t. Such averages are called *ensemble averages*, and we shall denote such an averaging process by the expectation symbol E. If the mean value $m = E(x_t)$ and the (unnormalized) autocorrelation coefficients $\phi(\tau) = E(x_{t+\tau} x_t)$ are finite and independent of t, the process is called *stationary* in the wide sense. Without loss of generality we assume $E(x_t)$ to be zero. There is another type of average known as a *time average* in which the averaging process is carried out with respect to all values of time t for a fixed realization x_t($-\infty < t < \infty$) of the stochastic process. For a large class of stationary processes, called *ergodic processes*, the Birkoff-Khintchine Ergodic Theorem tells us that ensemble averages and the corresponding time averages are equal with probability one. Consequently, the autocorrelation of an ergodic process may be expressed as the time average

$$\phi(\tau) = \lim_{T \to \infty} \frac{1}{2T+1} \sum_{t=-T}^{T} x_{t+\tau} x_t \tag{15}$$

taken over a single realization of the time series x_t. The autocorrelation function is a non-negative definite function, and hence it may be represented by the Fourier transform

$$\phi(\tau) = \frac{1}{2\pi} \int_{-\pi}^{\pi} e^{i\omega\tau}\, d\Lambda(\omega) = \frac{1}{\pi} \int_{0}^{\pi} \cos \omega\tau\, d\Lambda(\omega) \tag{16}$$

where $\Lambda(\omega)$, called the integrated spectrum, or the spectral distribution function, is a real, bounded, monotone non-decreasing function of ω.

Let us define the random variable $\hat{x}_{t+\alpha}$ to be the linear least-squares prediction of $x_{t+\alpha}$ in terms of the complete past $\ldots, x_{t-2}, x_{t-1}, x_t$ of the time series up to time t. That is, $\hat{x}_{t+\alpha}$ is the convolution of the one-sided operator k_t with the input x_t. We therefore have[2]

$$\hat{x}_{t+\alpha} = \sum_{s=0}^{\infty} k_s x_{t-s}, \quad K(\omega) = \sum_{s=0}^{\infty} k_s e^{-i\omega s} \tag{17}$$

which are determined by requiring the mean square error of the prediction

---

[2] The linear operator k_s in some cases must be interpreted as a generalized function (Robinson, 1963).

$$\sigma_\alpha^2 = E[(x_{t+\alpha} - \hat{x}_{t+\alpha})^2] = \frac{1}{2\pi} \int_{-\pi}^{\pi} \left| e^{i\omega\alpha} - K(\omega) \right|^2 \mathrm{d}\Lambda(\omega) \tag{18}$$

be a minimum. The x_t process is called *deterministic* provided $\sigma_1^2 = 0$, in which case the future $x_{t+\alpha}$ is completely determined from the remote past $x_s, x_{s-1}, x_{s-2}, \ldots$ where s approaches minus infinity. The process is called *non-deterministic* provided $\sigma_1^2 > 0$, in which case the future cannot be completely determined by a linear operation on the past.

If the spectral distribution $\Lambda(\omega)$ is absolutely continuous; that is, if $\Lambda(\omega)$ is equal to the integral of its derivative $\Lambda'(\omega)$, then the process is said to have a spectral density, or power spectrum, which is $\Phi(\omega) = \Lambda'(\omega)$. For real time series we see that $\Phi(\omega)$ is an even, non-negative function. For absolutely continuous $\Lambda(\omega)$ we may let $\mathrm{d}\Lambda(\omega) = \Phi(\omega)\,\mathrm{d}\omega$ in equations (16) and (18). If $\Sigma_{\tau=-\infty}^{\infty} |\phi(\tau)|$ converges, then $\Lambda(\omega)$ is absolutely continuous, and $\Phi(\omega)$ is continuous and is given by the inversion formula (Wold, 1938, Corollary to Theorem 5)

$$\Phi(\omega) = \sum_{\tau=-\infty}^{\infty} \phi(\tau) e^{-i\omega\tau} = \phi(0) + 2 \sum_{\tau=1}^{\infty} \phi(\tau) \cos \omega\tau . \tag{19}$$

Two real random variables x and y with finite variances are said to be *uncorrelated* provided $E(xy) = E(x)E(y)$, *orthogonal* provided $E(xy) = 0$, and *orthonormal* provided $E(xy) = 0$, $E(x^2) = 1$, and $E(y^2) = 1$. A *mutually uncorrelated process* is a stationary process for which the observations ε_t are uncorrelated in pairs; that is, $E(\varepsilon_t \varepsilon_s) = E(\varepsilon_t) E(\varepsilon_s)$ for t not equal to s. In what follows we shall consider uncorrelated variables to be normalized such that $E(\varepsilon_t) = 0$ and $E(\varepsilon_t^2) = 1$, in which case the ε_t forms an *orthonormal sequence of random variables*. Thus the autocorrelation of the ε_t process vanishes except for zero lag and the power spectrum is constant for the interval $(-\pi, \pi)$. These processes therefore have white light spectra and they may be called *white noise*.

For the fixed realization $\ldots, \varepsilon_{t-1}, \varepsilon_t, \varepsilon_{t+1}, \varepsilon_{t+2}, \ldots$ of a mutually uncorrelated process, the corresponding fixed realization of a *process of moving summation* is $\ldots, x_{t-1}, x_t, x_{t+1}, x_{t+2}, \ldots$ where

$$x_n = \sum_{s=-\infty}^{\infty} c_s \varepsilon_{n-s} , \qquad n = \ldots, t-1, t, t+1, t+2, \ldots . \tag{20}$$

It is supposed that the operator c_s is stable (in the mean). Processes of moving summation are ergodic. For orthonormal ε_t, the mean of the x_t is zero and the autocorrelation coefficients are given by

$$\Phi(\tau) = \sum_{t=-\infty}^{\infty} c_{t+\tau} c_t$$

Let $C(\omega)$ be the transfer function of the operator c_s. Then it may be shown that $\Phi(\omega) = C(\omega) C(-\omega)$.

The following theorem brings out the generality of processes of moving summation. *Any stationary time series with an absolutely continuous spectral distribution function is generated by a process of moving summation; and, conversely, any process of moving summation has an absolutely continuous spectral distribution function.* Let us interpret these results. Equation (20) is in the form of an infinite-length two-sided operator, where the white noise ε_t is the input and the time series x_t is the output. The transfer function of this linear system is $C(\omega)$. The power transfer function $C(\omega) C(-\omega)$ of this linear system is the power spectrum $\Phi(\omega)$ of the time series x_t. Since, in general, the c_s which are chosen to represent the time series x_t in equation (20) may be the coefficients of an infinite-length two-sided operator, this linear system need not necessarily be realizable. We now wish to investigate the conditions under which the two-sided set of operator coefficients c_s may be replaced in equation (20) by a unique stable one-sided set of operator coefficients b_s which have minimum phase-lag. As we shall see, the replacement of the c_s by the minimum-delay b_s leads to the so-called predictive decomposition of stationary time series x_t.

From a realization of an ergodic stationary time series x_t, we may compute the autocorrelation function $\phi(\tau)$ by means of equation (15), and then the power spectrum $\Phi(\omega)$ from the autocorrelation by means of equation (19). As we have seen, the power spectrum is the power transfer function of a linear system into which white noise ε_t is passed in order to obtain the time series x_t as output. Thus we know that the gain of this linear system is $\sqrt{\Phi(\omega)}$. Now any linear system with this gain, and with arbitrary phase-lag, would be an admissible linear system to describe the time series x_t. Nevertheless, let us specify that the particular linear system which we desire is one which is realizable and stable, with minimum phase-lag. In other words, we desire the linear system which has minimum phase-lag, denoted by $\theta(\omega)$, for the absolute gain $\sqrt{\Phi(\omega)}$. Thus the transfer function $B(\omega)$ of the desired minimum-delay linear system may be expressed as

$$B(\omega) = \sqrt{\Phi(\omega)} e^{-i\theta(\omega)} = \sum_{s=0}^{\infty} b_s e^{-i\omega s} \tag{21}$$

where $\theta(\omega)$ is the desired minimum phase-lag and the set $b_0, b_1, b_2, \ldots$ are the desired minimum-delay operator coefficients. The problem of factoring the power spectrum $\Phi(\omega)$ is the problem of expressing the power spectrum as $\Phi(\omega) = B(\omega)\overline{B(\omega)}$ where the factor $B(\lambda)$, $\lambda = \omega + i\sigma$, is free of singularities and zeros in the lower half λ-plane. Thus this factor $B(\omega)$ of the power spectrum has gain $\sqrt{\Phi(\omega)}$ and minimum phase-lag, and hence it is the transfer function $B(\omega)$, given by equation (21), of the desired minimum-delay system. Since $\overline{B(\omega)} = B(-\omega)$, we may alternatively write the factorization as $\Phi(\omega) = B(\omega)B(-\omega)$.

First let us find the minimum-delay system in the case of a finite moving-averages process by factoring its power spectrum. *A process of finite moving averages is a stationary process for which the autocorrelation $\phi(\tau)$ vanishes for τ greater than m.* Thus its power spectrum, $\Phi(\omega)$, given by equation (19), is seen to be a rational function in $z = \exp(-i\omega)$, and more specifically $z^m \Phi(z)$ is seen to be a polynomial of degree $2m$. Since $\Phi(\omega)$ is a real function of ω, it follows that if z_k is a root of this polynomial, then $\bar{z}_k^{-1}$ is also a root. Moreover, since $\Phi(\omega)$ is an even function of ω, it follows that if z_k is a root of this polynomial, then $\bar{z}_k$ is also a root. Since $\Phi(\omega)$ is a non-negative function of ω, it follows that any root of modulus one must appear an even number of times. Let γ_k and $\bar{\gamma}_k$ denote the complex roots of this polynomial with modulus greater than one, and also half of those complex roots with modulus equal to one. Similarly, let ρ_j denote the real roots of this polynomial with modulus greater than one, and also half of those real roots with modulus equal to one. Thus this polynomial may be factored into

$$z^m \Phi(z) = \left[a_m \prod_{k=1}^{h} (z - \gamma_k)(z - \bar{\gamma}_k) \prod_{j=1}^{l} (z - \rho_j) \right]$$

$$\left[\prod_{k=1}^{h} (z - \gamma_k^{-1})(z - \bar{\gamma}_k^{-1}) \prod_{j=1}^{l} (z - \rho_j^{-1}) \right] \quad (22)$$

where any root of order p is repeated p times. Let us denote the first factor in brackets in equation (22) by $B^*(z)$; then the second factor in brackets is seen to be $z^m B^*(z^{-1})$. We see that $B^*(z)$ is a polynomial in z with real coefficients, and so we may represent $B^*(z)$ by $b_0 + b_1 z + \ldots + b_m z^m$. Moreover, we see that $B^*(z)$ has no zeros within the unit circle. In those cases in which there are no γ_k and ρ_j of modulus one, the polynomial $B^*(z)$ has no zeros within or on the unit circle and hence is the transfer function in the z-plane of a (strictly) stable homogeneous difference equation with coefficients $b_0, b_1, \ldots, b_m$. Under the transformation $z = \exp(-i\omega)$, equation (22) becomes $\Phi(\omega) = B(\omega)B(-\omega)$ where $B(\lambda)$ is free of zeros and singularities in the lower half λ-plane; that is, $B(\omega)$ has minimum phase-lag. This equation then represents the Wold (1938) factorization of the power spectrum of the process of finite moving averages where $B(\omega)$ and $b_0, b_1, \ldots, b_m$ represent the transfer function and operator coefficients, respectively, of the desired minimum-delay system. On the other hand, if we did not choose the roots of the polynomial $z^m \Phi(z)$ in the above fashion, there being at most 2^m different ways of choosing the roots, then $B^*(z)$ would have roots, some of which have modulus greater than one, and some of which have modulus less than one. Consequently, the transfer function $B(\omega)$ would not be of the minimum phase-lag type.

The autoregressive process, like the process of finite moving averages, also has a power spectrum rational in $z = \exp(-i\omega)$. More specifically, *the reciprocal of the autoregressive spectrum has the same mathematical form as the spectrum of the process of finite moving averages.* Accordingly, Wold (1938) factored the reciprocal of the autoregressive spectrum into the product $A(\omega) A(-\omega)$, where the polynomial $A^*(z)$, $z = \exp(-i\omega)$, has no zeros within and on the unit circle. We note that $A^*(z)$ can have no zeros on the unit circle since we assume $\Phi(\omega)$ is integrable on the interval $(-\pi, \pi)$. Thus the spectrum $\Phi(\omega)$ of the autoregressive process may be factored as $\Phi(\omega) = B(\omega) B(-\omega)$ where the factor $B(\omega)$ is the reciprocal of $A(\omega)$, and, like $A(\omega)$, has minimum phase-lag. The factor $B(\omega)$ is then the transfer function of the desired minimum-delay system.

More generally, *any stationary process whose power spectrum is a rational function in $z = \exp(-i\omega)$ is a hybrid between the process of finite moving averages and the autoregressive process.* Accordingly, the numerator and denominator each may be factored by Wold's method to give

$$\Phi(\omega) = \frac{G(\omega)\, G(-\omega)}{H(\omega)\, H(-\omega)} = \frac{\left| \sum_{s=0}^{M} g_s\, e^{-i\omega s} \right|^2}{\left| \sum_{s=0}^{N} h_s\, e^{-i\omega s} \right|^2} \tag{23}$$

where, letting $z = \exp(-i\omega)$, the polynomials $G^*(z)$ and $H^*(z)$ have no common factors, the roots of $H^*(z)$ have modulus greater than one, and the roots of $G^*(z)$ have modulus greater than or equal to one. Thus the factor $B(\omega) = G(\omega)/H(\omega)$ is the transfer function of the desired minimum-delay system.

The *general solution of the factorization problem* for discrete stationary processes with arbitrary power spectra was given by Kolmogorov (1941) and independently by Wiener (1942). Here we wish to give an heuristic exposition. Let us first turn our attention to the properties of the desired realizable, stable, minimum phase-lag linear system with transfer function $B(\omega)=|B(\omega)|\exp[-i\theta(\omega)]$. Here $|B(\omega)|$ is the gain, and $\theta(\omega)$ is the minimum phase-lag associated with this gain. Since $B^*(z)$, $z=\exp(-i\omega)$, has no singularities or zeros within the unit circle, $\log B^*(z)$ is analytic within the unit circle. Consequently, $\log B^*(z)$ has a power series representation within the unit circle, which, as $|z|$ approaches one, we suppose converges to

$$\log B(\omega)=\beta_0+2\sum_{t=1}^{\infty}\beta_t\,e^{-i\omega t}=\beta_0+2\sum_{t=1}^{\infty}\beta_t\,\cos\omega t-2i\sum_{t=1}^{\infty}\beta_t\,\sin\omega t \quad (24)$$

where we have let $z=\exp(-i\omega)$. Let us now turn our attention to the power spectrum $\Phi(\omega)$. The following conditions on the power spectrum must be satisfied: (a) $\Phi(\omega)$ must be non-zero almost everywhere on the interval $(-\pi,\pi)$; (b) the integral of $\Phi(\omega)$ over the interval $(-\pi,\pi)$ must be finite; and (c) the integral of $\log\Phi(\omega)$ over the interval $(-\pi,\pi)$ must not diverge to minus infinity. Condition (c) is Kolmogorov's condition (9). Under these conditions, $\log\sqrt{\Phi(\omega)}$, which is an even real function of ω, may be expanded in a real, symmetric Fourier cosine series

$$\log\sqrt{\Phi(\omega)}=\tfrac{1}{2}\log\Phi(\omega)=\sum_{t=-\infty}^{\infty}\delta_t\,\cos\omega t=\delta_0+2\sum_{t=1}^{\infty}\delta_t\,\cos\omega t \quad (25)$$

where the Fourier coefficients δ_t are given by

$$\delta_t=\delta_{-t}=\frac{1}{2\pi}\int_0^{\pi}\cos\omega t\,\log\Phi(\omega)\,d\omega . \quad (26)$$

By taking the logarithm of each side of $B(\omega)=\sqrt{\Phi(\omega)}\,\exp[-i\theta(\omega)]$ and utilizing equation (25), we have

$$\log B(\omega)=\log\sqrt{\Phi(\omega)}-i\theta(\omega)=\delta_0+2\sum_{t=1}^{\infty}\delta_t\,\cos\omega t-i\theta(\omega) . \quad (27)$$

Now equation (24) gives an expression for $\log B(\omega)$ which was derived from the minimum phase-lag condition which is that $\log B^*(z)$ be analytic within the unit circle, whereas equation (27) gives an expression for $\log B(\omega)$ derived from the knowledge that the gain $|B(\omega)|$ be equal to $\sqrt{\Phi(\omega)}$. Setting these two equations equal to each other, we find that $\delta_t=\beta_t$. Thus the minimum phase-lag is given in terms of the power spectrum by

$$\theta(\omega)=2\sum_{t=1}^{\infty}\delta_t\,\sin\omega t=\frac{1}{\pi}\sum_{t=1}^{\infty}\sin\omega t\int_0^{\pi}\cos ut\,\log\Phi(u)\,du \quad (28)$$

which is the Hilbert transform of $\log\sqrt{\Phi(\omega)}$. Since $B^*(z)=\exp[\log B^*(z)]$, we have by letting $z = \exp(-i\omega)$ in equation (24) that

$$B^*(z) = \sum_{s=0}^{\infty} b_s z^s = \exp\left[\delta_0 + 2\sum_{t=1}^{\infty} \delta_t z^t\right], \quad |z| < 1. \tag{29}$$

By means of this equation we may solve for the desired minimum-delay operator b_s in terms of the δ_t given by equation (26). Letting $z = \exp(-i\omega)$ in $B^*(z)$, given by equation (29), we obtain $B(\omega)$ for which $\Phi(\omega) = B(\omega) B(-\omega)$. Thus the spectrum has been factored into two conjugate factors, $B(\omega)$ and $B(-\omega)$, where $B(\omega)$ has absolute gain $\sqrt{\Phi(\omega)}$ and minimum phase-lag $\theta(\omega)$. We shall refer to the minimum-delay transfer function $B(\omega)$ given by equation (21) as the Wold-Kolmogorov factor of the power spectrum, or briefly the W–K factor.

Let us now outline the proof of the Wold (1938) decomposition theorem. For a non-singular stationary process x_t we define the prediction error $\sigma_1 \varepsilon_t$ to be $\sigma_1 \varepsilon_t = x_t - \hat{x}_t$ where $\hat{x}_t$, given by equation (17) for $\alpha = 1$, is the best linear prediction (in the sense of the principle of least squares) of x_t from all the past values $x_{t-1}, x_{t-2}, \ldots$, and σ_1 is the positive square root of σ_1^2, the minimum mean square error (18). Equivalently, we may express this ε_t in terms of the one-sided operator (3) where $a_0 = \sigma_1^{-1}$, $a_1 = -\sigma_1^{-1} k_0$, $a_2 = -\sigma_1^{-1} k_1$, and so forth. Because of correlation properties of prediction errors (or regression residuals), Wold (1938, Theorem 6) shows that $E(x_{t-s}\varepsilon_t) = 0$ for $s > 0$, and consequently $E(\varepsilon_t \varepsilon_s) = 0$ for $t \neq s$. Thus ε_t forms an orthonormal set, and we may write the best linear regression of x_t from $\varepsilon_t, \varepsilon_{t-1}, \ldots$ as the sum of a Fourier series u_t and a regression residual v_t,

$$x_t = u_t + v_t, \quad u_t = \sum_{s=0}^{\infty} b_s \varepsilon_{t-s}, \quad \sum_{s=0}^{\infty} b_s^2 < \infty, \quad b_s = E(x_t \varepsilon_{t-s}), \quad b_0 > 0. \tag{30}$$

Because of the correlation properties of regressions residuals, we have $E(\varepsilon_t v_s) = E(u_t v_s) = 0$. Now the operators a_s and b_s are inverse to each other; so by substituting (30) into (3) we have $a_0 v_t + a_1 v_{t-1} + a_2 v_{t-2} + \ldots = 0$, which states that v_t is perfectly predictable from its past, and so the time series v_t is deterministic. Equation (30) represents the Predictive Decomposition Theorem of Herman Wold (1938, Theorem 7). This decomposition tells us that every non-deterministic stationary process may be decomposed into a purely non-deterministic component u_t plus a deterministic component v_t. By a purely non-deterministic component we mean a non-deterministic process with no deterministic component. The spectral distribution of the u_t process is the absolutely continuous component of the spectral distribution $\Lambda(\omega)$ of the x_t process. The spectral distribution of the v_t process is the sum of the step-function component and the continuous, non-absolutely-continuous component, of $\Lambda(\omega)$.

It is possible to put the decomposition (30) in analytic form. Wold (1938) showed that the b_s of the decomposition (30) in the case of the autoregressive process and the process of finite moving averages, both of which have rational spectral densities, may be found by his method of spectral factorization. Kolmogorov (1941) showed that for any stationary process with spectral distribution $\Lambda(\omega)$ the b_s of the decomposition may be found by his more general method of factorization, provided Kolmogorov's condition (9) is fulfilled for $\Lambda'(\omega)$. In our presentation of Kolmogorov's factorization method we tacitly assumed that $\Lambda(\omega)$ was absolutely continuous and utilized the spectral density $\Phi(\omega)$. In the case $\Lambda(\omega)$ is not absolutely continuous, the spectral density is not defined; so we should replace $\Phi(\omega)$ by $\Lambda'(\omega)$ in equations (25), (26), and (29). The desired factorization is then $\Lambda'(\omega) = B(\omega) B(-\omega)$, and from equation (29) the mean square prediction error, or prediction variance, is

$$\sigma_1^2 = b_0^2 = \exp\left(\frac{1}{2\pi} \int_{-\pi}^{\pi} \log \Lambda'(\omega)\,\mathrm{d}\omega\right), \qquad \alpha = 1 \tag{31}$$

(Kolmogorov, 1941; Wiener, 1942). Thus $\Lambda'(\omega)$ does not fulfill Kolmogorov's condition (9), $\sigma_1 = 0$, and the process is deterministic. As a result, a process is non-deterministic if and only if $\Lambda'(\omega) > 0$ almost everywhere for $(-\pi, \pi)$ and $\Lambda'(\omega)$ satisfies Kolmogorov's condition (9). In addition the decomposition (30) is unique, in that there is only one sequence of constants $b_0, b_1, b_2, \ldots$ and only one sequence of random variables ε_t satisfying the stated conditions.

In what follows we wish to consider stationary processes with absolutely continuous spectral distributions. We shall tacitly assume that Kolmogorov's condition (9) holds for $\Phi(\omega)$, so that these processes are purely non-deterministic. Thus we may replace the operator coefficients c_t in the moving summation (20) by the minimum-delay operator coefficients b_t found by the Wold-Kolmogorov factorization of the power spectrum. We thus obtain an expression for the purely non-deterministic time series x_t as the one-sided moving summation

$$x_t = \sum_{s=0}^{\infty} b_s \varepsilon_{t-s} = \sum_{s=-\infty}^{t} \varepsilon_s b_{t-s} = b_0 \varepsilon_t + b_1 \varepsilon_{t-1} + b_2 \varepsilon_{t-2} + \ldots\ . \tag{32}$$

This equation renders the time series x_t in terms of a minimum-delay operator $b_0, b_1, b_2, \ldots$ operating on the present and past values $\varepsilon_t, \varepsilon_{t-1}, \varepsilon_{t-2}, \ldots$ of a realization of an orthonormal process. That is, the value of x_t is expressed in terms of the present value ε_t and past values $\varepsilon_{t-1}, \varepsilon_{t-2}, \ldots$ but no future values $\varepsilon_{t+1}, \varepsilon_{t+2}, \ldots$.

Let us now consider the Predictive Decomposition Theorem (32) in the language of the engineer. A stationary time-series may be regarded as the output of a realizable and stable electric or mechanical network or filter with transfer function $B(\omega)$ which is the W–K factor of the spectrum. This filter has minimum phase-lag and absolute gain equal to $\sqrt{\Phi(\omega)}$. The minimum-delay operator b_t is the impulsive response of the filter. The input to the filter is ε_t $(-\infty < t < \infty)$, which may be considered to be the mutually uncorrelated impulses of wide-band resistance noise or "white" noise. In other words, the time series x_t $(-\infty < t < \infty)$ is the output of the minimum-delay filter in response to the white noise input ε_s $(-\infty < s \leq t)$. That is, ε_s may be regarded as an impulse of strength ε_s, which will produce a response $\varepsilon_s b_{t-s}$ at the subsequent time t. By adding the contributions of all the impulses ε_s $(-\infty < s \leq t)$, we obtain the total response, which is the time series x_t given by the Wold decomposition (32). Since the input ε_t is an orthonormal process, let us note that its power spectrum is equal to one and its autocorrelation vanishes except for lag zero. Then the W–K factorization, $B(\omega)B(-\omega) = \Phi(\omega)$, states that the power spectrum of the input ε_t multiplied by the power transfer function $B(\omega)B(-\omega)$ of the filter yields the power spectrum $\Phi(\omega)$ of the output x_t. Moreover, since the transfer function $B(\omega)$ is the Fourier transform of the realizable, stable linear operator b_t, we may call $B(\omega)$ the (complex) phase and amplitude spectrum of the damped transient time function b_t. The power transfer function $B(\omega)B(-\omega)$ may then be called the energy spectrum of the transient b_t. *Thus the W–K factorization $B(\omega)B(-\omega) = \Phi(\omega)$ states that the energy spectrum of the realizable, stable transient b_t is equal to the power spectrum of the time series x_t.* (See italicized theorem in section 1.5 of Wiener (1942).) In the time domain, the W–K factorization becomes

$$\sum_{t=0}^{\infty} b_{t+\tau} b_t = E[x_{t+\tau} x_t] = \phi(\tau) \tag{33}$$

which states that the autocorrelation of the transient b_t is equal to the autocorrelation of the time series x_t.

Let us now examine the predictive decomposition of the time series x_t with a rational power spectrum (23). The output x_t is obtained by passing white noise ε_t through the filter with transfer function $B(\omega) = G(\omega)/H(\omega)$. This linear operation is equivalent to first passing the white noise ε_t through the filter $G(\omega)$ and then passing the output of the $G(\omega)$ filter through the filter $1/H(\omega)$. Let $f_0, f_1, f_2, \ldots$ be the impulse response of the $1/H(\omega)$ filter; that is, $f_0, f_1, f_2, \ldots$ is the minimum-delay inverse operator to the minimum-delay operator $h_0, h_1, \ldots, h_N$. Then the predictive decomposition (32) becomes

$$x_t = \sum_{s=0}^{\infty} b_s \varepsilon_{t-s} = \sum_{s=0}^{\infty} f_s \sum_{r=0}^{\infty} g_r \varepsilon_{t-r-s} = \sum_{s=0}^{\infty} \left(\sum_{r=0}^{M} g_r f_{s-r} \right) \varepsilon_{t-s} \tag{34}$$

which may be written

$$\sum_{s=0}^{N} h_s x_{t-s} = \sum_{s=0}^{M} g_s \varepsilon_{t-s} \tag{35}$$

where ε_t is the orthonormal input and the time series x_t with spectrum (23) is the output. By setting $h_0 = 1$ and the other h's equal to zero, equation (35) represents the predictive decomposition of the process of moving averages. On the other hand, by setting $g_0 = 1$ and the other g's equal to zero, equation (35) represents the inverse decomposition of the autoregressive process.

Let us now derive the explicit prediction formula for purely non-deterministic stationary time series. Let t be the present time, and $\alpha > 0$ be the prediction distance. The value of $x_{t+\alpha}$ is given by the predictive decomposition

$$x_{t+\alpha} = (b_0 \varepsilon_{t+\alpha} + b_1 \varepsilon_{t+\alpha-1} + \ldots + b_{\alpha-1} \varepsilon_{t+1}) + (b_\alpha \varepsilon_t + b_{\alpha+1} \varepsilon_{t-1} + \ldots) . \tag{36}$$

At the present time t, the present and past values $x_t, x_{t-1}, x_{t-2}, \ldots$ are known. Consequently, the values $\varepsilon_t, \varepsilon_{t-1}, \varepsilon_{t-2}, \ldots$ at and prior to time t may be obtained by the inverse linear operator a_t of equation (14). Thus the component $(b_\alpha \varepsilon_t + b_{\alpha+1} \varepsilon_{t-1} + \ldots)$ of equation (36) can be computed at time t, and in fact this component is the optimum least squares prediction, $\hat{x}_{t+\alpha}$, of equation (17). Explicitly, we may write equation (17) as

$$\hat{x}_{t+\alpha} = \sum_{s=0}^{\infty} b_{\alpha+s} \varepsilon_{t-s} = \sum_{s=0}^{\infty} b_{\alpha+s} \sum_{n=0}^{\infty} a_n x_{t-n-s} = \sum_{r=0}^{\infty} \left[\sum_{s=0}^{\infty} b_{\alpha+s} a_{r-s} \right] x_{t-r} \tag{37}$$

by utilizing the a_s operator of equation (14). Comparing this with equation (17), we see that the explicit formula for the prediction operation coefficients is

$$k_r = \sum_{s=0}^{\infty} b_{\alpha+s} a_{r-s} . \tag{38}$$

By computing the transfer function $K(\omega)$ of the operator k_r of equation (38), we obtain Wiener's well-known formula

$$K(\omega) = \sum_{r=0}^{\infty} k_r\, e^{-i\omega r} = \frac{\displaystyle\sum_{s=0}^{\infty} b_{\alpha+s}\, e^{-i\omega s}}{\displaystyle\sum_{s=0}^{\infty} b_s\, e^{-i\omega s}} =$$

$$= \frac{1}{2\pi B(\omega)} \sum_{s=0}^{\infty} e^{-i\omega s} \int_{-\pi}^{\pi} B(u)\, e^{iu(\alpha+s)}\, du \tag{39}$$

for the filter characteristics of the optimum predictor. The other component of the decomposition (36), $(b_0 \varepsilon_{t+\alpha} + \dots + b_{\alpha-1}\varepsilon_{t+1})$, involves future values of the white noise and hence cannot be computed at time t.

This component is the prediction error for the prediction distance α, and its mean square value, $(b_0^2 + b_1^2 + \dots + b_{\alpha-1}^2)$, is the minimum mean square value σ_α^2 of equation (18).

Let us now consider the filtering problem, that is, separation of message and noise. The predictive decomposition of the time series x_t consisting of message m_t plus noise n_t is

$$x_t = m_t + n_t = \sum_{s=0}^{\infty} b_s \varepsilon_{t-s}. \tag{40}$$

Since the message has an absolutely continuous spectral distribution function, it may be represented by the process of moving summation given by

$$m_t = \sum_{s=-\infty}^{\infty} q_s \varepsilon_{t-s} + \sum_{s=-\infty}^{\infty} r_s \gamma_{t-s} \tag{41}$$

where ε_t and γ_t each represent an orthonormal sequence of random variables, and $E(\varepsilon_t \gamma_s) = 0$. Let $\Phi(\omega)$ be the power spectrum of x_t, $\Phi_{11}(\omega)$ be the power spectrum of the message, $\Phi_{12}(\omega)$ be the cross-spectrum of message and noise, and $\Phi_{22}(\omega)$ be the power spectrum of the noise. Using equation (40), we have

$$\Phi(\omega) = \Phi_{11}(\omega) + \Phi_{12}(\omega) + \Phi_{21}(\omega) + \Phi_{22}(\omega) = |B(\omega)|^2 \tag{42}$$

where $B(\omega)$ is the W–K factor of $\Phi(\omega)$. Letting $Q(\omega)$ and $R(\omega)$ be the transfer functions of the two-sided operators $q_s\,(-\infty < s < \infty)$ and $r_s\,(-\infty < s < \infty)$ respectively, it follows from equations (40) and (41) that $\Phi_{11}(\omega) = Q\bar{Q} + R\bar{R}$ and $\Phi_{12}(\omega) = B\bar{Q} - Q\bar{Q} - R\bar{R}$. Consequently, we have $B\bar{Q} = \Phi_{12}(\omega) + \Phi_{11}(\omega)$, so that

$$q_t = \frac{1}{2\pi} \int_{-\pi}^{\pi} \frac{\Phi_{11}(\omega) + \Phi_{21}(\omega)}{B(\omega)}\, e^{i\omega t}\, d\omega. \tag{43}$$

Following our usual notation, b_t is the minimum-delay operator obtained from the W–K factor $B(\omega)$, and a_t is the inverse, minimum-delay operator with transfer function $A(\omega) = 1/B(\omega)$. The operator a_t is obtained from b_t by means of equation (13). Since the present and past values, x_t, x_{t-1}, x_{t-2}, ... of the time series are known at time t, we may obtain the present and past values ε_t, ε_{t-1}, ε_{t-2}, ... by means of the a_s operator (14). By considering the moving summation (41) at the time $t + \alpha$, we see that the predictable part of the message $m_{t+\alpha}$, where α is the prediction distance or lead, is

$$\hat{m}_{t+\alpha} = \sum_{s=0}^{\infty} q_{s+\alpha}\varepsilon_{t-s} = \sum_{s=0}^{\infty} q_{s+\alpha} \sum_{n=0}^{\infty} a_n x_{t-n-s} = \sum_{r=0}^{\infty} \left[\sum_{s=0}^{\infty} q_{s+\alpha} a_{r-s} \right] x_{t-r} . \quad (44)$$

The non-predictable part, or filtering error, is $m_{t+\alpha} - \hat{m}_{t+\alpha}$, which has mean square value

$$E[(m_{t+\alpha} - \hat{m}_{t+\alpha})^2] = \sum_{s=-\infty}^{\alpha} q_s^2 + \sum_{s=-\infty}^{\infty} r_s^2 \quad (45)$$

which is a minimum. In equation (45) we see that the first term on the right depends on the lag $-\alpha$, whereas the second term does not. Thus the optimum linear operator in the sense of the principle of least squares to be used in filtering a non-deterministic time series x_t has coefficients h_r given by the expression in brackets on the right-hand side of equation (44); that is

$$h_r = \sum_{s=0}^{\infty} q_{s+\alpha} a_{r-s} . \quad (46)$$

The transfer function of the desired filter is then

$$H(\omega) = \sum_{r=0}^{\infty} h_r e^{-i\omega r} = \frac{\displaystyle\sum_{s=0}^{\infty} q_{s+\alpha} e^{-i\omega s}}{\displaystyle\sum_{s=0}^{\infty} b_s e^{-i\omega s}} =$$

$$= \frac{1}{2\pi B(\omega)} \sum_{s=0}^{\infty} e^{-i\omega s} \int_{-\pi}^{\pi} \frac{\Phi_{11}(u) + \Phi_{21}(u)}{B(u)} e^{iu(s+\alpha)} \, du .$$

2. Statistical Description of Time-Series

To formulate the concept of a statistical description of a time-series, we must consider in a conceptual and abstract way a great number of similar mechanisms, each of which generates a time-series. Such a group of mechanisms is called an *ensemble*. Let us denote the time-series generated by the various members of the ensemble by

$$x_t^{(1)}, x_t^{(2)}, x_t^{(3)}, \ldots, x_t^{(i)}, x_t^{(i+1)}, \ldots$$

where $x_t^{(i)}$ is the time-series generated by the ith member. The random character of the time-series is exhibited in this way: Although the mechanisms are similar, the value of the time-series $x_t^{(i)}$ generated by the ith member of of the ensemble at any specified time t is generally different from the value of the time-series $x_t^{(j)}$ generated by the jth member at the same time instant t. Here i and j represent any two different members of the ensemble. First let us consider for what fraction of the total number of mechanisms does $x_t^{(i)}$ (for a fixed value of t and for all values of i) lie in a given range $-\infty$ to u.

This fraction will depend upon u and t and, in fact, represents the probability that x_t (for some unspecified mechanism) will fall between $-\infty$ and u at time t. This probability may be denoted by

$$Pr\{-\infty < x_t \le u\}\ .$$

Hence for a fixed value of t and for an unspecified mechanism in the ensemble, x_t may be regarded as a random variable, whose distribution function $F_t(u)$, which is defined by

$$F_t(u) = Pr\{-\infty < x_t \le u\}\ ,$$

depends upon the fixed value of t. The distribution function $F_t(u)$ is called the *first probability distribution*. Next, for two fixed time instants t_1 and t_2, let us consider the pair of values

$$(x_{t_1}^{(i)}, x_{t_2}^{(i)})$$

of the time series generated by the ith mechanism. Then the fraction of the total number of pairs in the ensemble for which both $x_{t_1}^{(i)}$ lies in the range from $-\infty$ to u_1 and $x_{t_2}^{(i)}$ lies in the range from $-\infty$ to u_2 depends upon the fixed times t_1 and t_2 as well as u_1, u_2. This fraction represents the probability

$$F_{t_1,t_2}(u_1, u_2) = Pr\{-\infty < x_{t_1} \le u_1, -\infty < x_{t_2} \le u_2\}$$

of the two-dimensional random variable (x_{t_1}, x_{t_2}) where the time instants t_1 and t_2 are fixed but the particular mechanism in the ensemble is left unspecified. The distribution function $F_{t_1,t_2}(u_1, u_2)$ is called the *second probability distribution*. In a similar way, higher probability distribution functions may be constructed, the nth *distribution function* being denoted by

$$F_{t_1,t_2,\ldots,t_n}(u_1, u_2, \ldots, u_n),$$

where the order n of the distribution is a finite positive integer. These probability distributions for every finite set $t_1, t_2, \ldots, t_n$ embody all the information about the statistical properties of the family of random variables x_t. Such a family of random variables is called a *random process*, or in more sophisticated language a *stochastic process*.

To summarize our argument to this point, we may say that a *stochastic process*, denoted by x_t, is characterized by an ensemble of mechanisms. All the mechanisms in the ensemble are conceptually similar, but due to unspecified reasons the time-series generated by the various mechanisms are not the same but show statistical differences. Thus we cannot consider a time-series generated by one of the mechanisms as representative of the stochastic process, but must consider collectively the time-series generated by all the mechanisms. To do this, we let x_t represent a hypothetical time-series generated by an unspecified mechanism in the ensemble, and describe x_t by means of probability statements derived from the entire ensemble. These probability statements are summarized in the first probability distribution, the second probability distribution, the third probability distribution, and so on for every finite order, and the stochastic process x_t is the family of random variables described by these probability distributions.

On the other hand, a *time-series*, denoted by $x_t^{(i)}$, is the particular time-function generated by a single mechanism of the ensemble. Hence a time-series $x_t^{(i)}$ may be said to be a particular *realization* of the stochastic process x_t; more specifically the realization (i.e. time-function) generated by the ith mechanism. It is not unusual that only one time-series can ever be observed for a given stochastic process. Nevertheless, the basic idea of regarding a time-series as a realization of a stochastic process is a valuable one.

Before closing this section, let us point out that, in order for the distribution functions which characterize a stochastic process to be consistent with each other, they must satisfy

(1) *the permutation condition*, i.e. for any permutation $i_1, i_2, \ldots, i_n$ of the numbers $1, 2, \ldots, n$,

$$F_{t_{i_1},t_{i_2},\ldots,t_{i_n}}(u_{i_1}, u_{i_2}, \ldots, u_{i_n}) = F_{t_1,t_2,\ldots,t_n}(u_1, u_2, \ldots, u_n)$$

(2) *the nesting condition*, i.e. for $m < n$, and for any $t_{m+1}, \ldots, t_n$,

$$F_{t_1,t_2,\ldots,t_m,t_{m+1},\ldots,t_n}(u_1, u_2, \ldots, u_m, \infty, \ldots, \infty) =$$
$$= F_{t_1,t_2,\ldots,t_m}(u_1, u_2, \ldots, u_m).$$

Conversely, any system of distribution functions satisfying the above two conditions may be considered as defining a stochastic process.

In this section, we have characterized an arbitrary stochastic process, which may be either stationary or non-stationary, by a system of distribution functions.

3. Autoregressive Decomposition of Stochastic Processes

Let x_t be a stochastic process, where t is an integer denoting time. We do not assume that x_t is stationary. We assume that the first moment, or mean, of x_t is zero for all time t; that is,

$$E\{x_t\} = \int_{-\infty}^{\infty} u \, dF_t(u) = 0 .$$

Moreover we assume that the second moment, or variance, of x_t is finite and bounded for all time t; that is

$$\operatorname{var} x_t = E\{x_t^2\} = \int_{-\infty}^{\infty} u^2 \, dF_t(u) < \text{constant} .$$

We let $\phi(t, s)$ denote the covariance of x_t and x_s; that is

$$\phi(t,s) = E\{x_t x_s\} = \int_{-\infty}^{\infty} \int_{-\infty}^{\infty} uv \, dF_{t,s}(u;v) .$$

In particular, we note that the variance of x_t is

$$\operatorname{var} x_t = \phi(t, t) .$$

These variances are generally different for different values of t.

Let us now form the *least-squares linear regression* of x_t on the regressors $x_{t-1}, x_{t-2}, \ldots, x_{t-n}$. That is, we write the regression equation

$$x_t = c_1 x_{t-1} + c_2 x_{t-2} + \ldots + c_n x_{t-n} + \varepsilon_t$$

where the regression coefficients $c_1, c_2, \ldots, c_n$ are determined so that the variance $\operatorname{var} \varepsilon_t = E\{\varepsilon_t^2\}$ of the regression residual ε_t is a minimum.

The regression coefficients may be found as follows. Write

$$E\{\varepsilon_t^2\} = E\{(x_t - c_1 x_{t-1} - c_2 x_{t-2} - \ldots - c_n x_{t-n})^2\} .$$

Then the minimum of $E\{\varepsilon_t^2\}$ is found by setting its partial derivatives with respect to $c_1, c_2, \ldots, c_n$ equal to zero. We thus obtain the n equations (called the normal equations):

$$\frac{\partial E\{\varepsilon_t^2\}}{\partial c_1} = E\{2(x_t - c_1 x_{t-1} - c_2 x_{t-2} - \ldots - c_n x_{t-n})(-x_{t-1})\} = 0$$

$$\ldots \tag{47}$$

$$\frac{\partial E\{\varepsilon_t^2\}}{\partial c_n} = E\{2(x_t - c_1 x_{t-1} - c_2 x_{t-2} - \ldots - c_n x_{t-n})(-x_{t-n})\} = 0,$$

or

$$E\{\varepsilon_t x_{t-1}\} = 0$$

$$\ldots \tag{48}$$

$$E\{\varepsilon_t x_{t-n}\} = 0 .$$

Simplifying the normal equations as given in (47) we find:

$$E\{x_{t-1} x_t\} = c_1 E\{x_{t-1} x_{t-1}\} + c_2 E\{x_{t-1} x_{t-2}\} + \ldots + c_n E\{x_{t-1} x_{t-n}\}$$

$$\ldots \tag{49}$$

$$E\{x_{t-n} x_t\} = c_1 E\{x_{t-n} x_{t-1}\} + c_2 E\{x_{t-n} x_{t-2}\} + \ldots + c_n E\{x_{t-n} x_{t-n}\} ,$$

or

$$\phi(t-1, t) = c_1 \phi(t-1, t-1) + c_2 \phi(t-1, t-2) + \ldots + c_n \phi(t-1, t-n)$$

$$\ldots \tag{50}$$

$$\phi(t-n, t) = c_1 \phi(t-n, t-1) + c_2 \phi(t-n, t-2) + \ldots + c_n \phi(t-n, t-n) .$$

By solving these n normal equations, we obtain the n unknown coefficients $c_1, c_2, \ldots, c_n$. Moreover the *minimum value* of variance of the regression residual is given by

$$E\{\varepsilon_t^2\} = E\{\varepsilon_t(x_t - c_1 x_{t-1} - \ldots - c_n x_{t-n})\}$$

$$= E\{\varepsilon_t x_t\} - c_1 E\{\varepsilon_t x_{t-1}\} - \ldots - c_n E\{\varepsilon_t x_{t-n}\} .$$

Using the normal equations as given by (48) we thus have

$$E\{\varepsilon_t^2\} = E\{\varepsilon_t x_t\} .$$

Substituting for ε_t, we find

$$E\{\varepsilon_t^2\} = E\{(x_t - c_1 x_{t-1} - \ldots - c_n x_{t-n})x_t\} =$$

$$= E\{x_t^2\} - c_1 E\{x_t x_{t-1}\} - \ldots - c_n E\{x_{t-n} x_t\}$$

or

$$\operatorname{var} \varepsilon_t = \phi(t, t) - c_1 \phi(t, t-1) - \ldots - c_n \phi(t-n, t) .$$

Clearly, the coefficients $c_1, c_2, \ldots, c_n$ just found, as well as the corresponding regression residual

$$\varepsilon_t = x_t - c_1 x_{t-1} - c_2 x_{t-2} - \ldots - c_n x_{t-n}$$

depend upon the number n. To indicate this dependence we shall now use the more explicit notation

$$c_1(n), c_2(n), \ldots, c_n(n), \varepsilon_t(n)$$

for these coefficients and the residual.

Let us now compare the least-squares regression of x_t on the regressors $x_{t-1}, x_{t-2}, \ldots, x_{t-n}$ given by

$$x_t = c_1(n)x_{t-1} + c_2(n)x_{t-2} + \ldots + c_n(n)x_{t-n} + \varepsilon_t(n) ,$$

with the least-squares regression of x_t on the regressors $x_{t-1}, x_{t-2}, \ldots, x_{t-n}$, x_{t-n-1} given by

$$x_t = c_1(n+1)x_{t-1} + c_2(n+1)x_{t-2} + \ldots + c_n(n+1)x_{t-n} + c_{n+1}(n+1)x_{t-n-1} + \varepsilon_t(n+1) .$$

Clearly, by adding the regressor x_{t-n-1} the variance of the residual cannot increase, that is

$$\text{var } \varepsilon_t(n) \geq \text{var } \varepsilon_t(n+1) .$$

Letting $n = 1, 2, 3, \ldots$ we thus obtain

$$\text{var } \varepsilon_t(1) \geq \text{var } \varepsilon_t(2) \geq \text{var } \varepsilon_t(3) \geq \text{var } \varepsilon_t(4) \geq \ldots \geq 0 .$$

Hence var $\varepsilon_t(n)$ tends to a limiting value as n tends to infinity. Hence the residual $\varepsilon_t(n)$ tends to a limiting random variable that we denote by ε_t, that is

$$\varepsilon_t(n) \to \varepsilon_t \quad \text{as} \quad n \to \infty .$$

Also as $n \to \infty$, the coefficients

$$c_1(n), c_2(n), \ldots, c_n(n)$$

tend to a limiting infinite sequence[3]

$$c_1, c_2, c_3, \ldots$$

That is, by letting $n \to \infty$, we obtain the *least-squares regression of x_t on the*

---

[3] This infinite sequence in certain cases may have to be interpreted as a generalized function (Robinson, 1962, 1963).

infinite set of regressors $x_{t-1}, x_{t-2}, \ldots$; that is

$$x_t = c_1 x_{t-1} + c_2 x_{t-2} + \ldots + \varepsilon_t .$$

If we think of t as being the present time, then $x_{t-1}, x_{t-2}, \ldots$ represent the past of the stochastic process. Because the coefficients $c_1, c_2, \ldots$ depend upon the value of t, it is convenient to indicate this dependence by introducing new coefficients $a_{t0}, a_{t1}, a_{t2}, \ldots$ defined by the relations

$$a_{t0} = 1 , \quad a_{t1} = -c_1 , \quad a_{t2} = -c_2, \ldots$$

so that the least-squares regression

$$x_t = c_1 x_{t-1} + c_2 x_{t-2} + \ldots + \varepsilon_t$$

may be written instead as

$$\varepsilon_t = x_t + a_{t1} x_{t-1} + a_{t2} x_{t-2} + \ldots \tag{51}$$

This equation represents the *autoregressive decomposition* of the stochastic process x_t.

For $n \to \infty$ we note that the normal equations (48) become the infinite set of equations

$$E\{\varepsilon_t x_{t-1}\} = 0 , \quad E\{\varepsilon_t x_{t-2}\} = 0, \ldots \tag{52}$$

which say that the regression residual ε_t is uncorrelated with the entire past $x_{t-1}, x_{t-2}, \ldots$. Let us now consider the least-squares regression for time $t-1$. It is

$$\varepsilon_{t-1} = x_{t-1} + a_{t-1,1} x_{t-2} + \ldots$$

which shows that ε_{t-1} is a linear combination of $x_{t-1}, x_{t-2}, \ldots$. But as we have just seen ε_t is uncorrelated with each of $x_{t-1}, x_{t-2}, \ldots$. Therefore ε_t is uncorrelated with ε_{t-1}, that is

$$E\{\varepsilon_t \varepsilon_{t-1}\} = 0 .$$

Similarly it follows that ε_t is uncorrelated with ε_s provided $t \neq s$, that is

$$E\{\varepsilon_t \varepsilon_s\} = 0 \quad \text{for} \quad t \neq s . \tag{53}$$

The random variables ε_t for all values of t thus form a stochastic process with this uncorrelated property. The ε_t process is called the *innovation process*, and ε_t for a given value of t is called the *innovation* at time t in the x_t process.

4. Moving Average Composition of Stochastic Processes

In the foregoing section we decomposed the stochastic process x_t by the least-squares autoregression (51), and thereby obtained the innovation process ε_t which had the uncorrelated property (53). Our purpose in this section is to compose the x_t process, as well as we can by least-squares regression, from knowledge of the innovation process ε_t.

To do so, we form the *least-squares linear regression* of x_t on the regressors $\varepsilon_t, \varepsilon_{t-1}, \ldots, \varepsilon_{t-n}$. That is, we write the regression equation

$$x_t = b_0 \varepsilon_t + b_1 \varepsilon_{t-1} + \ldots + b_n \varepsilon_{t-n} + v_t ,$$

where the regression coefficients $b_0, b_1, \ldots, b_n$ are determined so that the variance var $v_t = E\{v_t^2\}$ of the regression residual v_t is a minimum. We find that the *normal equations* are

$$
\begin{aligned}
E\{v_t \varepsilon_t\} &= 0 \\
E\{v_t \varepsilon_{t-1}\} &= 0 \\
&\cdots \\
E\{v_t \varepsilon_{t-n}\} &= 0
\end{aligned}
\tag{54}
$$

or

$$E\{(x_t - b_0 \varepsilon_t - b_1 \varepsilon_{t-1} - \ldots - b_n \varepsilon_{t-n})\varepsilon_{t-s}\} = 0 \quad \text{for} \quad s = 0, 1, 2, \ldots, n .$$

Because of the uncorrelated property (53) of the ε_t process, the normal equations reduce to

$$E\{(x_t \varepsilon_{t-s}\} - b_s E\{\varepsilon_{t-s}^2\} = 0 \quad \text{for} \quad s = 0, 1, 2, \ldots, n$$

which give

$$b_s = \frac{E\{x_t \varepsilon_{t-s}\}}{E\{\varepsilon_{t-s}^2\}} \quad \text{for} \quad s = 0, 1, 2, \ldots, n . \tag{55}$$

This equation shows that b_s does not depend upon n (due to the uncorrelated property of the regressors), and, as we expect, b_s does depend upon t. Because it is convenient to indicate this dependence upon t, we introduce the new notation

$$b_{t1}, b_{t2}, \ldots, b_{tn} , \quad \text{and} \quad v_t(n)$$

instead of just

$$b_1, b_2, \ldots, b_n , \quad \text{and} \quad v_t$$

so the least-squares regression equation is

$$x_t = b_{t0} \varepsilon_t + b_{t1} \varepsilon_{t-1} + \ldots + b_{tn} \varepsilon_{t-n} + v_t(n) .$$

As before, we see that

$$\text{var } v_t(0) \geq \text{var } v_t(1) \geq \text{var } v_t(2) \geq \ldots \geq 0 \,,$$

so that $v_t(n)$ tends to a limiting random variable v_t as n tends to infinity. We thus obtain the *least-squares regression* of x_t on the infinite set of regressors $\varepsilon_t, \varepsilon_{t-1}, \ldots$; that is,

$$x_t = b_{t0}\varepsilon_t + b_{t1}\varepsilon_{t-1} + \ldots + v_t \,. \tag{56}$$

This equation represents the *moving-average composition* of the stochastic process x_t. Also we note that for $n \to \infty$ the normal equations (54) become the infinite set of equations

$$E\{v_t\varepsilon_t\} = 0 \,, \quad E\{v_t\varepsilon_{t-1}\} = 0 \,, \quad E\{v_t\varepsilon_{t-2}\} = 0 \,, \ldots \tag{57}$$

Let us introduce the notation

$$u_t = b_{t0}\varepsilon_t + b_{t1}\varepsilon_{t-1} + \ldots \tag{58}$$

so that

$$x_t = u_t + v_t \,.$$

Let us now consider the covariance of $v_t(n)$ and ε_{t-s}. It is

$$E\{v_t(n)\varepsilon_{t-s}\} = E\{(x_t - b_{t0}\varepsilon_t - \ldots - b_{tn}\varepsilon_{t-n})\varepsilon_{t-s} \,.$$

By (52) we have

$$E\{x_t\varepsilon_{t-s}\} = 0 \quad \text{for} \quad s = -1, -2, \ldots \,.$$

By (53) we have

$$E\{\varepsilon_{t-r}\varepsilon_{t-s}\} = 0 \quad \text{for} \quad r \neq s \,.$$

By (55) we have

$$E\{x_t\varepsilon_{t-s}\} = b_{ts}E\{\varepsilon_{t-s}^2\} \quad \text{for} \quad s = 0, 1, \ldots, n \,.$$

We therefore have

$$E\{v_t(n)\varepsilon_{t-s}\} = 0 \quad \text{for} \quad s = \ldots, -2, -1, 0, 1, 2, \ldots, n \,.$$

Letting $n \to \infty$ it follows that

$$E\{v_t\varepsilon_{t-s}\} = 0 \quad \text{for all} \quad s \,,$$

or, in words, v_t and ε_r are uncorrelated for all t and r. It therefore follows from (58) that v_t and u_r are uncorrelated for all t and r.

Moreover it may be shown that the v_t process is deterministic.

We may summarize in the following theorem (Wold, 1938; Cramér, 1960).

THEOREM 1. A non-deterministic stochastic process x_t with bounded variance has the uniquely determined decomposition

$$x_t = u_t + v_t, \quad \text{with} \quad u_t = b_{t0}\,\varepsilon_t + b_{t1}\,\varepsilon_{t-1} + \ldots$$

where the components u_t, v_t, ε_t are stochastic processes with bounded variance that satisfy the following properties:

1. Each of the random variables u_t, v_t, ε_t is a linear combination of x_t, $x_{t-1}, x_{t-2}, \ldots$
2. The process v_t is deterministic.
3. The random variables ε_t have zero mean, $E\{\varepsilon_t\} = 0$, and are mutually uncorrelated, $E\{\varepsilon_t \varepsilon_s\} = 0$ for $t \neq s$.
4. The process v_t is uncorrelated with the processes ε_t and u_t, that is

$$E\{v_t \varepsilon_s\} = E\{v_t u_s\} = 0 \quad \text{for} \quad t = s \text{ as well as } t \neq s\,.$$

5. $b_{t0} = 1$ for all t.

5. Recursive Decomposition of Vector Stationary Stochastic processes

We now let $\begin{pmatrix} x_t \\ y_t \end{pmatrix}$ be a (2×1) vector *stationary stochastic process* so that the autocovariance matrix

$$\begin{bmatrix} E\{x_{t+s} x_t\} & E\{x_{t+s} y_t\} \\ E\{y_{t+s} x_t\} & E\{y_{t+s} y_t\} \end{bmatrix}$$

does not depend upon t. Now consider the stochastic process given by

$$\ldots, y_{t-1}, x_{t-1}, y_t, x_t, y_{t+1}, x_{t+1}, \ldots,$$

where, for subject-matter reasons, y_t causes x_t, but not vice versa. Form the least-squares regression of x_t on the preceding random variables $y_t, x_{t-1}, y_{t-1}, x_{t-2}, \ldots$, and so obtain the autoregressive decomposition denoted by

$$\varepsilon_t = x_t + a_1 y_t + a_2 x_{t-1} + a_3 y_{t-1} + a_4 x_{t-2} + \ldots. \tag{59}$$

Also form the least-squares regression of y_t on the preceding random variables $x_{t-1}, y_{t-1}, x_{t-2}, y_{t-2}, \ldots$, and so obtain the autoregressive decomposition denoted by

$$\delta_t = y_t + \alpha_2 x_{t-1} + \alpha_3 y_{t-1} + \alpha_4 x_{t-2} + \alpha_5 y_{t-2} + \ldots. \tag{60}$$

Because of the stationarity, the coefficients

$$a_1, a_2, a_3, a_4, \ldots, \quad \text{and} \quad \alpha_2, \alpha_3, \alpha_4, \alpha_5, \ldots.$$

will not depend upon t. In matrix form, we may write (59) and (60) as

$$\begin{pmatrix} \varepsilon_t \\ \delta_t \end{pmatrix} = \begin{pmatrix} 1 & a_1 \\ 0 & 1 \end{pmatrix}\begin{pmatrix} x_t \\ y_t \end{pmatrix} + \begin{pmatrix} a_2 & a_3 \\ \alpha_2 & \alpha_3 \end{pmatrix}\begin{pmatrix} x_{t-1} \\ y_{t-1} \end{pmatrix} + \begin{pmatrix} a_4 & a_5 \\ \alpha_4 & \alpha_5 \end{pmatrix}\begin{pmatrix} x_{t-2} \\ y_{t-2} \end{pmatrix} + \ldots.$$

This equation is the recursive decomposition of the vector stationary stochastic process $\begin{pmatrix} x_t \\ y_t \end{pmatrix}$. By theorem 1 the random variables ε_t and δ_t are mutually uncorrelated, that is the $\begin{pmatrix} \varepsilon_t \\ \delta_t \end{pmatrix}$ process has the noncorrelation property:

$$\begin{bmatrix} E\{\varepsilon_t \varepsilon_s\} & E\{\varepsilon_t \delta_s\} \\ E\{\delta_t \varepsilon_s\} & E\{\delta_t \delta_s\} \end{bmatrix} = \begin{cases} \begin{bmatrix} \sigma_1^2 & 0 \\ 0 & \sigma_2^2 \end{bmatrix} & \text{when} \quad t=s \\ \begin{bmatrix} 0 & 0 \\ 0 & 0 \end{bmatrix} & \text{when} \quad t \neq s \end{cases}$$

where $\sigma_1^2 = E\{\varepsilon_t^2\}$ and $\sigma_2^2 = E\{\delta_t^2\}$.

The moving average composition is

$$\begin{pmatrix} x_t \\ y_t \end{pmatrix} = \begin{pmatrix} 1 & b_1 \\ 0 & 1 \end{pmatrix}\begin{pmatrix} \varepsilon_t \\ \delta_t \end{pmatrix} + \begin{pmatrix} b_2 & b_3 \\ \beta_2 & \beta_3 \end{pmatrix}\begin{pmatrix} \varepsilon_{t-1} \\ \delta_{t-1} \end{pmatrix} + \begin{pmatrix} b_4 & b_5 \\ \beta_4 & \beta_5 \end{pmatrix}\begin{pmatrix} \varepsilon_{t-2} \\ \delta_{t-2} \end{pmatrix} + \ldots + \begin{pmatrix} v_t \\ w_t \end{pmatrix}.$$

Theorem 5 of Robinson (1963) tells us that the set of matrix coefficients

$$\begin{pmatrix} 1 & b_1 \\ 0 & 1 \end{pmatrix}, \quad \begin{pmatrix} b_2 & b_3 \\ \beta_2 & \beta_3 \end{pmatrix}, \quad \begin{pmatrix} b_4 & b_5 \\ \beta_4 & \beta_5 \end{pmatrix}, \quad \ldots$$

is minimum-delay. Suppose now that $\begin{pmatrix} \varepsilon_t' \\ \delta_t' \end{pmatrix}$ is a stochastic process with the same non-correlation property as the $\begin{pmatrix} \varepsilon_t \\ \delta_t \end{pmatrix}$ process, and suppose that

$$\begin{pmatrix} x_t \\ y_t \end{pmatrix} = \begin{pmatrix} b_0' & b_1' \\ \beta_0' & \beta_1' \end{pmatrix}\begin{pmatrix} \varepsilon_t' \\ \delta_t' \end{pmatrix} + \begin{pmatrix} b_2' & b_3' \\ \beta_2' & \beta_3' \end{pmatrix}\begin{pmatrix} \varepsilon_{t-1}' \\ \delta_{t-1}' \end{pmatrix} + \ldots + \begin{pmatrix} v_t \\ w_t \end{pmatrix}$$

is a representation of $\begin{pmatrix} x_t \\ y_t \end{pmatrix}$ in terms of the $\begin{pmatrix} \varepsilon_t' \\ \delta_t' \end{pmatrix}$, $\begin{pmatrix} \varepsilon_{t-1}' \\ \delta_{t-1}' \end{pmatrix}$, $\ldots$.

By theorem 1 (Robinson, 1963) the following inequality holds

$$1 + b_1^2 + 1 \geq \left(b_0'\right)^2 + \left(b_1'\right)^2 + \left(\beta_0'\right)^2 + \left(\beta_1'\right)^2$$

with equality if and only if

$$\begin{pmatrix} b'_0 & b'_1 \\ \beta'_0 & \beta'_1 \end{pmatrix} = \begin{pmatrix} u_{11} & u_{12} \\ u_{21} & u_{22} \end{pmatrix}\begin{pmatrix} 1 & b_1 \\ 0 & 1 \end{pmatrix}$$

where the u-matrix is an orthogonal matrix.

6. Elementary Review of Information Theory

The measure of amount of information in communication theory is called *entropy*. It is measured in *bits*. Thus we may say that the entropy of a message source is so many bits per symbol. If the source produces symbols at a constant rate, then we can say that the source has an entropy of so many bits per second.

Suppose there are only two possible symbols, say X and Y, between which the message source chooses repeatedly and independently of previous choices. Let p_0 be the probability that X is chosen, and p_1 that Y is chosen. The entropy of the message source is defined as

$$H = -p_0 \log_2 p_0 - p_1 \log_2 p_1 \quad \text{(bits per symbol)} .$$

Let us now look at Shannon's theorem concerning messages produced by an ergodic source which selects symbols independently with certain probabilities.

Consider all messages the source can produce. Let each message consist of a large number, say M, of symbols. For example, M may be 10 000 symbols. Some of these messages are more probable than others. On the average in the message, symbol no. 1 will occur about Mp_1 times, symbol no. 2 about Mp_2 times, etc. The source might produce other sorts of messages, for example, a message consisting of symbol no. 1 repeated M times, or a message in which the numbers of the various symbols differ a great deal from M times their probabilities.

Shannon's theorem states: If H is the entropy of the source per symbol, then there are just about 2^{MH} probable messages, and the rest of the messages all have vanishingly small probabilities of ever occurring.

6.1. *Example*

Suppose the symbols produced are 1 or 0. If they are produced with equal probabilities, H is 1 bit per symbol. Suppose the source produced messages $M = 1\,000$ symbols long. Then $MH = 1\,000$ and there are $2^{1\,000}$ different probable messages.

Suppose now that the symbol 1 has a probability of 0.75 and the symbol 0 has a probability of 0.25. The entropy H is the 0.811 bit per symbol. Thus while there are $2^{1\,000}$ *possible* messages, there are only 2^{811} *probable* messages (end of Example).

Finally let us look at Shannon's fundamental theorem for a noisy channel: Let a channel have the capacity C and let a source have the entropy H per second. If $H < C$ there exists a coding system such that the output of the source can be transmitted over the channel with an arbitrary small frequency of errors (i.e. an arbitrarily small equivocation). If $H > C$ it is possible to encode the source so that the equivocation is less than $H - C + \varepsilon$, where ε is arbitrarily small. There is no method of encoding which gives an equivocation less than $H - C$.

7. Simulation and Theory Construction[4]

In this section, the impact of the electronic digital computing machine upon the social sciences in particular is examined. The most important developments seem to lie along two avenues, neither of which uses the computing machine as a calculator in the usual sense. These two approaches we will call simulation and theory construction. More familiar applications of computing machines in this area are also mentioned and the relationship of the applications is discussed.

All sciences involve the taking of data. In particular the social sciences and the aspects of the physical sciences that are most advanced involve a great deal of data reduction: the computation of statistics which both make its trends meaningful and allow exact conclusions to be drawn. The application of computing machines in this sort of work is obvious and accepted everywhere. Also well known is the fact that these computing machines have opened up the field of applied statistics for new development; the application of electronic computing machines even in this area is in its infancy.

But these applications, doing essentially what was done before in bigger, faster, and better ways, will probably not be the most important impact of the digital computing device upon the methodology of the social sciences. We believe these devices, taken together with development in modern logic and in the theory of programming these machines, will change the whole face of these disciplines.

---

[4] This section is written jointly with Calvin C. Mc Farling.

Turning now to the less familiar possibilities, consider simulation. This arises in connection with the different ways of generating an approximation to a given series of measurements on some process in time. The computer is programmed to generate an output in time as it is fed input data; the computer is said then to act like or simulate the process. The input can be varied and the output functions for expected inputs taken as predictions of the behavior of the process in the future.

There are two quite distinct aspects of simulation. The first is the idea of having the simulation serve as a model for the process. This may be a potent vehicle for yielding a better understanding of the process. The second is the notion of doing experiments (varying inputs) with the simulation. This second is the really critical application because there are many processes on which experiments cannot be performed for a variety of reasons. Experiments can always be performed, however, on a simulation. While most computers were designed for arithmetical calculations, for scientific problems, high speed digital machines may have a greater potential as simulators instead of as calculators.

In all simulation problems time enters into the picture in a unique way, that is in a one directional flow. Any process simulated on a computing machine must meet the restriction of physical realizability, that is that no stimulus follows its response and also the restriction of stability which is that no stimulus produces an unbounded response. Note that these are the fundamental assumptions behind the predictive decomposition of time series.

For the theory construction application we seek first a minimal definition of theory. It is not intended that this definition subsume all that the social scientist in general would intend to convey by that term. In fact, in the light of fundamental disagreements on this, such an undertaking is impossible. But this is not the only factor in our seeking a definition which is something less than the totality of what is meant by theory. We realize that there are valuable aspects of scientific theories quite apart from what can be done by computing machines or logical deductions in general. We know that when we are communicating with a machine the only contingency for success is that every symbol we use be correctly and uniquely defined. Theories however will always be communicated among people and here, how the story hangs together somehow may be more important in consideration of communicability than that every last detail be correct.

The definition here proposed follows naturally the evolution of the concept of a function in the contemporary pedagogy of mathematics. In elementary analysis, a function is defined as a rule which assigns to each of a set of objects called the domain another object from a set called the range of values of the function. An example of this is

$$f(x) = x^2$$

In more advanced analysis the student is informed that something more general is meant by functions than these rules. He is told that a function should be defined as a table or set of ordered pairs, the first member of which is from the domain and the second of which is from the range of values of the function. The rules then constitute the closed forms of a subset of these tables.

x	$f(x)$		x	$f(x)$
0	0		0	1 249
1	1		1	3
2	4		2	452
3	9		3	70 012
.	.		.	.
.	.		.	.
.	.		.	.

A function which ranges over objects which themselves could best be described as functions is sometimes called a function of second order or a functional.

In this definition of theory the functional whose domain is the set of all possible experiments and whose range of values is the set of all possible experimental outcomes is considered:

E	$F(E)$
E_1	0_1
E_2	0_2
E_3	0_3
.	.
.	.
.	.

A theory is defined as a closed form of a subset of this table. This definition is narrow enough that there should be general agreement among social scientists that theory is at least this for it is difficult to conceive of how anything could be called a theory unless it in some sense mapped experiments onto experimental outcomes, for how then could the theory be tested? And yet it is broad enough that it will probably encompass just about everything that can be done with computing machines.

Two problems arise immediately. The first is the problem that the unreliability of experimental data poses in theory construction. This question has long plagued the scientific theorist and one way of stating the usefulness of computing machines to social science is to state that they reduce this problem to the question of whether or not it makes any sense to experiment at all. The second question is whether an ordering of experiments can be found which will yield an orderliness in the range of values of the functional; in other words whether an economical theory can be found.

This second question concerns broadly speaking the existence of the functional called for in the definition. Explorations in modern logic, in particular in the theory of recursive functions, give hope for such structures contingent of course on some orderliness in the data. To the extent to which such a functional for any domain of possible experiments exists, it certainly is not unique, however.

Theorists in the social sciences have long tried to embody a property of completeness in their theories. By completeness they referred to the extent to which empirical results were covered by the theory. In the future, perhaps machines will be used along the lines laid down here to see how complete theories are in this sense. But putting a theory on a machine amounts to what is referred to in logic as formalization. Another notion of completeness arises here. Any consistent formalization for an extensive theory must, in the light of the Kleene generalization of Gödels famous result, necessarily be incomplete in the sense that all statements true in the theory would not be generated in the formal deductions. It would be quite interesting if an experimental result could be obtained which could not be generated due to this incompleteness property.

We envision an attitude among theorists of taking the formal structure as embodying the correctness or incorrectness of the theory. If a deduction is refuted by experiment then the formal structure must be altered. The net result could be a sharpening of the distinction between when one is claiming an experiment fits a theory and when one is changing the theory because it is inconsistent with an experimental result.

In connection with the practical problem of actually building the formal structure, we want a system which, given a description of an experiment together with a predicted result, would yield a decision as to whether or not the result is deducible from the theory. The programming problem is probably no more difficult than that associated with building a good compiler.

Once a formalization of a theory is given, the entire domain of meta-mathematics can be brought over as a tool to be used in studying its properties. Perhaps also attempts at formalization might be quite valuable in defining the state of theory in various areas.

To summarize: In this section, we have seen that the use of the machines as simulators and in theory construction will change the formulation and solution of scientific problems. Computing machines will be used for experimentation, that is for the extrapolation of experimentation into areas where experimenting has been heretofore impossible. They will also have a tendency to enlarge the role that theory plays in the experimental sciences.

8. Simultaneous Systems Subjected to Statistical Disturbances

Wold has made fundamental contributions to the theory and applications of simultaneous systems subjected to statistical disturbances. This section is written as an expository account of some of Wold's main results. Throughout we attempt to avoid involved mathematical arguments and use block-diagram models to show what the mathematics is doing.

The essential thing about any simultaneous model,—behavioral, physical, or biological,—is that it should be possible to *simulate* the model with a digital computer or electrical network. This assures first of all that the model be *physically realizable*, i.e. no response can occur before its stimulus, and and secondly that the model be *stable*, i.e. no stimulus can produce an unbounded response. This simulation concept is a basic principle behind the predictive decomposition of stationary time-series: that is, a purely non-deterministic stationary time-series can be simulated by passing white noise through a minimum-delay electrical filter.

Generally a system can be written in the form

$$y = Ax$$

where A is an operator. Simulation does not usually involve inverting operators; that is,

$$x = A^{-1} y$$

does not usually represent a *bona fide* simulation. Instead social, physical, and biological systems use the *iteration* or *feedback* principle, i.e. some x_0 is tried and the actuating error

$$y - Ax_0$$

is used to correct x_0 to

$$x_1 = x_0 + K(y - Ax_0)$$

by means of a motor (i.e. integrator). In differential form this becomes

$$\frac{dx}{dt} = K(y - Ax) .$$

The steady state solution of this differential equation is

$$\frac{dx}{dt} = 0$$

and the steady state x is the solution of the simultaneous system. The important thing is that it takes a certain non-zero response time for x to reach this steady state solution.

Causal chains are designed for direct simulation and so in this respect they are the most desirable type of model that can be obtained. If for some reason we cannot construct a pure causal chain, then we must consider some type of implicit system. In this category are the conditional causal chains and interdependent systems. The fundamental principles of simulation must be obeyed for systems with implicit relationships.

Let us consider one of Wold's basic illustrations: the *causal chain* given by the three equations

$$s_t = a p_{t-1} + u_t .$$
$$d_{t-1} = b p_{t-1} + v_{t-1}$$
$$p_t = p_{t-1} + K(d_{t-1} - s_t) + w_t .$$

The last equation is the discrete analogue of the feedback differential equa-

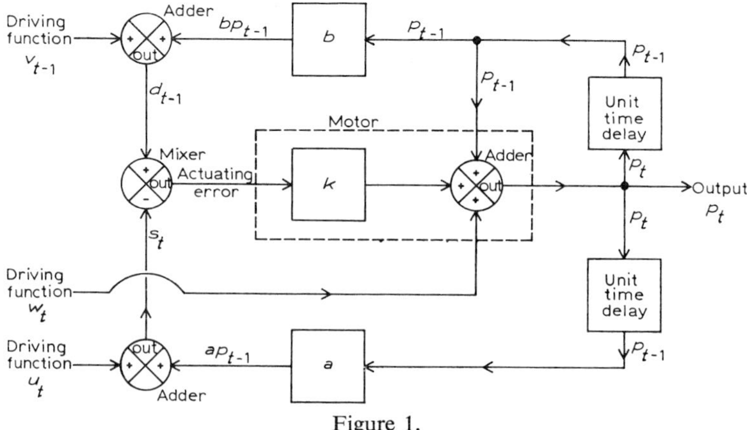

Figure 1.

tion involving the actuating error in what follows. Whereas in causal chains this feedback equation is specified, it is not specified in implicit models where p_t is treated as an equilibrating variable and instantaneous equilibrium is assumed. The block diagram for the causal chain is shown in figure 1. The important thing to recognize in figure 1 is that all the feedback loops contain a unit-time delay. As a result any given driving function, i.e. u_t, v_t, or w_t, at the time instant t, cannot influence the **value** at time t of the function to which it is added (i.e. the function which is coming into the same adder as the driving function). Consequently the driving functions u_t, v_t and w_t may be considered to be statistical errors in that they are never fed-back in the given time instant. Note that w_t represents a statistical error in the motor, and such an error is not taken into account in the implicit systems following.

Let us consider the *bicausal system*

$$s_t = q_t = a p_{t-1} + u_t$$
$$d_t = q_t = b p_t + v_t \, .$$

We suppose that u_t and v_t are the outside influences (i.e. the driving functions) and that p_t adjusts with time so that the system is satisfied. This adjustment time should be less than one time unit. If we do not assume instantaneous equilibrium (i.e. if we do not assume zero adjustment time), then a block diagram for the system is given in figure 2.

The differential equation is

$$p_t = \frac{K}{\dfrac{d}{dt}} \cdot \text{Actuating Error} = \frac{K}{\dfrac{d}{dt}} (d_t - s_t)$$

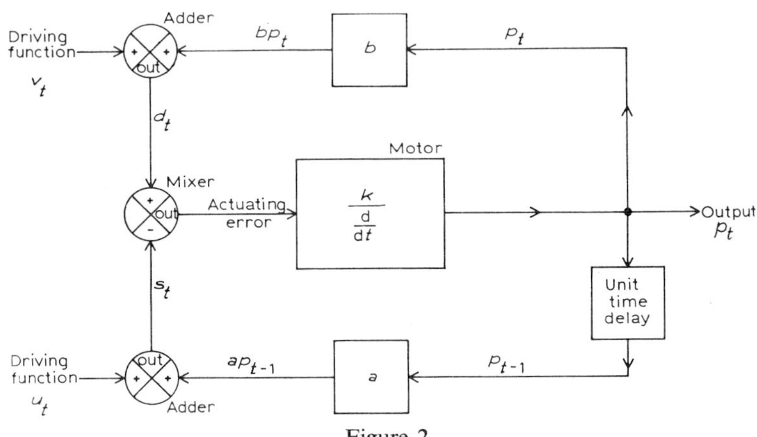

Figure 2.

which is

$$\frac{d}{dt}\, p_t = K \cdot \text{Actuating Error} = K[(b p_t + v_t) - (a p_{t-1} + u_t)]\,.$$

The steady state p_t and q_t will depend upon the driving functions u_t and v_t, as well as previous values u_{t-k}, v_{t-k}, $k > 0$. The driving function $u_t(-\infty < t < \infty)$ is a sequence of independent variates, so u_t is independent of p_{t-1} which depends upon u_{t-1}, v_{t-1}, etc. From the block diagram or differential equation, it is clear that p_t depends only on the difference variate $u_t - v_t$. Letting $w_t = u_t - v_t$ we have

$$q_t = a p_{t-1} + w_t + v_t$$
$$q_t = b p_t + v_t\,.$$

Suppose that w_t and v_t are independent variates, with zero means, then p_t will depend only on w_t, and p_t will be independent of v_t. In this case the system

$$q_t = a p_{t-1} + u_t$$
$$q_t = b p_t + v_t$$

is in Wold's terminology a *conditional causal chain*, since u_t and p_{t-1} are independent and v_t and p_t are independent. The driving functions u_t and v_t are correlated, as seen by

$$E[u_t v_t] = E[(w_t + v_t)v_t] = E[v_t^2]\,.$$

Let us consider the circular system or *causal circle* given by

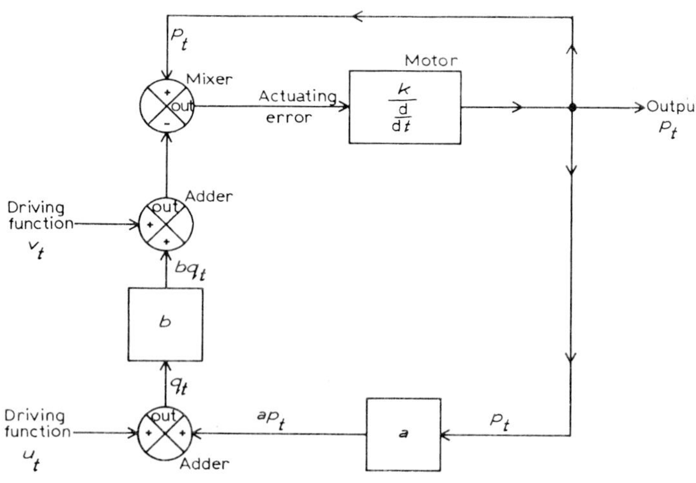

Figure 3.

$$q_t = a p_t + u_t$$
$$p_t = b q_t + v_t .$$

The block diagram is shown in figure 3., and the differential equation is

$$\frac{\mathrm{d}p_t}{\mathrm{d}t} = K[p_t - b(a p_t + u_t) - v_t] .$$

9. Estimation of the Trend Component

As a preliminary step in the analysis of time series, one might try various methods to estimate as well as possible the trend component, so that it may be separated from the trend-free component. One approach to this problem is to make use of the model specified by the following *assumptions*:

(a) The *trend component* s_t is given by

$$s_t = \sum_{i=1}^{p} c_i f_{it}$$

where $c_1, c_2, \ldots, c_p$ are *unknown* constants and $f_{1t}, f_{2t}, \ldots, f_{pt}$ are *known* fixed functions. The trend component cannot be directly observed.

(b) The *trend-free component* n_t is a normal stochastic process. The trend-free component cannot be directly observed.

(c) The mean value $E\{n_t\}$ is *known*, and is equal to zero.

(d) The covariances $\phi_{ij} = E\{n_i\, n_j\}$ of the trend-free component are *known*.

(e) We can directly observe the stochastic process x_t which is the sum of the trend component s_t and the trend-free component n_t, that is,

$$x_t = s_t + n_t .$$

(f) The sample time series $x_1, x_2, \ldots, x_N$ has been measured.

In this section we wish to consider various operations on the sample timeseries $x_1, x_2, \ldots, x_N$ which yield estimates of the trend component $s_1, s_2, \ldots, s_N$. Let us introduce the following matrix notation:

$$x = (x_1, x_2, \ldots, x_N): \quad 1 \times N \text{ row vector.}$$
$$n = (n_1, n_2, \ldots, n_N): \quad 1 \times N \text{ row vector.}$$
$$s = (s_1, s_2, \ldots, s_N): \quad 1 \times N \text{ row vector.}$$

$$\phi = \begin{bmatrix} \phi_{11} & \phi_{12} \cdots \phi_{1N} \\ \phi_{21} & \phi_{22} \cdots \phi_{2N} \\ \cdots & \cdots \\ \phi_{N1} & \phi_{N2} \cdots \phi_{NN} \end{bmatrix} = [\phi_{ij}]: \quad N \times N \text{ covariance matrix of the trend-free component.}$$

(ϕ is assumed to be non-singular.)

$\det \phi$: determinant of the covariance matrix ϕ.

$\mu = [\mu_{ij}] = \phi^{-1}$: inverse of the covariance matrix ϕ.

$c = (c_1, c_2, \ldots, c_p)$: $1 \times p$ row vector.

$$f = \begin{bmatrix} f_{11} & f_{12} & \cdots & f_{1N} \\ f_{21} & f_{22} & \cdots & f_{2N} \\ & \cdots & & \cdots \\ f_{p1} & f_{p2} & \cdots & f_{pN} \end{bmatrix} : p \times N \text{ matrix.}$$

$s = cf$.

$x = s + n = cf + n$.

$E\{n\} = 0$.

$E\{x\} = E\{cf + n\} = E\{cf\} + E\{n\} = cf = s$.

Note: A prime will indicate matrix transpose.

Let us now find the maximum-likelihood estimates of c and s. The density function of x, under the normality assumption, is

$$f(x) = \frac{1}{(2\pi)^{N/2}(\det \phi)^{1/2}} \exp[-\tfrac{1}{2}(x-s)\mu(x-s)'] \ .$$

The maximum-likelihood estimate of s, denoted by $\hat{s}$, is that value of $s = cf$ for which $f(x)$ is a maximum. The maximum occurs when the negative of the exponent of $f(x)$ is a minimum, that is, when

$$P = (x-s)\mu(x-s)' = \text{minimum} \ ,$$

subject to the constraint that $s = cf$. We shall now use the method of Lagrange multipliers to solve this constrained minimization problem. We therefore introduce the undetermined multipliers

$$\lambda = \begin{bmatrix} \lambda_1 \\ \lambda_2 \\ \cdot \\ \cdot \\ \cdot \\ \lambda_N \end{bmatrix} : N \times 1 \text{ column vector .}$$

We now wish to minimize

$$J = (x-s)\mu(x-s)' + (s-cf)\lambda + \lambda'(s-cf)'$$

with respect to s, λ, and c. Setting the partial derivatives equal to zero, we obtain:

$$\frac{\partial J}{\partial s} = 0 \quad \text{which gives:} \quad -(x-\hat{s})\mu + \hat{\lambda}' = 0 \,,$$

$$\frac{\partial J}{\partial \lambda} = 0 \quad \text{which gives:} \quad \hat{s} - \hat{c}f = 0 \,,$$

$$\frac{\partial J}{\partial c} = 0 \quad \text{which gives:} \quad \hat{\lambda}'f' = 0 \,.$$

Solving these equations, we have

$$\hat{\lambda}' = (x-\hat{s})\mu = (x-\hat{c}f)\mu$$

and

$$\hat{\lambda}'f' = (x-\hat{c}f)\mu f' = 0 \,.$$

These equations then give the equation

$$x\mu f' - \hat{c}f\mu f' = 0 \,.$$

Hence

$$\hat{c} = (x\mu f')(f\mu f')^{-1}$$

and

$$\hat{s} = \hat{c}f = (x\mu f')(f\mu f')^{-1}f \,,$$

where $\hat{s}$ is the *maximum-likelihood estimate* of s.

Let us now find the maximum-likelihood estimate of c. By the above reasoning, it is that value of c obtained by requiring

$$P = (x-s)\mu(x-s)' = (x-cf)\mu(x-cf)' = \text{minimum} \,.$$

We thus have

$$\frac{\partial P}{\partial c} = 0 \quad \text{which gives:} \quad -(x-\hat{c}f)\mu f' = 0 \,, \quad \text{or} \quad \hat{c} = x\mu f'(f\mu f')^{-1}$$

which is the same $\hat{c}$ as found above, as we would expect. That is, $\hat{c}$ is the *maximum-likelihood estimate* of c.

Let us find the expected value of $\hat{c}$. It is

$$E\{\hat{c}\} = E\{x\mu f'(f\mu f')^{-1}\} = E\{x\}\mu f'(f\mu f')^{-1}$$
$$= cf\mu f'(f\mu f')^{-1} = c \,.$$

Hence $E\{\hat{c}\} = c$ so $\hat{c}$ is an *unbiased estimate* of c.
Also

$$E\{\hat{s}\} = E\{\hat{c}f\} = cf = s$$

so $\hat{s}$ is an *unbiased estimate* of s.

The covariance matrix of $\hat{c}$ is

$$\Psi = E\{(\hat{c}-c)'(\hat{c}-c)\} = E\{[(x-cf)\mu f'(f\mu f')^{-1}]'[(x-cf)\mu f'(f\mu f')^{-1}]\}\;.$$

Because

$$x - cf = n$$

it follows that

$$\begin{aligned}
\Psi &= E\{[n\mu f'(f\mu f')^{-1}]'[n\mu f'(f\mu f')^{-1}]\}\\
&= E\{(f\mu f')^{-1}f\mu n'\,n\mu f'(f\mu f')^{-1}\}\\
&= (f\mu f')^{-1}f\mu E\{n'\,n\}\mu f'(f\mu f')^{-1}\\
&= (f\mu f')^{-1}f\mu\phi\mu f'(f\mu f')^{-1}\\
&= (f\mu f')^{-1}f\mu f'(f\mu f')^{-1} \quad \text{(since } \phi\mu = I \text{ where } I \text{ is the identity matrix)}\\
&= (f\mu f')^{-1}\;.
\end{aligned}$$

Hence the maximum-likelihood estimate $\hat{c}$ has mean c and covariance matrix $(f\mu f')^{-1}$.

The covariance matrix of $\hat{s}$ is

$$\begin{aligned}
E\{(\hat{s}-s)'(\hat{s}-s)\} &= E\{[(x-cf)\mu f'(f\mu f')^{-1}f]'[(x-cf)\mu f'(f\mu f')^{-1}f]\}\\
&= f'(f\mu f')^{-1}f\mu E\{n'\,n\}\mu f'(f\mu f')^{-1}f\\
&= f'(f\mu f')^{-1}f\mu\phi\mu f'(f\mu f')^{-1}f\\
&= f'(f\mu f')^{-1}f\;.
\end{aligned}$$

Hence the maximum-likelihood estimate $\hat{s} = \hat{c}f$ has mean $s = cf$ and covariance matrix $f'(f\mu f')^{-1}f$.

We note that the maximum likelihood estimates $\hat{c}$ and $\hat{s}$ are independent of the scale of the covariance matrix. That is, suppose $\phi = \sigma^2\psi$ and $\mu = \phi^{-1} = \sigma^{-2}\psi^{-1}$ where ψ is known but the scale factor σ^2 is unknown. Then

$$\hat{c} = x\psi^{-1}f'(f\psi^{-1}f')^{-1}\;, \quad \hat{s} = \hat{c}f\;,$$

independent of the numerical value of the scale factor σ^2.

For the remainder of this section, we want to relax the normality assumption (b). Instead of (b) we now assume:

(b_1)   The *trend-free component* n_t is a stochastic process with an *unknown* distribution. As before, n_t cannot be directly observed.
We have

$$x = cf + n$$

and we wish to estimate c. The least-squares estimate $\check{c}$ is that value of c such that the sum of squared errors

$$J = (x - cf)(x - cf)' = \text{minimum} .$$

We have

$$\frac{\partial J}{\partial c} = 0 \quad \text{which gives:} \quad (x - \hat{c}f)(-f') = 0 ,$$

or

$$\check{c}ff' = xf' .$$

Hence the least-squares estimate is

$$\check{c} = xf'(ff')^{-1} .$$

To compute $\check{c}$ we see that we do not need assumption (d); that is, we do not need to know the covariance matrix ϕ of n_t in order to calculate the least-squares estimate $\check{c}$. This fact makes the least-squares estimate extremely useful in practice.

The least-squares estimate is unbiased since

$$E\{\check{c}\} = E\{xf'(ff')^{-1}\} = E\{x\}f'(ff')^{-1}$$

(and using $E\{x\} = cf$)

$$E\{\check{c}\} = cff'(ff')^{-1} = c .$$

The covariance matrix of $\check{c}$ is defined to be

$$E\{(\check{c} - c)'(\check{c} - c)\}$$

Since

$$\begin{aligned}
\check{c} &= xf'(ff')^{-1} = (n + s)f'(ff')^{-1} \\
&= (n + cf)f'(ff')^{-1} \\
&= nf'(ff')^{-1} + cff'(ff)'^{-1} \\
&= nf'(ff')^{-1} + c
\end{aligned}$$

we have

$$\check{c} - c = nf'(ff')^{-1} .$$

Hence the covariance matrix of $\check{c}$ is

$$\begin{aligned}
E\{(\check{c} - c)'(\check{c} - c)\} &= E\{[nf'(ff')^{-1}]'[nf'(ff')^{-1}]\} \\
&= E\{(ff')^{-1}fn'\,nf'(ff')^{-1}\} \\
&= (ff')^{-1}fE\{n'\,n\}f'(ff')^{-1} .
\end{aligned}$$

Thus the covariance matrix of the least-squares estimate is

$$E\{(\check{c} - c)'(\check{c} - c)\} = (ff')^{-1}f\phi f'(ff')^{-1} .$$

Finally we wish to find an estimate $\hat{c}$ which

(1) is linear in the observations, i.e. $\hat{c}$ has the form

$$\hat{c} = xb$$

where the matrix

$$b \;=\; \begin{bmatrix} b_{11} & b_{12} & \ldots & b_{1p} \\ b_{21} & b_{22} & \ldots & b_{2p} \\ & \cdot\,\cdot\,\cdot & & \cdot\,\cdot\,\cdot \\ b_{N1} & b_{N2} & \ldots & b_{Np} \end{bmatrix}$$

is to be determined.

(2) is an unbiased estimate, i.e.

$$E\{\hat{c}\} = c$$

(3) is best in the sense that the covariance matrix of $\hat{c}$ is "less than" the covariance matrix of any other linear unbiased estimate $\tilde{c}$, i.e.

$$E\{(\hat{c}-c)'(\hat{c}-c)\} < E\{(\tilde{c}-c)'(\tilde{c}-c)\}\;.$$

(More precisely, the "less than" sign $<$ as used here means that

$$E\{(\tilde{c}-c)'(\tilde{c}-c)\} - E\{(\hat{c}-c)'(\hat{c}-c)\}$$

is a positive definite matrix.)

Using (1) we have

$$\hat{c} = xb = (s+n)b = (cf+n)b$$
$$= cfb + nb\;.$$

Using (2) we have

$$c = E\{\hat{c}\} = E\{cfb\} + E\{nb\}$$
$$= cfb + E\{n\}b$$
$$= cfb \quad \text{since} \quad E\{n\} = 0\;.$$

Thus we have the constraint

$$fb = I \;\text{(where } I \text{ is the } p \times p \text{ identity matrix)}\,.$$

Also we have $\hat{c} - c = nb$ so the covariance matrix of $\hat{c}$ is

$$E\{(\hat{c}-c)'(\hat{c}-c)\} = E\{(nb)'(nb)\}$$
$$= E\{b'\,n'\,nb\} \quad = b'\,E\{n'\,n\}b = b'\,\phi b\;.$$

Hence b may be determined as follows:

$$\text{Minimize the } p \times p \text{ matrix } b'\,\phi b$$
$$\text{subject to the constraint } fb = I\;.$$

We may use the method of Lagrange multipliers. Introduce the $p \times p$ matrix λ as an undetermined multiplier. We then wish to minimize

$$J = b'\phi b + 2\lambda(fb - I)$$

with respect to b and λ. We have

$$\frac{\partial J}{\partial b} = 0 \quad \text{which gives:} \quad b'\phi + \lambda f = 0$$

and

$$\frac{\partial J}{\partial \lambda} = 0 \quad \text{which gives:} \quad fb = I .$$

Solving these equations for b and λ we have

$$b' = -\lambda f \phi^{-1} .$$

Thus

$$b = -\phi^{-1} f' \lambda' .$$

Also $fb = I$ gives $-f\phi^{-1}f'\lambda' = I$, or $\lambda' = -(f\phi^{-1}f')^{-1}$. Hence

$$b = -\phi^{-1}f'\lambda' = \phi^{-1}f'(f\phi^{-1}f')^{-1} .$$

We recall that the notation for ϕ^{-1} was

$$\mu = \phi^{-1} .$$

Hence

$$b = \mu f'(f\mu f')^{-1} .$$

Thus the best unbiased linear estimate $\hat{c}$ of c is

$$\hat{c} = xb = x\mu f'(f\mu f')^{-1}$$

which, we see, is the same as the maximum-likelihood estimate obtained under the normality assumption (b). The covariance matrix of $\hat{c}$ is

$$E\{(\hat{c}-c)'(\hat{c}-c)\} = b'\phi b = (f\mu f')^{-1} f\mu\phi\mu f'(f\mu f')^{-1} = (f\mu f')^{-1} .$$

10. Spectral Band Power Estimates

The problem we wish to consider in this section is the following: Let x_t be a real normal stationary stochastic process with autocorrelation

$$\phi(s) = E\{x_{t+s}x_t\}$$

and spectral density $\Phi(\omega)$. Given the sample time series

$$x_1, x_2, \ldots, x_N$$

consisting of N observations, we wish to estimate the spectral density, or some function of the spectral density.

First of all, we may compute the *sample autocorrelation* which is

$$\phi_N(s) = \frac{1}{N} \sum_{t=1}^{N-s} x_{t+s} x_t = \phi_N(-s) \quad \text{for} \quad s = 0, 1, 2, \ldots, N-1\,,$$

and which is zero for $|s| \geq N$. Then we compute the *periodogram* which is

$$\Phi_N(\omega) = \sum_{s=-N+1}^{N-1} \phi_N(s)\, e^{-i\omega s} = \phi_N(0) + 2 \sum_{t=1}^{N-1} \phi_N(t) \cos \omega t$$

$$= \frac{1}{N} \sum_{s=1}^{N} \sum_{t=1}^{N} x_s x_t\, e^{-i\omega(s-t)} = \frac{1}{N} \left| \sum_{t=1}^{N} x_t\, e^{-i\omega t} \right|^2.$$

Let us write the spectral representation of the stationary process (Robinson, 1959, 1962). The spectral representation is

$$x_t = \frac{1}{2\pi} \int_{-\pi}^{\pi} e^{i\omega t}\, X(\omega)\, d\omega$$

where $X(\omega)$ satisfies

$$E\{X(\mu)\,\overline{X(\lambda)}\} = 2\pi\delta(\mu - \lambda)\, \Phi(\mu)\,.$$

Here $\delta(\mu - \lambda)$ is the Dirac delta function for the argument $\mu - \lambda$. Let us define $X_N(\omega)$ to be

$$X_N(\omega) = \sum_{s=1}^{N} x_s\, e^{-i\omega s}\,.$$

By the spectral representation, we have

$$X_N(\omega) = \sum_{s=1}^{N} \left[\frac{1}{2\pi} \int_{-\pi}^{\pi} e^{i\mu s}\, X(\mu)\, d\mu \right] e^{-i\omega s}$$

$$= \frac{1}{2\pi} \int_{-\pi}^{\pi} d\mu\, X(\mu) \sum_{s=1}^{N} e^{-i(\omega-\mu)s}\,.$$

Now define the function $D(\omega)$ to be

$$D(\omega) = \sum_{s=1}^{N} e^{-i\omega s}\,.$$

Hence

$$X_N(\omega) = \frac{1}{2\pi} \int_{-\pi}^{\pi} X(\mu) D(\omega-\mu)\,d\mu$$

$$= \frac{1}{2\pi} X(\omega) * D(\omega)$$

where the asterisk $*$ indicates convolution (see Robinson, 1962). Thus the periodogram may be written

$$\Phi_N(\omega) = \frac{1}{N} \left| X_N(\omega) \right|^2 = \frac{1}{N} \frac{1}{4\pi^2} \int_{-\pi}^{\pi} d\mu\, X(\mu) D(\omega-\mu) \int_{-\pi}^{\pi} d\lambda\, \overline{X(\lambda)}\, \overline{D(\omega-\lambda)}$$

where the bar indicates the complex conjugate, as usual.

Now let us find the expected value of the periodogram. The expected value is

$$E\{\Phi_N(\omega)\} = \frac{1}{4\pi^2 N} \int_{-\pi}^{\pi} d\mu\, D(\omega-\mu) \int_{-\pi}^{\pi} d\lambda\, \overline{D(\omega-\lambda)}\, E\{X(\mu)\,\overline{X(\lambda)}\}$$

$$= \frac{1}{4\pi^2 N} \int_{-\pi}^{\pi} d\mu\, D(\omega-\mu) \int_{-\pi}^{\pi} d\lambda\, \overline{D(\omega-\lambda)}\, 2\pi\delta(\mu-\lambda)\, \Phi(\mu)$$

$$= \frac{1}{2\pi N} \int_{-\pi}^{\pi} \Phi(\mu)|D(\omega-\mu)|^2\, d\mu\ .$$

Define $H(\omega)$ to be

$$H(\omega) = |D(\omega)|^2 = \sum_{s=1}^{N} \sum_{t=1}^{N} e^{i\omega(t-s)} = \left[\frac{\sin N\omega/2}{\sin \omega/2}\right]^2.$$

Then the expected value is

$$E\{\Phi_N(\omega)\} = \frac{1}{2\pi N} \int_{-\pi}^{\pi} \Phi(\mu) H(\omega-\mu)\,d\mu$$

$$= \frac{1}{2\pi N} \Phi(\omega) * H(\omega)\ .$$

Because

$$\frac{1}{2\pi N} H(\omega) \to \delta(\omega) \quad \text{as} \quad N \to \infty$$

we have

$$\lim_{N \to \infty} E\{\Phi_N(\omega)\} = \Phi(\omega) * \delta(\omega) = \Phi(\omega)\ .$$

This equation says that *the periodogram $\Phi_N(\omega)$ is an asymptotically unbiased estimate of the spectral density.*

Because x_t is a normal process, the *covariance of the periodogram* may be derived in a straight forward manner. The covariance is

$$\text{cov}\{\Phi_N(\omega), \Phi_N(\mu)\} = \frac{1}{N^2} \left[|E\{X_N(\omega) X_N(\mu)\}|^2 + |E\{X_N(\omega) \overline{X_N(\mu)}\}|^2 \right]$$

$$= \left| \frac{1}{2\pi N} \int_{-\pi}^{\pi} \Phi(\lambda) D(\omega - \lambda) D(\mu + \lambda) d\lambda \right|^2$$

$$+ \left| \frac{1}{2\pi N} \int_{-\pi}^{\pi} \Phi(\lambda) D(\omega - \lambda) \overline{D(\mu - \lambda)} d\lambda \right|^2 .$$

If $\omega \neq \pm \mu$, then it may be shown that the covariance tends to zero as $N \to \infty$. If $\omega = \mu$, then the covariance becomes the variance. Thus the *variance of the periodogram* is

$$\text{var}\{\Phi_N(\omega)\} = \left| \frac{1}{2\pi N} \int_{-\pi}^{\pi} \Phi(\lambda) D(\omega - \lambda) D(\omega + \lambda) d\lambda \right|^2 +$$

$$+ \left(\frac{1}{2\pi N} \int_{-\pi}^{\pi} \Phi(\lambda) H(\omega - \lambda) d\lambda \right)^2 .$$

As $N \to \infty$, the limit of the first term on the right hand side is

$$\begin{cases} \Phi^2(-\pi) & \text{if} \quad \omega = -\pi \\ \Phi^2(0) & \text{if} \quad \omega = 0 \\ \Phi^2(\pi) & \text{if} \quad \omega = \pi \\ 0 & \text{if} \quad \omega \neq -\pi, 0, \pi , \end{cases}$$

whereas the limit of the second term on the right hand side is

$$\Phi^2(\omega) .$$

Therefore the limit of the variance is

$$\lim_{N \to \infty} \text{var}\{\Phi_N(\omega)\} = \begin{cases} 2\Phi^2(\omega) & \text{if} \quad \omega = -\pi, 0, \pi \\ \Phi^2(\omega) & \text{otherwise} . \end{cases}$$

That is, the variance of the periodogram does not tend to zero as the sample size N tends to infinity. Thus, *the periodogram $\Phi_N(\omega)$ is not a consistent estimate of the spectral density $\Phi(\omega)$.*

Hence, let us not try to estimate the spectral density $\Phi(\omega)$, but instead *let us try to estimate a function of the spectral density.* Because the x_t process is real, the spectral density is an even function, that is

$$\Phi(\omega) = \Phi(-\omega) .$$

Let us try to estimate a portion of the area under the spectral density curve. Let S_1 denote the area under the $\Phi(\omega)$ curve between the lower frequency ω_0-h and the upper frequency ω_0+h, where ω_0 and h are fixed numbers. That is,

$$S_1 = \int_{\omega_0-h}^{\omega_0+h} \Phi(\omega)\,d\omega \, .$$

Similarly, let S_{-1} denote the area under the $\Phi(\omega)$ curve between the lower frequency $-\omega_0-h$ and the upper frequency $-\omega_0+h$. That is

$$S'_{-1} = \int_{-\omega_0-h}^{-\omega_0+h} \Phi(\omega)\,d\omega \, .$$

Because $\Phi(\omega)$ is an even function, it follows that

$$S_{-1}=S_1 \, .$$

Let us try to estimate the following function of the spectral density:

$$F(\omega_0) = \frac{1}{2h}\frac{1}{2\pi}\int_{\omega_0-h}^{\omega_0+h} \Phi(\omega)\,d\omega \, .$$

We see that $F(\omega_0)$ is the spectral power in a band of width $2h$ centered at the frequency ω_0, and so let us call $F(\omega)_0$ the *band power* about ω_0. The band power may be written

$$F(\omega_0) = \frac{1}{4\pi h}\,S_1 = \frac{1}{8\pi h}(S_1+S_{-1}) \, .$$

Let us define the transfer function $B(\omega)$ of an *ideal band-pass filter* by[5]

$$B(\omega) = \begin{cases} \dfrac{1}{4h} & \text{if} \quad -\omega_0-h\le \omega \le -\omega_0+h \\[2ex] \dfrac{1}{4h} & \text{if} \quad \omega_0-h\le \omega \le \omega_0+h \\[1ex] 0 & \text{otherwise} \, . \end{cases}$$

Thus the band power is

$$F(\omega_0) = \frac{1}{2\pi}\int_{-\pi}^{\pi} B(\omega)\,\Phi(\omega)\,d\omega \, .$$

The logical estimate of this band power is therefore $F_N(\omega_0)$ defined by

---

[5] In case ω_0 lies within an h-neighborhood of $-\pi$, 0, or π the formula must be accordingly modified.

$$F_N(\omega_0) = \frac{1}{2\pi} \int_{-\pi}^{\pi} B(\omega)\,\Phi_N(\omega)\,d\omega\ ,$$

where $\Phi_N(\omega)$ is the periodogram. Let us call $F_N(\omega_0)$ the *band power estimate*.

The band power estimate can be written in terms of the sample autocorrelation $\phi_N(t)$ as follows. The impulse response of the ideal bandpass-filter is the two-sided operator

$$b_t = \frac{1}{2\pi} \int_{-\pi}^{\pi} B(\omega)\,e^{i\omega t}\,d\omega = \frac{1}{2\pi}\frac{\sin ht}{ht}\cos \omega_0 t$$

where $-\infty < t < \infty$. Hence the band power estimate is

$$F_N(\omega_0) = \frac{1}{2\pi} \int_{-\pi}^{\pi} B(\omega)\,\Phi_N(\omega)\,d\omega = \sum_{t=-N+1}^{N-1} b_t \phi_N(t)\ .$$

Substituting for b_t, the band power estimate becomes

$$F_N(\omega_0) = \sum_{t=-N+1}^{N-1} \frac{\sin ht}{2\pi ht}\cos \omega_0 t\ \phi_N(t)$$

$$= \frac{\phi_N(0)}{2\pi} + 2\sum_{t=1}^{N-1} \frac{\sin ht}{2\pi ht}\cos \omega_0 t\ \phi_N(t)$$

which is the desired expression for the band power estimate in terms of the sample autocorrelation.

The band power estimate can also be written as a quadratic form. The band power estimate is

$$F_N(\omega_0) = \sum_{t=-N+1}^{N-1} b_t \phi_N(t)$$

where the sample autocorrelation is given by

$$\phi_N(t) = \frac{1}{N} \sum_{k=1}^{N-|t|} x_{k+|t|}\,x_k\ .$$

The band power estimate may then be written

$$F_N(\omega_0) = \frac{1}{N} \sum_{t=-N+1}^{N-1} b_t \sum_{k=1}^{N-|t|} x_{k+|t|} x_k$$

$$= \frac{1}{N} \sum_{i=1}^{N} \sum_{j=1}^{N} x_i\,b_{i-j}\,x_j\ .$$

Hence the band power estimate is the quadratic form given by

$$F_N(\omega_0) = \sum_{i=1}^{N} \sum_{j=1}^{N} x_i(b_{i-j}/N)x_j = x' a x$$

where we have let x be the $N \times 1$ column vector with elements $x_1, x_2, \ldots, x_N$ and a be the $N \times N$ matrix whose $i^{\text{th}}, j^{\text{th}}$ element is b_{i-j}/N.

For brevity, let us use the symbol Q (standing for "quadratic form") to denote the band power estimate, that is

$$Q = F_N(\omega_0) = x' a x.$$

We now wish to examine the statistical properties of Q. Because Q is a quadratic form in normal random variables, we may find its characteristic function as follows.

The joint probability density function of x is

$$(2\pi)^{-N/2}|\phi^{-1}|^{1/2} \exp\left[-\tfrac{1}{2}x' \phi^{-1} x\right]$$

where

$$\phi = \begin{bmatrix} \phi(0) & \phi(1) & \ldots & \phi(N-1) \\ \phi(1) & \phi(0) & & \phi(N-2) \\ & \ldots & & \ldots \\ \phi(N-1) & \phi(N-2) & \ldots & \phi(0) \end{bmatrix}$$

is the autocorrelation matrix, and where $|\quad|$ now indicates the determinant. The characteristic function of Q is

$$E\{\exp(itQ)\} = \int_{-\infty}^{\infty} \ldots \int_{-\infty}^{\infty} \exp[itx' a x - \tfrac{1}{2}x' \phi^{-1} x](2\pi)^{-N/2}|\phi^{-1}|^{\frac{1}{2}}dx_1 \ldots dx_N$$

$$= (2\pi)^{-N/2}|\phi^{-1}|^{\frac{1}{2}} \int_{-\infty}^{\infty} \ldots \int_{-\infty}^{\infty} \exp[-\tfrac{1}{2}x'(\phi^{-1} - 2ita)x]dx_1 \ldots dx_N.$$

Using formula (11.12.2) on page 120 of Cramér (1945) we see that

$$\int_{-\infty}^{\infty} \ldots \int_{-\infty}^{\infty} \exp[-\tfrac{1}{2}x'(\phi^{-1} - 2ita)x]dx_1 \ldots dx_N = \frac{(2\pi)^{N/2}}{|\phi^{-1} - 2ita|^{\frac{1}{2}}}.$$

Therefore the characteristic function of Q is

$$E\{\exp(itQ)\} = \frac{1}{|\phi^{-1}|^{-\frac{1}{2}}|\phi^{-1} - 2ita|^{\frac{1}{2}}} = \frac{1}{|I - 2it\phi a|^{\frac{1}{2}}}$$

where I is the identity matrix. Let us denote the eigenvalues of the matrix product ϕa by

$$\lambda_1^{(N)}, \lambda_2^{(N)}, \ldots, \lambda_N^{(N)} .$$

Then the logarithm of the characteristic function of Q is

$$\psi(t) = -\tfrac{1}{2} \sum_{j=1}^{N} \log(1 - 2it\lambda_j^{(N)})$$

and hence the cumulants of Q are

$$\kappa_s = 2^{s-1}(s-1)! \sum_{j=1}^{N} [\lambda_j^{(N)}]^s .$$

Thus, in particular, the band power estimate Q has mean

$$E\{Q\} = \kappa_1 = \sum_{j=1}^{N} \lambda_j^{(N)} = \lambda_1^{(N)} + \lambda_2^{(N)} + \ldots + \lambda_N^{(N)}$$

and variance

$$\text{var}\{Q\} = E\{[Q - E\{Q\}]^2\} = \kappa_2 = 2 \sum_{j=1}^{N} [\lambda_j^{(N)}]^2$$

$$= 2[\lambda_1^{(N)}]^2 + 2[\lambda_2^{(N)}]^2 + \ldots + 2[\lambda_N^{(N)}]^2 .$$

We have thus found expressions for the mean and the variance of the band power estimate $Q = F_N(\omega_0)$.

Let us now consider the problem of finding *asymptotic approximations for the mean and variance of the spectral band estimate.*

Each of the two $N \times N$ matrices

$$\boldsymbol{\phi} = [\phi_{s-t}]$$

and

$$\boldsymbol{a} = [b_{s-t}/N]$$

has equal elements along any diagonal. But it is easy to verify that their product $\boldsymbol{\phi a}$ does *not* have equal elements along any diagonal. Nevertheless, as N tends to infinity, it turns out that the product $\boldsymbol{\phi a}$ has *approximately* equal elements along any diagonal.

We recall that the sth, t^{th} element of the matrix $\boldsymbol{\phi}$ has the representation

$$\phi_{s-t} = \frac{1}{2\pi} \int_{-\pi}^{\pi} e^{i\omega(s-t)} \Phi(\omega) \, d\omega$$

and that the sth, t^{th} element of the matrix $\boldsymbol{a}$ has the representation

$$\frac{b_{s-t}}{N} = \frac{1}{2\pi} \int_{-\pi}^{\pi} e^{i\omega(s-t)} \frac{B(\omega)}{N} \, d\omega .$$

Hence the sth, t^{th} element of the matrix product $\boldsymbol{\phi a}$ has the approximate representation

$$c_{s-t} = \frac{1}{2\pi} \int_{-\pi}^{\pi} e^{i\omega(s-t)} \Phi(\omega) \frac{B(\omega)}{N} \, d\omega .$$

Let us now consider the $N \times N$ matrix

$$c = \begin{bmatrix} c_0 & c_{-1} & c_{-2} & \cdots & c_{1-N} \\ c_1 & c_0 & c_{-1} & & c_{2-N} \\ c_2 & c_1 & c_0 & & c_{3-N} \\ & & & & \\ c_{N-1} & c_{N-2} & c_{N-3} & & c_0 \end{bmatrix}$$

whose elements along any diagonal are given by

$$c_s = \frac{1}{2\pi N} \int_{-\pi}^{\pi} e^{i\omega s} \Phi(\omega) B(\omega) \, d\omega = \frac{1}{2\pi} \int_{-\pi}^{\pi} e^{i\omega s} C(\omega) \, d\omega$$

where we have defined $C(\omega)$ by

$$C(\omega) = \frac{\Phi(\omega) B(\omega)}{N} .$$

Let us assume that $C(\omega)$ satisfies

$$m \leq C(\omega) \leq M$$

for finite constants m and M. If we let

$$\lambda_1^{(N)}, \lambda_2^{(N)}, \ldots, \lambda_N^{(N)}$$

be the eigenvalues of the matrix c, it follows that

$$m \leq \lambda_j^{(N)} \leq M$$

for $j = 1, 2, 3, \ldots, N$ and for $N = 1, 2, 3, \ldots$. Let $G(\lambda)$ be an arbitrary continuous function of λ in the interval $m \leq \lambda \leq M$. Then it may be shown that

$$\lim_{N \to \infty} \frac{1}{N} [G(\lambda_1^{(N)}) + G(\lambda_2^{(N)}) + \ldots + G(\lambda_N^{(N)})] = \frac{1}{2\pi} \int_{-\pi}^{\pi} G[C(\omega)] \, d\omega .$$

But

$$\frac{1}{2\pi} \int_{-\pi}^{\pi} G[C(\omega)] \, d\omega = \frac{1}{2\pi} \int_{-\pi}^{\pi} G\left[\frac{\Phi(\omega) B(\omega)}{N}\right] d\omega .$$

Hence we have the following asymptotic relation

$$G(\lambda_1^{(N)}) + G(\lambda_2^{(N)}) + \ldots + G(\lambda_N^{(N)}) \to \frac{N}{2\pi} \int_{-\pi}^{\pi} G\left[\frac{\Phi(\omega)B(\omega)}{N}\right] d\omega$$

for N tending to infinity.

Let us now apply this formula to our expressions for the mean and variance of the band power estimate Q. We recall that the mean is

$$E\{Q\} = \lambda_1^{(N)} + \lambda_2^{(N)} + \ldots + \lambda_N^{(N)}$$

so, for this case,

$$G(\lambda) = \lambda \ .$$

Hence

$$E\{Q\} \to \frac{N}{2\pi} \int_{-\pi}^{\pi} \frac{\Phi(\omega)B(\omega)}{N} \, d\omega = \frac{1}{2\pi} \int_{-\pi}^{\pi} \Phi(\omega)B(\omega) d\omega$$

as $N \to \infty$. But we recognize the expression on the right to be the band power $F(\omega_0)$. Hence the above equation is

$$E\{F_N(\omega_0)\} \to F(\omega_0) \quad as \quad N \to \infty$$

which says that *the band power estimate $F_N(\omega_0)$ is an asymptotically unbiased estimate of the band power $F(\omega)_0$.*

We recall that the variance of the band power estimate is

$$\mathrm{var}\{Q\} = 2[\lambda_1^{(N)}]^2 + 2[\lambda_2^{(N)}]^2 + \ldots + 2[\lambda_N^{(N)}]^2$$

so, for this case,

$$G(\lambda) = 2\lambda^2 \ .$$

Hence

$$\mathrm{var}\{Q\} \to \frac{N}{2\pi} \int_{-\pi}^{\pi} 2\left[\frac{\Phi(\omega)B(\omega)}{N}\right]^2 d\omega = \frac{1}{\pi N} \int_{-\pi}^{\pi} \Phi^2(\omega)B^2(\omega) d\omega$$

as $N \to \infty$. Recalling that $B(\omega)$ is an ideal band-pass transfer function, we see that the expression on the right is equal to[5]

$$\frac{1}{16h^2 \pi N}\left[\int_{-\omega_0-h}^{-\omega_0+h} \Phi^2(\omega)d\omega + \int_{\omega_0-h}^{\omega_0+h} \Phi^2(\omega)d\omega\right] = \frac{2}{16h^2 \pi N} \int_{\omega_0-h}^{\omega_0+h} \Phi^2(\omega)d\omega \ .$$

Thus we have the following asymptotic equation[5] for the variance of the band power estimate:

---

[5] Tee note 5 on page 161.

$$\text{var}\{F_N(\omega_0)\} \to \frac{1}{8h^2\pi N}\int_{\omega_0-h}^{\omega_0+h}\Phi^2(\omega)\,d\omega \approx \frac{\Phi^2(\omega_0)}{4h\pi N} \quad \text{as} \quad N \to \infty .$$

Because of the N in the denominator in the expression on the right hand side, we see that the variance of the band power estimate tends to zero as the sample size N tends to infinity. Thus, *the band power estimate $F_N(\omega_0)$ is a consistent estimate of the band power $F(\omega_0)$.*

References

BLACKMAN, R. B., and TUKEY, J. W. (1959). *The Measurement of Power Spectra.* Dover, New York.

BODE, H. W., and SHANNON, C. E. (1950). A simplified derivation of linear least-square smoothing and prediction theory. *Proc. IRE,* **38,** 417–425.

CRAMÉR, H. (1945). *Mathematical Methods of Statistics.* Princeton University Press, Princeton.

—. (1960). On the structure of purely non-deterministic stochastic processes. *Arkiv för Matematik,* **4,** 249–266.

GRENANDER, U., and ROSENBLATT, M. (1957). *Statistical Analysis of Stationary Time-Series.* John Wiley, New York.

HANNAN, E. J. (1960). *Time Series Analysis.* Methuen, London.

KARHUNEN, K. (1947). Über lineare Methoden in der Wahrscheinlichkeitsrechnung. *Ann. Acad. Sci. Fenn.,* **37.**

KENDALL, M. G., and STUART, A. (1958). *The Advanced Theory of Statistics* (three volumes). Charles Griffin and Co., London.

KLEENE, S. C. (1952). *Introduction to Metamathematics.* Van Nostrand, New York.

KOLMOGOROV, A. N. (1941). Stationary sequences in Hilbert space (Russian). *Bull. Moscow State University, Math.,* **2,** 40 pp.

LANING, J. H., and BATTIN, R. H. (1956). *Random Processes in Automatic Control.* McGraw-Hill, New York.

ROBINSON, E. A. (1954). *Predictive Decomposition of Time Series with Applications to Seismic Exploration.* MIT Ph.D. thesis, Cambridge, Mass.

—. (1957). Predictive decomposition of seismic traces. *Geophysics,* **22,** 767–778.

—. (1959). *An Introduction to Infinitely Many Variates.* Charles Griffin and Co., London, and Stechert-Hafner, New York.

—. (1962). *Random Wavelets and Cybernetic Systems.* Charles Griffin and Co., London, and Stechert-Hafner, New York.

—. (1963). Structural properties of stationary stochastic processes with applications. *Brown University Symposium on Time Series Analysis,* 170–192, John Wiley, New York.

— and WOLD, H. (1963). Minimum-delay structure of least-squares and *eo-ipso* predicting systems for stationary stochastic processes. *Brown University Symposium on Time Series Analysis,* 192–196, John Wiley, New York.

SIMPSON, S. M. (1953). *Statistical Approaches to Certain Problems in Geophysics.* MIT Ph.D. thesis, Cambridge, Mass.

—. (1954). *A Multiple Trace Criterion for Linear Operator Selection*. MIT GAG Report **8**, Cambridge, Mass.

—. (1961). *Initial Studies on Underground Nuclear Detection with Seismic Data Prepared by a Novel Digitization System*. Scientific Report **1**, Geophysics Research Directorate, Air Force Cambridge Research Center, Office of Aerospace Research, Bedford, Massachusetts.

—. (1961). *Time Series Techniques Applied to Underground Nuclear Detection and Further Digitized Seismic Data*. Scientific Report **2** of *ibid*.

—. (1962). *Continued Numerical Studies on Underground Nuclear Detection and Further Digitized Seismic Data*. Scientific Report **3** of *ibid*.

WADSWORTH, G. P., ROBINSON, E. A., BRYAN, J. G., and HURLEY, P. M. (1953). Detection of reflections on seismic records by linear operators. *Geophysics*, **18**, 539–586.

WIENER, N. (1942). *The Extrapolation, Interpolation, and Smoothing of Stationary Time Series with Engineering Applications*. MIT DIC Contract 6037, National Defense Research Council, Section D2, Cambridge, Mass. (Reprinted (1949). John Wiley, New York).

WOLD, H. (1938). *A Study in the Analysis of Stationary Time-series*. Thesis, University of Stockholm. (Second Edition (1954). Almqvist & Wiksell, Uppsala.)

— and JURÉEN, L. (1953). *Demand Analysis*. John Wiley, New York.

—. (1963). Forecasting by the chain principle. *Brown University Symposium on Time Series Analysis*, 471–497, John Wiley, New York.

ZADEH, L. A., and RAGAZZINI, J. R. (1950). An extension of Wiener's theory of prediction. *Journal Appl. Physics*, **21**, 645–655.

ZASUHIN, V. (1941). On the theory of multidimensional stationary random processes. *C.R. (Doklady) Acad. Sci. U.R.S.S., N.S.* **33**, 435–437.

CHAPTER 6

REALIZABILITY AND MINIMUM-DELAY ASPECTS OF MULTICHANNEL MODELS

BY

ENDERS A. ROBINSON
University of Wisconsin, N. A. S. A., and Digital Consultants, Inc.

CONTENTS

REALIZABILITY AND MINIMUM-DELAY ASPECTS OF MULTICHANNEL MODELS

BY

ENDERS A. ROBINSON

1. MATHEMATICAL REVIEW

1.1 The Characteristic Equation of a Matrix

The motivation for introducing the concept of the characteristic equation of a square matrix a of order n may be described in the following way[1]. Let $y=ax$ be a linear transformation which carries a vector x into a vector y. We want to investigate the possibility of certain vectors x being carried by the transformation into λx, where λ is a scalar. Any such vector x which is transformed into λx, that is, any vector x for which

$$(1.11) \qquad\qquad ax=\lambda x$$

is called an *invariant vector* or *characteristic vector* or *eigenvector* under the transformation.

Let us now introduce the characteristic equation. From equation (1.11) we may write

$$(1.111) \qquad\qquad \lambda x-ax=0$$

which may be factored as

$$(1.112) \qquad\qquad (\lambda I-a)x=0 \; .$$

---

[1] Before proceeding any further, let us explain some of the mathematical symbols used. An asterisk * in the superscript position, as well as a horizontal bar over a symbol, each denote complex conjugate. However, an asterisk * on line between two symbols denotes convolution. A superscript T indicates matrix transpose. I denotes the identity matrix, namely a square matrix with ones on its diagonal and zeros elsewhere. adj a denotes the adjugate of the matrix a; the adjugate is defined as adj $a = a^{-1} (\det a) = (\det a)\, a^{-1}$. The superscript -1 on a matrix indicates matrix inversion.

The matrix $(\lambda I - a)$ is called the *characteristic matrix* of a. The system of homogenous equations (1.112) has non-trivial solutions if and only if the determinant

$$(1.12) \qquad \det\,(\lambda I - a) = 0 \ .$$

The expansion of this determinant yields a polynomial $\varphi(\lambda)$ of degree n in λ, where n is the order of the matrix a. This polynomial is known as the *characteristic polynomial* of the matrix a. Equation (1.12), which is

$$(1.121) \qquad \varphi(\lambda) = 0 \ ,$$

is called the characteristic equation of a and its roots $\lambda_1, \lambda_2, \ldots, \lambda_n$ are called the *characteristic roots*, *eigenroots*, or *eigenvalues* of a. If $\lambda = \lambda_i$ is an eigenroot then equation (1.112) has a non-trivial solution x_i which is the eigenvector associated with the eigenroot λ_i, that is

$$(1.13) \qquad (\lambda_i I - a)x_i = 0 \ .$$

The following are some of the general theorems about eigenroots and eigenvectors.

1.   If $\lambda_1, \lambda_2, \ldots, \lambda_n$ are distinct eigenroots of a and if $x_1, x_2, \ldots, x_n$ are non-zero eigenvectors respectively associated with these eigenroots, then the $x_1, x_2, \ldots, x_n$ are linearly independent.

2.   The eigenroots of a and a^T are the same.

3.   The eigenroots of a^* are the conjugates of the eigenroots of a.

4.   If a is non-singular and if λ_i is an eigenroot of a, then λ_i^{-1} is an eigenroot of a^{-1} and $\lambda_i^{-1} \det a$ is an eigenroot of adj a.

5.   The characteristic equation of an orthogonal matrix p is a reciprocal equation. Proof:

$$
\begin{aligned}
\varphi(\lambda) &= \det\,(\lambda I - p) = \det\,(\lambda p I p^T - p) = \det\left(-p\lambda(\lambda^{-1} I - p^T)\right) \\[4pt]
&= \pm \lambda^n \det\,(\lambda^{-1} I - p) = \pm \lambda^n \varphi(\lambda^{-1}) \ .
\end{aligned}
$$

(1.14)

Instead of using the variable λ in what follows, we will find it more convenient to use the variable z defined as the reciprocal of λ, that is

$$(1.15) \qquad z = \lambda^{-1} \ .$$

Thus the characteristic equation (1.112) becomes

$$(1.16) \qquad (z^{-1}I-a)x=0$$

which upon multiplying through by z becomes

$$(1.161) \qquad (I-az)x=0 \ .$$

The matrix $(I-az)$ is the *characteristic matrix*. The system of homogenous equations (1.161) has a non-trivial solution provided the determinant

$$(1.17) \qquad \det (I-az)=0 \ .$$

The expansion of this determinant yields a polynomial $\Psi(z)$ of degree n in z. This polynomial may be called the *characteristic polynomial* (in the variable z) of the matrix a. The roots of this polynomial $z_1,z_2,...,z_n$ may be called the *eigenroots* or *eigenvalues* (again in terms of the variable z). These eigenroots are related to those in terms of the variable λ by

$$(1.18) \qquad z_1=\lambda_1^{-1} , z_2=\lambda_2^{-1} , ... , z_n=\lambda_n^{-1}$$

since

$$(1.181) \qquad \Psi(z)=z^n\varphi(z^{-1}) \ .$$

If $z=z_i$ is an eigenroot, then equation (1.161) has a non-trivial solution x_i which is the eigenvector associated with the eigenroot z_i, that is

$$(1.182) \qquad (I-az_i)x_i=0 \ .$$

The eigenvector x_i is the same eigenvector as the one associated with the eigenroot $\lambda_i=z_i^{-1}$.

1.2 Similar Matrices

Two square matrices a and b are called *similar* provided there exists a non-singular matrix r such that

$$(1.21) \qquad b=r^{-1} a \, r \ .$$

The following two theorems hold for similar matrices a and b.

1.   If z_i is an eigenroot of a, then z_i is also an eigenroot of b.

2.   If y_i is an eigenvector of b corresponding to the eigenroot z_i of b, then $x_i = r y_i$ is an eigenvector of a corresponding to the same eigenroot z_i of a.

A matrix d is called a diagonal matrix provided that all its elements vanish except those upon its main diagonal. If the elements upon its main diagonal are denoted by $d_{11}, d_{22}, \ldots, d_{nn}$ then we shall denote the diagonal matrix d by

$$(1.22) \qquad d = \mathrm{diag}\,(d_{11}, d_{22}, \ldots, d_{nn}) \ .$$

The characteristic roots z_i of a diagonal matrix are simply the reciprocals of its diagonal elements, that is $z_i = d_{ii}^{-1}$ for $i = 1, 2, \ldots, n$. A set of eigenvectors of a diagonal matrix is the set of elementary vectors e_i (where e_i is defined as the vector with all zero elements, except the ith element which is unity). As a consequence, it follows that a diagonal matrix always has n linearly independent eigenvectors. Thus a square matrix a is similar to a diagonal matrix if and only if a has n linearly independent eigenvectors.

It follows therefore that not every square matrix is similar to a diagonal matrix. However, every square matrix a is similar to a triangular matrix whose diagonal elements are the eigenroots of a.

1.3 Polynomials

A polynomial is an expression of the form

$$(1.31) \qquad A(z) = a_0 + a_1 z + a_2 z^2 + \ldots + a_m z^m$$

where the coefficients, in particular, may be real or complex constants or real or complex matrices and where z is any symbol. Since nothing is assumed known about z, the symbol z is usually called an *indeterminate*. Using only the rules for adding, subtracting, and multiplying the coefficients, one can add, subtract, and multiply any two polynomials. Thus the polynomial written above can be multiplied with the polynomial

$$(1.32) \qquad B(z) = b_0 + b_1 z + b_2 z^2 + \ldots + b_n z^n$$

to yield the polynomial

$$(1.33) \qquad A(z)B(z) = \sum_{i=0}^{m} \sum_{j=0}^{n} a_i b_j z^{i+j} \ .$$

Letting $k=i+j$ this expression becomes

$$(1.331) \qquad A(z)B(z) = \sum_{k=0}^{m+n} \left(\sum_i a_i b_{k-i} \right) z^k .$$

In this result the coefficient

$$(1.34) \qquad c_k = \sum_i a_i b_{k-i}$$

of z^k is a sum for all i with $0 \leqslant k-i \leqslant n$. The coefficients

$$(c_0, c_1, c_2, \ldots, c_{m+n})$$

represent the complete (transient) convolution of the coefficients

$$(a_0, a_1, a_2, \ldots, a_m)$$

and

$$(b_0, b_1, b_2, \ldots, b_n) .$$

This terminology comes from the fact that the convolution of two series a_i and b_i of indefinite extent (i. e., $-\infty < i < \infty$) is defined as

$$(1.35) \qquad c_k = \sum_{i=-\infty}^{\infty} a_i b_{k-i} .$$

In the case when the coefficients a_i vanish for values of i outside of the range $0 \leqslant i \leqslant m$ and the coefficients b_j vanish for values of j outside of the range $0 < j < n$, then the coefficients c_k given by equation (1.35) vanish for values of k outside of the range $0 < k < m+n$, and are the same as those given by the polynomial multiplication algorithm, equation (1.34).

1.4   Matrix Polynomials

Let z denote an abstract symbol (an indeterminate) which is assumed to be commutative with itself and with matrices. Thus if a is a matrix then

$$zaz = z^2 a = a z^2 .$$

Let $a_0, a_1, \ldots, a_{p-1}, a_p$ be square matrices of order n. Then $A(z)$ defined as

$$A(z) = a_0 + a_1 z + \ldots + a_{p-1} z^{p-1} + a_p z^p$$

is called a *polynomial matrix* or alternatively a *matrix polynomial*. A matrix polynomial $A(z)$ is called singular or non-singular according to whether det $A(z)$ is or is not zero.

Let $A(z)$ and $B(z)$ be two matrix polynomials:

$$(1.41) \qquad A(z) = a_0 + a_1 z + \ldots + a_{p-1} z^{p-1} + a_p z^p$$

$$(1.42) \qquad B(z) = b_0 + b_1 z + \ldots + b_{q-1} z^{q-1} + b_q z^q \; .$$

We shall say that these two polynomial matrices are equal, that is, $A(z) = B(z)$, provided that $p = q$ and $a_i = b_i$ for $i = 0, 1, 2, \ldots, p$.

The sum of two polynomial matrices $A(z)$ and $B(z)$ is a polynomial matrix $C(z) = A(z) + B(z)$ whose coefficients are obtained by adding the corresponding coefficients of $A(z)$ and $B(z)$. The product of two polynomial matrices $A(z)$ and $B(z)$ is a polynomial matrix $C(z) = A(z)B(z)$ whose coefficients are obtained by convolving the coefficients of $A(z)$ with $B(z)$. In general, two polynomial matrices are not commutative, that is $A(z)B(z)$ is not in general equal to $B(z)A(z)$. The product $A(z)B(z)$ is a polynomial matrix of degree of most $p+q$. If either $A(z)$ or $B(z)$ is non-singular, the degree of $A(z)B(z)$ and also of $B(z)A(z)$ is exactly $p+q$.

The equality

$$A(z) = a_0 + a_1 z + \ldots + a_{p-1} z^{p-1} + a_p z^p$$

is not disturbed when z is considered as a (scalar) complex number. In our work we shall generally assume that z is a (scalar) complex variable. However, when z is replaced by a square matrix C, several results can be obtained depending upon how we distribute the C's on each side of the coefficients A_i. In particular we define

$$(1.43) \qquad A_R(C) = a_0 + a_1 C + \ldots + a_{p-1} C^{p-1} + a_p C^p$$

and

$$(1.431) \qquad A_L(C) = a_0 + C a_1 + \ldots + C^{p-1} a_{p-1} + C^p a_p$$

as respectively the right and left functional values of $A(C)$.

Let $A(z)$ and $B(z)$ be the matrix polynomials (1.41) and (1.42) and suppose b_q is non-singular. Then there exists unique matrix polynomials $Q_1(z)$ and $R_1(z)$, where $R_1(z)$ is either zero or of degree less than $B(z)$, such that

$$(1.44) \qquad A(z) = Q_1(z)\, B(z) + R_1(z) \ .$$

If $R_1(z)=0$, then $B(z)$ is called a *right divisor* of $A(z)$. Similarly there exists unique matrix polynomials $Q_2(z)$ and $R_2(z)$, where $R_2(z)$ is either zero or of degree less than that of $B(z)$, such that

$$(1.441) \qquad A(z) = B(z)\, Q_2(z) + R_2(z) \ .$$

If $R_2(z)=0$, then $B(z)$ is called a *left divisor* of $A(z)$.

A matrix polynomial of the form

$$
\begin{aligned}
(1.45) \qquad S(z) &= s_0 I + s_1 z I + \ldots + s_{q-1} z^{q-1} I + s_q z^q I \\
&= (s_0 + s_1 z + \ldots + s_{q-1} z^{q-1} + s_q z^q) I
\end{aligned}
$$

(where $s_0, s_1, \ldots, s_{q-1}, s_q$ are scalars and I is the identity matrix) is called a *scalar matrix polynomial*. That is, a scalar matrix polynomial $S(z)$ is equal to a scalar polynomial times the identity matrix I. A scalar matrix polynomial $S(z)$ commutes with every matrix polynomial, that is,

$$(1.451) \qquad A(z)S(z) = S(z)A(z) \ .$$

If in equations (1.44) and (1.441), $B(z)$ is a scalar matrix polynomial $S(z)$, then

$$(1.452) \qquad A(z) = Q_1(z)S(z) + R_1(z) = S(z)Q_1(z) + R_1(z) \ .$$

A matrix polynomial $A(z) = \big[a_{ij}(z)\big]$ of degree n is divisible by a scalar matrix polynomial $S(z)$ given by equation (1.45) if and only if every $a_{ij}(z)$ is divisible by the scalar polynomial $s_0 + s_1 z + \ldots + s_{q-1} z^{q-1} + s_q z^q$.

Let us now outline the remainder theorem. Let $A(z)$ be the matrix polynomial given by equation (1.41) and let $b=[b_{ij}]$ be a constant non-singular square matrix. Since $I - bz$ is non-singular, we may write

$$(1.46) \qquad A(z) = Q_1(z)(I - bz) + R_1$$

and

$$(1.461) \qquad A(z) = (I - bz)Q_2(z) + R_2$$

where R_1 and R_2 are free of z. The *remainder theorem* states that the remainder R_1 is given by

$$(1.462) \qquad R_1 = A_R(b^{-1}) = a_0 + a_1 b^{-1} + \ldots + a_{p-1} b^{-p+1} + a_p b^{-p}$$

and the remainder R_2 is given by

$$(1.463) \qquad R_2 = A_L(b^{-1}) = a_0 + b^{-1} a_1 + \ldots + b^{-p+1} a_{p-1} + b^{-p} a_p \, .$$

In the special case when $A(z)$ is a scalar matrix polynomial, say

$$(1.47) \qquad \begin{aligned} A(z) &= S(z)I \\ &= (s_0 + s_1 z + \ldots + s_{p-1} z^{p-1} + s_p z^p)I \end{aligned}$$

where $s_0, s_1, \ldots, s_{p-1}, s_p$ are scalars, then the remainders R_1 and R_2 are identical, so that

$$(1.471) \qquad R_1 = R_2 = s_0 I + s_1 b^{-1} + \ldots + s_{p-1} b^{-p+1} + s_p b^{-p} = S(b^{-1}) \, .$$

As a consequence we have that a scalar matrix polynomial $S(z)I$ is divisible by $I - bz$ if and only if $S(b^{-1}) = 0$.

From these results the Cayley-Hamilton theorem (for a non-singular matrix) may readily be proved. Consider the non-singular square matrix $a = [a_{ij}]$ having characteristic matrix

$$(1.48) \qquad\qquad I - az$$

and characteristic equation

$$(1.481) \qquad\qquad \Psi(z) = \det (I - az) = 0 \, .$$

By definition of the adjugate, adj $(I - az)$, we have

$$(1.482) \qquad\qquad (I - az) \text{ adj } (I - az) = \Psi(z)I \, .$$

Hence the scalar polynomial $\Psi(z)I$ is divisible by $I - az$, and hence $\Psi(a^{-1}) = 0$. Thus we have proved the *Cayley-Hamilton theorem* for a non-singular square matrix a, namely:

The inverse a^{-1} of a non-singular square matrix a satisfies its characteristic equation

$$(1.49) \qquad\qquad \Psi(a^{-1}) = 0$$

where $\Psi(z)$ is the characteristic polynomial in z defined by

$$(1.491) \qquad\qquad \Psi(z) = \det (I - az) \, .$$

1.5   Factorization of Matrix Polynomials

Suppose that we are given the matrix polynomial

$$(1.51) \qquad A(z) = a_0 + a_1 z + a_2 z^2 + \ldots + a_m z^m$$

where $a_0, a_1, a_2, \ldots, a_m$ are each $n \times n$ matrices. The problem is:

Problem.   To find the $n \times n$ matrices b_k such that

$$(1.52) \qquad A(z) = b_0(I - zb_1)(I - zb_2) \ldots (I - zb_m) .$$

The solution to this problem is as follows:
The adjugate $\mathrm{adj} A(z)$ is defined by

$$(1.53) \qquad \mathrm{adj}\ A(z) = A(z)^{-1}\ \mathrm{det}\ A(z)$$

where $A(z)^{-1}$ is the inverse of $A(z)$ and where det $A(z)$ is the determinant of $A(z)$. Postmultiply equation (1.52) by $\mathrm{adj} A(z)$. We have

$$(1.531) \qquad I\ \mathrm{det}\ A(z) = b_0(I - zb_1)(I - zb_2) \ldots (I - zb_m)\ \mathrm{adj}\ A(z) .$$

Let z_k (when $k = 1, 2, \ldots, mn$) be the zeros of $\mathrm{det} A(z)$. That is

$$(1.54) \qquad \mathrm{det}\ A(z_k) = 0 .$$

(Assume the zeros are non-zero, distinct, and finite). Thus the matrix $A(z_k)$ is singular. By a theorem in Frazer, Duncan, and Collar (1938, page 61) it follows that adj $A(z_k)$ has rank at most 1 and can be factored as

$$(1.541) \qquad \mathrm{adj}\ A(z_k) = c_k r_k$$

where c_k is a column and r_k is a row. For $z = z_k$, equation (1.531) becomes

$$(1.55) \qquad \begin{aligned} 0 &= b_0(I - z_k b_1)\ (I - z_k b_2) \ldots (I - z_k b_m)\ \mathrm{adj} A(z_k) \\ &= b_0(I - z_k b_1)\ (I - z_k b_2) \ldots (I - z_k b_m) c_k r_k . \end{aligned}$$

Equation (1.55) is satisfied provided

$$(1.56) \qquad (I - z_k b_m) c_k = 0 .$$

But equation (1.56) states that c_k is an eigenvector of b_m with eigenvalue $\lambda_k = 1/z_k$. Pick n roots, say $z_1, z_2, \ldots, z_n$, from the mn roots of det $A(z)$ that are available. Then equation (1.56) is

$$(1.561) \qquad c_k = z_k b_m c_k \quad k = 1, 2, \ldots, n \,.$$

If we let $C = (c_1, c_2, \ldots, c_n)$ and $D =$ the diagonal matrix with $z_1, z_2, \ldots, z_n$ on the diagonal, then equation (1.561) becomes

$$(1.562) \qquad b_m c_k = c_k \, \frac{1}{z_k} \quad k = 1, 2, \ldots, n$$

which gives

$$(1.563) \qquad b_m C = C\, D^{-1}$$

or

$$(1.564) \qquad b_m = C\, D^{-1} C^{-1} \,.$$

Thus the matrix b_m has been determined.

Consider now

$$(1.57) \qquad I \det A(z) = b_0 (I - z b_1) \ldots (I - z b_{m-1}) \left[(I - z b_m) \text{ adj } A(z) \right] \,.$$

The factor

$$(1.571) \qquad (I - z b_m) \text{ adj} A(z)$$

also becomes the product of a column and a row at the remaining zeros of $\det A(z)$. Since $\det A(z)$ has mn zeros, we can continue until all the b_i are determined.

Since $A(z=0) = a_0 = b_0$, we obtain $b_0 = a_0$.

Thus we have found $B(z)$ satisfying

$$(1.58) \qquad I \det A(z) = B(z) \text{ adj } A(z)$$

for $z = z_k$ $(k = 1, \ldots, mn)$ and for z_0. But

$$(1.581) \qquad I \det A(z) = A(z) \text{ adj } A(z)$$

for all z. Each of the elements of the matrix equations (1.58) and (1.581) is a polynomial of degree at most mn. From a theorem in algebra, it is known that if any two polynomials of degree mn are equal to each other at $mn+1$ different values of the independent variable z, then they are equal over the entire range of z. Thus

(1.582)
$$I \det A(z) = B(z) \text{ adj } A(z)$$

for all z or

(1.583)
$$A(z) = B(z) \ .$$

Since the choice of the n zeros for each of the m factors is arbitrary, it follows that there are $(mn)!/(n!)^m$ ways to factor $A(z)$.

Example:

$$A(z) = a_0 + a_1 z + a_2 z^2 \qquad m=2, \ n=2$$

$$a_0 = \begin{bmatrix} 1 & 0 \\ 0 & 1 \end{bmatrix} \qquad a_1 = \begin{bmatrix} -3 & -1 \\ 14 & -11 \end{bmatrix} \qquad a_2 = \begin{bmatrix} -4 & 4 \\ -58 & 28 \end{bmatrix}$$

factors in $\dfrac{(mn)!}{(n!)^m} = \dfrac{4!}{(2!)^2} = 6$ ways as

$$\left\{ I + z \begin{bmatrix} 2 & -1 \\ 20 & -7 \end{bmatrix} \right\} \qquad \left\{ I + z \begin{bmatrix} -5 & 0 \\ -6 & -4 \end{bmatrix} \right\}$$

$$\left\{ I + z \begin{bmatrix} -4 & 0 \\ -10 & -2 \end{bmatrix} \right\} \qquad \left\{ I + z \begin{bmatrix} 1 & -1 \\ 24 & -9 \end{bmatrix} \right\}$$

$$\left\{ I + z \begin{bmatrix} 0 & -1 \\ 10 & -7 \end{bmatrix} \right\} \qquad \left\{ I + z \begin{bmatrix} -3 & 0 \\ 4 & -4 \end{bmatrix} \right\}$$

$$\left\{ I + \begin{bmatrix} -4 & 0 \\ -4 & -3 \end{bmatrix} \right\} \qquad \left\{ I + z \begin{bmatrix} 1 & -1 \\ 18 & -8 \end{bmatrix} \right\}$$

$$\left\{ I + z \begin{bmatrix} -1 & -1 \\ 8 & -7 \end{bmatrix} \right\} \qquad \left\{ I + z \begin{bmatrix} -2 & 0 \\ 6 & -4 \end{bmatrix} \right\}$$

$$\left\{ I + z \begin{bmatrix} -4 & 0 \\ 2 & -5 \end{bmatrix} \right\} \qquad \left\{ I + z \begin{bmatrix} 1 & -1 \\ 12 & -6 \end{bmatrix} \right\}$$

2. THEORY OF FINITE DISCRETE LINEAR OPERATORS

2.1 The Finite Discrete Linear Operator

In this chapter we wish to consider properties of the finite discrete linear operator for a single time series, that is, a linear operator which has a finite number of discrete operator coefficients which perform linear operations on single discrete time series.

In this paper we deal exclusively with discrete time series. A discrete time series is a sequence of equidistant observations x_t which are associated with the discrete time parameter t. Without loss of generality we may take the spacing between each successive observation to be one unit of time, and thus we may represent the time series as

$$(2.11) \qquad \ldots, x_{t-2}, x_{t-1}, x_t, x_{t+1}, x_{t+2} \ldots$$

where t takes on integer values. As a result the minimum period which may be observed is equal to two units, and consequently the maximum frequency which may be observed is equal to $1/2$, which is an angular frequency of π. Thus, we may require all frequencies f of the discrete time series to lie between $-1/2$ and $1/2$, and all angular frequencies $\omega = 2\pi f$ to lie between $-\pi$ and π.

Time series x_t may be finite or infinite in extent, and may be treated from a deterministic or statistical point of view. In this chapter we shall develop those properties of the finite discrete linear operator which are independent of the nature of the x_t time series, whereas in subsequent chapters we shall be mainly concerned with the nature of the x_t time series.

Following Kolmogorov (1939, 1941) we shall distinguish between extrapolation or prediction type operators on the one hand, and interpolation or smoothing type operators on the other hand.

2.2 Prediction Operators

The extrapolation or prediction operator (Kolmogorov, 1939, 1941) is given by

$$(2.21) \qquad \hat{x}_{t+\alpha} = k_o x_t + k_1 x_{t-1} + \ldots + k_M x_{t-M}$$

$$= \sum_{s=0}^{M} k_s x_{t-s}, \ \alpha \geqslant 0$$

where α is the prediction distance. The operator coefficients, $k_0, k_1, \ldots, k_M$ are chosen so that the actual output, $\hat{x}_{t+\alpha}$, approximates the desired output, $x_{t+\alpha}$, according to some criterion. The operator is discrete since its coefficients are discrete values, and the operator is finite since there are only a finite number M of such coefficients.

In Section 5.2 we discuss the solution of the prediction problem in which the operator coefficients are determined by the least squares criterion. In general, we shall see that an infinite number of operator coefficients are required, although for autoregressive time series only a finite number of operator coefficients are required.

The prediction operator (2.21) is one from which a specific time function may be generated if a sufficient number of initial values of the time function are specified. That is, if we are given the initial values

$$(2.22) \qquad x_0, x_1, x_2, \ldots, x_{M+\alpha-1},$$

and if we define

$$(2.221) \qquad \hat{x}_t = x_t \ \text{for} \ t = M+\alpha, \ M+\alpha+1, \ M+\alpha+2, \ \ldots$$

then we may generate the time function x_t by means of equation (2.21). For example, ·let us consider the case given by

$$(2.23) \qquad \hat{x}_{t+1} = .5 x_t, \ x_0 = 1.$$

Then we may generate the time function

$$(2.24) \qquad x_t = \hat{x}_t = .5 x_{t-1} \ \text{for} \ t = 1, 2, 3, \ldots$$

which is the sequence

$$(2.241) \qquad (.5), (.5)^2, (.5)^3, (.5)^4, (.5^5), \ldots, (.5)^t, (.5)^{t+1}, \ldots$$

In Section 2.6 we shall see that such a sequence must form a damped motion in order for the operator to be minimum-delay.

Further, the prediction operator (2.21) has the property that only the values x_t, x_{t-1}, x_{t-2},... of the time series at time t and prior to time t, and no values x_{t+1},x_{t+2},..., subsequent to time t, are required in order to compute the actual output $\hat{x}_{t+\alpha}$. Thus a prediction operator has an inherent one-sidedness in that it operates on present and past values, but no future values, of the time series. As a result if time t represents the present calendar time, as it is the case for a meteorologist who makes daily weather predictions, then only observations of the time series at the present time and at past times, and no observations at future times are required to carry out the necessary computation.

As we shall see in Section 2.5, the prediction operator coefficients represent the impulsive response of an equivalent electric network. Thus those impulses which have arrived at time t or prior to time t will make the network respond, whereas those impulses which have not yet arrived at time t, i. e., those impulses which arrive subsequent to time t, cannot make the network respond. In summary, then, we may say that a finite prediction operator is computationally realizable to the statistician, and physically realizable to the engineer.

Instead of considering the predicted values $\hat{x}_{t+\alpha}$ as the output of the prediction operator, one may consider the prediction error $\xi_{t+\alpha}=x_{t+\alpha}-\hat{x}_{t+\alpha}$ as the output (Wadsworth, *et al*, 1953). Then we have

$$(2.25) \qquad \xi_{t+\alpha}=x_{t+\alpha}-\hat{x}_{t+\alpha}=x_{t+\alpha}-\sum_{s=0}^{M} k_s x_{t-s}$$

which may be rewritten

$$(2.26) \qquad \xi_t=x_t-\sum_{s=0}^{M} k_s x_{t-s-\alpha}$$

$$=x_t-k_0 x_{t-\alpha}-k_1 x_{t-\alpha-1}-\ldots-k_M x_{t-\alpha-M}\ .$$

Let us define $m=M+\alpha$ and

$$(2.27) \qquad a_0=1, a_1=0, a_2=0,\ldots,a_{\alpha-1}=0,\ a_\alpha=-k_0,$$

$$a_{\alpha+1}=-k_1,\ldots,\ a_{\alpha+M}=a_m=-k_M\ .$$

Then the prediction error ξ_t may be rewritten as the convolution of the operator coefficients with the time series, that is

$$\xi_t=a_0 x_t+a_1 x_{t-1}+\ldots+a_m x_{t-m}$$

$$(2.271) \qquad =\sum_{s=0}^{m} a_s x_{t-s},\ a_0=1\ .$$

In the sequel we shall be primarily concerned with the prediction operator for prediction distance α equal to one. Then equation (2.271) for the prediction error $(\alpha=1)$ becomes

$$(2.28) \qquad \xi_t = \sum_{s=0}^{m} a_s x_{t-s}$$

$$= x_t + a_1 x_{t-1} + \dots + a_m x_{t-m}, \quad a_0 = 1$$

and equation (2.27) becomes

$$(2.281) \qquad a_0 = 1, \; a_1 = -k_0, \; a_2 = -k_1, \; \dots, \; a_m = -k_M \, .$$

We shall regard equation (2.28) as the basic form of the prediction type operator, the prediction error operator, or more generally the realizable, or one-sided memory type, operator, with the coefficients $a_0, a_1, a_2, \dots, a_m$. The prediction error ξ_t shall be regarded as the actual output at time t of the operator (2.28).

In general, the operator coefficients are chosen so that the actual output approximates a certain desired output according to some criterion. In Section 5.2 we discuss the solution to the general filtering problem in which realizable operators are used, the coefficients of which are chosen according to the least squares criterion.

In equation (2.28) we may let the operator coefficients represent the impulsive response of a network and the time series x_t the input. Then ξ_t is the actual output of the network.

2.3 Smoothing Operators

The interpolation or smoothing operator (Kolmogorov, 1939, 1941) is given by

$$\hat{x}_t = d_{-M} x_{t+M} + d_{-M+1} x_{t+M-1} + \dots + d_{-1} x_{t-1}$$

$$+ d_1 x_{t-1} + d_2 x_{t-2} + \dots + d_M x_{t-M}$$

$$(2.31) \qquad = \sum_{\substack{s=-M \\ s \neq 0}}^{M} d_s x_{t-s} \, .$$

We may consider the smoothing error given by

$$(2.32) \qquad \gamma_t = x_t - \hat{x}_t = -\sum_{s=-M}^{1} d_s x_{t-s} + x_t - \sum_{s=1}^{M} d_s x_{t-s}$$

and by letting $m = M$, $c_0 = 1$, and $c_s = -d_s$ $(s = \pm 1, \pm 2, ..., \pm M)$ we have

$$(2.33) \qquad \gamma_t = \sum_{s=-m}^{m} c_s x_{t-s}, \; c_0 = 1 .$$

In the sequel we shall regard this equation as the basic form of the smoothing type operator, the smoothing error operator, or more generally the non-realizable, or two-sided (memory and anticipation) type, operator, with operator coefficients $c_{-m}, c_{-m+1}, ..., c_0 = 1, c_1, ..., c_m$. The smoothing error γ_t is the actual output of the operator.

The smoothing operator does not have the same properties in regard to computational or physical realizability as the prediction operator. In particular, it is not possible to generate a time function from specified initial values by means of equation (2.31) as was done in the case of the prediction operator. The smoothing operator (2.33) has the property that values of the time series $x_{t+m}, ..., x_{t+2}, x_{t+1}$ at times subsequent to time t, as well as values $x_t, x_{t-1}, x_{t-2}, ..., x_{t-m}$ at time t and prior to time t, are required to compute γ_t. Consequently, if time t represents the present calendar time, as in the example of the meteorologist, then γ_t given by equation (2.33) cannot be computed since it involves observations $x_{t+1}, x_{t+2}, ..., x_{t+m}$ at future times and which thereby are not observable at the present time t. Similarly, the network which would be equivalent to a smoothing type operator would be one which would respond to impulses which have not yet arrived at the present time t. Consequently, smoothing type operators are not computationally realizable to the statistician, or physically realizable to the engineer.

Nevertheless, a very simple trick makes finite smoothing type operators computationally and physically realizable. That is, in order to compute

$$(2.33) \qquad \gamma_t = \sum_{s=-m}^{m} c_s x_{t-s} = c_{-m} x_{t+m} + c_{-m+1} x_{t+m-1} + ... + c_{-1} x_{t+1}$$

$$+ c_0 x_t + c_1 x_{t-1} + ... + c_m x_{t-m}$$

the statistician must delay his computations until time $t+m$, or later, which is a time delay of m or greater, at which time all the values needed in the computation will have occurred. That is, the statistician delays his computations at least until time $t'=t+m$, and then computes

(2.34)

$$\gamma_t = \sum_{s=-m}^{m} c_s x_{t-s} = \sum_{s=-m}^{m} c_s x_{t'-m-s}$$

$$= c_{-m} x_{t'} + c_{-m+1} x_{t'-1} + \ldots + c_{-1} x_{t'-m-1}$$

$$+ c_0 x_{t'-m} + c_1 x_{t'-m-1} + \ldots + c_m x_{t'-2m}$$

which we shall call the time-delay form of the smoothing error operator with coefficients $c_s(s=0, \pm 1, \ldots, \pm m)$. Such an operator is in the form of the realizable operator (2.28) with the present time now being t'. That is, t' is the time at which computations are to be carried out.

Similarly the engineer may introduce a time delay m, or greater, into his network (Bode and Shannon, 1950) to transform the non-physically realizable system (at the time instant t)

(2.35)

$$\sum_{s=-m}^{m} c_s x_{t-s}$$

into the physically realizable system (at the time $t'=t+m$)

(2.36)

$$\sum_{s=-m}^{m} c_s x_{t'-m-s} \; .$$

2.4   The Transfer Function or Filter Characteristics

As Wiener (1942) points out, the linear operator is the approach from the standpoint of time to a filter which is essentially an instrument for the separation of different frequency ranges. The filtering action of the linear operator is brought out by its transfer function, which is the analogue of the transfer or system function of the linear system of which the linear operator is the unit impulse response.

Smith (1954) gives the following interpretation to the transfer function. Let the input information x_t be points from a sine wave of angular frequency ω. Since the system is linear, the output will be a sine wave of the same frequency but, in general, will differ in phase and amplitude. Using the complex notation for a sine wave, $x_t = e^{i\omega t}$, of angular frequency ω, the transfer ratio at angular frequency ω is the output of the linear operator, which is a complex sine wave of angular frequency ω, divided by the input $x_t = e^{i\omega t}$. The transfer function is the totality of these transfer ratios for $-\pi \leqslant \omega \leqslant \pi$, and represents the filter characteristics of the linear operator. As we shall now see the transfer function is the Fourier transform of the linear operator. Since the operator is discrete, the transfer function is in the form of a Fourier series rather than a Fourier integral.

For the prediction operator, equation (2.21),

$$(2.21) \qquad \hat{x}_{t+\alpha} = \sum_{s=0}^{M} k_s x_{t-s}, \quad \alpha \geqslant 0,$$

by letting $x_t = e^{i\omega t}$ we obtain the transfer ratio

$$(2.41) \qquad \frac{\hat{x}_{t+\alpha}}{x_t} = \frac{\displaystyle\sum_{s=0}^{M} k_s x_{t-s}}{x_t} = \frac{\displaystyle\sum_{s=0}^{M} k_s e^{i\omega(t-s)}}{e^{i\omega t}} = \sum_{s=0}^{M} k_s e^{-i\omega s}.$$

The totality of these transfer ratios yields the transfer function

$$(2.441) \qquad K(\omega) = \sum_{s=0}^{M} k_s e^{-i\omega s}$$

which is the Fourier transform of the operator coefficients k_s.

By letting $x_t = e^{i\omega t}$ be the input for the operator for the prediction error, equation (2.28),

$$(2.28) \qquad \xi_t = \sum_{s=0}^{m} a_s x_{t-s}, \quad a_0 = 1,$$

we obtain the transfer ratio

$$(2.42) \qquad \frac{\xi_t}{x_t} = e^{-i\omega t} \sum_{s=0}^{m} a_s e^{i\omega(t-s)} = \sum_{s=0}^{m} a_s e^{-i\omega s}.$$

The transfer function is then

$$(2.421) \qquad A(\omega) = \sum_{s=0}^{m} a_s e^{-i\omega s}$$

which is the Fourier transform of the operator coefficients.
For the smoothing operator, equation (2.31),

$$(2.31) \qquad \hat{x}_t = \sum_{\substack{s=-M \\ s \neq 0}}^{M} d_s x_{t-s} \, ,$$

by letting the input be $x_t = e^{i\omega t}$, the transfer ratio is

$$(2.43) \qquad \frac{\hat{x}_t}{x_t} = e^{-i\omega t} \sum_{\substack{s=-M \\ s \neq 0}}^{M} d_s e^{i\omega(t-s)} = \sum_{\substack{s=-M \\ s \neq 0}}^{M} d_s e^{-i\omega s}$$

and the transfer function is

$$(2.431) \qquad D(\omega) = \sum_{\substack{s=-M \\ s \neq 0}}^{M} d_s e^{-i\omega s} \, .$$

Similarly letting $x_t = e^{i\omega t}$ be the input for the operator for the smoothing error

$$(2.33) \qquad \Upsilon_t = \sum_{s=-m}^{m} c_s x_{t-s}$$

we obtain the transfer ratio

$$(2.44) \qquad \frac{\Upsilon_t}{x_t} = e^{-i\omega t} \sum_{s=-m}^{m} c_s e^{i\omega(t-s)} = \sum_{s=-m}^{m} c_s e^{-i\omega s}$$

so that the transfer function is

$$(2.441) \qquad C(\omega) = \sum_{s=-m}^{m} c_s e^{-i\omega s}$$

Using the same operator coefficients c_s $(s=0, \pm 1, \ldots, \pm m)$ but introducing a time delay m, the time-delay smoothing error operator, equation (2.34),

$$(2.34) \qquad \gamma_t = \sum_{s=-m}^{m} c_s x_{t'-m-s}$$

is realizable at the time instant $t'=t+m$. Since t' now represents the time instant at which computations are to be carried out the input is $x_{t'} = e^{i\omega t'}$. The transfer ratio is

$$(2.45) \qquad \frac{\sum_{s=-m}^{m} c_s e^{i\omega(t'-m-s)}}{e^{i\omega t'}} = \sum_{s=-m}^{m} c_s e^{-i\omega(s+m)}$$

and its transfer function is

$$(2.451) \qquad \sum_{s=-m}^{m} c_s e^{-i\omega(s+m)} = e^{-i\omega m} \sum_{s=-m}^{m} c_s e^{-i\omega s} = e^{-i\omega m} C(\omega) \ .$$

A summary of the various linear operator forms and their transfer functions is given in Tables 2.1 and 2.2.

2.5   Realizability of Linear Operators and their Relationship to Electric Networks

We now wish to summarize those parts of the preceding sections concerning the realizability of linear operators. It was seen that operators of the one-sided (memory) type, for example, as represented by the prediction error operator (2.28)

$$(2.28) \qquad \xi_t = \sum_{s=0}^{m} a_s x_{t-s} \ ,$$

are computationally realizable at the time instant t. On the other hand, operators of the two-sided (memory and anticipation) type, for example, as represented by the smoothing error operator (2.33)

$$(2.33) \qquad \gamma_t = \sum_{s=-m}^{m} c_s x_{t-s}$$

TABLE 2.1

VARIOUS TYPES OF FINITE DISCRETE LINEAR OPERATORS FOR SINGLE TIME SERIES

	Operator Form
Prediction Operator (2.21) (Computationally realizable at time instant t)	$x_{t+\alpha} = \sum_{s=0}^{M} k_s x_{t-s}, \ \alpha \geqslant 0, \ M \geqslant 0$
Prediction Error Operator (2.28) (Computationally realizable at time instant t)	$\xi_t = \sum_{s=0}^{m} a_s x_{t-s}, \ m \geqslant 0, \ a_0 = 1$
Smoothing Operator (2.31) (Not computationally realizable at time instant t)	$x_t = \sum_{\substack{s=-M \\ s \neq 0}}^{M} d_s x_{t-s}, \ M \geqslant 0$
Time Delay Smoothing Operator (Computationally realizable at time instant $t'=t+M$)	$x_t = \sum_{\substack{s=-M \\ s \neq 0}}^{M} d_s x_{t'-M-s}, \ M \geqslant 0$
Smoothing Error Operator (2.33) (Not computationally realizable at time instant t)	$\gamma_t = \sum_{s=-m}^{m} c_s x_{t-s}, \ m \geqslant 0, \ c_0 = 1$
Time Delay Smoothing Error Operator (2.34) (Computationally realizable at time instant $t'=t+m$)	$\gamma_t = \sum_{s=-m}^{m} c_s x_{t'-m-s}, \ c_0 = 1$

are not computationally realizable at the time instant t. Nevertheless by delaying the computations at least to the time instant $t'=t+m$, one may compute

$$(2.34) \qquad \gamma_t = \sum_{s=-m}^{m} c_s x_{t'-m-s} \ .$$

TABLE 2.2

TRANSFER FUNCTIONS OF VARIOUS TYPES OF FINITE LINEAR OPERATORS
GIVEN IN TABLE 2.1

	Transfer Function
Prediction Operator (2.21)	$K(\omega) = \displaystyle\sum_{s=0}^{M} k_s \mathrm{e}^{-i\omega s}$
Prediction Error Operator (2.28)	$A(\omega) = \displaystyle\sum_{s=0}^{m} a_s \mathrm{e}^{-i\omega s}$
Smoothing Operator (2.31)	$D(\omega) = \displaystyle\sum_{\substack{s=-M \\ s \neq 0}}^{M} d_s \mathrm{e}^{-i\omega s}$
Time-Delay Smoothing Operator	$\displaystyle\sum_{\substack{s=-M \\ s \neq 0}}^{M} d_s \mathrm{e}^{-i\omega(s+M)} = \mathrm{e}^{-i\omega M} D(\omega)$
Smoothing Error Operator (2.33)	$C(\omega) = \displaystyle\sum_{s=-m}^{m} c_s \mathrm{e}^{-i\omega s}$
Time-Delay Smoothing Error Operator (2.34)	$\displaystyle\sum_{s=-M}^{M} c_s \mathrm{e}^{-i\omega(s+m)} = \mathrm{e}^{-i\omega m} C(\omega)$

This later form of the smoothing error operator, which is called the time-delad smoothing error operator, is therefore computationally realizable, and indeey has the same form as a one-sided memory type operator (2.28).

The realizability of these operator forms is reflected in their respective transfer functions as follows. Let us consider the complex plane $\lambda = \omega + i\sigma$, where we let the real ω axis denote the angular frequency ω. By analytic continuation, the transfer function of the prediction type operator becomes in the complex λ-plane

$$(2.51) \qquad A(\lambda) = \sum_{s=0}^{m} a_s e^{-i\lambda s} = \sum_{s=0}^{m} a_s e^{\sigma s}\, e^{-i\omega s}\, .$$

By examining this equation, we see that $A(\lambda)$ has no singularities in the lower half λ-plane, that is, for $\sigma < 0$, which reflects the realizability of the prediction type operator.

On the other hand, the transfer function of the smoothing type operator in the complex plane is

$$(2.52) \qquad C(\lambda) = \sum_{s=-m}^{m} c_s e^{-i\lambda s} = \sum_{s=-m}^{-1} c_s e^{-i\lambda s} + \sum_{s=0}^{m} c_s e^{-i\lambda s}\, .$$

By letting $r = -s$ for $s = -m, -m+1, \ldots, -2, -1$, the transfer function becomes

$$(2.521) \qquad C(\lambda) = \sum_{r=1}^{m} c_r e^{-\sigma r}\, e^{i\omega r} + \sum_{s=0}^{m} c_s e^{\sigma s}\, e^{-i\omega s}\, ,$$

which has singularities in the upper half λ-plane and in the lower half λ-plane, thereby reflecting the non-realizability of the smoothing type operator.

The transfer function of the time-delay smoothing error operator,

$$(2.53) \qquad \begin{aligned} e^{-i\lambda m}\, C(\lambda) &= \sum_{s=-m}^{m} c_s e^{-i\lambda(s+m)} \\ &= \sum_{s+m=0}^{2m} c_s e^{\sigma(s+m)}\, e^{-i\omega(s+m)} \end{aligned}$$

has no singularities in the lower half λ-plane ($\sigma < 0$), which reflects the realizability of such an operator.

The computationally realizable linear operator corresponds to a physically realizable passive lumped element network together with a single amplifier (Bode and Shannon, 1950). The linear operator coefficients represent the impulsive response of the network, and the transfer function represents the transfer or system function of the network (Levinson, 1947a; Smith, 1954). For example, in equation (2.28) the operator coefficient a_s may be interpreted as the output obtained from an electric filter at time $t+s$ in response to a unit impulse impressed upon the input of the filter at time t. Since $a_s=0$ for $s<0$ the output obtained from the filter is zero for times less than t, and since $a_s=0$ for $s>t+m$ the output is zero for times greater than $t+m$, the filter is physically realizable. The transfer or system function of the electric filter is the transfer function of the linear operator,

$$(2.421) \qquad A(\omega) = \sum_{s=0}^{m} a_s e^{-i\omega s}.$$

Smoothing type operators, on the other hand, are not realizable to the statistician unless he introduces a time delay in his computations, or to the engineer unless he introduces a time delay in his network. Nevertheless, for problems in which physical calendar time is not important, for example, as in the analysis of seismic records, one may make computations based on the time-delay smoothing operator, but consider these computations from the point of view of the smoothing operator itself. In other words, computations may be carried out with respect to the realizable transfer function

$$(2.53) \qquad e^{-i\lambda m}\, C(\lambda)$$

of the time-delay smoothing error operator, but considered as if they were carried out with respect to the non-realizable transfer function

$$(2.521) \qquad C(\lambda)$$

of the smoothing error operator with same the operator coefficients. This same procedure is available to the engineer in such cases (Bode and Shannon, 1950).

The realizable operator

$$(2.28) \qquad \xi_t = \sum_{s=0}^{m} a_s x_{t-s}$$

has the transfer function

$$(2.421) \qquad A(\omega) = \sum_{s=0}^{m} a_s e^{-i\omega s}$$

with real part

$$(2.541) \qquad \text{Re}\,[A(\omega)] = \sum_{s=0}^{m} a_s \cos \omega s$$

and imaginary part

$$(2.542) \qquad \text{Im}\,[A(\omega)] = \sum_{s=0}^{m} a_s \sin \omega s.$$

Since both the real and imaginary parts of the transfer function depend on the same variables, namely the operator coefficients, a_0, a_1,..., a_m, the real part $\text{Re}[A(\omega)]$ and the imaginary part $\text{Im}[A(\omega)]$ cannot be chosen independently (Smith, 1954). In other words, knowledge of $\text{Re}[A(\omega)]$ lets us compute the values of the operator coefficients a_s $(s=0,1,...,m)$ by means of the equation

$$
\begin{aligned}
(2.543) \qquad & \frac{2}{\pi} \int_0^{\pi} \text{Re}[A(\omega)]\, e^{i\omega t}\, d\omega \\
&= \frac{2}{\pi} \int_0^{\pi} \sum_{s=0}^{m} a_s \cos \omega s \cos \omega t\, d\omega \\
&= a_t,\ \text{for}\ t = 0, 1, 2, ..., m\,.
\end{aligned}
$$

With the values of the operator coefficients $a_0, a_1,...,a_m$ thus found, the imaginary part $\text{Im}[A(\omega)]$ may be computed by means of equation (2.542).

On the other hand the smoothing error operator, say

$$(2.33) \qquad \gamma_t = \sum_{s=-m}^{m} c_s x_{t-s}\,,\ m \geqslant 0$$

has the transfer function

$$(2.441) \qquad C(\omega) = \sum_{s=-m}^{m} c_s e^{-i\omega s},$$

with real part

$$(2.551) \qquad \mathrm{Re}\,[C(\omega)] = \sum_{s=-m}^{m} c_s \cos \omega s = c_0 + \sum_{s=1}^{m} (c_s + c_{-s}) \cos \omega s \;,$$

and imaginary part

$$(2.552) \qquad \mathrm{Im}[C(\omega)] = \sum_{s=-m}^{m} c_s \sin \omega s = \sum_{s=1}^{m} (c_s - c_{-s}) \sin \omega s .$$

Thus the real part of the transfer function depends only on the symmetric component, $(c_s + c_{-s})$, of the smoothing error operator, and the imaginary part of the transfer function depends only on the antisymmetric component, $(c_s - c_{-s})$, of the smoothing error operator. Since the symmetric component, $(c_s + c_{-s})$, is linearly independent of the antisymmetric component, $(c_s - c_{-s})$, for each value of $s = 0,1,2,...,m$, we see that the real part, $\mathrm{Re}[C(\omega)]$, may be chosen independently of the imaginary part, $\mathrm{Im}[C(\omega)]$, of the transfer function (Smith, 1954).

Now for the smoothing error operator (2.33), let us suppose that the computations are to be carried out with respect to the time t' where $t' = t + p$. That is, since $t = t' - p$, we have

$$(2.56) \qquad \gamma_t = \sum_{s=-m}^{m} c_s x_{t'-p-s} = \sum_{s=-m}^{m} c_s x_{t'-(p+s)} .$$

Letting $x_{t'} = e^{i\omega t'}$, the transfer ratio is

$$(2.561) \qquad \frac{1}{e^{i\omega t'}} \sum_{s=-m}^{m} c_s e^{i\omega[t'-(p+s)]} = \sum_{s=-m}^{m} c_s e^{-i\omega(p+s)}$$

and so the transfer function of the operator (2.56), denoted by $C_p(\omega)$, is

$$(2.562) \qquad C_p(\omega) = \sum_{s=-m}^{m} c_s e^{-i\omega(s+p)} .$$

Now when $p=0$, the operator (2.56) is the smoothing error operator (2.33), with transfer function (2.441), the real and imaginary parts of which are independent of each other. On the other hand, when $p=m$, the operator (2.56) is the time-delay smoothing error operator (2.34), which is a realizable operator, with transfer function (2.451), the real and imaginary parts of which are entirely dependent upon each other. For those p for which $0<p<m$, the operator (2.56), has transfer function (2.562), the real and imaginary parts of which are partially dependent upon each

other, and partially independent of each other. We shall now examine to what extent they are dependent and independent.

We see that $C_p(\omega)$, given by equation (2.562), is

$$(2.563) \qquad C_p(\omega) = \sum_{s=-m}^{m} c_s e^{-i\omega(s+p)} = e^{-i\omega p} \sum_{s=-m}^{m} c_s e^{-i\omega s} = e^{-i\omega p}\, C(\omega)$$

where $C(\omega)$ is the transfer function (2.441) of the smoothing error operator (2.33). For $0<p<m$, the real part of $C_p(\omega)$ is given by

$$(2.564) \quad \mathrm{Re}[C_p(\omega)] = c_{-p} + \sum_{l=1}^{m-p} (c_{-l-p} + c_{l-p}) \cos \omega l + \sum_{l=m-p+1}^{m+p} c_{l-p} \cos \omega l$$

and the imaginary part is given by

$$(2.565) \quad \mathrm{Im}[C_p(\omega)] = \sum_{l=1}^{m-p} (c_{-l-p} - c_{l-p}) \sin \omega l + \sum_{l=m-p+1}^{m+p} c_{l-p} \sin \omega l\,.$$

Thus, given $\mathrm{Re}[C_p(\omega)]$, we may compute c_{-p}, $(c_{-l-p} + c_{l-p})$ for $l=1,2,\ldots,m-p$, and c_{l-p} for $l=m-p+1$, $m-p+2,\ldots,m+p$. The values $(c_{-l-p} - c_{l-p})$ for $l=1,2,\ldots,m-p$, which enter into equation (2.565) for the imaginary part $\mathrm{Im}[C_p(\omega)]$ are independent of the values $(c_{-l-p}+c_{l-p})$ for $l=1,2,\ldots,m-p$, which were computed from the real part $\mathrm{Re}[C_p(\omega)]$, and thus reflect the partial independence of the real and imaginary parts of the transfer function. On the other hand, the values c_{l-p} for $l=m-p+1$, $m-p+2,\ldots,m+p$, which enter into equation (2.565) for the imaginary part $\mathrm{Im}[C_p(\omega)]$ are the same values c_{l-p} for $l=m-p+1$, $m-p+2,\ldots,m+p$ which were computed from the real part $\mathrm{Re}[C_p(\omega)]$, and thus reflect the partial dependence of the imaginary part on the real part of the transfer function.

2.6 The Minimum-Delay Operator

In the remaining sections of this chapter we shall consider only the operator form of the realizable type, namely

$$(2.28) \qquad \xi_t = \sum_{s=0}^{m} a_s x_{t-s}\,.$$

Let us now consider the linear difference equation

$$(2.61) \qquad \sum_{s=0}^{m} a_s x_{t-s} = 0$$

obtained from equation (2.28) by requiring $\xi_t=0$. We see that the constant coefficients a_s of the difference equation are the operator coefficients. There is no loss of generality in assuming that a_0 is equal to one, in conformity with our usual convention.

The theory of difference equations is presented by various authors, and the reader is referred especially to Wold (1938) and Samuelson (1947). In the first part of this section we state the condition that the coeficients of the difference equation (2.61) be minimum-delay, that is, the condition that the solution x_t describes a damped oscillation. Then in the last part of this section we show that this minimum-delay condition is precisely the condition that the Fourier transform of the operator (i. e. the transfer function $A(\lambda)$) has zeros and singularities, all of which lie in the upper half λ-plane, where $\lambda=\omega+i\sigma$. In other words, we show that the difference equation (2.61) has a stable solution if the transfer function of the operator has minimum phase shift characteristic. Let us now examine the condition that the difference equation (2.61) have a stable solution. The characteristic equation of the difference equation (2.61) is defined to be

$$(2.62) \qquad P(\zeta) = a_0\zeta^m + a_1\zeta^{m-1} + \ldots + a_{m-1}\zeta + a_m = \sum_{s=0}^{m} a_s\zeta^{m-s} .$$

Since $P(\zeta)$ is a polynomial of degree m, it follows from the fundamental theorem of algebra that $P(\zeta)$ has m roots or zeros ζ_j such that

$$(2.621) \qquad P(\zeta_j)=0 \ , \ \text{for } j=1,2,\ldots,m .$$

As a result $P(\zeta)$ may be written in the form

$$(2.622) \qquad P(\zeta) = a_0(\zeta-\zeta_1) (\zeta-\zeta_2) \ldots (\zeta-\zeta_m) .$$

Since the operator coefficients $a_s(s=0,1,\ldots,m)$ are real, the roots or zeros $\zeta_1,\zeta_2,\ldots,\zeta_m$ must be real or occur in complex conjugate pairs. Let the distinct real roots of $P(\zeta)$ be represented by $\alpha_j,j=1,2,\ldots,h$ where each distinct root α_j is repeated γ_j times $(j=1,2,\ldots,h)$; that is, the zero α_j is a zero of order γ_j. Let the distinct complex roots and their conjugate roots be represented by $\beta_j\,e^{i\theta_j}$ and $\beta_j\,e^{-i\theta_j}$, $(j=1,2,\ldots,k)$ where each distinct complex root is repeated ρ_j times

$(j=1,2,\ldots,k)$; that is, the zero $\beta_j e^{i\theta_j}$ is a zero of order ρ_j, and the zero $\beta_j e^{-i\theta_j}$ is also a zero of order ρ_j. Here β_j represents the modulus, and θ_j or $-\theta_j$ represents the argument, of the complex root. Consequently, equation (2.62) becomes

$$P(\zeta)=a_0\zeta^m + a_1\zeta^{m-1} + \ldots + a_{m-1}\zeta + a_m$$

$$= a_0(\zeta - \zeta_1)(\zeta - \zeta_2)\ldots(\zeta - \zeta_m) = a\prod_{j=1}^{m}(\zeta - \zeta_j)$$

(2.623)

$$= a_0\prod_{j=1}^{h}(\zeta-\alpha_j)^{\gamma_j}\prod_{j=1}^{k}(\zeta-\beta_j e^{i\theta_j})^{\rho_j}(\zeta-\beta_j e^{-i\theta_j})^{\rho_j}$$

$$= a_0\prod_{j=1}^{h}(\zeta-\alpha_j)^{\gamma_j}\prod_{j=1}^{k}(\zeta^2 - 2\beta_j\cos\theta_j\,\zeta + \beta_j^2)^{\rho_j}\,.$$

For an arbitrary set of initial values

$$(2.63)\qquad\qquad x_0, x_1, x_2, \ldots, x_{t-1},$$

the series $x_t, x_{t+1}, x_{t+2}, \ldots,$ may be generated by recursive deductions from the difference equation (2.61), and this series will form the general solution of the difference equation. Explicitly the general solution is given by

$$(2.631)\quad x_t = \sum_{j=1}^{h} P^{(1)}_{\gamma_j-1}(t)\,\alpha_j^t + \sum_{j=1}^{k}[P^{(2)}_{\rho_j-1}(t)\,\cos\theta_j t + P^{(3)}_{\rho_j-1}(t)\,\sin\theta_j t]\,\beta_j^t$$

where the $P_r^{(s)}(t)$ denotes a polynomial of order r with arbitrary coefficients, and the h, k, α_j, β_j, γ_j, and ρ_j are given by the characteristic equation (2.623). The asymptotic behavior of the general solution x_t in equation (2.631) is dependent on the exponential factors α_j^t and β_j^t. A necessary and sufficient condition that

$$(2.64)\qquad\qquad \sum_{t=0}^{\infty} x_t^2 \text{ and } \sum_{t=0}^{\infty}|x_t|$$

converge for any values taken on by the $P_r^{(s)}(t)$ is that the magnitude of the roots $|\zeta_j|$, $j=1,2,\ldots,m$, of the characteristic equation (2.623) be less than one, that is, $|\zeta_j|<1$, $j=1,2,\ldots,m$, which is

$$|\alpha_j|<1, j=1,2,\ldots,h$$

(2.641)

$$\beta_j<1, j=1,2,\ldots,k\,.$$

In this case the solution x_t of the difference equation (2.61) describes a damped oscillation, and we shall call the corresponding linear operator

$$(2.28) \qquad \xi_t = \sum_{s=0}^{m} a_s x_{t-s}, \quad a_0 = 1$$

minimum-delay. That is, a linear operator is minimum-delay if the zeros ζ_j of its associated characteristic equation $P(\zeta)$ lie within the periphery $|\zeta|=1$ of the unit circle, that is $|\zeta_j|<1$.

Let us now turn our attention to the Fourier transform of the operator coefficients, that is, the transfer function,

$$(2.65) \qquad A(\omega) = \sum_{s=0}^{m} a_s e^{-i\omega s}$$

of the linear operator (2.28). By analytic continuation to the complex plane $\lambda = \omega + i\sigma$, where the real ω axis denotes angular frequency, the transfer function becomes

$$(2.651) \qquad A(\lambda) = \sum_{s=0}^{m} a_s e^{-i\lambda s} = \sum_{s=0}^{m} a_s e^{\sigma s} \, e^{-i\omega s} .$$

Let us apply the transformation $z = e^{-i\lambda}$, for $-\pi < \omega \leqslant \pi$, to the transfer function $A(\lambda)$. The transformation $z = e^{-i\lambda}$ maps the strip between $\omega = -\pi$ to $\omega = \pi$ of the upper half λ plane $(\sigma > 0)$ into the exterior $|z| > 1$ of the unit circle in the z plane; it maps the strip between $\omega = -\pi$ to $\omega = \pi$ of the lower half of the λ plane, $(\sigma < 0)$ into the interior $|z| < 1$ of the unit circle in the z plane; and it maps the real axis $-\pi < \omega \leqslant \pi$ of the λ plane into the periphery $|z| = 1$ of the unit circle in the z plane. See Figure 2.1. Under this transformation the transfer function $A(\lambda)$, equation (2.651) becomes the polynomial $A(z)$ where

$$(2.652) \qquad A(z) = \sum_{s=0}^{m} a_s z^s = a_0 + a_1 z + a_2 z^2 + \ldots + a_m z^m .$$

This polynomial or entire rational function, is analytic in the whole plane and has a pole at infinity. Let us call $A(z)$ the transfer function in the z-plane, or the z-transform, of the operator.

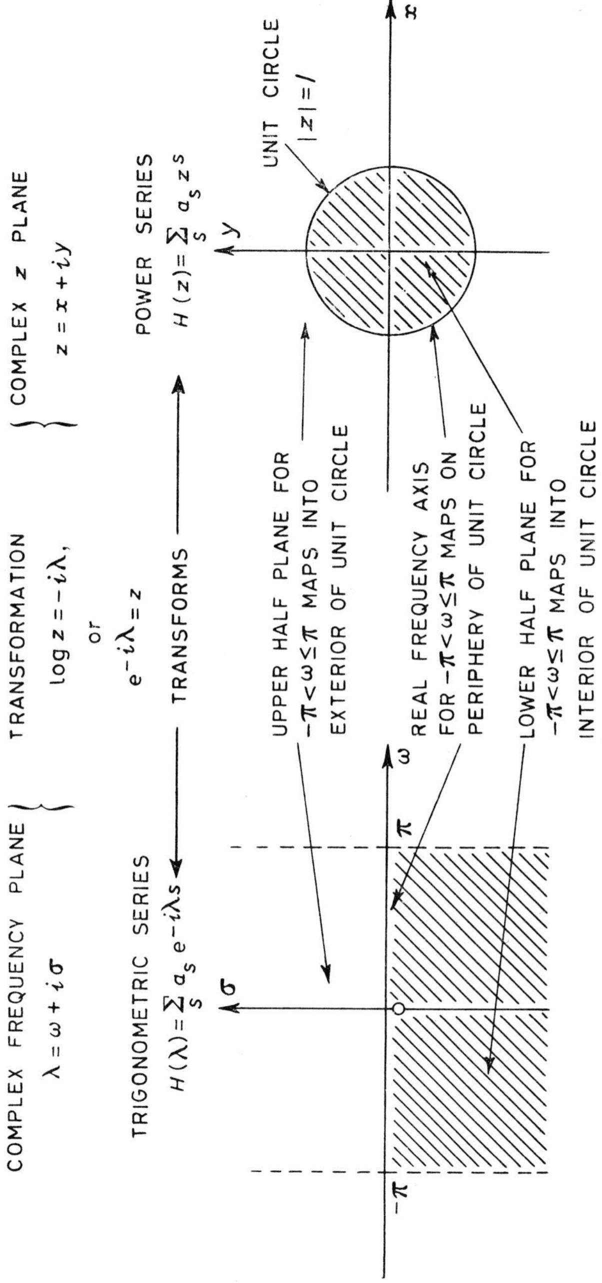

Figure 2.1 — Transformation of trigonometric series into power series.

For a minimum-delay operator we have seen that the characteristic equation

$$(2.623) \qquad P(\zeta) = \sum_{s=0}^{m} a_s \zeta^{m-s} = a_0 \zeta^m + a_1 \zeta^{m-1} + \ldots + a_m, \quad a_0 = 1$$

has roots $\zeta_1, \zeta_2, \ldots, \zeta_m$, all of which have modulus $|\zeta_k|$ less than one. Without loss of generality, we assume $a_m \neq 0$, so that $|\zeta_k| \neq 0$ for $k = 1, 2, \ldots, m$. Thus the characteristic equation (2.623) may be written

$$
\begin{aligned}
P(\zeta) &= (\zeta - \zeta_1)(\zeta - \zeta_2)\ldots(\zeta - \zeta_m) \\[2mm]
&= \zeta^m \zeta_1 \zeta_2 \ldots \zeta_m \, (\zeta_1^{-1} - \zeta^{-1})\,(\zeta_2^{-1} - \zeta^{-1})\ldots(\zeta_m^{-1} - \zeta^{-1}) \\[2mm]
&= \zeta^m \zeta_1 \zeta_2 \ldots \zeta_m \, (-1)^m \, (\zeta^{-1} - \zeta_1^{-1})\,(\zeta^{-1} - \zeta_2^{-1})\ldots(\zeta^{-1} - \zeta_m^{-1}) \\[2mm]
&= a_m \zeta^m \, (\zeta^{-1} - \zeta_1^{-1})\,(\zeta^{-1} - \zeta_2^{-1})\ldots(\zeta^{-1} - \zeta_m^{-1})
\end{aligned}
$$

(2.653)

since

$$(2.654) \qquad a_m = (-1)^m \, \zeta_1 \zeta_2 \ldots \zeta_m .$$

Under the transformation $z = \zeta^{-1}$, the function $P(\zeta)$, given by equation (2.653), becomes

$$(2.655) \qquad P(\zeta = z^{-1}) = \sum_{s=0}^{m} a_s z^{-m+s} = z^{-m} a_m (z - z_1)(z - z_2) \ldots (z - z_m)$$

where we define

$$(2.656) \qquad z_1 = \zeta_1^{-1}, \; z_2 = \zeta_2^{-1}, \; \ldots, z_m = \zeta_m^{-1} .$$

Thus the function

$$(2.657) \qquad z^m P(\zeta = z^{-1}) = \sum_{s=0}^{m} a_s z^s = a_m (z - z_1)\,(z - z_2)\ldots(z - z_m)$$

is the function $A(z)$ in equation (2.652) which was obtained from the transfer function $A(\lambda)$ by the transformation $z = e^{-i\lambda}$. That is, the transfer function in the z-plane is

$$(2.658) \qquad A(z) = \sum_{s=0}^{m} a_s z^s = a_m (z - z_1)\,(z - z_2)\ldots(z - z_m)$$

where the roots $z_k(k=1,2,\ldots,m)$ are given by equation (2.656). Since the minimum-delay condition is that the roots ζ_k have modulus $|\zeta_k|$ less than one, we see that this condition is that the roots z_k of $A(z)$ have modulus

$$(2.659) \qquad |z_k| = |\zeta_k^{-1}| > 1$$

greater than one.

Thus a finite, realizable operator is minimum-delay if the roots z_k of the polynomial $A(z)$ all have modulus greater than one. That is, the operator with coefficients $a_0, a_1, \ldots, a_m$ is minimum-delay if the transfer function in the z-plane,

$$A(z) = a_0 + a_1 z + a_2 z^2 + \ldots + a_m z^m$$

$$(2.66) \qquad = a_m (z - z_1)\,(z - z_2)\,\ldots\,(z - z_m)$$

has roots z_k where

$$(2.661) \qquad |z_k| > 1,\ k = 1,2,\ldots,m\,.$$

In other words, the minimum-delay requirement is that all the zeros of $A(z)$ must lie exterior to the periphery $|z| = 1$ of the unit circle in the z plane.

For example, let us consider the so-called cosine operator (Simpson, 1953)

$$(2.67) \qquad \xi_t = x_t + a_1 x_{t-1} + a_2 x_{t-2}$$

$$= x_t - (2\cos\omega_0)\,x_{t-1} + x_{t-2}\,.$$

The coefficients of this operator were chosen by the requirement that the prediction error $\xi_t = 0$ for $x_t = \cos\omega_0 t$. The transfer function in the z-plane is then

$$A(z) = 1 + a_1 z + a_2 z^2$$

$$(2.671) \qquad = 1 - (2\cos\omega_0)\,z + z^2$$

$$= (e^{i\omega_0} - z)\,(e^{-i\omega_0} - z)$$

so that the roots of $A(z)$ are

$$(2.672) \qquad z_1 = e^{i\omega_0},\ z_2 = e^{-i\omega_0}\,.$$

Since $|z_1| = |z_2| = 1$, the cosine operator is just on the bordeline of minimum-delay.

Related to the concept of minimum-delay are the associated concepts of mixed-delay and maximum-delay. An operator is called *maximum-delay* provided that all of the zeros of its z-transform $A(z)$ lie within the periphery $|z|=1$ of the unit circle in the z-plane. An operator which is neither minimum-delay nor maximum-delay is called *mixed-delay*; a mixed-delay operator is one for which some of the zeros of its z-transform $A(z)$ lie within the unit circle and others of the zeros lie without the unit circle.

The usage of any of the terms «minimum-delay operator», «maximum-delay operator», or «mixed-delay operator» always implies that the operator in question is realizable.

Let us consider the minimum-delay condition (2.661) in terms of the transfer function $A(\lambda)$ in the complex λ-plane. The stability condition (2.661) becomes, under the transformation $e^{-i\omega} = z$, that the transfer function

$$(2.651) \qquad A(\lambda) = \sum_{s=0}^{m} a_s e^{-i\lambda s}$$

has zeros only in the upper half λ-plane. See Figure 2.1. Then $A(\lambda)$ has no singularities or zeros below the axis of real frequency ω and its logarithm in that half plane is as small as possible at infinity.

A linear system has minimum phase-shift characteristic if its transfer function has no singularities or zeros in the lower half λ-plane (Bode and Shannon, 1950). Thus if we consider $A(\lambda)$ to be a transfer function of a linear system and if $A(\lambda)$ satisfies the conditions of a minimum phase-shift characteristic, then the linear operator

$$(2.67) \qquad a_t = \frac{1}{2\pi} \int_{-\pi}^{\pi} A(\omega)e^{i\omega t} d\omega$$

is minimum-delay.

In summary, then, we see that the minimum-delay condition is that the difference equation (2.28) formed by the operator coefficients $a_s(s=0,1,\ldots,m)$ is to have a solution which describes a damped oscillation. This minimum-delay condition is also that the transfer function $A(\lambda)$ of the linear operator is to have no singularities or zeros below the axis of real frequency ω, and that its logarithm in that half plane is to be as small as possible at infinity. Briefly, a minimum-delay operator represents a minimum phase network, and conversely.

2.7   The Inverse Linear Operator

Let us consider the finite minimum-delay operator

$$(2.28) \qquad \xi_t = \sum_{s=0}^{m} a_s x_{t-s}, \quad a_0 = 1$$

so that its transfer function in the z-plane

$$(2.652) \qquad A(z) = \sum_{s=0}^{m} a_s z^s$$

has zeros, all of which are exterior to the periphery $|z|=1$ of the unit circle. In other words, the function $A(z)$ is analytic for $|z| \leqslant 1$, and

$$(2.71) \qquad A(z) \neq 0 \text{ for } |z| \leqslant 1, \quad \sum_{s=0}^{m} a_s^2 < \infty .$$

Consequently the function

$$(2.711) \qquad A^{-1}(z) = \frac{1}{A(z)} = \frac{1}{\displaystyle\sum_{s=0}^{m} a_s z^s} = B(z)$$

is a function which is analytic for $|z| \leqslant 1$, and has no zeros for $|z| \leqslant 1$. As a result we may expand this function in the power series

$$(2.712) \qquad \frac{1}{A(z)} = \frac{1}{\displaystyle\sum_{s=0}^{m} a_s z^s} = \sum_{t=0}^{\infty} b_t z^t = B(z)$$

which converges for $|z| \leqslant 1$, and has no zeros for $|z| \leqslant 1$. The values of b_t for $t=0,1,2,\ldots$, may be found by direct division of the polynomial $a_0 + a_1 z + a_2 z^2 + \ldots + a_m z^m$ into unity. We shall define b_t to be those values given by equation (2.712) and define

$$(2.713) \qquad b_t = 0, \text{ for } t < 0 .$$

From the equation (2.712) we have

$$(2.72) \qquad A(z)B(z) = 1 = \sum_{s=0}^{m} a_s z^s \sum_{t=0}^{\infty} b_t z^t$$

$$= \sum_{s=0}^{m} a_s \sum_{t=0}^{\infty} b_t z^{s+t} = \sum_{s=0}^{m} a_s \sum_{n=s}^{\infty} b_{n-s} z^n$$

where we have let $n=s+t$. Recalling that b_t is equal to zero for $t<0$ we have

$$(2.721) \qquad A(z)B(z) = 1 = \sum_{s=0}^{m} a_s \sum_{n=0}^{\infty} b_{n-s} \zeta^n = \sum_{n=0}^{\infty} \left(\sum_{s=0}^{m} a_s b_{n-s} \right) \zeta^n .$$

In order for this equation to hold, we see that

$$(2.722) \qquad a_0 b_0 = 1$$

$$(2.723) \qquad \sum_{s=0}^{m} a_s b_{n-s} = 0 \ , \ \text{ for } \ n = 1,2,3,\dots$$

Since, we let $a_0=1$, we have $b_0=1$. Thus, given the operator coefficients $a_0, a_1, \dots, a_m$, the $b_1, b_2, \dots$, may be uniquely determined by recursive deductions from the difference equation

$$(2.724) \qquad \sum_{s=0}^{m} a_s b_{t-s} = 0, \ t = 1,2,3,\dots$$

subject to the initial values $b_t=0$ for $t<0$ and $b_0 = (a_0)^{-1} = 1$.

Since we have assumed that the linear operator with the coefficients $a_0, a_1, \dots, a_m$ is minimum-delay, the difference equation (2.724) with the constant coefficients $a_0, a_1, \dots, a_m$ has the characteristic equation

$$(2.73) \qquad P(\zeta) = \sum_{s=0}^{m} a_s \zeta^{m-s}$$

all the roots ζ_k of which lie within the periphery of the unit circle, that is $|\zeta_k|<1$ for $k=1,2,\ldots,m$. As a result, in virtue of equation (2.631) for the general solution of this difference equation, the solution $b_t(t=0,1,2,\ldots)$ describes a damped oscillation. In particular, the series

$$(2.731) \qquad \sum_{t=0}^{\infty} |b_t| \quad \text{and} \quad \sum_{t=0}^{\infty} b_t^2$$

converge, so the inverse operator $b_0,b_1,b_2,\ldots$, is stable.

Let us now examine the linear operator, with coefficients $b_0,b_1,b_2,\ldots$, given by

$$(2.74) \qquad \sum_{r=0}^{\infty} b_r \zeta_{t-r}.$$

Since $b_t=0$ for $t<0$, this operator is of the one-sided memory type, and since the operator has an infinite number of discrete coefficients it may be called an infinite discrete linear operator. Strictly speaking infinite linear operations may not be computed because an infinite number of multiplications would be required, and hence in this strict sense they are not computationally realizable. Nevertheless, because of equation (2.731), the operator equation (2.74) may be approximated by the partial sum

$$(2.741) \qquad \sum_{r=0}^{M} b_r \zeta_{t-r}$$

to any degree of computational accuracy by choosing M sufficiently large, and in this sense the operator equation (2.74) is computationally realizable.

Since $B(z)$, given by equation (2.711), has no singularities or zeros for $|z|\leqslant 1$, the transfer function of the b_t operator

$$(2.742) \qquad B(\omega) = \sum_{r=0}^{\infty} b_r e^{-i\omega r}$$

has no singularities or zeros in the lower half λ-plane where $\lambda=\omega+i\sigma$, (see Figure 2.1). Thus the linear operator b_t has minimum phase shift characteristic, and is minimum-delay.

In summary, then, the infinite linear operator b_t has the following properties:

$$b_t = 0 \ \text{for} \ t < 0$$

$$b_0 > 0$$

(2.75)

$$\sum_{t=0}^{\infty} b_t^2 < \infty$$

and the

(2.751)
$$B(z) = \sum_{t=0}^{\infty} b_t z^t \neq 0 \ \text{for} \ |z| \leqslant 1 \,,$$

and we shall call such infinite one-sided operators computationally realizable and *minimum-delay*.

Let us now examine the significance of the linear operator

(2.74)
$$\sum_{r=0}^{\infty} b_r \xi_{t-r} \,.$$

Inserting into this equation the prediction errors ξ_t, ξ_{t-1}, $\xi_{t-2}, \dots$ given by the prediction error operator (2.28), we have

(2.76)
$$\sum_{r=0}^{\infty} b_r \xi_{t-r} = \sum_{r=0}^{\infty} b_r \sum_{s=0}^{m} a_s x_{t-r-s}$$
$$= \sum_{s=0}^{m} a_s \sum_{r=0}^{\infty} b_r x_{t-r-s} \,.$$

Letting $n = r+s$, we have

(2.761)
$$\sum_{r=0}^{\infty} b_r \xi_{t-r} = \sum_{s=0}^{m} a_s \sum_{n=s}^{\infty} b_{n-s} x_{t-n}$$

and recalling $b_r = 0$ for $r < 0$, we have

$$(2.762) \qquad \sum_{r=0}^{\infty} b_r \xi_{t-r} = \sum_{s=0}^{m} a_s \sum_{n=0}^{\infty} b_{n-s} x_{t-n} = \sum_{n=0}^{\infty} \left(\sum_{s=0}^{m} a_s b_{n-s} \right) x_{t-n} .$$

Therefore, because of equations (2.722) and (2.723) we have

$$(2.77) \qquad \sum_{r=0}^{\infty} b_r \xi_{t-r} = x_t .$$

That is, the linear operator (2.77) with coefficients $b_0, b_1, b_2, \ldots$ operates on the prediction errors $\xi_t, \xi_{t-1}, \xi_{t-2}, \ldots$, at time t and prior to time t in order to yield the value of the time series x_t. Thus we see that the operator (2.28) with coefficients $a_m, a_{m-1}, \ldots, a_2, a_1, a_0$ operates on the time series $\ldots, x_{t-2}, x_{t-1}, x_t$ to yield the prediction errors $\ldots, \xi_{t-2}, \xi_{t-1}, \xi_t$, whereas the operator (2.77) with coefficients $\ldots, b_2, b_1, b_0$, performs the inverse operation. Therefore the a_s operator (2.28) is called the inverse to the b_s operator (2.77), and conversely.

More generally, the infinite linear operators

$$(2.78) \qquad \xi_t = \sum_{s=0}^{\infty} a_s x_{t-s}$$

$$(2.781) \qquad x_t = \sum_{s=0}^{\infty} b_s \xi_{t-s}$$

are realizable, minimum-delay, and inverse to each other if the coefficients a_t satisfy

$$a_t = 0 \text{ for } t < 0$$

$$a_0 > 0$$

$$(2.782) \qquad \sum_{t=0}^{\infty} a_t^2 < \infty$$

$$A(z) = \sum_{t=0}^{\infty} a_t z^t \neq 0 \text{ for } |z| \leqslant 1,$$

if the coefficients b_t satisfy

$$b_t = 0 \ \text{ for } \ t < 0$$

$$b_0 > 0$$

(2.783)
$$\sum_{t=0}^{\infty} b_t^2 < \infty$$

$$B(z) = \sum_{t=0}^{\infty} b_t z^t \neq 0 \ \text{ for } \ |z| \leqslant 1$$

and if

(2.784)
$$A(z)B(z) = 1 \ .$$

Hence the a_t and b_t are related by

$$a_0 b_0 = 1$$

(2.785)
$$\sum_{s=0}^{t} a_s b_{t-s} = 0, \ \text{ for } \ t = 1,2,\ldots$$

Thus, given the set a_s, the set b_s may be uniquely determined, and vice versa. For example equation (2.785) for $t=1,2,3$, yields

$$a_1 = -b_1$$

(2.786)
$$a_2 = -b_2 + b_1^2$$

$$a_3 = -b_3 + 2b_1 b_2 - b_1^3$$

on the one hand, and

$$b_1 = -a_1$$

(2.787)
$$b_2 = -a_2 + a_1^2$$

$$b_3 = -a_3 + 2a_1 a_2 - a_1^3$$

on the other hand. Both the a_t series and the b_t series $(t=0,1,2,...)$ form damped oscillations.

The transfer functions

$$(2.791) \qquad A(\omega) = \sum_{s=0}^{\infty} a_s e^{-\omega s}$$

and

$$(2.792) \qquad B(\omega) = \sum_{s=0}^{\infty} b_s e^{-i\omega s}$$

are free from singularities and zeros in the lower half λ-plane, $\lambda=\omega+i\sigma$, and have minimum phase shift characteristic. They are related by

$$(2.793) \qquad A(\omega)B(\omega) = 1,$$

$$(2.794) \qquad \frac{1}{A(\omega)} = A^{-1}(\omega) = B(\omega),$$

$$(2.795) \qquad \frac{1}{B(\omega)} = B^{-1}(\omega) = A(\omega),$$

$$(2.796) \qquad A^{-1}(\omega)\ A(\omega) = 1,$$

and

$$(2.797) \qquad B^{-1}(\omega)\ B(\omega) = 1.$$

2.8   The Power Transfer Function and its Minimum-Delay Operator

The power transfer function $\Psi(\omega)$ is defined to be the square of the absolute value $|A(\omega)|$ of the transfer function $A(\omega)$; that is, the power transfer function is given by

$$\Psi(\omega) = |A(\omega)|^2 = A(\omega)\ \overline{A(\omega)}$$

$$(2.81)$$

$$= \mathrm{Re}[A(\omega)]^2 + \mathrm{Im}[A(\omega)]^2 \geqslant 0$$

where the bar indicates the complex conjugate. Here $|A(\omega)|$ is called the gain of the linear operator, and it is given by the square root of the power transfer function; that is

$$(2.811) \qquad |A(\omega)| = \sqrt{\Psi(\omega)}$$

so that knowledge of the gain of the linear operator and knowledge of the power transfer function are equivalent.

The power transfer function $\Psi(\omega)$ of a finite linear operator $a_0, a_1, \ldots, a_m$ may be expressed by the finite trigonometric series

$$(2.812) \qquad \Psi(\omega) = \sum_{s=0}^{m} a_s e^{-i\omega s} \sum_{t=0}^{m} a_t e^{i\omega t} = |A(\omega)|^2 \geqslant 0$$

which is non-negative for $-\pi < \omega \leqslant \pi$. We may rewrite this expression in the following way

$$\Psi(\omega) = \sum_{s=0}^{m} \sum_{t=0}^{m} a_s a_t e^{-i\omega(s-t)}$$

$$(2.813)$$

$$= \sum_{\tau=-m}^{m} e^{-i\omega\tau} \sum_{t=0}^{m} a_t a_{t+\tau}$$

where $\tau = s - t$. Let us define r_τ to be

$$(2.814) \qquad r_\tau = \sum_{t=0}^{m} a_t a_{t+\tau}; \quad r_\tau = r_{-\tau}$$

that is,

$$r_0 = a_0^2 + a_1^2 + a_2^2 + \ldots + a_{m-2}^2 + a_{m-1}^2 + a_m^2$$

$$r_1 = r_{-1} = a_0 a_1 + a_1 a_2 + a_2 a_3 + \ldots + a_{m-2} a_{m-1} + a_{m-1} a_m$$

$$r_2 = r_{-2} = a_0 a_2 + a_1 a_3 + a_2 a_4 + \ldots + a_{m-2} a_m$$

$$(2.814)$$

$$\cdot \quad \cdot \quad \cdot$$

$$r_{m-1} = r_{1-m} = a_0 a_{m-1} + a_1 a_m$$

$$r_m = r_{-m} = a_0 a_m$$

Therefore if we are given the linear operator with coefficients $a_0, a_1, \ldots, a_m$, we may find the r_τ by means of equation (2.814) and thereby determine the power transfer function

$$\Psi(\omega) = \left| \sum_{s=0}^{m} a_s e^{-i\omega s} \right|^2 = \sum_{\tau=-m}^{m} r_\tau e^{-i\omega\tau}$$

(2.815)

$$= r_0 + 2 \sum_{\tau=0}^{m} r_\tau \cos \omega\tau \geqslant 0 \,.$$

In this section we wish to consider the inverse problem; that is, given the power transfer function $\Psi(\omega)$, find the coefficients $a_0, a_1, \ldots, a_m$ of the linear operator which yield this power transfer function. This inverse problem, as it stands, is not unique in that several different linear operators may yield the same power transfer function $\Psi(\omega)$. As we shall see, however, all of these linear operators are non-minimum-delay, except one. In other words, given the power transfer function $\Psi(\omega)$ we wish to find the one, and only one, realizable, minimum-delay operator which yields the power transfer function $\Psi(\omega)$. This minimum--delay operator has the transfer function with gain equal to $\sqrt{\Psi(\omega)}$ and minimum phase characteristic. The import of this section resides in the fact that if one wishes to design a minimum-delay operator, he needs only to have information about the desired absolute gain characteristics $| A(\omega) | = \sqrt{\Psi(\omega)}$, and needs no information concerning the phase characteristics. In this section, we give a direct procedure for the determination of such finite, minimum-delay linear operators. This procedure may be readily programmed for automatic computation on a digital computer.

Since we wish to consider finite linear operators with the coefficients $a_0, a_1, \ldots, a_m$, it is necessary to express the power transfer function in terms of a finite trigonometric series

(2.82)
$$\Psi(\omega) = \sum_{\tau=-m}^{m} r_\tau e^{-i\omega\tau} = r_0 + 2 \sum_{\tau=1}^{m} r_\tau \cos \omega\tau \,, \, r_0 > 0 \,, \, r_\tau = r_{-\tau}$$

which is non-negative for $-\pi < \omega \leqslant \pi$, and the r_τ are real.

If the power transfer function is given by the infinite Fourier series

(2.821)
$$\Psi(\omega) = \rho_0 + 2 \sum_{\tau=1}^{\infty} \rho_\tau \cos \omega\tau \geqslant 0 \,, \, -\pi < \omega \leqslant \pi \,,$$

which is non-negative for $-\pi < \omega \leqslant \pi$, then the Cesaro partial sum

$$(2.822) \qquad \rho_0 + 2 \sum_{\tau=1}^{N} \left(1 - \frac{\tau}{N}\right) \rho_\tau \cos \omega\tau$$

which is also non-negative for $-\pi < \omega \leqslant \pi$, may be used as the finite series approximation to $\Psi(\omega)$. We then have

$$(2.823) \qquad \Psi(\omega) \approx r_0 + 2 \sum_{\tau=1}^{m} r_\tau \cos \omega\tau, \quad r_\tau = \left(1 - \frac{\tau}{N}\right) \rho_\tau$$

which may be used for equation (2.82).

Let $Q(u)$ be the polynomial of order m obtained from

$$(2.824) \qquad r_0 + 2 \sum_{\tau=1}^{m} r_\tau \left(z^{-\tau} + z^{\tau}\right)$$

by the substitution $z^{-1} + z = u$. Wold (1938), in connection with his study of the process of moving averages, shows that a necessary and sufficient condition that the finite series

$$(2.825) \qquad r_0 + 2 \sum_{\tau=1}^{m} r_\tau \left(e^{i\omega\tau} + e^{-i\omega\tau}\right)$$

be non-negative for $-\pi \leqslant \omega \leqslant \pi$ is that the equation $Q(u) = 0$ should have no real root of odd multiplicity in the interval $-2 < u < 2$.

The method which we give in this section was used by Fejér (1915) and Wold (1938), working in different but mathematically equivalent settings, Fejér in terms of trigonometric series, Wold in terms of stationary stochastic processes.

Returning now to the power transfer function let us suppose that the $r_\tau = r_{-\tau}$ are such that $\Psi(\omega)$ given by equation (2.82) is non-negative for $-\pi < \omega \leqslant \pi$. Under the transformation $z = e^{-i\lambda}$, where $\lambda = \omega + i\sigma$, (see Figure 2.1), the power transfer function $\Psi(\omega)$ becomes

$$(2.83) \qquad \Psi(z) = \sum_{\tau=-m}^{m} r_\tau z^\tau$$

where $\Psi(z)$ is a rational function in z.

We see that

$$(2.831) \qquad z^m \, \Psi(z) = \sum_{\tau=-m}^{m} r_\tau z^{\tau+m}$$

is a polynomial of order $2m$. Expressing this polynomial in terms of its roots z_k $(k=1,2,...,2m)$ we have

$$(2.832) \qquad z^m \, \Psi(z) = r_m(z - z_1) \, (z - z_2) \ldots (z - z_{2m})$$

where we assume $r_m \neq 0$ so that $| \, z_k \, | \neq 0$ for $k=1,2,...,2m$. Hence the rational function $\Psi(z)$ may be expressed as

$$(2.833) \qquad \Psi(z) = z^{-m} r_m \, (z - z_1) \, (z - z_2) \ldots (z - z_{2m})$$

and the power transfer function is therefore

$$(2.834) \qquad \Psi(\omega) = e^{i\omega m} \, r_m \, (e^{-i\omega} - z_1) \, (e^{-i\omega} - z_2) \ldots (e^{-i\omega} - z_{2m}) \, .$$

Since the power transfer function $\Psi(\omega)$ is a real function of ω, we have

$$\Psi(\omega) = \overline{\Psi(\omega)} = r_m e^{-i\omega m} \, (e^{i\omega} - \bar{z}_1) \, (e^{i\omega} - \bar{z}_2) \ldots (e^{i\omega} - \bar{z}_{2m})$$

$$= r_m e^{i\omega m} \, (1 - \bar{z}_1 e^{-i\omega}) \, (1 - \bar{z}_2 e^{-i\omega}) \ldots (1 - \bar{z}_{2m} e^{-i\omega})$$

$$(2.835)$$

$$= \bar{z}_1 \bar{z}_2 \ldots \bar{z}_{2m} r_m e^{i\omega m} \, (\frac{1}{\bar{z}_1} - e^{-i\omega}) \, (\frac{1}{\bar{z}_2} - e^{-i\omega}) \ldots (\frac{1}{\bar{z}_{2m}} - e^{-i\omega}) \, .$$

Letting $z = e^{-i\lambda}$ where $\lambda = \omega + i\sigma$ we have

$$(2.836) \qquad \Psi(z) = \bar{z}_1 \bar{z}_2 \ldots \bar{z}_{2m} r_m z^{-m} (\bar{z}_1^{-1} - z) \, (\bar{z}_2^{-1} - z) \ldots (\bar{z}_{2m}^{-1} - z) \, .$$

Comparing equations (2.824) and (2.831) we see that if z_k is a root of the polynomial $z^m \, \Psi(z)$, then $\bar{z}_k^{-1}$ is also a root.

Moreover since the power transfer function $\Psi(\omega)$ is a real even function of ω, we have

$$(2.84) \qquad \Psi(-\omega) = \overline{\Psi(\omega)} = \Psi(\omega)$$

which is

$$(2.841) \qquad \begin{aligned} \Psi(-\omega) &= r_m e^{i\omega m} \, (e^{-i\omega} - \bar{z}_1) \, (e^{-i\omega} - \bar{z}_2) \ldots (e^{-i\omega} - \bar{z}_{2m}) \\ &= \overline{\Psi(\omega)} = r_m e^{i\omega m} \, (e^{-i\omega} - z_1) \, (e^{-i\omega} - z_2) \ldots (e^{-i\omega} - z_{2m}) = \Psi(\omega) \, . \end{aligned}$$

Letting $z = e^{-i\lambda}$, where $\lambda = \omega + i\sigma$, we have

$$(2.842) \qquad \begin{aligned} \Psi(z) &= r_m z^{-m} (z - \bar{z}_1) \, (z - \bar{z}_2) \ldots (z - \bar{z}_{2m}) \\ &= r_m z^{-m} (z - z_1) \, (z - z_2) \ldots (z - z_{2m}) \end{aligned}$$

so that if z_k is a root of the polynomial $z^m \, \Psi(z)$ then $\bar{z}_k$ is also a root.

In summary, then, if z_k is a root of $z^m \Psi(z)$ then $\bar{z}_k, z_k^{-1}$, and $\bar{z}_k^{-1}$ are also roots. Thus if α_k is a complex root of $z^m \Psi(z)$ with modulus $|\alpha_k| \neq 1$, then $\alpha_k, \alpha_k^{-1}, \bar{\alpha}_k^{-1}$ are distinct from each other, and are all roots of $z^m \Psi(z)$. If β_k is a real root of $z^m \Psi(z)$ with modulus $|\beta_k| \neq 1$, then β_k and β_k^{-1} are distinct from each other, and are both roots of $z^m \Psi(z)$. If γ_k is a complex root of $z^m \Psi(z)$ with modulus $|\gamma_k| = 1$, then γ_k and $\bar{\gamma}_k$ are distinct from each other, and are both roots of $z^m \Psi(z)$. Let ρ_k represent the real roots of $z^m \Psi(z)$ with modulus $|\rho_k| = 1$.

Accordingly, the polynomial $z^m \Psi(z)$ may be expressed as

$$z^m \Psi(z) = r_m \prod_{k=1}^{h} (z - \alpha_k)(z - \bar{\alpha}_k)(z - \alpha_k^{-1})(z - \bar{\alpha}_k^{-1}) \prod_{k=1}^{j} (z - \beta_k)(z - \beta_k^{-1})$$

(2.85)

$$\prod_{k=1}^{l} (z - \gamma_k)(a - \bar{\gamma}_k) \prod_{k=1}^{n} (z - \rho_k)$$

where any root of order p is repeated p times.

Let us now turn our attention to equation (2.812),

(2.812)
$$\Psi(\omega) = \sum_{s=0}^{m} a_s e^{-i\omega s} \sum_{t=0}^{m} a_t e^{i\omega t}$$

which expresses the relationship of the power transfer function $\Psi(\omega)$ with the coefficients $a_0, a_1, \ldots, a_m$ of the realizable operator which yields $\Psi(\omega)$. Under the transformation $z = e^{-i\lambda}$, $\lambda = \omega + i\sigma$, (see Figure 2.1), we have

(2.86)
$$\Psi(z) = \sum_{s=0}^{m} a_s z^s \sum_{t=0}^{m} a_t z^{-} .$$

We thus have

(2.861)
$$z^m \Psi(z) = \sum_{\tau=-m}^{m} r_\tau z^{\tau+m} = \sum_{s=0}^{m} a_s z^s \sum_{t=0}^{m} a_t z^{m-t} .$$

In section 2.6 we defined $A(z)$, called the transfer function in the z plane, to be

(2.652)
$$A(z) = \sum_{s=0}^{m} a_s z^s$$

so equation (2.86) becomes

(2.862)
$$\Psi(z) = A(z) A(z^{-1})$$

and equation (2.861) becomes

$$(2.863) \qquad z^m \, \Psi(z) = [A(z)][z^m A(z^{-1})]$$

which is

$$(2.864) \quad r_m z^{2m} + r_{m-1} z^{2m-1} + \ldots + r_1 z^{m+1} + r_0 z^m + r_1 z^{m-1} + \ldots + r_{m-1} z + r_m$$

$$= (a_0 + a_1 z + \ldots + a_{m-1} z^{m-1} + a_m z^m)(a_0 z^m + a_1 z^{m-1} + \ldots + a_{m-1} z + a_m).$$

In order to factor $z^m \, \Psi(z)$ into the two real polynomials $A(z)$ and $z^m A(z^{-1})$, we see that one of the real polynomials $(z-\alpha_k)(z-\bar{\alpha}_k)$ or $(z-\alpha_k^{-1})(z-\bar{\alpha}_k^{-1})$ must be a factor in the polynomial $A(z)$. Since β_k is real, then either $(z-\beta_k)$ or $(z-\beta_k^{-1})$ is a factor in $A(z)$. On the other hand, since the factors $(z-\gamma_k)$ and $(z-\bar{\gamma}_k)$ are complex, both of them must be contained in $A(z)$. Likewise $(z-\rho_k)$ must appear in $A(z)$. Thus it is necessary that the roots γ_k and ρ_k, which have modulus 1, appear an even number of times, that is $l=2l'$ and $n=2n'$. This condition that roots of modulus one appear an even number of times is satisfied since $\Psi(\omega) \geqslant 0$. Thus we have

$$(2.87) \qquad A(z) = a_m \prod_{k=1}^{h} (z-\alpha_k)(z-\bar{\alpha}_k) \prod_{k=1}^{j} (z-\beta_k) \prod_{k=1}^{l'} (z-\gamma_k)(z-\bar{\gamma}_k) \prod_{k=1}^{n'} (z-\rho_k)$$

and

$$z^m A(z^{-1}) = \prod_{k=1}^{h} (z-\alpha_k^{-1})(z-\bar{\alpha}_k^{-1}) \prod_{k=1}^{j} (z-\beta_k^{-1})$$

$$(2.871)$$

$$\prod_{k=1}^{l'} (z-\gamma_k^{-1})(z-\bar{\gamma}_k^{-1}) \prod_{k=1}^{n'} (z-\rho_k^{-1})$$

$$\text{since } \bar{\gamma}_k = \gamma_k^{-1}, \; \gamma_k = \bar{\gamma}_k^{-1}, \text{ and } \rho_k = \rho_k^{-1}.$$

Thus if the zeros of $A(z)$ are z_k, then the zeros of $z^m A(z^{-1})$ are z_k^{-1}. In order for the linear operator with coefficients $a_0, a_1, \ldots, a_m$ to be strictly minimum-delay, then all the roots of

$$(2.652) \qquad A(z) = \sum_{s=0}^{m} a_s z^s$$

must have modulus $|z_k| > 1$. Thus if the transfer function $\Psi(\omega)$ yields roots γ_k and ρ_k which do have modulus $|\gamma_k| = |\rho_k|$ equal to one, then there is no

12

strictly minimum-delay operator which yields this transfer function, although there is a linear operator on the borderline of minimum-delay, such as the cosine operator (2.67).

Let us suppose that $\Psi'(\omega)$ yields no roots γ_k and ρ_k with modulus equal to one. Then we have

$$(2.88) \qquad A(z) = a_m \prod_{k=1}^{h} (z-\alpha_k)\,(z-\alpha_k) \prod_{k=1}^{j} (z-\beta_k) = \sum_{s=0}^{m} a_s z^s$$

and

$$(2.881) \qquad z^m A(z^{-1}) = \prod_{k=1}^{h} (z-\alpha_k^{-1})\,(z-\bar{\alpha}_k^{-1}) \prod_{k=1}^{j} (z-\beta_k^{-1}) = \sum_{s=0}^{m} a_s z^{m-s}$$

where $|\alpha_k| \neq 1$ for $k = 1,2,...,h$ and $|\beta_k| \neq 1$ for $k = 1,2,...,j$. If the zeros of $A(z)$ are z_k for $k = 1,2,...,m$, then the zeros of $z^m A(z^{-1})$ are z_k^{-1} for $k = 1,2,...,m$, and since $|z_k| \neq 1$, it follows that half of the $2m$ roots of

$$(2.863) \qquad z^m \Psi'(z) = A(z) z^m A(z^{-1})$$

have modulus greater than one, and the other half of the $2m$ roots has modulus less than one.

Thus if we choose those m roots of $z^m \Psi'(z)$ which have modulus greater than one, and call these roots the α_k $(k = 1,2,...,h)$, the $\bar{\alpha}_k$ $(k = 1,2,...,h)$, and the $\beta_k (k = 1,2,...,h)$ which appear in equation (2.88), then $A(z)$ will represent the transfer function in the z plane, of a minimum-delay operator. That is, since $A(z)$ has roots, all of which have modulus greater than one, the transfer function

$$(2.421) \qquad A(\omega) = \sum_{s=0}^{m} a_s e^{-i\omega s}$$

will be free from zeros in the lower half plane of $\lambda = \omega + i\sigma$, and be of the minimum phase shift type. The realizable operator, with coefficients

$$(2.864) \qquad a_t = \frac{1}{2\pi} \int_{-\pi}^{\pi} A(\omega) e^{i\omega t}\, dt$$

is then minimum-delay.

On the other hand, if we did not choose the roots in the above fashion, there being at most 2^m different ways of choosing the roots, then $A(z)$ would have roots, some of which have modulus greater than one, and some of which have modulus less than one. Consequently the transfer function $A(\omega)$ would not be of the minimum phase shift type and its linear operator would not be minimum-delay.

Let us summarize the computational procedure required to determine the stable operator coefficients $a_0, a_1, \ldots, a_m$ from the power transfer function

$$(2.82) \qquad \Psi(\omega) = \sum_{\tau=-m}^{m} r_\tau e^{-i\omega\tau} \geqslant 0, \; r_0 > 0 .$$

Form the polynomial

$$(2.831) \qquad z^m \, \Psi(z) = \sum_{\tau=-m}^{m} r_\tau z^{\tau+m}$$

and solve for its roots $z_1, z_2, \ldots, z_{2m}$. Let $z'_1, z'_2, \ldots, z'_m$ be those $z_k (k=1,2,\ldots,2m)$ of modulus greater than one and also those z_k of modulus one counted half as many times. (In order for there to be a strictly minimum-delay operator, there may be no z_k of modulus one.) Then we form the polynomial

$$(2.89) \qquad A(z) = (z - z'_1) \, (z - z'_2) \ldots (z - z'_m) = \sum_{s=0}^{m} a_s z^s ,$$

and the operator coefficients are given by the a_s. They represent a minimum-delay operator, the transfer function of which has minimum phase characteristic. Thus we have shown that the power transfer function $\Psi(\omega)$, equation (2.82), may be factored into $\Psi(\omega) = A(\omega)\overline{A(\omega)}$ where the transfer function $A(\omega)$ is minimum-delay.

3. THEORY OF FINITE DISCRETE MULTICHANNEL OPERATORS

3.1 Minimum–Delay Multichannel Operators

A multiple process consists of many time series; these time series are interrelated. A multichannel operator acts upon all these time series simultaneously, and thus it can take advantage of the structure between different time series as well as along a single time series. In the foregoing chapter (Section 2.7) we have shown how the inverse to a single channel minimum-delay operator may be computed. The purpose of the next section (Section 3.2) is to extend this result to the multichannel case. In this section we want to define the concept of a minimum-delay multichannel operator.

The theory of multichannel discrete operators conceptually is the same as the theory of single channel discrete operators; mathematically it represents the extension from ordinary (scalar) algebra to matrix algebra. That is, multichannel theory can be obtained from single channel theory by the appropriate substitution of matrices for scalars.

A multichannel operator is one with several input channels and several output channels. In the case when there is the same number of input channels as output channels, then we need deal only with square matrices and vectors, where the order of a matrix or a vector is equal to the number of channels. If the number of input channels is different from the number of output channels then we must deal with rectangular matrices.

As an example let us consider a discrete multichannel operator with two input channels and two output channels. In this case the input and output may be represented by the 2×1 column vectors:

$$\text{input} = x_t = \begin{bmatrix} x_{1t} \\ x_{2t} \end{bmatrix} \quad , \quad \text{output} = y_t = \begin{bmatrix} y_{1t} \\ y_{2t} \end{bmatrix}$$

and the operator coefficients may be represented by the 2×2 square matrices:

$$a_0 = \begin{bmatrix} 2 & 1 \\ 0 & 6 \end{bmatrix} \quad , \quad a_1 = \begin{bmatrix} 1 & 0 \\ 1 & 1 \end{bmatrix} .$$

The convolution formula relating input x_t and output y_t is in this case:

$$y_t = a_0 x_t + a_1 x_{t-1}$$

or

$$\begin{bmatrix} y_{1t} \\ y_{2t} \end{bmatrix} = \begin{bmatrix} 2 & 1 \\ 0 & 6 \end{bmatrix} \begin{bmatrix} x_{1t} \\ x_{2t} \end{bmatrix} + \begin{bmatrix} 1 & 0 \\ 1 & 1 \end{bmatrix} \begin{bmatrix} x_{1,\,t-1} \\ x_{2,\,t-1} \end{bmatrix}.$$

Hence the theory of multichannel discrete operators can be regarded as the matrix-valued counterpart of single-channel discrete operator theory. The essential difference in programming multichannel operations on a digital computer is that we now deal with discrete operators whose coefficients are matrices instead of scalars. Single channel concepts, such as autocorrelation, cross-correlation, and convolution are the same in the multichannel case, except that now the digital computer must be programmed to carry out matrix multiplication instead of scalar multiplication, matrix addition instead of scalar addition, matrix inversion instead of scalar division, etc., and that we must be careful to take into account the special features of matrix algebra, such as the fact that matrix multiplication is not commutative as is scalar multiplication.

The z-transform of a multichannel operator is formed in the same way as in the single channel case, except that now the z-transform is a polynomial in z with matrix-valued coefficients instead of scalar coefficients. Alternatively we may describe the z-transform as a matrix with scalar polynomials as entries.

In the classification of the delay properties of such a multichannel operator, the determinant of its matrix-valued z-transform plays a central role. This determinant is a scalar-valued polynomial in z. *If the coefficients of this polynomial represent a single-channel minimum-delay operator, then the parent multichannel operator is also called minimum-delay. On the other hand, if they represent a single-channel maximum-delay operator, then the parent multichannel operator is also called maximum-delay. Finally, if they represent a single-channel mixed-delay operator, then the parent multichannel operator is also called mixed-delay.*

Exactly as we do in the case of single-channel theory, we can make the following statements for the multichannel case: A minimum-delay multichannel discrete operator has a multichannel inverse that is also a minimum-delay (and hence realizable) multichannel operator. A maximum-delay multichannel discrete operator has a multichannel inverse that is a stable but purely non-realizable function. A mixed-delay multichannel discrete operator has a stable inverse, this inverse being made up of a realizable component and a non-realizable component.

In the next section we will give the mathematical justification for the above statements.

3.2   The Inverse of a Minimum–Delay Multichannel Operator

This section will be mathematical in nature and somewhat condensed; we wish to establish the basic conditions for the stability of the inverse of a multichannel finite-length discrete operator. We will treat the case where there are the same number of output channels as input channels; we designate the number of channels by the letter p . The basic convolution formula for the action of the multichannel operator is

$$(3.21) \qquad\qquad y_t = a_0 x_t + a_1 x_{t-1} + \ldots + a_m x_{t-m}$$

where t is an integer denoting the discrete time index, x_t is the input at time t, y_t is the output at time t, and

$$(a_0, a_1, \ldots, a_m)$$

denotes the set of operator coefficients. The input x_t and output y_t for a fixed t each are $p \times 1$ column vectors, and any given filter coefficient a_s is a $p \times p$ constant square matrix. The z-transform of the operator is defined as

$$(3.22) \qquad\qquad A(z) = a_0 + a_1 z + \ldots + a_m z^m$$

for the scalar complex variable z. For example, suppose the operator coefficients are the matrices

$$a_0 = \begin{bmatrix} 2 & 1 \\ 0 & 6 \end{bmatrix} \,, \quad a_1 = \begin{bmatrix} 1 & 0 \\ 1 & 1 \end{bmatrix} .$$

Then the z-transform of the operator is

$$(3.23) \qquad\qquad A(z) = \begin{bmatrix} 2 & 1 \\ 0 & 6 \end{bmatrix} + \begin{bmatrix} 1 & 0 \\ 1 & 1 \end{bmatrix} z .$$

This equation exhibits the z-transform as a polynomial with matrix coefficients. Combining terms, we may write $A(z)$ as

$$(3.231) \qquad\qquad A(z) = \begin{bmatrix} 2+z & 1 \\ z & 6+z \end{bmatrix} .$$

This equation exhibits the z-transform as a matrix with polynomial entries. Both of these ways of looking at $A(z)$ will prove valuable at one time or another.

The determinant, adjugate, and inverse of a matrix with polynomial entries can be defined in a way entirely analogous to scalar valued matrices. See Frazer, Duncan, and Collar (1938, Chapter 3). Given $A(z)$, we denote its determinant by $\det A(z)$, its adjugate by $\operatorname{adj} A(z)$, and its inverse by $A^{-1}(z)$. For the above example, these quantities may be computed easily and are

$$(3.24) \qquad \det\ A(z) = 12 + 7z + z^2$$

$$(3.241) \qquad \operatorname{adj}\ A(z) = \begin{bmatrix} 6+z & -1 \\ -z & 2+z \end{bmatrix}$$

$$(3.242) \qquad A^{-1}(z) = \frac{\operatorname{adj}\ A(z)}{\det\ A(z)} = \frac{1}{12+7z+z^2} \begin{bmatrix} 6+z & -1 \\ -z & 2+z \end{bmatrix}.$$

Now $A^{-1}(z)$ represents the transfer function of the inverse to the multichannel operator with transfer function $A(z)$. Since $\operatorname{adj}\ A(z)$ is necessarily a matrix with polynomial entries, it follows that $\operatorname{adj}\ A(z)$ represents a finite-length realizable discrete operator. To obtain $A^{-1}(z)$ we see that each term in $\operatorname{adj}\ A(z)$ is multiplied by $1/\det A(z)$. Hence the stability of the inverse operator $A^{-1}(z)$ depends entirely on the stability of the single-channel operator $1/\det A(z)$. But the single-channel operator $1/\det A(z)$ is the inverse of the single-channel operator with z-transform $\det A(z)$. Now from Section 2.7, we know that the inverse of a single-channel operator can be represented by a stable minimum-delay realizable function provided the single-channel operator is minimum-delay. Likewise the inverse can be represented by a stable purely non-realizable function provided the single-channel operator is maximum-delay; the inverse can be represented by a stable function involving both a realizable component and a purely non-realizable component provided the single-channel operator is mixed-delay.

From this we can conclude that the inverse of a multichannel operator $A(z)$ can be represented by a stable matrix-valued minimum-delay realizable function provided the single-channel filter $\det\ A(z)$ is minimum-delay; the inverse of $A(z)$ can be represented by a stable matrix-valued purely non-realizable function provided $\det A(z)$ is maximum-delay; and the inverse of $A(z)$ can be represented by a stable matrix-valued function with both a realizable component and a purely non-realizable component provided $\det A(z)$ is mixed-delay. Hence we call the multichannel filter $A(z)$ minimum-delay, maximum-delay, or mixed-delay provided that the single-channel filter $\det\ A(z)$ is minimum-delay, maximum-delay, or mixed-delay respectively. In our example, we see that

$$\det A(z) = (3+z)(4+z)$$

which has zeros $z_1 = -3$ and $z_2 = -4$, both of which lie outside the unit circle; hence $\det A(z)$ and therefore $A(z)$ are each minimum-delay. Thus the inverse single-channel operator $1/\det A(z)$ can be represented by a minimum-delay stable realizable function, and the inverse multichannel operator $A^{-1}(z)$ can be represented by a minimum-delay stable matrix-valued realizable function.

It is worthwhile to exhibit the general structure of the inverse of a multichannel finite-length operator $A(z)$. To do so we write the relationship between $A(z)$, $\operatorname{adj}A(z)$, and $\det A(z)$ as

$$(3.25) \qquad\qquad [\operatorname{adj}\ A(z)]A(z)=I\ \det A(z)$$

where I is the $p \times p$ identity matrix. The characteristic equation of the polynomial matrix $A(z)$ is defined as

$$\det\ A(z)=0\ .$$

Now $\det A(z)$ is a polynomial of degree of at most n, where n is defined as equal to mp. For the sake of simplicity of notation and clarity of exposition, we assume that the charateristic equation has n distinct roots, which we denote by

$$z_1, z_2, \ldots, z_n\ .$$

Hence $\det A(z)$ may be written in factored form as

$$(3.251) \qquad\qquad \det\ A(z)=c(z-z_1)(z-z_2)\ldots(z-z_n)$$

where c is a constant. From equation (3.25) we obtain

$$(3.252) \qquad\qquad [\operatorname{adj}\ A(z_i)]A(z_i)=0$$

where 0 in this equation denotes a $p \times p$ matrix of zeros. Now $\operatorname{adj}\ A(z_i)$ can be written as the product of a column vector c_i, and a row vector r_i (Frazer, Duncan, and Collar, 1938, p. 61); that is

$$\operatorname{adj}\ A(z_i)=c_i r_i\ .$$

Hence equation (3.252) becomes

$$c_i r_i A(z_i)=0$$

where again 0 denotes a $p \times p$ zero matrix. From this equation it follows that

$$r_i A(z_i)=0$$

where here 0 denotes a $1 \times p$ row vector of zeros. Because of this equation we may call r_i the eigen-row-vector of the polynomial matrix of $A(z)$ corresponding to the eigenvalue z_i.

Again returning to the relationship between $A(z)$, adj $A(z)$, and det $A(z)$ we may write

$$(3.253) \qquad A(z)\bigl[\text{adj } A(z)\bigr] = I \det A(z) \, ,$$

from which it follows that

$$A(z_i)\bigl[\text{adj } A(z_i)\bigr] = 0$$

where 0 denotes a $p \times p$ zero matrix. Again writing adj $A(z_i)$ as $c_i r_i$ we obtain

$$A(z_i) c_i r_i = 0$$

where 0 denotes the $p \times p$ zero matrix. From this equation it follows as before that

$$A(z_i) c_i = 0$$

where 0 denotes a $p \times 1$ column vector of zeros. Hence we may call c_i the eigen-column-vector of the polynomial matrix $A(z)$ corresponding to the eigenvalue z_i.

Returning to our numerical example, let us substitute the eigenvalue $z_1 = -3$ for z in adj $A(z)$. We obtain

$$\text{adj } A(z_1) = \text{adj } A(-3) = \begin{bmatrix} 3 & -1 \\ 3 & -1 \end{bmatrix} = \begin{bmatrix} 1 \\ 1 \end{bmatrix} \begin{bmatrix} 3 & -1 \end{bmatrix}$$

so the eigenvectors are

$$c_1 = \begin{bmatrix} 1 \\ 1 \end{bmatrix} \, , \quad r_1 = \begin{bmatrix} 3 & -1 \end{bmatrix} \, .$$

Likewise for the eigenvalue $z_2 = -4$ we have that

$$\text{adj } A(z_2) = \text{adj } A(-4) = \begin{bmatrix} 2 & -1 \\ 4 & -2 \end{bmatrix} = \begin{bmatrix} 1 \\ 2 \end{bmatrix} \begin{bmatrix} 2 & -1 \end{bmatrix}$$

so the eigenvectors are

$$c_2 = \begin{bmatrix} 1 \\ 2 \end{bmatrix} \, , \quad r_2 = \begin{bmatrix} 2 & -1 \end{bmatrix} \, .$$

Recalling that $A^{-1}(z) = \operatorname{adj} A(z)/\det A(z)$ and using equation (3.251) we may write the z-transfom of the inverse multichannel discrete operator as

$$(3.26) \qquad A^{-1}(z) = \frac{\operatorname{adj} A(z)}{c(z-z_1)(z-z_2)\ldots(z-z_n)} \, .$$

This may be expanded in partial fractions as

$$(3.261) \qquad A^{-1}(z) = \frac{u_1}{z-z_1} + \frac{u_2}{z-z_2} + \ldots + \frac{u_n}{z-z_n}$$

where $u_1, u_2, \ldots, u_n$ are each $p \times p$ matrices. To determine u_1, we multiply both sides of the equation by $z-z_1$, and then set $z=z_1$, thereby obtaining

$$u_1 = \frac{\operatorname{adj} A(z_1)}{c(z_1-z_2)(z_1-z_3)\ldots(z_1-z_n)} \, .$$

If we define α_1 as

$$\alpha_1 = \frac{1}{c(z_1-z_2)(z_1-z_3)\ldots(z_1-z_n)}$$

and recall that $\operatorname{adj} A(z_1) = c_1 r_1$ then we have

$$u_1 = \alpha_1 c_1 r_1 \, .$$

In a similar fashion we may define $\alpha_2, \alpha_3, \ldots, \alpha_n$. Hence the partial fraction expansion of the inverse filter is

$$(3.262) \qquad A^{-1}(z) = \frac{\alpha_1 c_1 r_1}{z-z_1} + \frac{\alpha_2 c_2 r_2}{z-z_2} + \ldots + \frac{\alpha_n c_n r_n}{z-z_n}$$

where z_i, c_i, and r_i are the respective eigenvalues, eigen-column-vectors, and eigen-row-vectors of the polynomial matrix $A(z)$.

In our numerical example, we have

$$\alpha_1 = \frac{1}{z_1-z_2} = 1 \qquad \alpha_2 = \frac{1}{z_2-z_1} = -1$$

so the partial fraction expansion of the inverse operator is

$$A^{-1}(z) = \frac{1}{z+3}\begin{bmatrix} 1 \\ 1 \end{bmatrix}\begin{bmatrix} 3 & -1 \end{bmatrix} - \frac{1}{z+4}\begin{bmatrix} 1 \\ 2 \end{bmatrix}\begin{bmatrix} 2 & -1 \end{bmatrix}.$$

If $|z_i| > 1$, the fraction $1/(z-z_i)$ can be expanded in a stable series of non-negative powers of z, namely the series

$$\frac{1}{z-z_i} = -z_i^{-1} - z_i^{-2}\,z - z_i^{-3}\,z^2 - \ldots .$$

Hence if all the eigenvalues z_i are greater than unity in magnitude, then the inverse system $A^{-1}(z)$ can be expanded in a stable series of non-negative powers of z; that is,

$$A^{-1}(z) = \sum_{i=1}^{n} \alpha_i c_i r_i (-z_i^{-1} - z_i^{-2}\,z - z_i^{-3}\,z^2 - \ldots) .$$

But the eigenvalues z_i are the roots of the characteristic equation $\det A(z)=0$, and hence these eigenvalues will all be greater than unity in magnitude provided that $\det A(z)$ represents a single-channel minimum-delay operator. Thus this is precisely the case when $\det A(z)$ and hence $A(z)$ represent minimum-delay operators with inverses that can be represented by stable realizable functions. In our numerical example both eigenvalues exceed unity in magnitude, so that the operator is minimum-delay. Its inverse has the representation

$$A^{-1}(z) = \begin{bmatrix} 1 \\ 1 \end{bmatrix}\begin{bmatrix} 3 & -1 \end{bmatrix}\left(\frac{1}{3} - \frac{1}{9}\,z + \frac{1}{27}\,z^2 - \ldots \right)$$

$$- \begin{bmatrix} 1 \\ 2 \end{bmatrix}\begin{bmatrix} 2 & -1 \end{bmatrix}\left(\frac{1}{4} - \frac{1}{16}\,z + \frac{1}{64}\,z^2 - \ldots \right) .$$

If $|z_i| > 1$, the fraction $1/(z-z_i)$ can be expanded in a stable series of negative powers of z, namely the series

$$\frac{1}{z-z_i} = z^{-1} + z_i\,z^{-2} + z_i^2\,z^{-3} + \ldots .$$

If all the eigenvalues are less than unity in magnitude, then the multichannel filter is maximum-delay; its inverse can be represented by a stable purely non-realizable function; here we make use of the expansion of $1/(z-z_i)$ given above for each z_i. If some of the eigenvalues are greater than unity in magnitude, while the others are less than unity in magnitude, then we must make use of one or the other of the two expansions of $1/(z-z_i)$ for each z_i, depending on the magnitude of z_i. This is the case of $A(z)$ being mixed-delay; its inverse can be expressed as a stable function involving both a realizable component and a purely non-realizable component.

3.3   The Multichannel Power Transfer Function and its Multichannel Minimum–Delay Operator

In this section we extend the method given in Section 2.8 for the single channel case to the multichannel case. Corresponding to the single-channel equations (2.83) and (2.82) given in Section 2.8, we have the multichannel counterpart of the power transfer function $\Psi(z)$ given by

$$\Psi(z) = \sum_{s=-m}^{m} r_s z^s = A(z) A_*(z)$$

where

$$A(z) = a_0 + a_1 z + \ldots + a_m z^m$$

is the z-transform of the p-channel operator $(a_0, a_1, \ldots, a_m)$, as previously given by equation (3.22) in Section 3.2, and where we have used the subscript $*$ notation, defined as

$$A_*(z) = \left[A(1/z^*) \right]^{*T}$$

$$= \left[a_0 + a_1(1/z^*) + \ldots + a_m(1/z^*) \right]^{*T}$$

$$= a_0^{*T} + a_1^{*T}(1/z) + \ldots + a_m^{*T}(1/z)^m$$

Let us note that the following equalities hold for the subscript $*$ notation:

$$A_{**}(z) = A(z)$$

$$[A(z)B(z)]_* = B_*(z) A_*(z) \ .$$

In particular we see that the power transfer function satisfies

$$\Psi_*(z) = \left[A(z)A_*(z)\right]_* = A_{**}(z)A_*(z)$$

$$= A(z)A_*(z) = \Psi(z) \ .$$

This property, namely that $\Psi(z) = \Psi_*(z)$, may be described as the *reverse-hermitian* property of $\Psi(z)$. Hence on the unit circle, where $z = e^{i\omega}$, we have

$$\Psi(\omega) = \left[\Psi(\omega)\right]^{*T}$$

which shows that $\Psi(\omega)$ is hermitian in the ordinary sense. In the special case the coefficients r_s of $\Psi(z)$ are real matrices, then the reverse-hermitian property of $\Psi(z)$ simplifies to

$$\Psi(z) = \left[\Psi(1/z)\right]^T$$

which may be described as a *reverse-symmetrical* property. From the reverse-hermitian property of $\Psi(z)$ we have

$$\sum_{s=-m}^{m} r_s z^s = \left[\sum_{s=-m}^{m} r_s(1/z^*)^s\right]^{*T}$$

$$= \sum_{s=-m}^{m} r_s^{*T} z^{-s}$$

which gives

$$r_{-s} = r_s^{*T} \ .$$

In the special case when the r_s are real matrices, this reduces simply to

$$r_{-s} = r_s^{T} \ .$$

Before discussing the general case of factoring the multichannel power transfer function so as to obtain its corresponding multichannel minimum-delay operator, let us consider a numerical example. Let us suppose $\Psi(z)$ is given as

$$\Psi(z) = \begin{bmatrix} 2z^{-1}+6+2z & 3+7z \\ \\ 7z^{-1}+3 & 6z^{-1}+38+6z \end{bmatrix} \ .$$

We wish to find a minimum-delay operator $[a_0 \; a_1]$ such that

$$\Psi'(z) = (a_0 + a_1 z)(a_0^{*T} + a_1^{*T} z^{-1}) \; .$$

From $\Psi'(z)$ we may readily compute the adjugate and determinant as

$$\text{adj } \Psi'(z) = \begin{bmatrix} 6z^{-1}+38+6z & -3-7z \\ -7z^{-1}-3 & 2z^{-1}+6+2z \end{bmatrix}$$

$$\det \Psi'(z) = 12z^{-2}+91z^{-1}+194+91z+12z^2 \; .$$

Next we wish to factor $\det \Psi'(z)$ into binomial factors. (Note: The quasi-polynomial $\det \Psi'(z)$ can be reduced to a polynomial of one-half the degree by the substitution $z+z^{-1}=u$, thereby reducing the computation required for the factorization). The result of the factorization is

$$\det \Psi'(z) = (z+3)(z+4)(3+z^{-1})(4+z^{-1})$$

so that the roots of $\det \Psi'(z) = 0$ are

$$z_1 = -3, \quad z_2 = -4, \quad z_3 = -\frac{1}{3}, \quad z_4 = -\frac{1}{4} \; .$$

In order to obtain the minimum-delay factorization we are required to choose those roots which lie outside the unit circle, namely $z_1 = -3$ and $z_2 = -4$. Substituting, one by one, these roots into adj $\Psi'(z)$ we obtain the equations

$$\text{adj } \Psi'(-3) = \begin{bmatrix} 18 & 18 \\ -2/3 & -2/3 \end{bmatrix} = \begin{bmatrix} 1 \\ -1/27 \end{bmatrix} \begin{bmatrix} 18 & 18 \end{bmatrix} = c_1 r_1$$

$$\text{adj } \Psi'(-4) = \begin{bmatrix} 12.50 & 25.00 \\ -1.25 & -2.50 \end{bmatrix} = \begin{bmatrix} 1 \\ -0.1 \end{bmatrix} \begin{bmatrix} 12.5 & 25 \end{bmatrix} = c_2 r_2$$

which define the column vectors c_1 and c_2 and the row vectors r_1 and r_2. We then form the matrix

$$R = \begin{bmatrix} r_1 \\ r_2 \end{bmatrix} = \begin{bmatrix} 18 & 18 \\ 12.5 & 25 \end{bmatrix}$$

and the diagonal matrix

$$D = \begin{bmatrix} 1/z_1 & 0 \\ 0 & 1/z_2 \end{bmatrix} = \begin{bmatrix} -1/3 & 0 \\ 0 & -1/4 \end{bmatrix}$$

from which we compute

$$u = R^{-1} D\, R = \begin{bmatrix} -5/12 & -1/6 \\ 1/12 & -1/6 \end{bmatrix} .$$

By construction the matrix u satisfies

$$\Psi(z) = (I - uz)\, a_0\, a_0^{*T}\, (I - u^{*T}z) .$$

(Since u is real, we may drop the superscript $*$ on the u). Next we want to determine $a_0 a_0^{*T}$. By letting $z = 1$ we have

$$\Psi(1) = (I - u)\, a_0\, a_0^{*T}\, (I - u^{T})$$

which gives

$$a_0 a_0^{*T} = (I - u)^{-1} \Psi(1)(I - u^{T})^{-1} = \begin{bmatrix} 4 & 2 \\ 2 & 37 \end{bmatrix} .$$

If c is any arbitrary unitary matrix (i. e. $cc^{*T} = I$) then we see that

$$(a_0 c)\, (a_0 c)^{*T} = \begin{bmatrix} 4 & 2 \\ 2 & 37 \end{bmatrix}$$

and hence the matrix a_0 is determined except for the arbitrary unitary postmultiplier c. In particular, therefore, the matrix a_0 can be chosen to be a triangular matrix, thereby leading to a causal-chain system (Wold, 1953). Such a choice gives

$$a_0 = \begin{bmatrix} 2 & 0 \\ 1 & 6 \end{bmatrix} .$$

Hence we may compute

$$a_0 + a_1 z = (I - uz)a_0$$

so that the required minimum-delay causal-chain operator is

$$a_0 = \begin{bmatrix} 2 & 0 \\ 1 & 6 \end{bmatrix}, \quad a_1 = -ua_0 = \begin{bmatrix} 1 & 1 \\ 0 & 1 \end{bmatrix}.$$

In summary, then, the multichannel factorization method is as follows. Given $\Psi(z)$ we wish to find a minimum-delay $A(z)$ such that

$$\Psi(z) = A(z)A_*(z) .$$

Write $A(z)$ in the factored form as

$$A(z) = (I - u_1 z)(I - u_2 z) \ldots (I - u_m z)a_0 .$$

Hence

$$\Psi(z) = (I - u_1 z) \ldots (I - u_m z)\, a_0\, a_0^{*T}(I - u_m^{*T}z) \ldots (I - u_1^{*T}z) .$$

Next post-multiply by $\Psi^{-1}(z)\det\Psi(z) = \mathrm{adj}\,\Psi(z)$; this yields

$$I \det \Psi(z) = \mathrm{adj}\,\Psi(z)\,(I - u_1 z) \ldots (I - u_m z)\, a_0\, a_0^{*T}(I - u_m^{*T}z) \ldots (I - u_1^{*T}z) .$$

Now the number of zeros of the quasi-polynomial $\det\Psi(z)$ is $2mp$. Moreover because

$$\det \Psi(z) = \det A(z) \; \det A_*(z)$$

we see that if z_i is a zero of $\det \Psi(z)$ then $1/z_i^*$ is also a zero. As in the single channel case discussed in Section 2.8, one of this pair of zeros goes into the construction of $A(z)$ and the other into the construction of $A_*(z)$, and the choice is arbitrary for each of the mp pairs of zeros of $\det\Psi(z)$. As a result there will be 2^{mp} multichannel factorizations if we exclude the case of zeros on the unit circle and other multiple zeros. In order to obtain the minimum-delay factorization we choose those zeros which are greater than unity in magnitude. At a zero z_i the adjugate factors into the product of a column c_i and a row r_i, that is

$$\mathrm{adj}\,\Psi(z_i) = c_i r_i$$

which gives

$$0 = c_i r_i (I - u_1 z_i) \dots (I - u_m z_i)\, a_0\, a_0^{*T}(I - u_m^{*T} z_i)\dots(I - u_1^{*T} z_i)\ .$$

Thus we have

$$r_i (I - u_1 z_i) = 0\ ,$$

and using this equation for the first p zeros (i. e. $i = 1,2,\dots,p$) we may determine u_1, as we did in the numerical example. At the remaining zeros $z = z_i$, the factor adj $\Psi(z)$ $(I - u_1 z)$ also becomes the product of a column and a row and hence we may use the same procedure with the next p zeros (i. e. $i = p+1, p+2,\dots,2p$) and thus determine u_2. Likewise we can then consecutively determine $u_3,\dots,u_m$, thereby exhausting the mp zeros which have magnitude greater than one. Finally we determine a_0 as we did in the numerical example, and hence we obtain the required causal-chain minimum-delay operator:

$$A(z) = (I - u_1 z)(I - u_2 z) \dots (I - u_m z) a_0$$

$$= a_0 + a_1 z + a_2 z^2 + \dots + a_m z^m\ .$$

4. THEORY OF DISCRETE STATIONARY TIME SERIES

4.1 Random Processes

In this chapter we wish to develop the theory of discrete stationary time series. For more comprehensive presentations of the theory of discrete stationary time series, the reader is referred in particular to Wiener (1930), Khintchine (1934), Wold (1938), Kolmogorov (1941), Wiener (1942), Doob (1953), and Robinson (1959, 1962).

A discrete time series is a sequence of equidistant observations x_t which are associated with the discrete time parameter t. Without loss of generality we may take the spacing between each successive observation to be one unit of time, and thus we may represent the time series as

$$(4.11) \qquad \ldots, x_{t-2}, x_{t-1}, x_t, x_{t+1}, x_{t+2}, \ldots$$

where t takes on all integer values from minus infinity to infinity ($-\infty < t < \infty$). Thus all angular frequencies ω may be required to lie between $-\pi$ and π.

Any observational time series x_t ($-\infty < t < \infty$) may be considered as a realization of a so-called random process, or stochastic process, which is a mathematical abstraction defined with respect to a probability field. In many phenomena, the number of observational time series supplied by a random process is limited. It is often the case, especially in economic applications, that only one time series is generated by a random process . Such a case, nevertheless, is in full accord with the frequency interpretation of probability.

4.2 Stationary Time Series

A time series is said to be stationary if the probabilities involved in the stochastic process are not tied down to a specific origin in time; that is, the probability of any event associated with the time t is equal to the probability of the corresponding event associated with the time $t+\tau$ where t and τ are any integer values.

For any stochastic process, one may form averages with respect to the statistical population or «ensemble» of realizations x_t for a fixed value of time t. Such averages are called ensemble averages or space averages, and we shall denote such an averaging process by the expectation symbol E. In particular the mean value $m = \mathrm{E}[x_t]$ and the variance $\sigma^2 = \mathrm{E}[(x_t - m)^2]$ of a stationary stochastic process are independent of time t. Likewise, the (unnormalized) autocorrelation coefficients

$$(4.21) \qquad \varphi(\tau) = \mathrm{E}[x_t\, x_{t+\tau}]$$

are independent of t, and constitute an even function of the time lag τ, that is

$$(4.22) \qquad \varphi(\tau) = \varphi(-\tau) \ .$$

Also we have

$$(4.221) \qquad |\,\varphi(\tau)\,| \leqslant \varphi(0) \ .$$

The normalized autocorrelation function is defined to be

$$(4.222) \qquad \varphi(\tau) = \frac{\mathrm{E}[(x_t - m)\,(x_{t+\tau} - m)]}{\mathrm{E}[(x_t - m)]^2}$$

so that

$$(4.223) \qquad \varphi(0) = 1,\ |\,\varphi(\tau)\,| \leqslant 1 \ .$$

In what follows we shall assume that the mean value m is equal to zero which we may do without loss of generality. Also we shall utilize the unnormalized autocorrelation (4.21).

There is another type of average known as time average in which the averaging process is carried out with respect to all values of time t for a fixed realization $x_t(-\infty < t < \infty)$ of the stochastic process. A stationary process is called an ergodic process if the ensemble averages and time averages are equal with probability one. As a result the autocorrelation of an ergodic process may be expressed as the time average

$$(4.23) \qquad \varphi(\tau) = \lim_{T \to \infty} \frac{1}{2\,T + 1} \sum_{t=-T}^{T} x_{t+\tau}\, x_t \ .$$

4.3 The Autocorrelation

The autocorrelation function $\varphi(\tau)$ is a non-negative definite function, that is,

$$\varphi(\tau) = \varphi(-\tau) \, ,$$

(4.31)
$$\sum_{j=1}^{N} \sum_{k=1}^{N} \varphi(j - k)\, a_j a_k \geqslant 0 \, , \quad N = 1,2,\dots$$

for every real set of $a_1, a_2, \dots, a_N$. Thus, the autocorrelation matrix $(N{=}1,2,\dots)$

(4.32)
$$\begin{bmatrix} \varphi(0) & \varphi(1) & \dots & \varphi(N) \\ \varphi(-1) & \varphi(0) & \dots & \varphi(N-1) \\ & \dots & & \dots \\ \varphi(-N) & \varphi(-N+1) & \dots & \varphi(0) \end{bmatrix}$$

is symmetric, has its elements equal along its diagonal and along any super or sub diagonal, has non-negative eigenroots $\lambda_j \geqslant 0 (j{=}1,2,\dots,N)$, has a non-negative determinant, and has a non-negative definite quadratic form given by equation (4.31). The non-negative definiteness of the autocorrelation follows from the inequality

$$\sum_{j=1}^{N} \sum_{k=1}^{N} \varphi(j - k) a_j a_k = \sum_{j=1}^{N} \sum_{k=1}^{N} \varphi(|\, j - k\,|) a_j a_k$$

(4.33)
$$= \sum_{j=1}^{N} \sum_{k=1}^{N} \mathrm{E}[a_j a_k x_j x_k]$$

$$= \mathrm{E}\left[\left(\sum_{j=1}^{N} a_j x_j \right)^2\right] \geqslant 0$$

4.4 The Spectrum

The property that the autocorrelation function $\varphi(\tau)$ is non-negative definite is equivalent to its representation by the Fourier transform

$$(4.41) \qquad \varphi(\tau) = \frac{1}{\pi} \int_0^\pi \cos \omega\tau \, d\Lambda(\omega)$$

where $\Lambda(\omega)$, called the integrated spectrum or the spectral distribution function, is a real monotone non-decreasing function of $\omega(0 \leqslant \omega \leqslant \pi)$ with $\Lambda(0) = 0$ and $\Lambda(\pi) = \pi$. This theorem was given by Khintchine (1934) in his development of the theory of continuous stationary time series, and by Wold (1938) for discrete stationary time series.

The inversion formula expresses the integrated spectrum in terms of the autocorrelation, that is,

$$(4.42) \qquad \Lambda(\omega) = \omega + 2 \sum_{\tau=1}^{\infty} \frac{\varphi(\tau)}{\tau} \sin \omega\tau, \ 0 \leqslant \omega \leqslant \pi \ .$$

Moreover, if $\displaystyle\sum_{\tau=0}^{\infty} |\varphi(\tau)|$ is convergent, then $\Lambda(\omega)$ will be absolutely continuous, with the continuous derivative

$$\Lambda'(\omega) = \frac{d\Lambda(\omega)}{d\omega} = \Phi(\omega) = 1 + 2 \sum_{\tau=1}^{\infty} \varphi(\tau) \cos \omega\tau$$

$$(4.43)$$

$$= \sum_{\tau=-\infty}^{\infty} \varphi(\tau) \cos \omega\tau \ .$$

In the remaining part of this paper, we shall confine ourselves to stochastic proceses for which the spectral distribution function $\Lambda(\omega)$ is absolutely continuous and for which $\Sigma \, |\varphi(\tau)| < \infty$, unless it is otherwise stated. Wiener (1942) restricts himself to those processes where the spectral distribution function $\Lambda(\omega)$ is absolutely continuous.

The derivative $\Phi(\omega)=\Lambda'(\omega)$ is called the spectral density function, the power spectrum, or simply the spectrum. Since $\Phi(\omega)$ is the slope of a real monotone non-decreasing function $\Lambda(\omega)$, we have

$$(4.44) \qquad \Phi(\omega) \geqslant 0 ,$$

and $\Phi(\omega)$ may be considered to represent the power density in the time series $x_t \, (-\infty < t < \infty)$. In order to have equal power at ω and $-\omega$, let us define $\Phi(\omega)$ to be an even function of ω, that is,

$$(4.45) \qquad \Phi(-\omega) = \Phi(\omega) , \; -\pi \leqslant \omega \leqslant \pi .$$

Equation (4.41) thus becomes

$$(4.46) \qquad \varphi(\tau) = \frac{1}{\pi} \int_0^\pi \cos \omega\tau \; \Phi(\omega) \; d\omega = \frac{1}{2\pi} \int_{-\pi}^\pi e^{i\omega\tau} \; \Phi(\omega) \; d\omega$$

and, in particular, we have

$$(4.461) \qquad \varphi(0) = \frac{1}{2\pi} \int_{-\pi}^\pi \Phi(\omega) d\omega .$$

To show that the autocorrelations $\varphi(\tau)$ as given by equation (4.46) is a non-negative definite function, we let

$$(4.47) \qquad H_N(\omega) = \sum_{k=1}^N a_k e^{-i\omega k}$$

where $a_k(k=1,2,...,N)$ is any arbitrary set of real numbers. Then the quadratic form of the autocorrelation matrix is

$$(4.48) \qquad \sum_{j=1}^N \sum_{k=1}^N \varphi(j-k) \; a_j a_k = \int_{-\pi}^\pi \sum_{j=1}^N \sum_{k=1}^N a_j a_k e^{i\omega(j-k)} \; \Phi(\omega) \; d\omega$$

which is equal to

$$(4.49) \qquad \int_{-\pi}^\pi H_N(\omega) \, \overline{H_N(\omega)} \; \Phi(\omega) \; d\omega = \int_{-\pi}^\pi | \, H_N(\omega) \, |^2 \, \Phi(\omega) \; d\omega \geqslant 0 , \; N=1,2,...$$

(where the bar indicates the complex conjugate). Thus equation (4.31) is verified for $\varphi(\tau)$ given by equation (4.46).

The important role played by the harmonic analysis of a stationary time series is brought out by the Spectral Representation Theorem due to Harald Cramér (1942) and others. The theorem allows the time series x_t to be represented by a stochastic integral which involves the harmonic components of the time series. For a statement and discussion of this theorem, the reader is referred to Doob (1953, p. 481).

4.5   Processes with White Light Spectra

A process is said to have a white light spectrum if its power spectrum has a constant value, that is,

$$(4.51) \qquad \Phi(\omega) = \text{constant}, \quad -\pi \leqslant \omega \leqslant \pi .$$

In this section we wish to consider two types of processes which have white light spectra, namely, the purely random process and the mutually uncorrelated process. For a further discussion of these processes, see Doob (1953) and Wold (1953). As a matter of terminology, one should distinguish between the «purely random process» and a «random process». The purely random process is defined in this section. On the other hand, a random or stochastic process designates any process which generates one or more observational time series, and such processes range from purely random processes to non-random or deterministic processes.

A realization from a purely random process is the time series $\xi_t(-\infty < t < \infty)$ where each ξ_t is an independent random variate from a single probability distribution function. Therefore the joint distribution functions of the ξ_t's are simple products of this probability distribution function. Consequently, we have

$$(4.52) \qquad E[\xi_t \xi_s] = E[\xi_t] \, E[\xi_s], \quad (s \neq t) .$$

Such a process is stationary and ergodic. In order to normalize the process so that it will have zero mean and unit variance, we let

$$(4.53) \qquad E[\xi_t] = 0, \quad E[\xi_t^2] = 1 .$$

The autocorrelation function is then given by

$$
\begin{aligned}
\varphi(0) &= E[\xi_t^2] = 1 \\
(4.531) \\
\varphi(\tau) &= E[\xi_t \xi_{t+\tau}] = E[\xi_t] E[\xi_{t+\tau}] = 0, \quad \tau = \pm 1, \pm 2, \pm 3, \ldots
\end{aligned}
$$

An alternative assumption as to the nature of the ξ_t leads to the definition of the so-called mutually uncorrelated process. That is, if instead of assuming the ξ_t and $\xi_{t+\tau}$ are independent, we assume for a mutually uncorrelated process that they are uncorrelated in pairs, that is

$$(4.54) \qquad E[\xi_t\xi_s] = E[\xi_t]\,E[\xi_s], \; (s \neq t)$$

which is the same as equation (4.52). That is, independent random variables are uncorrelated, but the converse is not necessarily true. Again we shall normalize the ξ_t as in equation (4.53), in which case the uncorrelated random variables ξ_t are called orthogonal random variables. The orthogonality is illustrated by

$$(4.55) \qquad \begin{aligned} E[\xi_t^2] &= 1 \,, \\[4pt] E[\xi_t\xi_s] &= 0 \,, \; (t \neq s) \,. \end{aligned}$$

In what follows, we shall assume all uncorrelated random variables are normalized in this manner (equation (4.53)) so, alternatively, we may call them orthogonal random variables. Since equation (4.531) holds for a (normalized) uncorrelated process we see that this process has the same autocorrelation coefficients as the purely random process. Also, an uncorrelated process is stationary and ergodic.

The spectral distribution function, equation (4.42), of a purely random process or of an uncorrelated process is given by

$$(4.56) \qquad \Lambda(\omega) = \omega \,, \quad 0 \leqslant \omega \leqslant \pi$$

and the power spectrum, by equation (4.43), is a constant given by

$$(4.57) \qquad \Phi(\omega) = 1 \,, \; -\pi \leqslant \omega \leqslant \pi \,.$$

These processes therefore have white light spectra, and the ξ_t may be called «white» noise (Wiener, 1930).

In summary, then, the purely random process and the uncorrelated process both have white light spectra and have autocorrelation coefficients which vanish except for lag zero.

4.6   Processes of Moving Summation

An important theorem states that a time series with an absolute continuous spectral distribution function is generated by a process of moving summation, and conversely. In this section we shall define what is meant by a process of moving summation, and then indicate in an heuristic way why this theorem holds. For a rigorous proof, the reader is referred to Doob (1953).

For the fixed realization

$$(4.61) \qquad \ldots, \xi_{t-1}, \ \xi_t, \ \xi_{t+1}, \ \xi_{t+2}, \ldots$$

of a purely random process or of a mutually uncorrelated process, the corresponding fixed realization of a process of moving summation is

$$(4.62) \qquad \ldots, x_{t-1}, \ x_t, \ x_{t+1}, \ x_{t+2}, \ldots$$

where

$$(4.621) \qquad x_n = \sum_{k=-\infty}^{\infty} c_k \xi_{n-k}, \ \ n = t, \ t \pm 1, \ t \pm 2, \ldots$$

Since two random variables are uncorrelated if they are independent, whereas the converse is not always true, in what follows we shall impose only the weaker restriction on the ξ_t and thus only assume they are mutually uncorrelated (i. e. orthogonal) in the definition of the process of moving summation. (See Section 4.5). In particular, we assume

$$(4.63) \qquad E[\xi_t] = 0, \ \ E[\xi_t^2] = 1, \ \ E[\xi_t \xi_s] = E[\xi_t] \ E[\xi_s] = 0 \ \text{for} \ t \neq s .$$

The mean of the x_t process is

$$(4.64) \qquad E[x_t] = \sum_{k=-\infty}^{\infty} c_k \ E[\xi_{t-k}] = 0$$

and the variance is

$$(4.641) \qquad E[x_t^2] = \sum_{k=-\infty}^{\infty} c_k^2 \ E[\xi_{t-k}^2] = \sum_{k=-\infty}^{\infty} c_k^2 \ .$$

The autocorrelation coefficients are

$$(4.642) \qquad \varphi(\tau) - E[x_t x_{t+\tau}] = \sum_{t=-\infty}^{\infty} c_{t-\tau} c_t = \sum_{t=-\infty}^{\infty} c_t c_{t+\tau} \ .$$

We now wish to indicate why a process of moving summation has an absolutely continuous spectral distribution function. Let x_t be a process of moving summation given by equation (4.621). Define $F(\omega)$ by

$$(4.65) \qquad F(\omega) = \sum_{k=-\infty}^{\infty} c_k e^{-i\omega k}$$

which is the transfer function of the infinite smoothing operator c_k. Then, by equation (4.642) we have

$$(4.651) \qquad \varphi(\tau) = \sum_{j=-\infty}^{\infty} c_j c_{j-\tau} = \frac{1}{2\pi} \int_{-\pi}^{\pi} \sum_{j=-\infty}^{\infty} c_{j-\tau} e^{i\omega j} \sum_{k=-\infty}^{\infty} c_k e^{-i\omega k} d\omega$$

since

$$(4.652) \qquad \int_{-\pi}^{\pi} e^{i\omega j} e^{-i\omega k} d\omega = \begin{cases} 2\pi, & j=k \\ 0, & j\neq k \end{cases}$$

for integer j and k, and by letting $j-\tau=l$, we have

$$(4.653) \qquad \varphi(\tau) = \frac{1}{2\pi} \int_{-\pi}^{\pi} e^{i\omega\tau} \sum_{l=-\infty}^{\infty} c_l e^{i\omega l} \sum_{k=-\infty}^{\infty} c_k e^{-i\omega k} d\omega \ .$$

Then, using equation (4.65), we have

$$(4.654) \qquad \varphi(\tau) = \frac{1}{2\pi} \int_{-\pi}^{\pi} F(\omega) \overline{F(\omega)} \, e^{i\omega\tau} d\omega = \frac{1}{2\pi} \int_{-\pi}^{\pi} |\, F(\omega) \,|^2 e^{i\omega\tau} d\omega$$

which is in the form of equation (4.46) with the power spectrum

$$(4.655) \qquad \Phi(\omega) = |\, F(\omega) \,|^2 \ .$$

Thus the spectral distribution function

$$(4.656) \qquad \Lambda(\omega) = \int_{0}^{\omega} \Phi(u) du$$

is absolutely continuous.

　　　　Conversely, any process with absolutely continuous spectral distribution is a process of moving summation, and in this paragraph we wish to indicate some reasons for this theorem. Because of equation (4.44) which states that the spectrum $\Phi(\omega)$ is non-negative for every value of ω, we may set

$$(4.66) \qquad \Phi(\omega) = |\, F(\omega) \,|^2 = F(\omega) \, \overline{F(\omega)}$$

where $F(\omega)$ is the Fourier series of any square root of $\Phi(\omega)$. Let us represent this Fourier series by

$$(4.661) \qquad F(\omega) = \sum_{k=-\infty}^{\infty} c_k e^{-i\omega k} .$$

Using the coefficients c_k we may define the process of moving summation

$$(4.662) \qquad x_t = \sum_{k=-\infty}^{\infty} c_k \xi_{t-k} .$$

The autocorrelation of the process (4.662) will be given by equation (4.642). The Fourier transform of this autocorrelation function gives the power spectrum of the process (4.662) which is the same as the original power spectrum in equation (4.66). Thus the process x_t given by equation (4.662) is a process of moving summation which has the given power spectrum (4.66).

Hence, for the process of moving summation represented by

$$(4.621) \qquad x_t = \sum_{k=-\infty}^{\infty} c_k \xi_{t-k} ,$$

the c_k represents a linear operator, and the ξ_{t-k} represents white «noise». The transfer function is given by $F(\omega)$ in equation (4.65), and the power transfer function is then the power spectrum of the process, that is

$$(4.655) \qquad \Phi(\omega) = |F(\omega)|^2 .$$

Thus the time series x_t is the output of a linear system, with power transfer function $\Phi(\omega)$, into which white «noise» is passed. Since the c_k may be an infinite smoothing operator, this system need not necessarily be realizable or minimum-delay. However a unique realizable and minimum-delay operator may be found, which leads to the predictive decomposition of stationary time series, initiated by Wold (1938). For a development of the theory of predictive decompositon, see Robinson (1954, 1959).

Finally, let us note that processes of moving summation are ergodic, and for a proof the reader is referred to Doob (1953).

4.7 Multichannel Time Series

In this section we wish to consider multiple discrete stationary ergodic time series with absolutely continuous spectral distribution functions. We shall have to modify our previous notation to some extent in order to accommodate the bulk of notation required. Let us consider the set of stationary processes $x_j(t)$, $(j=1,2,...,n)$ which we take to be real functions. From now on, where two or more secondary symbols appear, subscripts will denote the particular time series under consideration, whereas the time parameter will appear in the parentheses following the symbol for the function.

We define the correlation functions (Cramér, 1940) to be

$$(4.71) \qquad \varphi_{jk}(\tau) = E\{x_j(t+\tau)x_k(t)\} = \lim_{T \to \infty} \frac{1}{2T+1} \sum_{t=-T}^{T} x_j(t+\tau)x_k(t) \ .$$

For $j=k$ this equation gives the autocorrelation function of $x_j(t)$, whereas for $j \neq k$ it gives the cross-correlation function of $x_j(t)$ and $x_k(t)$. We have

$$(4.711) \qquad \varphi_{jj}(\tau) = \varphi_{jj}(-\tau) , \quad j=1,2,...,n$$

which states that the autocorrelation function is an even function of τ. For the cross-correlation functions, we have

$$(4.712) \qquad \varphi_{jk}(\tau) = \varphi_{kj}(-\tau) , \quad k \neq j , \quad j,k=1,2,...,n \ .$$

Since the time series $x_j(t)$ are real functions of time t, the correlation functions are real functions of τ. From their definition (4.71), the Schwarz inequality gives

$$(4.713) \qquad |\varphi_{jk}(\tau)| \leqslant \sqrt{\varphi_{jj}(0)\varphi_{kk}(0)}$$

which provides a basis for normalizing the correlation functions.

Cramér (1940) shows that these correlation functions may be expressed as Fourier-Stieltjes integrals of the form:

$$(4.72) \qquad \varphi_{jk}(\tau) = \frac{1}{2\pi} \int_{-\pi}^{\pi} e^{i\omega\tau}\Phi_{jk}(\omega)d\omega, \quad \text{for } j,k=1,2,...,n \ .$$

The inverse transforms may be written as

$$(4.73) \qquad \Phi_{jk}(\omega) = \sum_{\tau=-\infty}^{\infty} e^{-i\omega\tau}\varphi_{jk}(\tau) \ .$$

Here the $\Phi_{jk}(\omega)$ are the spectra of the set of stationary processes. For $j=k$, we have the power spectrum of $x_j(t)$:

$$(4.731) \qquad \Phi_{jj}(\omega)=\Phi_{jj}(-\omega) \ , \quad j=1,2,\ldots,n$$

which is a positive real function of ω. For $j \neq k$, we have the cross-power spectrum $\Phi_{jk}(\omega)$ of $x_j(t)$ and $x_k(t)$. The cross-power spectrum, which is a complex valued function of the real variable ω, satisfies

$$(4.732) \qquad \Phi_{jk}(\omega)=\Phi_{kj}(-\omega)$$

and

$$(4.733) \qquad \Phi_{jk}(\omega)=\overline{\Phi_{kj}(\omega)}$$

where the bar indicates the complex conjugate. Consequently, we have

$$(4.734) \qquad \Phi_{jk}(\omega)=\overline{\Phi_{jk}(-\omega)} \ .$$

Thus we see that the real part of the cross spectrum

$$(4.735) \qquad \mathrm{Re}\big[\Phi_{jk}(\omega)\big]=\mathrm{Re}\big[\Phi_{jk}(-\omega)\big]$$

is an even function of the real variable ω, whereas the imaginary part

$$(4.736) \qquad \mathrm{Im}\big[\Phi_{jk}(\omega)\big]=-\,\mathrm{Im}\big[\Phi_{jk}(-\omega)\big]$$

is an odd function of the real variable ω.

By letting $\tau=0$ in equation (4.72), we see that

$$(4.74) \qquad \varphi_{jk}(0) = \frac{1}{2\pi}\int_{-\pi}^{\pi}\Phi_{jk}(\omega)d\omega = \frac{1}{\pi}\int_{0}^{\pi}\mathrm{Re}[\Phi_{jk}(\omega)]d\omega \ .$$

Let us consider the linear combination

$$(4.75) \qquad x(t) = \sum_{1}^{n} a_j x_j(t)$$

which defines the time series $x(t)$. Here the weighting coefficients a_j are real. The autocorrelation of $x(t)$ is

$$\varphi(\tau) = \lim_{T \to \infty} \frac{1}{2T+1} \sum_{t=-T}^{T} x(t+\tau)\,x(t)$$

$$(4.76)$$

$$= \sum_{j=1}^{n} \sum_{k=1}^{n} a_j a_k \lim_{T \to \infty} \frac{1}{2T+1} \sum_{t=-T}^{T} x_j(t+\tau) x_k(t)$$

which is

$$(4.761) \qquad \varphi(\tau) = \sum_{j=1}^{n} \sum_{k=1}^{n} a_j a_k \varphi_{jk}(\tau) \ .$$

Since $\varphi(\tau)$ is the autocorrelation function of the time series $x(t)$ it is an even function of τ:

$$(4.762) \qquad \varphi(\tau) = \varphi(-\tau) \ .$$

The spectrum of $x(t)$ is given by the hermitian form

$$(4.77) \qquad \Phi(\omega) = \sum_{\tau=-\infty}^{\infty} \varphi(\tau) e^{-i\omega\tau} d\tau = \sum_{j=1}^{n} \sum_{k=1}^{n} a_j a_k \Phi_{jk}(\omega) \geqslant 0$$

which is a non-negative function of ω (Cramér, 1940). The matrix of the hermitian form (4.77) may be represented by the hermitian matrix

$$(4.771) \qquad \left[\Phi_{jk}(\omega)\right], \quad \text{for } j,k=1,2,\ldots,n$$

which is

$$(4.772) \qquad \begin{bmatrix} \Phi_{11}(\omega) & \Phi_{12}(\omega) & \ldots & \Phi_{1n}(\omega) \\ \Phi_{21}(\omega) & \Phi_{22}(\omega) & \ldots & \Phi_{2n}(\omega) \\ & \cdots & & \\ \Phi_{n1}(\omega) & \Phi_{n2}(\omega) & \ldots & \Phi_{nn}(\omega) \end{bmatrix} \ .$$

We shall call this matrix, which determines the spectra of all possible linear combinations of $x_1(t),\ldots,x_n(t)$, the *coherency matrix*. The elements of our coherency matrix are the derivatives of the elements of Wiener's coherency matrix (Wiener, 1930, 1942).

Further, for the time series $x_1(t)$ and $x_2(t)$ with the coherency matrix

$$(4.773) \qquad \begin{bmatrix} \Phi_{11}(\omega) & \Phi_{12}(\omega) \\ \Phi_{21}(\omega) & \Phi_{22}(\omega) \end{bmatrix}$$

the significant invariants of this hermitian matrix are

$$(4.774) \qquad \mathrm{coh}_{12}(\omega) = \frac{\Phi_{12}(\omega)}{\left[\Phi_{11}(\omega)\Phi_{22}(\omega)\right]^{\frac{1}{2}}}$$

which Wiener (1930) calls the coefficient of coherency of $x_1(t)$ and $x_2(t)$ for frequency ω, and

$$(4.775) \qquad \sigma_1(\omega) = \frac{\Phi_{12}(\omega)\sqrt{\Phi_{11}(\omega)}}{\Phi_{22}(\omega)}$$

and

$$(4.776) \qquad \sigma_2(\omega) = \frac{\Phi_{21}(\omega)\sqrt{\Phi_{22}(\omega)}}{\Phi_{11}(\omega)}$$

which Wiener (1930) calls the coefficients of regression respectively of x_1 on x_2 and of x_2 on x_1. Wiener (1930) points out that the modulus of the coefficient of coherency represents the amount of linear coherency between $x_1(t)$ and $x_2(t)$ and the argument, the phase-lag of this coherency. The coefficients of regression determine in addition the relative scale for equivalent changes of $x_1(t)$ and $x_2(t)$.

Cramér (1940) shows that the determinant

$$(4.777) \qquad \Phi_{11}(\omega)\Phi_{22}(\omega)-\Phi_{12}(\omega)\Phi_{21}(\omega)$$

of the coherency matrix (4.773) is non-negative. Therefore we have

$$(4.778) \qquad \Phi_{12}(\omega)\Phi_{21}(\omega) = |\,\Phi_{12}(\omega)\,|^2 \leqslant \Phi_{11}(\omega)\Phi_{22}(\omega)$$

so that the magnitude of the coefficient of coherency,

$$(4.779) \qquad |\mathrm{coh}_{12}(\omega)| = \frac{|\Phi_{12}(\omega)|}{\sqrt{\Phi_{11}(\omega)\Phi_{22}(\omega)}}$$

lies between zero and one.

Inequality (4.778) may be written

$$(4.78) \qquad \sum_{\tau=-\infty}^{\infty} \varphi_{12}(\tau)e^{-i\omega\tau} \sum_{s=-\infty}^{\infty} \varphi_{21}(s)e^{-i\omega s} \leqslant \sum_{\tau=-\infty}^{\infty} \varphi_{11}(\tau)e^{-i\omega\tau} \sum_{s=-\infty}^{\infty} \varphi_{22}(s)e^{-i\omega s}$$

which is

$$(4.781) \qquad \sum_{r=-\infty}^{\infty} \sum_{\tau=-\infty}^{\infty} \varphi_{12}(\tau)\varphi_{12}(r)e^{-i\omega(\tau-r)} \leqslant \sum_{r=-\infty}^{\infty} \sum_{\tau=-\infty}^{\infty} \varphi_{11}(\tau)\varphi_{22}(r)e^{-i\omega(\tau-r)}$$

or

$$(4.782) \qquad \sum_{n=-\infty}^{\infty} e^{-i\omega n} \sum_{r=-\infty}^{\infty} \varphi_{12}(n+r)\varphi_{12}(r) \leqslant \sum_{n=-\infty}^{\infty} e^{-i\omega n} \sum_{r=-\infty}^{\infty} \varphi_{11}(n+r)\varphi_{22}(r) \, .$$

This inequality states that for two stationary time series the Fourier transform of the autocorrelation of their cross-correlation cannot exceed the Fourier transform of the cross-correlation of their autocorrelations.

As Wold (1953, Chapter 12.7, p. 202) observes, in view of the abundance of possible variations and combinations available in the extention of the theory of one dimensional stochastic processes to multiple stochastic processes, the main difficulty does not lie in developing formulae of great generality, but rather in picking out those processes for further study that merit interest from the point of view of applications.

5. MULTICHANNEL SIGNAL ENHANCEMENT AND PREDICTION

5.1 General Discussion of Signal Enhancement and Prediction

The possibility of physical prediction depends upon the process having statistical regularity, and on the existence of correlations between past values of the known data and future values of the signal. If the prediction is to be accomplished by a linear operation, then the only type of correlation that can be used is linear correlation. Thus a linear predictor has the disadvantage of not making use of relations contained in the higher moments of the process. It does have the advantage of simplifying the analysis. Moreover, if the process is Gaussian, then all the information is contained in the linear correlations, and so in this case the linear predictor is the absolute optimum.

Many factors ought to be considered when «optimizing» the performance of a filter. An optimization depends upon the purpose of the filter, the nature of the input data, the criterion used for evaluating performance, and allowable tolerances in accuracy.

We will consider the problems involved in preserving or enhancing the waveshape of a signal immersed in noise, and in predicting the future values of the signal. A filter which performs the operation of signal enhancement is called an *enhancement filter*, or a *smoothing filter*. The reason for the use of the word «smoothing» is that such a filter generally produces the effect of «smoothing the data» in performing its task of removing the unwanted random roughness caused by the noise. A filter which performs the operation of signal prediction or forecasting is called a *prediction filter*. A filter may combine both operations; a *smoothing and predicting filter* separates the wanted signal from a signal-plus-noise complex and yields future values of the signal.

The simplest case of *prediction* is that in which the pure signal s_t is processed. That is, we suppose that the signal time-series s_t is available without any contamination. We know the signal s_t for past values of time, and we require a device to yield, as well as possible, the value at a future, or advanced, time. In other words, we wish to feed the signal time-series s_t into a filter. The filter is designed so that its actual output y_t is the best approximation to the desired output $s_{t+\alpha}$, which is the signal advanced by the positive time advance α.

Signal enhancement may be described as the problem of separating as well as possible a signal time-series from an unwanted noise time-series. The time-series x_t which we have available is a contaminated time-series. The contaminated or perturbed time-series x_t consists of a pure component s_t, called the signal, and an impure component n_t, called the noise, that is $x_t = s_t + n_t$. We are only interested in the signal s_t, the n_t being unwanted. We therefore want to design a filter $(f_0, f_1, f_2, \ldots)$, which will filter out, as well as possible, the noise which is contaminating the signal, thereby enhancing the signal in relation to the noise. The filter should be designed so that its actual output y_t is as close as possible to the desired output s_t, the signal.

The signal enhancement operation may be modified in several ways. For example, the desired output may be the signal delayed by a certain time constant β. This is the case of *signal enhancement with delay*. The contaminated time-series x_t is fed into the filter. The filter is designed so that its actual output y_t is the best approximation to the desired output $s_{t-\beta}$, which is the signal delayed by the positive time delay (or time lag) β.

Another example is the case in which the desired output is the signal advanced by a certain time constant α. The introduction of this time advance α means that the filter is required to predict future values of the signal. This case which incorporates both signal enhancement and prediction is called *signal enhancement with advance*. Here we feed the contaminated time-series x_t into the filter. The filter is designed so that its actual output y_t is the best approximation to the desired output $s_{t+\alpha}$, which is the signal advanced by the positive time advance α.

Finally, let us formulate the general case, which includes each of the above cases as a special case. The desired output may be any time-series z_t. We wish to feed a time-series x_t into a filter. The filter is designed so that its actual output y_t is the best approximation to the desired output z_t.

The basic problem in the theory of signal enhancement and prediction is that of determining the numerical values of the filter coefficientes $(f_0, f_1, f_2, \ldots)$. The solution of this problem rests on three main assumptions which determine the range of application of the results. The three assumptions are:

(1) The time-series representing the input x_t and the desired output z_t are stationary. As we have seen in Chapter IV, this means essentially that the statistical properties of the input and desired output do not change with time. The theory as given here cannot properly be applied, for example, to long-term economic effects, for the statistics of economic time-series is not the same today as one hundred years ago.

(2) The approximation criterion is taken to be the mean-square-error between the desired output and the actual output. This means that we determine the operator $(f_0, f_1, f_2, \ldots)$ in such a way as to minimize the trace of the mean-

square-error matrix between the desired output z_t and the actual output y_t. This trace is given by

$$(5.11) \qquad I = \operatorname{tr} \operatorname{E}\left\{ (z_t - y_t)(z_t - y_t)^T \right\}.$$

The average is the ensemble average E, which is taken over all possible inputs and desired outputs with each weighted according to its probability of occurrence. In case the time-series are ergodic, the ensemble averages can be replaced by the corresponding time averages.

(3) The operation to be used for signal enhancement and prediction is assumed to be a linear operation on the available information, or in engineering terms, a time-invariant linear filter has to be used. The available information consists of the past history of the perturbed signal, i. e. the time-series x_s with $-\infty < s \leqslant t$ where t is the present time. A physically realizable filter (i. e. one with a memory function $(f_0, f_1, f_2, \ldots)$ but no anticipation function) performs a linear operation on x_s over just this range. For digital computer applications, we want to further restrict ourselves to the case when the operator is of finite length, say $(f_0, f_1, \ldots, f_m)$, as done in Section 5.2.

The theory may therefore be described as linear least-square smoothing and prediction of stationary time-series by means of a realizable, time-invariant linear operator. It is recognized that the theory applies only when the above three assumptions are fulfilled, or at least are approximately satisfied. If any one of the conditions is eliminated or changed, then the change must be taken into account, and the problem can become very difficult mathematically.

Linear least-square smoothing and prediction theory is based on statistical regularity. It is on this basis that it is possible to predict, albeit not perfectly the future behavior of a time-series when all that is known is a perturbed version of its past history. In general, physical prediction depends fundamentally on the assumption that regularities that have occurred in the past will recur in the future. Such an assumption, of course, can never be proved, but should be regarded as a postulate. The basic assumption that statistical regularities of the past will hold in the future appears in the mathematics as the assumption that the various time-series are stationary. This implies that a value of a parameter obtained by averaging the time-series over the past will be the same as the value obtained by averaging over the future. More specifically, prediction depends essentially on the existence of correlations between the desired output z_t and the input x_t. The assumption that the smoothing and/or prediction is to be done by a linear operation implies that the only type of correlation that can be used is linear correlation, namely the autocorrelation function

$$(5.12) \qquad \varphi_{xx}(\tau) = \varphi_{xx}^T(-\tau) = \operatorname{E}\left\{ x_t\, x_{t-\tau}^T \right\}, \qquad \tau = 0, 1, 2, \ldots,$$

of the input, and the cross-correlation function

$$(5.13) \qquad \varphi_{zx}(\tau) = \mathrm{E}\left\{ z_t \, x_{t-\tau}^T \right\}, \qquad \tau = 0, 1, 2, \ldots,$$

between desired output and input. If the cross-correlation function between desired output and input (which in the case of pure prediction reduces to the autocorrelation function of the input) were zero, then no significant linear prediction would be possible, and so the best mean-square estimate of the desired output would be its mean value.

5.2   Multichannel Signal Enhancement and Prediction with Finite Operators

For digital computing applications it is important to develop the theory of smoothing and prediction in terms of finite-length operators. The method is, actually, classical least-squares in the context of statistical communication theory.

With the proper use of vectors and matrices the case of multichannel filtering is only a simple extension of the case of single channel filtering.

Let the *filter* be represented by the $n \times n$ matrix-valued coefficients

$$(5.21) \qquad f_x \equiv f(s) = \begin{bmatrix} f_{11}(s) & f_{12}(s) & \cdots & f_{1n}(s) \\ & \cdots & & \\ f_{n1}(s) & f_{n2}(s) & \cdots & f_{nn}(s) \end{bmatrix}.$$

Let the input time series be given by the $n \times 1$ vector-valued time series

$$(5.211) \qquad x_t \equiv x(t) = \begin{bmatrix} x_1(t) \\ x_2(t) \\ \vdots \\ x_n(t) \end{bmatrix}$$

and let the desired output time series be given by the $n \times 1$ vector-valued time series

$$(5.212) \qquad z_t \equiv z(t) = \begin{bmatrix} z_1(t) \\ z_2(t) \\ \vdots \\ z_n(t) \end{bmatrix}.$$

Let us now derive the optimum finite-length operator for smoothing and/or prediction. Without loss of generality, we may assume that the mean values have been removed from the input and desired output, so $E\{x_t\}=0$ and $E\{z_t\}=0$. The actual output, which is the convolution of the input with the operator, is given by the matrix equation

$$(5.22) \qquad y_t = f_0 x_t + f_1 x_{t-1} + \cdots + f_m x_{t-m} .$$

The structure of these mathematical operations is shown schematically by the figure:

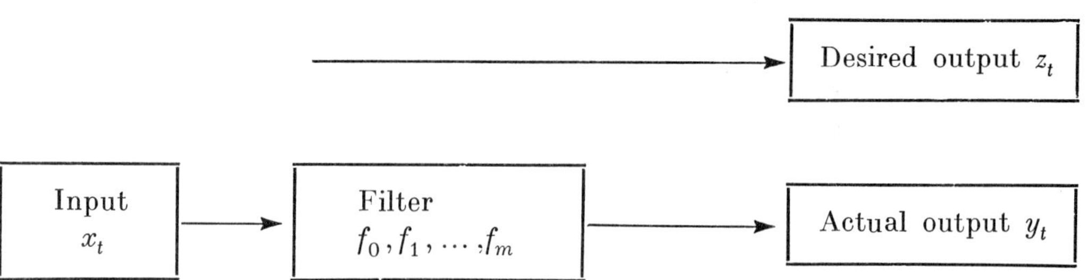

We note that the actual output is given by the $n \times 1$ vector-valued time series

$$(5.221) \qquad y_t \equiv y(t) = \begin{bmatrix} y_1(t) \\ y_2(t) \\ \vdots \\ y_n(t) \end{bmatrix} .$$

The error e_t between desired output and actual output is then

$$(5.23) \qquad e_t = z_t - y_t = z_t - (f_0 x_t + f_1 x_{t-1} + \cdots + f_m x_{t-m}) .$$

This error is, in fact, the $n \times 1$ vector-valued time series

$$(5.231) \qquad e_t = z_t - y_t = \begin{bmatrix} z_1(t) - y_1(t) \\ \cdots \\ z_n(t) - y_n(t) \end{bmatrix} \equiv \begin{bmatrix} e_1(t) \\ \cdots \\ e_n(t) \end{bmatrix} .$$

As a matter of notation, we let T in the superscript position indicate matrix transposition. Now we consider the *mean-square-error matrix* defined by

$$(5.232) \qquad \mathrm{E}\{e_t e_t^T\} = \begin{bmatrix} \mathrm{E}\{e_1^2(t)\} & \mathrm{E}\{e_1(t)e_2(t)\} \dots \mathrm{E}\{e_1(t)e_n(t)\} \\ \dots & \\ \mathrm{E}\{e_n(t)e_1(t)\} & \mathrm{E}\{e_n(t)e_2(t)\} \dots \quad \mathrm{E}\{e_n^2(t)\} \end{bmatrix}.$$

The trace (abbreviated tr) of a square matrix is defined as the sum of its diagonal entries. Hence the trace of the mean-square-error matrix is

$$(5.233) \qquad I \equiv \text{trace } \mathrm{E}\{e_t e_t^T\} = \mathrm{E}\{e_1{}^2(t)\} + \mathrm{E}\{e_2{}^2(t)\} + \dots + \mathrm{E}\{e_n{}^2(t)\} .$$

We wish to determine the values of the filter coefficients f_s such that the trace I of the mean-square-error matrix is a minimum. Now the error e_t may be written as

$$(5.24) \qquad e_t = z_t = y_t = z_t - \sum_s f_s x_{t-s} \quad \text{where} \quad \sum_s \equiv \sum_{s=0}^{m} .$$

Hence

$$e_t e_t^T = \left(z_t - \sum f_s x_{t-s}\right)\left(z_t - \sum f_r x_{t-r}\right)^T$$

$$(5.241) \qquad = z_t z_t^T - z_t\left(\sum f_r x_{t-r}\right)^T - \sum f_s x_{t-s} z_t^T + \left(\sum f_s x_{t-s}\right)\left(\sum f_r x_{t-r}\right)^T$$

$$= z_t z_t^T - z_t \sum x_{t-r}^T f_r^T - \sum f_s x_{t-s} z_t^T + \sum \sum f_s x_{t-s} x_{t-r}^T f_r^T .$$

Thus, taking the ensemble average, we have

$$\mathrm{E}\{e_t e_t^T\} = \mathrm{E}\{z_t z_t^T\} - \sum \mathrm{E}\{z_t x_{t-r}^T\} f_r^T - \sum f_s \mathrm{E}\{x_{t-s} z_t^T\}$$

$$(5.242)$$

$$+ \sum \sum f_s \mathrm{E}\{x_{t-s} x_{t-r}^T\} f_r .$$

Define the cross-correlation matrix between desired output and input as

$$(5.25) \qquad \mathrm{E}\{z_t x_{t-r}^T\} = \mathrm{E}\{z_{t-r} x_t^T\} = \varphi_{zx}(r) \qquad (n \times n \text{ matrix})$$

where the first equal sign in this equation follows because of stationarity. Also define the cross-correlation matrix between input and desired output as

$$(5.251) \qquad \mathrm{E}\{x_{t-s}z_t^T\} = \varphi_{xz}(-s) \qquad (n \times n \text{ matrix}) .$$

At this point let us note that these two cross-correlation functions are related by

$$(5.252) \qquad \varphi_{xz}^T(-s) = \mathrm{E}\{z_t x_{t-s}^T\} = \varphi_{zx}(s) .$$

Finally define the autocorrelation function of the input as

$$(5.253) \qquad \mathrm{E}\{x_{t-s}x_{t-r}^T\} = \varphi_{xx}(r-s) \qquad (n \times n \text{ matrix}) .$$

Hence we see that $\mathrm{E}\{e_t e_t^T\}$ may be written in terms of these correlation functions as

$$(5.254) \qquad \mathrm{E}\{e_t e_t^T\} = \mathrm{E}\{z_t z_t^T\} - \sum \varphi_{zx}(r)f_r^T - \sum f_s \varphi_{xz}(-s) + \sum \sum f_s \varphi_{xx}(r-s)f_r^T .$$

Setting the partial derivative of

$$(5.26) \qquad I = \mathrm{tr}\, \mathrm{E}\{e_t e_t^T\}$$

with respect to each filter coefficient equal to zero, we obtain the set of simultaneous equations given by

$$(5.261) \qquad \begin{aligned}
f_0 \varphi_{xx}(0) + f_1 \varphi_{xx}(-1) + \dots + f_m \varphi_{xx}(-m) &= \varphi_{zx}(0) \\
f_0 \varphi_{xx}(1) + f_1 \varphi_{xx}(0) + \dots + f_m \varphi_{xx}(1-m) &= \varphi_{zx}(1) \\
\dots \\
f_0 \varphi_{xx}(m) + f_1 \varphi_{xx}(m-1) + \dots + f_m \varphi_{xx}(0) &= \varphi_{zx}(m) .
\end{aligned}$$

The solution of these equations yields the optimum $(m+1)$-length operator. These equations are called the *normal equations*. The known quantities are the autocorrelation coefficients $\varphi_{xx}(\tau)$ of the input x_t and the cross-correlation coefficients $\varphi_{zx}(\tau)$ of the desired output z_t with the input x_t. The unknown quantities are the operator coefficients $(f_0, f_1, \dots, f_m)$. Because the matrix of this set of equations is an autocorrelation matrix, the solution may be obtained by making use of the computationally efficient recursion given in Robinson (1964, pp. 98-105). In practice one will usually assume that the time series are ergodic, and estimate the required autocorrelation and cross-correlation coefficients as time averages rather than as ensemble averages.

Let us now investigate the structure of the normal equations with respect to the minimization problem. If we make the following definitions

$$\text{Term } 1 = \mathrm{E}\{z_t z_t^T\} = \varphi_{zz}(0)$$

$$\text{Term } 2 = \sum \varphi_{zx}(r) f_r^T$$

(5.27)

$$\text{Term } 3 = \sum f_s \varphi_{xz}(-s) = \sum f_s \varphi_{zx}^T(s)$$

$$\text{Term } 4 = \sum \sum f_s \varphi_{xx}(r-s) f_r^T$$

then we see that

(5.271) $$\mathrm{E}\{e_t e_t^T\} = \text{Term } 1 - \text{Term } 2 - \text{Term } 3 + \text{Term } 4 .$$

When we minimize $I = \mathrm{tr}\, \mathrm{E}\{e_t e_t^T\}$ the result is that

(5.272) $$\text{Term } 4 = \text{Term } 2 = \text{Term } 3 .$$

Hence from

(5.273) $$\text{Term } 4 = \text{Term } 2$$

we have

(5.274) $$\sum_r \sum_s f_s \varphi_{xx}(r-s) f_r^T = \sum_r \varphi_{zx}(r) f_r^T .$$

Cancelling the $\sum_r$ and f_r^T we have

(5.275) $$\sum_{s=0}^{m} f_s \varphi_{xx}(r-s) = \varphi_{zx}(r) \qquad \text{for } r = 0,1,2,\ldots,m$$

which are the normal equations. As a verification, let us consider the equation

(5.28) $$\text{Term } 4 = \text{Term } 3 .$$

This equation gives

$$(5.281) \qquad \sum \sum f_s \varphi_{xx}(r-s) f_r^T = \sum f_s \varphi_{xz}(-s)$$

or

$$(5.282) \qquad \sum \varphi_{xx}(r-s) f_r^T = \varphi_{xz}(-s) \ .$$

But

$$(5.283) \qquad \varphi_{xz}^T(-s) = \varphi_{zx}(s) \ .$$

Hence

$$(5.284) \qquad \varphi_{xz}(-s) = \varphi_{zx}^T(s) \ .$$

Thus

$$(5.285) \qquad \sum \varphi_{xx}(r-s) f_r^T = \varphi_{zx}^T(s) \ .$$

Taking transposes, we have $\left(\text{since } \varphi_{xx}^T(r-s) = \varphi_{xx}(s-r)\right)$

$$(5.286) \qquad \sum_{r=0}^{m} f_r \varphi_{xx}(s-r) = \varphi_{zx}(s) \qquad \text{for } s = 0,1,2,\ldots,m$$

which again are the same normal equations, but in this case the role of s and r is reversed from that of the foregoing case.

The minimum value of the trace of the mean-square-error matrix is

$$(5.29) \qquad I_{\min} = \text{tr } (\text{Term } 1 - \text{Term } 2) = \text{tr } \varphi_{zz}(0) - \text{tr } \sum_{r=0}^{m} \varphi_{zx}(r) f_r^T \ .$$

The value of the minimum trace $I_{\min}$ of the mean-square-error matrix depends upon the parameter m (i. e. depends upon the filter length). If m is increased (i. e. if we use a longer filter) the value of $I_{\min}$ must either remain stationary or decrease. In practice one usually computes filters of successively increasing lengths, making use of the recursion given in Robinson (1964, pp. 98-105). The values of minimum mean-square-error traces $I_{\min}$ will diminish as the length of the filter increases. The recursion from m to $m+1$ can be stopped when one of a set of pre-set conditions is first met, such as:

(1)   the filter reaches some maximum pre-set length,
(2)   the minimum mean-square-error trace $I_{\min}$ reaches some minimum pre-set value,
(3)   the minimum mean-square-error trace curve levels out and shows little probability of further appreciable decrease.

6. APPLICATIONS OF MULTICHANNEL PREDICTION THEORY

6.1 Prediction of Commodity Futures

As a practical application of the methods given in the foregoing chapters, the commodity futures on the Chicago grain market were investigated. Table 6.1 gives the weekly prices of May oats and May corn from June 1965 to May 1966. Each time series is made up of 50 observations spaced one week apart.

The mean value of the oats time series is 70.098 and that of the corn time series is 125.922. These means were removed from the time series before the computations for the predictions were made and then added back into the final predicted values, so as to make the predicted values compatible with the original time series.

TABLE 6.1

TWO TIME SERIES OF WEEKLY PRICES IN CENTS. EACH TIME SERIES
HAS 50 OBSERVATIONS FROM JUNE 1965 TO MAY 1966

May Oats	May Corn	May Oats (Continued)	May Corn (Continued)	May Oats (Continued)	May Corn (Continued)
72.5	124.6	67.9	124.2	72.2	131.9
72.5	125.0	67.9	123.0	71.2	131.2
72.1	126.0	68.1	122.1	71.3	131.0
72.3	128.0	68.2	121.5	72.0	131.0
72.3	127.0	68.3	121.2	71.5	128.9
71.5	127.2	68.6	121.3	71.0	127.0
70.8	125.8	69.2	122.0	70.0	125.0
70.7	125.1	69.5	123.2	70.0	125.0
70.6	125.2	70.0	122.7	70.2	125.0
70.5	126.5	70.0	123.0	70.0	125.0
71.5	124.9	70.7	126.0	68.5	125.2
70.0	124.2	70.6	128.0	69.6	126.0
68.5	123.9	70.1	128.0	70.6	127.0
68.5	125.0	69.6	127.2	70.6	127.7
68.5	125.7	69.1	127.7	68.6	127.2
68.0	124.0	70.5	130.0	67.9	125.9
67.9	125.0	72.7	131.9		

In each case we predicted one observation ahead, that is, we used a prediction distance of one time unit, i. e. one week. The necessary correlations were computed over the entire known 50 observations of each time series.

The predicted values of interest are the next (unavailable) observations, namely the 51st observation of each of the two time series. That is, the 50th observation represents the latest available observation of each time series, and the 51st observation is the future unavailable observation which we desire to predict.

Tables 6.2 and 6.3 give the results of predicting the 51st observation of each time series. «Single channel prediction» refers to predicting the given time series from only its own past, whereas «double channel prediction» refers to predicting the given time series from its own past as well as from the past of the other time series. In the notation of Section 5.2 the operator length is the parameter $M=m+1$, the percent error is the parameter $I_{\min}/\mathrm{tr}\,\varphi_{zz}(0)$ (multiplied by 100 in order to convert it to percent), and the predicted value is y_{51}. The desired output z_t, of course, is the given input x_t (the oats and corn time series) advanced by one time unit.

Table 6.2

PREDICTION OF THE 51st OBSERVATION OF THE OATS TIME SERIES

SINGLE-CHANNEL PREDICTION			DOUBLE-CHANNEL PREDICTION		
Operator length	Percent error	Predicted value	Operator length	Percent error	Predicted value
1	34	68.3	1	21	68.4
2	32	68.2	2	18	68.2
3	29	68.7	3	17	68.6

Table 6.3

PREDICTION OF THE 51st OBSERVATION OF THE CORN TIME SERIES

SINGLE-CHANNEL PREDICTION			DOUBLE-CHANNEL PREDICTION		
Operator length	Percent error	Predicted value	Operator length	Percent error	Predicted value
1	18	125.9	1	21	125.7
2	15	125.4	2	18	125.2
3	15	125.4	3	17	125.4

From Tables 6.2 and 6.3 we see that the oats time series is less predictable than the corn time series (as evidenced by the single-channel prediction error range of 34 percent to 29 percent for oats versus 18 percent to 15 percent for corn).

The latest observed value for oats (i. e. the 50th value in Table 6.1) is 67.9. The predictions of the 51st value (given in Table 6.2) range from 68.2 to 68.7, and hence our predictions indicate that the price of oats will rise. Moreover, acting on the basis of the longer operator ($M=3$) the predicted rise would be to the value of 68.7 or 68.6.

Likewise the latest observed value for corn (i. e. the 50th value in Table 6.1) is 125.9. The predictions of the 51st value (given in Table 6.3) range from 125.2 to 125.9, and hence our predictions indicate that the price of corn will fall. Moreover, acting on the basis of the longer operator ($M=3$) the predicted fall would be to the value of 125.4.

6.2 Predictive Deconvolution of Seismic Traces in Oil Exploration

There are two basic approaches to seismology, namely the deterministic approach and the statistical approach. The deterministic approach is concerned with the building of mathematical and physical models of the layered earth in order to better understand seismic wave propagation. These models involve no random elements; they are completely deterministic. The statistical approach is concerned with the building of seismic models involving random components. For example, in the statistical model which we will discuss the depths and reflectivities of the deep reflecting horizons are considered to have a random distribution. A major justification for using the statistical approach in seismology is due to the fact that large amounts of data must be processed; any data in large enough quantities takes on a statistical character, even if each individual piece of data is of a deterministic nature.

The water reverberation problem in marine exploration seismic operations may be described as follows. The water-air interface is a strong reflector, with reflection coefficient nearly -1. If the water-bottom interface is also a strong reflector, then the water layer represents a non-attenuating medium bounded by two strong reflecting interfaces and hence represents an energy trap. A seismic pulse generated in this energy trap will be successively reflected between the two interfaces. Consequently, reflections from deep horizons below the water layer will be obscured by the water reverberations.

Let us consider a downward travelling impulse plane wave (i. e. we assume that the source wavelet is a unit spike) and let us suppose that we have a transducer which only measures downward motion in the water layer. The signal received by the transducer represents the successive reflections from the air-water interface. We assume that we are dealing with discrete time so that successive sampled data points are separated by a fixed sampling time-increment; for example, the sampling time unit might be 4 milliseconds. We let the integer n represent the two-way travel time in the water measured in terms of the sampling time unit; for example, if the water depth is 100 feet and the water velocity is 5000 feet per second, then the two way travel time would be 40 milliseconds so for a 4 millisecond sampling increment the value of n would be 10. The *water-confined reverberation spike-train* is of the form

$$(6.21) \qquad (1,0,0,\ldots,0,-c,0,0,\ldots,0,c^2,0,0,\ldots,0,-c^3,0,0,\ldots) \, .$$

$$\text{time:} \quad 0,1,2,\ldots \quad\quad n \quad\quad\quad 2n \quad\quad\quad 3n$$

In this spike-train, we see that the coefficient 1 occurs at time 0 and represents the initial downgoing spike. The coefficient $-c$ occurs at discrete time n and represents the second downgoing spike, which has suffered a reflection at the bottom (reflection coefficient c) and a reflection at the surface (reflection coefficient -1). The coefficient c^2 occurs at discrete time $2n$ and represents the third downgoing spike, which has suffered two reflections at the bottom (c^2) and two reflections at the surface $((-1)^2)$. The coefficient $-c^3$ occurs at discrete time $3n$ and represents the fourth downgoing spike, which has suffered three reflections at the bottom (c^3) and three reflections at the surface $((-1)^3)$, etc. The reflection coefficient c cannot exceed unity in magnitude; generally c will be less than unity in magnitude, for otherwise the water layer would represent a perfect energy trap and no energy would penetrate to the deeper layers. Thus in the case where $|c| < 1$, the water-confined reverberation spike-train (6.21) represents a convergent sequence, and is a wavelet (according to the mathematical definition of a discrete wavelet as being a one-sided stable time sequence). The z-transform of the water-confined reverberation spike-train is

$$1 - cz^n + c^2 z^{2n} - c^3 z^{3n} + \ldots$$

which is in the form of a geometric series, which can be summed to give

$$\frac{1}{1 + cz^n} \, .$$

The *water-confined reverberation elimination filter* would be the inverse filter, and thus would have the z-transform

$$1 + cz^n$$

so that the impulse response function would be the couplet

$$(6.22) \qquad\qquad (1,0,0,\ldots,c)$$

$$\uparrow\uparrow\uparrow \quad\ \uparrow$$

$$\text{time:}\quad 0,1,2 \quad\ n$$

Because $|c| < 1$, this couplet is minimum-delay, and it also follows that the water-confined reverberation spike-train (6.21) is minimum-delay because the inverse of a minimum-delay wavelet is also minimum-delay. To verify that the water-confined reverberation elimination filter (6.22) does indeed convert the water-confined reverberation spike-train (6.21) into a spike (and thereby eliminates the reverberation) we may convolve (6.21) with (6.22) and thereby obtain the spike $(1, 0, 0, 0, \ldots)$ which represents the impulse plane wave without the water reverberation.

In our analysis so far we have only considered the water reverberation effect itself, and have not considered the effect of the water layer on deep subsurface reflections. To do so, we must take into account that the water layer acts as a filter, and the seismic energy passes through this filter once as it goes down to the deep reflecting horizon and again as it returns to the surface. That is, the water layer acts on the deep reflection data twice, once going down to the deep strata and once again coming back to the surface. Hence we have, in effect, a situation where the deep reflection data passes through two cascaded sections of the water layer, so that the overall z-transform of the filtering introduced by the water layer is the square of the z-transform due to a single pass, as given in the foregoing paragraph. That is, the z-transform of the reverberation spike-train resulting from a deep reflecting horizon is

$$(1 - cz^n + c^2z^{2n} - c^3z^{3n} + \ldots)^2$$

or

$$\frac{1}{(1 + cz^n)^2} \, .$$

In the time domain, we see that the *deep-reflection reverberation spike-train* is given by

$$(6.23) \qquad (1,0,0,\ldots,0,-2c,0,0,\ldots,0,3c^2,0,0,\ldots,0,\ -4c^3,0,0,\ldots)$$

$$\uparrow\uparrow\uparrow \qquad\ \uparrow \qquad\quad\ \uparrow \qquad\qquad\ \uparrow$$

$$\text{time:}\quad 0,1,2,\ldots \quad\ n \qquad\quad 2n \qquad\qquad 3n$$

This expression may be obtained by convolving the water-confined reverberation spike-train (6.21) with itself. Since spike-train (6.21) is minimum-delay it follows that spike-train (6.23) is also minimum-delay. The reverberation spike-train (6.23) may be interpreted as being made up of the deep reflection spike at time 0, and the attached reverberation spikes at times n, $2n$, $3n$, $4n$,... . To eliminate the reverberation part of spike-train (6.23) we must pass spike-train (6.23) through the *inverse filter* whose z-transform is given by

$$(1+cz^n)^2$$

or

$$1+2cz^n+c^2z^{2n} \ .$$

The impulse response function of this inverse filter is

$$(6.24) \qquad\qquad (1,0,0,...,0,2c,0,0,...,0,c^2)$$

$$\uparrow\uparrow\uparrow \qquad \uparrow \qquad\quad \uparrow \quad .$$
$$\text{time:} \quad 0,1,2 \qquad n \qquad\quad 2n$$

Convolving spike-train (6.23) with spike-train (6.24), we see that the output is indeed a spike $(1,0,0,0,...)$ which represents the direct deep reflection spike (at time 0) with the attached reverberations spike eliminated.

In practice the reverberations are often generated by a more complicated physical situation than the one we have described, so the reverberation elimination operator (6.24) is not adequate in many field applications. However, there is an important principle which we can extract from the foregoing deterministic reverberation model, namely the fact that the deep-reflection reverberation spike-train (6.23) is minimum-delay. This minimum-delay property of reverberation spike-trains is also valid in other more complicated reverberation models and moreover it is upheld on recorded data from various field situations. As a result we would want to use a deconvolution method which exploits this minimum-delay property.

Up to now, for simplicity, we have assumed that the seismic source pulse is a spike. In practice it will not be a spike but will have some more complicated shape. Let us designate the pulse-train waveform due to a deep reflection by the symbol b_t. The pulse-train waveform b_t is given by the convolution of the deep reflection reverberation spike-train (whatever it is) with the seismic source wavelet (whatever it is). For example, in the reverberation model which we have described, the deep-reflection reverberation spike-train was given by (6.23). As we have just noted, the deep-reflection reverberation spike-train in many instances is minimum-delay, or at least approximately so. Hence, if the seismic source is sufficiently sharp so that the source wavelet is also approximately minimum-delay then it follows that the waveform b_t is also approximately minimum-delay. (This follows from the fact that the signal resulting from the convolution of two minimum-delay signals is minimum-delay).

Let us designate the received seismic trace by x_t (where it is assumed that the filtering effect of the recording instruments has already been removed from x_t by some restoring filtering operation). The received seismic trace x_t is the resultant of many deep reflections; each reflection contributes a pulse-train waveform of shape b_t. However, because of the fact that all the pulse-train waveforms are overlapping to various degrees with each other, it is not possible to obtain a direct measurement of the individual waveform shape b_t. In such a physical situation, the method of predictive decomposition of seismic traces (Robinson, 1954, 1959) can be used to eliminate the reverberations, and thus better bring out the true reflection pulses. (The word «composition» is the old term for «convolution»; because of the prevalence of the coined word «deconvolution» it is appropriate now to use the term «predictive deconvolution» in place of «predictive decomposition»). The predictive deconvolution method represents a geophysical application of the *predictive decomposition theorem* of Wold (1938).

The predictive deconvolution method is based on the following statistical model. The received seismic trace x_t (included within an appropriately chosen time gate) is considered to be the result of convolving the waveform b_t with a random spike series ε_t; that is, $x_t = b_t * \varepsilon_t$. The spike series ε_t represents the reflections from the deep reflecting horizons in the sense that the timing of a spike represents the direct arrival time of a reflection and the amplitude of the spike represents the strength of the reflection. For data-processing purposes this spike series ε_t is considered as a random uncorrelated (i. e. white noise) series. Hence the autocorrelation of the received seismic trace x_t is the same as the autocorrelation of the individual waveform b_t, except for a constant scale factor. This scale factor will not affect the final results, so it may be neglected. We can therefore compute the autocorrelation of the waveform b_t from the received seismic trace x_t. From the autocorrelation we can compute the prediction operator for the waveform b_t. Because the waveform b_t is minimum-delay, a prediction operator for prediction distance n ($=$ two-way travel time in the water) will predict the reverberation component of the waveform b_t. (This would not be true if the waveform were not minimum-delay. For an illustration of this principle, see Robinson (1962, pp. 90-91)). A delay of n time units will line up this predicted reverberation with the reverberation portion of the waveform b_t; by subtracting this delayed predicted reverberation from the waveform b_t we obtain the prediction error. Because the delayed predicted reverberation cancels the reverberation part of the waveform b_t, it follows that the prediction error represents the non-reverberation part of the waveform b_t. Hence the prediction error operator eliminates the reverberation part of the waveform b_t. Since the prediction error operator is linear, we can apply it to the received seismic trace $x_t = b_t * \varepsilon_t$ (which represents many overlapping waveforms b_t with arrival times and strengths given according to the spike series ε_t). In so doing, we eliminate the reverberations from each of the waveforms b_t but leave intact the initial non-reverberation portions, thereby increasing seismic resolution. If more resolution is desired, the prediction distance can be lessened which will have the effect of further compressing the energy in the waveform. Also some type of band-pass filtering or shaping operation can be applied as a post-filtering operation to yield a cleaned-up deconvolved seismic trace.

As an extra result, we can obtain the shape of the waveform b_t. See Robinson (1954, p. 248).

The model required for the application of the predictive deconvolution is a statistical model. This model depends upon two basic hypotheses namely, (1) the statistical hypothesis that the strengths and arrival times of the information-bearing events on seismic trace can be represented as a random spike series, and (2) the deterministic hypothesis that the basic waveform associated with each of these events is minimum-phase. There are various ways of checking a model to see if it conforms with the physical situation. In the writing of my thesis at M.I.T. during 1953 and 1954 some checks of the above model were made but were necessarily of a limited nature because of the data and computing facilities (the M.I.T. Whirlwind digital computer) then available. Despite these limitations, however, this model seemed a reasonable one in many instances.

From a theoretical point of view the concept of minimum-phase provided much fertile ground for mathematical research. We recall that minimum-phase was originally introduced as a frequency-domain concept; the definition is that a system is minimum-phase provided it produces the minimum possible phase shift for its gain. This concept has proven to be vital in the design of feedback control systems. Our research was aimed at taking the concept of minimum-phase out of the frequency domain and transferring it into the time-domain; by so doing we arrived at the identical concept in the time-domain, which we called «minimum-delay» (Robinson, 1962, 1963). (That is, the terms «minimum-phase» and «minimum-delay» are synonymous, except that the former connotes frequency domain and the latter time domain). A minimum-delay waveform is one which has its largest concentration of energy in the early part of the waveform. In this section we have given a fundamental example of a minimum-delay pulse-train waveform, namely that occurring in the model of water reverberation generation. The reason why we can often expect minimum-delay pulse-train waveforms in the layered earth situation is that reflection coefficients are less than unity in magnitude. Hence the more times a seismic pulse is reflected and transmitted, the more it is delayed and attenuated. Thus the concentration of energy in a seismic pulse-train waveform appears at its beginning rather than at its end; this is the condition that the pulse-train waveform be minimum-delay.

15

The reverberation model which we described represents a special case of much more general deterministic layered-earth models. From these models it is not difficult to show that in many instances certain basic seismic pulse-train waveforms generated by such models are minimum-delay. These results have done much from a theoretical point of view to justify the deterministic hypothesis required by our statistical model, namely that the basic waveform is minimum-delay. With the advent of routine digital field recording and extensive digital computer processing of seismic data in the last few years it has been possible to test statistical models on a large-scale basis with high quality data. Because a seismic trace is made up of many overlapping waveforms, it is usually not possible to obtain a direct measurement of the individual waveform shape. Hence the minimum-delay nature of a pulse-train waveform must be verified indirectly. This indirect verification can be carried out by applying the method of predictive deconvolution to seismic data; if the method works satisfactorily, then we can conclude that the minimum-delay hypothesis is upheld.

The method of predictive deconvolution has been highly successful in deconvolving seismic exploration records, especially marine records with reverberations. The general success of the method shows that the minimum-delay hypothesis is valid under a wide range of field situations. The next step therefore is to categorize these various field situations so as to be able to predict those cases where the minimum-delay hypothesis does not apply. For such non-minimum-delay cases, adaptive procedures must be devised whereby some measure of the deviations from minimum-delay can be determined. With such information it is then possible to design the necessary deconvolution operators.

In summary, we may say that in the standard deterministic model of water reverberation generation, the reverberation pulse-train resulting from a deep reflection is minimum-delay. Even in the more complex physical situations encountered in the field, there is evidence that in many cases the reverberation pulse-train waveforms are minimum-delay, or at least approximately so. The reason for this minimum-delay property is that a pulse-train waveform results from multiple reflections and transmissions within the layered earth; because reflection coefficients are less than unity in magnitude, the concentration of energy in a pulse-train must appear at its beginning rather than its end; this early concentration of energy is the condition that pulse-train waveform be minimum-delay. Each deep reflection horizon contributes a minimum-delay reverberation pulse-train waveform to a seismic trace. If we let a spike series represent the deep horizons in the sense that the timing of a spike represents the direct arrival time of a reflection and the amplitude of the spike represents the strength of the reflection, then the seismic trace may be considered as the convolution of the spike series with the reverberation pulse-train waveform. Because the reverberation pulse-train waveform is minimum-delay, and because, at least approximately, the deep horizon spike series represents a statistically uncorrelated series, the two conditions required for the application of the method of predictive deconvolution (Robinson, 1954, 1959) are met, and hence this method can be used as a practical digital data processing method to eliminate water reverberations on exploration seismic traces. As a result large water-covered areas of the globe can now be explored for oil, which hitherto had to be classified as «no-good» seismic regions because of the destructive interference of the water reverberations with the desired deep reflection signals.

REFERENCES

Bode, H. W., and Shannon, C. E. (1950). A simplified derivation of linear least square smoothing and prediction theory. *Proc. Inst. Rad. Eng.*, vol. 38, 417-425.

Cochrane, D., and Orcutt, G. H. (1949). Application of least-squares regression to relationships containing autocorrelated error terms. *J. Am. Stat. Assoc.*, vol. 44, 32-61.

Cramér, H. (1940). On the theory of stationary random processes. *Ann. Math.*, vol. 41, 215-230.

——— (1942). On harmonic analysis in certain functional spaces. *Ark. Mat. Astr. Fys.*, vol. 28 B, N.º 12, 17 pp.

——— (1945). *Mathematical methods of statistics*. Almqvist & Wiksells, Uppsala. Parallel ed. (1946) Princeton University Press, Princeton, N. J., 575 pp.

——— (1947). Problems in probability theory. *Annals Math. Stat.*, vol. 18, 165-193.

——— (1951). A contribution to the theory of stochastic processes. *Proc. Sec. Berkeley Symp. Math. Statistics and Prob.*, Berkeley, 329-339.

Doob, J. L. (1949). Time series and harmonic analysis. *Proc. Berkeley Symp. Math. Statistics and Prob.*, Berkeley, 303-343.

——— (1953). *Stochastic processes*. John Wiley & Sons, Inc., New York.

Fejér, L. (1915). Uber trigonometrische Polynome. *J. reine ang. Mat.*, vol. 146, 53-82.

Frazer, R. A., Duncan, W. J., and Collar, A. R. (1938). *Elementary matrices*. Cambridge University Press, Cambridge, 416 pp.

Gantmacher, F. R. (1959). *The theory of matrices*, vol. 1. Chelsea Publishing Co., New York, 374 pp.

Khintchine, A. (1933). Asymptotische Gesetze der Wahrscheinlichkeitsrechnung. *Ergebnisse der Mathe. und ihrer Grenzgebiete*, vol. 2, no. 4.

——— (1934). Korrelationstheorie der stationaren stochastischen Prozesse. *Mathematische Annalen*, vol. 109, 604-615.

Kolmogorov, A. (1933). Grundbegriffe der Warscheinlichkeitsrechnung. *Ergebnisse der Math. und ihrer Grenzgebiete*, vol. 2, n.º 3. (Also separately at Springer, Berlin. American edition (1950), Chelsea Publishing Co., New York), 74 pp.

——— (1939). Sur l'interpolation et extrapolation des suites stationnaires. *C. R. Acad. Sci.*, Paris, vol. 208, 2 043-2 045.

——— (1941). Interpolation und Extrapolation von stationären zufalligen Folgen. *Bull. Acad. Sci. U. R. S. S., Ser. Math.*, vol. 5, 3-14.

——— (1941). Stationary sequences in Hilbert space. (Russian.) *Bull. Moscow State Univ.*, vol. 2. (Spanish trans., *Trab. Estad.*, vol. 4, 55 and 243.)

LEVINSON, Norman (1947a). The Wiener RMS (root mean square) error criterion in filter design and prediction. *J. Math. Phys.* (M.I.T.), vol. 25, 261-278.

——————— (1947b). A heuristic exposition of Wiener's mathematical theory of prediction and filtering. *J. Math. Phys.*, vol. 26, 110-119.

ROBINSON, E. A. (1950). *The theories of fiducial limits and confidence limits.* M.I.T. S.B. Thesis, Cambridge, Mass.

——————— (1952). *Eight new applications of the theory of stochastic processes to economic analysis.* M.I.T. S.M. Thesis, Cambridge, Mass., 54 pp.

——————— (1954). *Predictive decomposition of time series with applications to seismic exploration.* M.I.T. Ph. D. Thesis, and M.I.T. Geophysical Group, 7, Cambridge, Mass., and VESIAC No. 1104, 265 pp.

——————— (1959). *An introduction to infinitely many variates.* Charles Griffin & Co., London, and Stechert-Hafner, New York, 132 pp.

——————— (1962). *Random wavelets and cybernetic systems.* Charles Griffin & Co., London, and Stechert-Hafner, New York, 132 pp.

——————— (1963). Properties of the Wold decomposition of stationary stochastic processes. *Teoriya Veroyatnostei i ee Primeneniya* (U. S. S. R. Academy of Sciences), vol. 7, no. 2, 201-211; also in *The theory of probability and its applications*, vol. 7, 187-194.

——————— (1964). Wavelet composition of time series. *Econometric model building. Essays on the causal chain approach* (edited by Herman Wold), North-Holland Publishing Co., Amsterdam, 37-106.

——————— (1966). *Statistical communication and detection with special reference to digital data processing of radar and seismic signals.* Charles Griffin & Co., London, and Stechert-Hafner, New York (about 400 pp., in press).

SAMUELSON, P. A. (1941a). The stability of equilibrium: Comparative statics and dynamics. *Econometrica*, vol. 9, 97-120.

——————— (1941b). Conditions that the roots of a polynomial be less than unity in absolute value. *Annals of Mathematical Statistics*, vol. 12, 360-366.

——————— (1941c). A method of determining explicity the coefficients of the characteristic equation. *Annals of Mathematical Statistics*, vol. 12, 424-429.

——————— (1947). *Foundations of economic analysis.* Harvard University Press, Cambridge, Mass., 447 pp.

SIMPSON, S. M. (1953). *Statistical approaches to certain problems in geophysics.* M.I.T. Ph. D. Thesis, Cambridge, Mass.

SMITH, M. K. (1954). *Filter theory of linear operators with seismic applications.* M.I.T. Ph. D. Thesis, Cambridge, Mass. (Sections to which references are made are reprinted in M.I.T. GAG Report No. 6).

WADSWORTH, G. P., ROBINSON, E. A., BRYAN, J. G., and HURLEY, P. M. (1953). Detection of reflections on seismic records by linear operators. *Geophysics*, vol. 18, 539-586.

WIENER, N. (1923). Differential space. *J. Math. Phys.* (M.I.T.), vol. 2, 131-174.

WIENER, N. (1930). Generalized harmonic analysis. *Acta Math.*, vol. 55, 117-258.

———————— (1933). *The Fourier integral and certain of its applications.* Cambridge University Press, Cambridge.

———————— (1942). The extrapolation, interpolation, and smoothing of stationary time series with engineering applications. M.I.T. DIC Contract 6037, National Defense Research Council (Section D2), Cambridge, Mass. (Reprinted (1949), John Wiley & Sons, Inc., New York, to which references to page numbers refer).

———————— (1948). *Cybernetics.* John Wiley & Sons, Inc., New York.

WOLD, H. (1938). *A study in the analysis of stationary time series.* Almqvist & Wiksells, 214 pp.

———————— (1948). On prediction in stationary time series. *Annals of Math. Statistics*, vol. 19, 559-567.

———————— (1953) in association with L. JURÉEN. *Demand analysis.* John Wiley & Sons, Inc., New York, 358 pp.

CHAPTER 7

STATIONARY PROCESSES

7.1. Stationary processes and quantum mechanics

Definition 7.10   A sequence $x(t)$, $-\infty < t < \infty$, in Hilbert space $\mathbf{V}_\infty$ is called a *stationary process* if the inner product

$$R(s) = ((x(t+s), x(t)))$$

does not depend on t. The set of sequences

$$x_1(t), ..., x_n(t), \quad -\infty < t < \infty$$

is called a (*multiple*) *stationary process* if the inner products

$$R_{ij}(s) = ((x_i(t+s), x_j(t))) \quad (i, j = 1, ..., n)$$

do not depend on t. Both s and t are assumed to be integers. The variable s is called the *lag*. (Note: $R_{ij}(s)$ may also be denoted by $R_{x_i x_j}(s)$.)

If $\mathbf{V}_\infty$ is variate space $\mathbf{L}^2(P)$, then $x_i(t)$ is called a (second-order) *stationary random* (or *stochastic*) *process*, and if t represents time, a (second-order) *stationary time-series*. If the variates of a stationary stochastic process are centred at their means, then $R_{ii}(s)$ is called the *autocovariance*; $R_{ii}(s)/R_{ii}(0)$ the *autocorrelation*; $R_{ij}(s)$ the *cross-covariance*; and $R_{ij}(s)/[R_{ii}(0) R_{jj}(0)]^{1/2}$ the *cross-correlation*. We shall assume that all variates are centred, unless otherwise stated.

There is an elegant relationship between stationary processes and unitary operators due to André Kolmogorov.

Theorem 7.10 (Kolmogorov, 1941)   Let

$$x_1(t), ..., x_n(t), \quad -\infty < t < \infty$$

be a multiple stationary process in Hilbert space, and let $\mathbf{X}_{12 \dots n}$ be the closed linear manifold spanned by all the elements of the stationary process. Then the equation

$$x_j(t) U = x_j(t+1), \quad (j = 1, ..., n), \quad -\infty < t < \infty$$

uniquely determines the unitary operator $U \equiv U_{12 \dots n}$ with domain and range $\mathbf{X}_{12 \dots n}$.

This theorem shows that each stationary process is characterized by a uniquely determined unitary operator. As we have seen in Chapter VI, there is a one-to-one correspondence between unitary operators and hypermaximal Hermitian operators, and hence each stationary process is characterized by a uniquely determined hypermaximal Hermitian operator. John von Neumann (1932) has shown that the spectral representation enters so essentially into all quantum mechanical concepts that its existence cannot be dispensed with. Accordingly, in the mathematical foundations of quantum mechanics he admits only hypermaximal Hermitian operators. Thus quantum mechanical concepts are associated with hypermaximal Hermitian operators. Therefore we see that stationary processes and quantum mechanics have a common mathematical framework, namely the hypermaximal Hermitian operators.

7.2 Time-domain and frequency-domain

Let $x(t)$, $-\infty < t < \infty$, be a stationary time-series. The closed linear manifold $\mathbf{X}$ spanned by $x(t)$, $-\infty < t < \infty$, is called the *time-domain* of the time-series. By Kolmogorov's Theorem 7.10, the time-series is characterized by the unitary operator U with domain $\mathbf{X}$ and range $\mathbf{X}$ where $x(t+1) = x(t)U$ for all integers t. Using the von Neumann spectral representation of U (Theorem 6.50), we obtain

$$x(t+1) = x(t)U = \int_{-0.5}^{0.5} e^{2\pi i \nu}\, dx(t)\, E(\nu)$$

Hence the spectral representation of the time-series is

$$x(t) = \int_{-0.5}^{0.5} e^{2\pi i \nu t}\, dx(0)\, E(\nu) \tag{7.20}$$

a result given by Kolmogorov (1941). This integral is a stochastic integral (Section 5.5). The stochastic process $f(\nu)$ defined by $f(\nu) = x(0)E(\nu)$, $-0.5 \leqslant \nu \leqslant 0.5$, has orthogonal increments. By considering the properties of $f(\nu)$ (instead of the Spectral Representation Theorem, 6.50), Cramér (1942) independently arrived at the spectral representation (7.20) (with $x(0)E(\nu)$ replaced by $f(\nu)$).

Define the function $F(\nu)$ to be

$$F(\nu) = \|x(0)E(\nu)\|^2, \qquad -0.5 \leqslant \nu \leqslant 0.5 \tag{7.21}$$

Because $E(\nu)$ is a resolution of the identity, it follows that

$$F(-0.5) = \|0\|^2 = 0, \quad F(0.5) = \|x(0)I\|^2 = R(0)$$

and $F(\nu)$ is a real, monotone, non-decreasing function of ν that is continuous on the right. Since $F(\nu)$ has the properties of a distribution function (except that $F(0.5)$, although finite, is not necessarily one), the function $F(\nu)$ is called the *spectral distribution function* of the time-series. The variable ν is called the *frequency*.

From the spectral representation (7.20) we may deduce the *Khintchine–Wold Theorem*:

Theorem 7.20 (Khintchine, 1934; Wold, 1938) Let $x(t)$ be a stationary time-series with autocovariance $R(s)$ and spectral distribution function $F(\nu)$. Then $R(s)$ may be determined from $F(\nu)$ by

$$R(s) = \int_{-0\cdot5}^{0\cdot5} e^{2\pi i\nu s}\, dF(\nu) \qquad (7.22)$$

and, conversely, $F(\nu)$ may be determined from $R(s)$ by $F(\nu) = W(\nu+0)+C$ where

$$W(\nu) = R(0)\nu - \sum_{s\neq0} \frac{R(s)}{2\pi is}\, e^{-2\pi i\nu s}$$

and the constant C is calculated from $F(-0\cdot5) = 0$.

Proof. From the spectral representation (7.20) we see that

$$R(s) = ((x(s), x(0))) = \left(\left(\int_{-0\cdot5}^{0\cdot5} e^{2\pi i\nu s}\, dx(0)\, E(\nu),\quad x(0)\right)\right)$$

$$= \int_{-0\cdot5}^{0\cdot5} e^{2\pi i\nu s}\, d((x(0)\, E(\nu), x(0))) = \int_{-0\cdot5}^{0\cdot5} e^{2\pi i\nu s}\, dF(\nu)$$

Since $F(0\cdot5)-F(-0\cdot5)$ is finite, $F(\nu)$ has the Fourier series expansion $F(\nu) = W(\nu+0)+C$, where $W(\nu)$ and C are determined as above. Thus the spectral distribution function determines the autocorrelation, and conversely. Q.E.D.

The spectral distribution function $F(\nu)$ defines the Lebesgue–Stieltjes measure on the real line segment $-0\cdot5 \leqslant \nu \leqslant 0\cdot5$ (see Example 1.20). Then the function space $\mathbf{L}^2(F)$, for the line segment $-0\cdot5 \leqslant \nu \leqslant 0\cdot5$ with F-measure, is called the *frequency-domain* of the time-series. That is, the frequency-domain $\mathbf{L}^2(F)$ is the space of all complex-valued, measurable functions $\Phi(\nu)$ of the frequency ν for which

$$\int_{-0\cdot5}^{0\cdot5} |\Phi(\nu)|^2\, dF(\nu)$$

is finite.

Theorem 7.21 *The Isomorphism Theorem* (Stone, 1932; Kolmogorov, 1941) Given a stationary time-series

$$x(t), \quad -\infty < t < \infty$$

define a correspondence between an element x of the time-domain $\mathbf{X}$ and an element $\Phi(\nu)$ of the frequency-domain $\mathbf{L}^2(F)$ by

$$x = \int_{-0\cdot5}^{0\cdot5} \Phi(\nu)\, dx(0)\, E(\nu) \leftrightarrow \Phi(\nu) \qquad (7.23)$$

Then under this correspondence the time-domain and the frequency-domain are isomorphic in the sense that:

(1) The correspondence is one-to-one between the whole $\mathbf{X}$ space and the whole $\mathbf{L}^2(F)$ space;

(2) $cx \leftrightarrow c\Phi$ if $x \leftrightarrow \Phi$ (c a complex number);

(3) $x + y \leftrightarrow \Phi + \Psi$ if $x \leftrightarrow \Phi$ and $y \leftrightarrow \Psi$;

(4) $((x, y)) = ((\Phi, \Psi)) = \displaystyle\int_{-0\cdot5}^{0\cdot5} \Phi\overline{\Psi}\, dF(\nu)$;

(5) A sequence of elements in $\mathbf{X}$ converges and has the limit

$$x = \int_{-0\cdot5}^{0\cdot5} \Phi(\nu)\, dx(0)\, E(\nu)$$

in $\mathbf{X}$ if and only if the corresponding sequence of functions in $\mathbf{L}^2(F)$ converges and has the limit $\Phi(\nu)$ in $\mathbf{L}^2(F)$.

Proof. The set $\mathbf{M}$ of all elements of $\mathbf{X}$ of the form

$$x = \int_{-0\cdot5}^{0\cdot5} \Phi(\nu)\, dx(0)\, E(\nu)$$

where Φ belongs to $\mathbf{L}^2(F)$ is a closed linear manifold in $\mathbf{X}$, by Theorem 6.2 of Stone (1932). Since $x(t)$ is given by the spectral representation (7.20) and since the functions $e^{2\pi i\nu t}$, $-\infty < t < \infty$, belong to $\mathbf{L}^2(F)$, we see that $\mathbf{M}$ contains all the $x(t)$, $-\infty < t < \infty$. But $\mathbf{X}$ is the closed linear manifold spanned by all the $x(t)$, $-\infty < t < \infty$, so therefore $\mathbf{M} \supset \mathbf{X}$. But, by the definition of $\mathbf{M}$, $\mathbf{M} \subset \mathbf{X}$. Therefore $\mathbf{M} = \mathbf{X}$. The remainder of the theorem follows directly from Theorem 6.2 of Stone (1932). The verity of statements (2), (3), and (4) may be seen by formal manipulation; for example, in the case of (4), we have:

$$((x, y)) = \int_{-0\cdot5}^{0\cdot5} \Phi(\nu)\, d((x(0)\, E(\nu), y))$$

$$= \int_{-0\cdot5}^{0\cdot5} \Phi(\nu)\, d((x(0), yE(\nu)))$$

$$= \int_{-0\cdot5}^{0\cdot5} \Phi(\nu)\, d\overline{((yE(\nu), x(0)))}$$

$$= \int_{-0\cdot5}^{0\cdot5} \Phi(\nu)\, d\int_{-0\cdot5}^{0\cdot5} \overline{\Psi(\sigma)\, d\overline{((x(0)\, E(\sigma)\, E(\nu), x(0)))}}$$

$$= \int_{-0\cdot5}^{0\cdot5} \Phi(\nu)\, d\int_{-0\cdot5}^{\nu} \overline{\Psi(\sigma)\, d\overline{((x(0)\, E(\sigma), x(0)))}}$$

$$= \int_{-0\cdot5}^{0\cdot5} \Phi(\nu)\overline{\Psi(\nu)}\, dF(\nu) \qquad\qquad \text{Q.E.D.}$$

We recall Definition 5.30, which states that two functions of $\mathbf{L}^2(F)$ are considered as identical if and only if they are equal for all sets of frequencies ν that have positive measure. That is, two functions that differ only on a ν-set of F-measure zero are identical in $\mathbf{L}^2(F)$. From the definition of the F-measure (Example 1.20), we see that a ν-set has positive measure if and only if it is a set on which $F(\nu)$ increases. That is, any set of frequencies on which $F(\nu)$ does not increase has zero F-measure, and conversely. For

$$\| x(0)[E(\nu)-E(\sigma)]\|^2 = ((x(0)[E(\nu)-E(\sigma)],\, x(0)[E(\nu)-E(\sigma)]))$$
$$= ((x(0)E(\nu),\, x(0)E(\nu))) - ((x(0)E(\nu),\, x(0)E(\sigma)))$$
$$\quad - ((x(0)E(\sigma),\, x(0)E(\nu))) + ((x(0)E(\sigma),\, x(0)E(\sigma)))$$
$$= \| x(0)E(\nu)\|^2 - \| x(0)E(\sigma)\|^2 = F(\nu)-F(\sigma)$$

we write symbolically

$$\| dx(0)E(\nu)\|^2 = dF(\nu) \tag{7.24}$$

Definition 7.20   Let $x(t)$ be a stationary time-series with unitary operator U which has the resolution of the identity, $E(\nu)$, $-0.5 \leqslant \nu \leqslant 0.5$. Those frequencies ν in whose neighbourhoods $E(\nu)$ is not constant form the *spectrum* S of the time-series $x(t)$. Those frequencies σ for which $E(\nu)$ is discontinuous form the *point* (or *discrete*) *spectrum*. The spectrum consists of the point spectrum and the *continuous spectrum*. The set of frequencies that do not belong to the spectrum forms the *resolvent* of the time-series.

From equation (7.24) we see that the spectrum S is the set of those frequencies ν $(-0.5 \leqslant \nu \leqslant 0.5)$ for which the spectral distribution function $F(\nu)$ increases, and the resolvent is the set of those frequencies for which the spectral distribution function $F(\nu)$ does not increase. Accordingly, two functions $\Phi(\nu)$ and $\Psi(\nu)$, $-0.5 \leqslant \nu \leqslant 0.5$, that are equal for frequencies $\nu \in S$, but which may not be equal for frequencies ν in the resolvent, are considered as identical functions in $\mathbf{L}^2(F)$. Also, all integrals on ν for $\nu = -0.5$ to $\nu = 0.5$ may be replaced by integrals on ν for $\nu \in S$. Because it is common practice to use the terms "spectrum" and "spectral distribution function" interchangeably, the reader should note the distinction between these terms in our usage: the spectrum S is a point-set of frequencies ν on the real line segment $-0.5 \leqslant \nu \leqslant 0.5$, whereas the spectral distribution function $F(\nu)$ is a distribution function on the real line segment $-0.5 \leqslant \nu \leqslant 0.5$, the set of increase of $F(\nu)$ being S.

The Isomorphism Theorem 7.21 allows us to work with functions $\Phi(\nu)$ defined on the spectrum S instead of the variates x defined on the abstract space Ω. For example, the time-series $x(t)$ at time t corresponds to a complex sinusoidal function; i.e.

$$x(t) \leftrightarrow e^{2\pi i \nu t} = \cos 2\pi \nu t + i \sin 2\pi \nu t, \quad \nu \in S$$

In particular, we see that

$$x(0) \leftrightarrow 1, \quad \nu \in S \tag{7.25}$$

The stochastic process $x(0)E(\sigma)$ (which has orthogonal increments) corresponds to the characteristic function $\gamma_\sigma(\nu)$ of the set $-0{\cdot}5 \leqslant \nu \leqslant \sigma$;

i.e.
$$x(0)E(\sigma) \leftrightarrow \gamma_\sigma(\nu) = \begin{cases} 1, & -0{\cdot}5 \leqslant \nu \leqslant \sigma, & \nu \in S \\ 0, & \sigma < \nu \leqslant 0{\cdot}5, & \nu \in S \end{cases} \tag{7.26}$$

Moreover, if $x \leftrightarrow \Phi$, then

$$xU^t \leftrightarrow \Phi(\nu)\, e^{2\pi i \nu t}, \quad \nu \in S$$

so that operation on the variate x in the time-domain by the unitary operator U corresponds to multiplication of the function $\Phi(\nu)$ in the frequency-domain by the complex sinusoidal function $e^{2\pi i \nu}$. If $x \leftrightarrow \Phi$, then

$$xE(\sigma) = \int_{-0{\cdot}5}^{0{\cdot}5} \Phi(\nu)\, dx(0)E(\nu)E(\sigma) = \int_{-0{\cdot}5}^{\sigma} \Phi(\nu)\, dx(0)E(\nu)$$

so $xE(\sigma) \leftrightarrow \Phi(\nu)\gamma_\sigma(\nu)$. Because, if $x \leftrightarrow \Phi$ and $y \leftrightarrow \Psi$,

$$((x,y)) = \int_\Omega x(\omega)\overline{y(\omega)}\, P(d\omega) = \int_S \Phi(\nu)\overline{\Psi(\nu)}\, dF(\nu) = ((\Phi,\Psi))$$

we see that

$$((xE(\sigma),y)) = \int_S \Phi\gamma_\sigma \overline{\Psi}\, dF(\nu) = \int_{-0{\cdot}5}^{\sigma} \Phi(\nu)\overline{\Psi(\nu)}\, dF(\nu) \tag{7.27}$$

7.3   The Gibbs ergodic theorem

The first systematic exposition of the foundations of statistical mechanics and its applications to thermodynamics was given by Willard Gibbs (1902). It is a tribute to the originality of his ideas and the magnitude of his contribution that his ergodic theorem remained unproven for thirty years. B. O. Koopman (1931) perceived the relationship between the ergodic theorem and function space, which led John von Neumann (1932a), using Hilbert function space $\mathbf{L}^2$, to give the first proof of the ergodic theorem. Carleman did the fundamental work for the $\mathbf{L}^2$ version independently, and Birkoff gave a proof for function space $\mathbf{L}^1$. Khintchine (1934) applied the ergodic theorem to stationary processes. For second-order stationary processes, von Neumann's $\mathbf{L}^2$ version, in the framework of Khintchine, is required.

Theorem 7.30   The Gibbs ergodic theorem Let $x(t)$ be a stationary time-series with spectral representation (7.20). Then as $m-n \to \infty$, the time average

$$\frac{1}{m-n} \sum_{t=n+1}^{m} x(t)\, e^{-2\pi i \sigma t} \Rightarrow x(0)[E(\sigma) - E(\sigma-0)] \tag{7.30}$$

Proof. Since $cx \leftrightarrow c\Phi$ if $x \leftrightarrow \Phi$, we have

$$x(t)\, \mathrm{e}^{-2\pi i \sigma t} = \int_{-0.5}^{0.5} \mathrm{e}^{2\pi i\,(\nu-\sigma)\,t}\, dx(0)\, E(\nu)$$

Hence

$$\frac{1}{m-n} \sum_{n+1}^{m} x(t)\, \mathrm{e}^{-2\pi i \sigma t} = \int_{-0.5}^{0.5} \left[\frac{1}{m-n} \sum_{n+1}^{m} \mathrm{e}^{2\pi i\,(\nu-\sigma)\,t} \right] dx(0)\, E(\nu)$$

Consider the integrand in the above integral (i.e. the expression in the brackets). Clearly, when $\nu = \sigma$, the integrand is equal to $(1+1+\ldots+1)/(m-n) = 1$, whatever be the value of $m-n$. That is, when $\nu = \sigma$, the components of the complex sinusoidal functions $\cos 2\pi(\nu-\sigma)t + i \sin 2\pi(\nu-\sigma)t$ are exactly in phase, so their sum (for $t = n+1$ to m) divided by $m-n$ is precisely equal to 1. On the other hand, when $\nu \neq \sigma$, the components of the complex sinusoidal functions are out of phase, and hence their sum (for $t = n+1$ to m) suffers interference effects. Consequently, their sum divided by $m-n$ tends to zero as $m-n$ tends to infinity. Thus when $m-n \to \infty$, the integrand tends to zero for all ν except for $\nu = \sigma$, where the integrand is 1, and equation (7.30) results.   Q.E.D.

Definition 7.30   An *eigenvalue* of a unitary operator U in Hilbert space is a number $\mathrm{e}^{2\pi i \sigma} (-0.5 \leqslant \sigma \leqslant 0.5)$ for which the characteristic equation

$$\phi U = \mathrm{e}^{2\pi i \sigma} \phi \tag{7.31}$$

has non-null solutions ϕ in the Hilbert space. Orthonormal ϕ are called *eigensolutions*.

Theorem 7.31   Let $x(t)$ be a stationary time-series with unitary operator U. Then the time average

$$\frac{1}{m-n} \sum_{t=n+1}^{m} x(t)\, \mathrm{e}^{-2\pi i \sigma t} \Rightarrow 0$$

as $m-n \to \infty$ if and only if $\mathrm{e}^{2\pi i \sigma}$ is not an eigenvalue of U.

Proof. The unitary-operator counterpart of Theorem 6.60 states that $\mathrm{e}^{2\pi i \sigma}$ is an eigenvalue of U if and only if σ is a point of discontinuity of the resolution of the identity $E(\nu)$ belonging to U; i.e. if and only if $E(\sigma) - E(\sigma - 0) \neq 0$. Application of Theorem 7.30 then gives the desired result.   Q.E.D.

Theorem 7.32   Let $x(t)$ be a stationary time-series with unitary operator U. If $\mathrm{e}^{2\pi i \sigma}$ is an eigenvalue of U, then the non-null solutions ϕ (in $\mathbf{X}$) of the characteristic equation (7.31) form a closed linear manifold, which is the eigenmanifold of the projection operator $E(\sigma) - E(\sigma - 0)$.

Theorem 7.32 is the unitary-operator counterpart of Theorem 6.61. We see that $x(0)[E(\sigma) - E(\sigma - 0)]$ of expression (7.30) is the projection of $x(0)$ on the eigenmanifold of $e^{2\pi i\sigma}$. Corresponding to Theorem 6.62, eigenmanifolds of distinct eigenvalues of U are orthogonal.

Theorem 7.33 Let $x(t)$ be a stationary time-series with auto-covariance $R(t)$ and spectral distribution function $F(\nu)$. Then, as $m-n\to\infty$, the time-average

$$\frac{1}{m-n}\sum_{t=n+1}^{m} R(t)\,e^{-2\pi i\sigma t} \to F(\sigma)-F(\sigma-0) \qquad (7.32)$$

Proof. By using equation (7.22) the proof is similar to the proof of Theorem 7.30, except that we now have point-wise convergence ($\to$) instead of metric space convergence ($\Rightarrow$). Since

$$\| x(0)\,[E(\sigma)-E(\sigma-0)]\|^2 = F(\sigma)-F(\sigma-0) \qquad (7.33)$$

it follows that the right-hand side of expression (7.32) is non-zero if and only if $e^{2\pi i\sigma}$ is an eigenvalue.   Q.E.D.

In practice the autocorrelation or the autocovariance is estimated by time-averages over a *fixed realization* (i.e. a given observation) of a stationary time-series. We recall that a variate is a numerical-valued function of the points ω of Ω, and so, more specifically, we should write $x(t;\omega)$ instead of $x(t)$ in order to show the dependence of the variate $x(t)$ on the point ω. A fixed realization, then, is $x(t;\omega_0)$, $-\infty<t<\infty$, where ω_0 is a single fixed point. Thus we may think of each point ω as a (two-sided) sequence

$$\omega = (\ldots, u_{-1}, u_0, u_1, u_2, \ldots)$$

i.e. each ω represents a fixed realization u_t, $-\infty<t<\infty$ (where u_t are complex numbers). Then the random variable $x(t;\omega)$ is the *co-ordinate function* that assigns to ω the tth co-ordinate of ω, i.e. $x(t;\omega) = u_t$.

By inspection we see that the time-average given in expression (7.34) below is unbiased (i.e. has mean equal to $R(u)$). The following theorem gives the condition for the time-average to be consistent (i.e. to be asymptotically equal to $R(u)$).

Theorem 7.34 (Wiener, 1942; Doob, 1953) Let $x(t)$, $-\infty<t<\infty$, be a stationary time-series in $\mathbf{L}^2(P)$ with autocovariance $R(s)$. Define $y(t) = x(t+u)\,\overline{x(t)}$, $-\infty<t<\infty$, where u is a fixed integer. (Note: The variates $y(t)$ will not in general be centred at their means.) If $y(t)$ is a stationary time-series in $\mathbf{L}^2(P)$, i.e. if the inner products

$$S(s) = ((x(t+u+s)\,\overline{x(t+s)},\, x(t+u)\,\overline{x(t)}))$$

do not depend on t, then the time-average

$$\frac{1}{m-n}\sum_{t=n+1}^{m} x(t+u)\,\overline{x(t)} \Rightarrow R(u)+0 \qquad (7.34)$$

(where 0 is the null variate) as $m - n \to \infty$ if and only if

$$\frac{1}{m-n} \sum_{s=n+1}^{m} S(s) \to |R(u)|^2$$

as $m - n \to \infty$.

Proof. We have

$$E[y(t)] = ((y(t), 1)) = \int_{\Omega} x(t+u)\overline{x(t)}\, P(d\omega) = R(u)$$

Hence the time-series $y(t) - R(u)$ is centred (i.e. has zero mean) and so application of Theorem 7.30 (with $\sigma = 0$) shows that, as $m - n \to \infty$, the time-average

$$\frac{1}{m-n} \sum_{t=n+1}^{m} [y(t) - R(u)] = \frac{1}{m-n} \sum_{t=n+1}^{m} x(t+u)\overline{x(t)} - R(u)$$

converges ($\Rightarrow$) to a limit variate in $\mathbf{L}^2(P)$. By Theorem 7.33 (equations (7.32) and (7.33)) this limit is the null variate of $\mathbf{L}^2(P)$ if and only if

$$\frac{1}{m-n} \sum_{s=n+1}^{m} ((y(t+s) - R(u), y(t) - R(u)))$$

$$= \frac{1}{m-n} \sum_{s=n+1}^{m} [((y(t+s), y(t))) - \overline{R(u)}\,((y(t+s), 1))$$

$$\qquad\qquad\qquad\qquad - R(u)((1, y(t))) + |R(u)|^2]$$

$$= \frac{1}{m-n} \sum_{s=n+1}^{m} S(s) - |R(u)|^2$$

converges ($\to$) to zero as $m - n \to \infty$. We recall that $[R(u) + \text{null}$ variate$]$ is a function equal to $R(u)$ with probability 1 (i.e. equal to $R(u)$ except on an ω-set of probability zero).   Q.E.D.

7.4 Multiple stationary time-series

Let $x_1(t)$ and $x_2(t)$, $-\infty < t < \infty$, be a multiple stationary time-series. Then, by Theorem 7.10,

$$x_j(t) = \int_{-0\cdot5}^{0\cdot5} e^{2\pi i v t}\, dx_j(0)\, E(v) \quad (j = 1, 2) \tag{7.40}$$

where $E \equiv E_{12}$ is the resolution of the identity belonging to the unitary operator $U \equiv U_{12}$. The *cross-spectral function* is defined to be

$$F_{12}(v) = ((x_1(0)\, E(v), x_2(0))), \quad -0\cdot5 \leqslant v \leqslant 0\cdot5 \tag{7.41}$$

It follows that $F_{12}(v)$ is a function continuous on the right and has bounded variation. From $E(-0\cdot5) = 0$ we see that $F_{12}(-0\cdot5) = 0$. The counterpart of Theorem 7.20 is that

$$R_{12}(s) = \int_{-0\cdot5}^{0\cdot5} e^{2\pi i v s}\, dF_{12}(v)$$

and, conversely, $F_{12}(\nu) = W_{12}(\nu) + C$ where

$$W_{12}(\nu) = R_{12}(s)\nu - \sum_{s \neq 0} \frac{R_{12}(s)}{2\pi i s}\, e^{-2\pi i \nu s}$$

and C is calculated from $F_{12}(-0{\cdot}5) = 0$. We see that

$$R_{12}(s) = \bar{R}_{21}(-s), \quad F_{12}(\nu) = \bar{F}_{21}(\nu)$$

Moreover, it is true that the increments $\Delta F_{ij} = F_{ij}(\sigma) - F_{ij}(\nu), \sigma > \nu$, satisfy $|\Delta F_{12}|^2 \leqslant \Delta F_{11}\Delta F_{22}$.

Definition 7.40   A stationary time-series $x_2(t)$, $-\infty < t < \infty$, is called *subordinated* to a stationary time-series $x_1(t)$, $-\infty < t < \infty$, if $x_1(t)$ and $x_2(t)$ form a multiple stationary time-series and if $x_2(t)$, $-\infty < t < \infty$, belongs to the time-domain $\mathbf{X}_1$ of $x_1(t)$.

If $x_2(t)$ is subordinated to $x_1(t)$, then $\mathbf{X}_{12} = \mathbf{X}_1$, $U_{12} = U_1$, and $E_{12} = E_1$.

Theorem 7.40 (Kolmogorov, 1941)   For each stationary time-series $x_2(t)$ subordinated to $x_1(t)$ there corresponds one and only one function $\Phi_2^{(1)}(\nu)$ in the frequency-domain $\mathbf{L}^2(F_{11})$ for which

$$F_{22}(\nu) = \int_{-0{\cdot}5}^{\nu} |\Phi_2^{(1)}(\nu)|^2\, dF_{11}(\nu) \tag{7.42}$$

$$F_{21}(\nu) = \int_{-0{\cdot}5}^{\nu} \Phi_2^{(1)}(\nu)\, dF_{11}(\nu) \tag{7.43}$$

The correspondence $x_2(0) \leftrightarrow \Phi_2^{(1)}(\nu)$ is one-to-one between the class of all stationary time-series $x_2(t)$ subordinated to $x_1(t)$ and the frequency-domain $\mathbf{L}^2(F_{11})$ of $x_1(t)$. For any $x_2(t)$ and $x_3(t)$ in the class of stationary time-series subordinated to $x_1(t)$ we have

$$F_{23}(\nu) = \int_{-0{\cdot}5}^{\nu} \Phi_2^{(1)}\, \overline{\Phi_3^{(1)}}\, dF_{11}(\nu) \tag{7.44}$$

Proof. If $x_2(t)$ is a stationary time-series subordinated to $x_1(t)$, then $x_2(0)$ belongs to $\mathbf{X}_1$. According to the Isomorphism Theorem 7.21, there is a function $\Phi_2^{(1)}$ of $\mathbf{L}^2(F_{11})$ for which $y(0) \leftrightarrow \Phi_2^{(1)}$. Hence, by (7.41) we have

$$F_{23}(\nu) = ((x_2(0)\, E_{23}(\nu),\, x_3(0))) = ((x_2(0)\, E_1(\nu),\, x_3(0)))$$

which, by equation (7.27), gives equation (7.44). In the particular case when $x_3(t) = x_1(t)$, then $x_3(0) = x_1(0) \leftrightarrow 1 = \Phi_3^{(1)}$ by equation (7.25), and hence (7.43) results from (7.44). In the particular case when $x_3(t) = x_2(t)$, then $x_3(0) = x_2(0) \leftrightarrow \Phi_2^{(1)}$ and hence equation (7.42) results from (7.44). That $x_2(t)$ determines $\Phi_2^{(1)}$ uniquely follows from equation (7.43), by which

$$\Phi_2^{(1)}(\nu) = \frac{dF_{21}(\nu)}{dF_{11}(\nu)} \qquad \qquad \text{Q.E.D.}$$

Theorem 7.41 (Kolmogorov, 1941) Let $x_1(t)$ and $x_2(t)$ be a multiple stationary time-series. Then $x_2(t)$ is subordinated to $x_1(t)$ if and only if there is a function Φ in $\mathbf{L}^2(F_{11})$ for which

$$F_{22}(v) = \int_{-0\cdot5}^{v} |\Phi(v)|^2 \, dF_{11}(v) \qquad (7.45)$$

$$F_{21}(v) = \int_{-0\cdot5}^{v} \Phi(v) \, dF_{11}(v) \qquad (7.46)$$

Proof. The "only if" statement follows from Theorem 7.40. To establish the "if" statement, construct the time-series $x_3(t)$ by the correspondence $x_3(0) \leftrightarrow \Phi$. Since, by hypothesis, Φ is in $\mathbf{L}^2(F_{11})$ and equations (7.45) and (7.46) are satisfied, the one-to-one correspondence of Theorem 7.40 shows that $x_3(t)$ is subordinated to $x_1(t)$. Then it can be shown that $x_3(t)$ coincides with $x_2(t)$.   Q.E.D.

Definition 7.41   Two stationary time-series $x_1(t)$ and $x_2(t)$ are called *equivalent* if they are mutually subordinated, i.e. if $x_1(t)$ and $x_2(t)$ form a multiple stationary time-series and $\mathbf{X}_1 = \mathbf{X}_2$.

Theorem 7.42   The stationary time-series $x_2(t)$, subordinated to $x_1(t)$, is equivalent to $x_1(t)$ if and only if $\Phi_2^{(1)} \neq 0$ for v in the spectrum S_1 of $x_1(t)$ (i.e. if and only if $\Phi_2^{(1)} = 0$ only on a v-set of F_{11}-measure zero).

Proof. The proof can be established from the following observation: If $x_2(t)$ is equivalent to $x_1(t)$ then by equation (7.42) F_{22}-measure is absolutely continuous with respect to F_{11}-measure, and (by interchanging indices) the converse is also true. Hence the spectra S_1 and S_2 of equivalent time-series $x_1(t)$ and $x_2(t)$ coincide, and

$$\Phi_2^{(1)}(v) = \frac{1}{\Phi_1^{(2)}(v)}, \quad v \epsilon S_1 = S_2 \qquad (7.47)$$

Q.E.D.

7.5   The spectral decomposition

Let S be the spectrum, and $F(v)$ the spectral distribution function, of the stationary time-series $x(t)$. Let S be divided into two non-overlapping sets S_1 and S_2 (i.e. $S = S_1 \cup S_2$, $S_1 \cap S_2 = $ null set) each of which is measurable with respect to F-measure. If the spectral representation of $x(t)$ is given by equation (7.20), define $x_j(t)$ $(j = 1, 2)$ to be

$$x_j(t) = \int_{S_j} e^{2\pi i v t} \, dx(0) E(v) = \int_{S} e^{2\pi i v t} \Lambda_j(v) \, dx(0) E(v) \qquad (7.50)$$

where $\Lambda_j(v)$ is the characteristic function of the set S_j (i.e. $\Lambda_j(v) = 1$ for v in S_j, and $= 0$ for v not in S_j). We see that $x_j(t) \leftrightarrow e^{2\pi i v t} \Lambda_j(v)$ in the isomorphism of $\mathbf{X}$ and $\mathbf{L}^2(F)$ (Theorem 7.21). We note that

$\Lambda_1(\nu) + \Lambda_2(\nu) = 1$ for ν in S, and $\Lambda_1(\nu)\Lambda_2(\nu) = 0$ for ν in S. By Theorem 7.21 we have

$$((x_j(t+s), x_k(t))) = \int_S e^{2\pi i\nu s}\,\Lambda_j(\nu)\,\Lambda_k(\nu)\,dF(\nu)$$

which $= 0$ for $j \neq k$ and which does not depend on t for $j = k$. Hence $x_1(t)$ and $x_2(t)$ form a multiple stationary time-series, and $x_1(t)$ and $x_2(s)$ are orthogonal for all t and s. The spectrum of $x_1(t)$ is S_1 and its spectral distribution function is

$$F_{11}(\nu) = \int_{-0.5}^{\nu} \Lambda_1(\nu)\,dF(\nu) \tag{7.51}$$

and similarly for $F_{22}(\nu)$. The cross-spectral function $F_{12}(\nu)$ is equal to zero. Thus the decomposition

$$x(t) = \int_{S_1} e^{2\pi i\nu t}\,dx(0)\,E(\nu) + \int_{S_2} e^{2\pi i\nu t}\,dx(0)\,E(\nu) = x_1(t) + x_2(t)$$

divides $x(t)$ into two orthogonal time-series, where

$$F(\nu) = F_1(\nu) + F_2(\nu)$$

If $F_{1x}(\nu)$ is the cross-spectral function of $x_1(t)$ and $x(t)$, then from the decomposition we see that $F_{1x}(\nu) = F_{11}(\nu)$. Hence from equations (7.43) and (7.51) we have

$$\Phi_1^{(x)}(\nu) = \frac{dF_{1x}(\nu)}{dF(\nu)} = \frac{dF_{11}(\nu)}{dF(\nu)} = \Lambda_1(\nu)$$

and similarly for $\Phi_2^{(x)}(\nu)$. Corresponding orthogonal decompositions can be made by dividing S into any finite or countable number of non-overlapping sets $S_1, S_2, S_3, \ldots$.

If σ is in the *point spectrum* of $x(t)$ then $e^{2\pi i\sigma}$ is an eigenvalue of U, and conversely. For each eigenvalue of U, choose one and only one eigensolution. In this way we obtain an orthonormal set in $\mathbf{X}$. By Theorem 4.70 this set must be either finite or countable. Since, by construction, there is a one-to-one correspondence between the eigensolutions in this set and the eigenvalues of U, the eigenvalues must be either finite or countable. Hence the point spectrum of a time-series must be either a finite or countable set, and so has Lebesgue measure zero.

Suppose that the *continuous spectrum* of $x(t)$ exists, and let the interval $\nu_1 \leqslant \nu \leqslant \nu_2$ be in the continuous spectrum of $x(t)$. That is, in this interval $E(\nu)$ increases continuously, so $E(\nu_1) < E(\nu_2)$. For $\sigma \leqslant \nu_1$, we have $E(\sigma) - E(\sigma - 0) \leqslant E(\sigma) \leqslant E(\nu_1)$; for $\nu_1 < \sigma \leqslant \nu_2$,

$$E(\sigma) - E(\sigma - 0) = 0$$

by the continuity; and for $\nu_2 < \sigma$,

$$E(\sigma) - E(\sigma - 0) \leqslant I - E(\sigma - 0) \leqslant I - E(\nu_2)$$

Hence $E(\sigma) - E(\sigma - 0)$ is always orthogonal to $E(\nu_2) - E(\nu_1)$. Let $\mathbf{N}$ be the eigenmanifold of the projection operator $E(\nu_2) - E(\nu_1)$. Thus the eigenmanifolds of all the eigenvalues are orthogonal to $\mathbf{N}$. Now $\mathbf{N}$ contains elements other than the null element (because, by hypothesis, $E(\nu_2) - E(\nu_1) \neq 0$). Hence the eigensolutions (which, by definition, belong to the eigenmanifolds of the eigenvalues) cannot form a maximal orthonormal set in $\mathbf{X}$. In other words, if the continuous spectrum of $x(t)$ exists, then $x(t)$ cannot be expressed by a Fourier series in the eigensolutions.

Let $\sigma_1, \sigma_2, \ldots$ be the point spectrum of $x(t)$. Then, by definition, the eigenmanifolds of the eigenvalues are the eigenmanifolds of the projection operators $E(\sigma_1) - E(\sigma_1 - 0)$, $E(\sigma_2) - E(\sigma_2 - 0)$, $\ldots$. Let $\mathbf{M}$ be the closed linear manifold spanned by these eigenmanifolds. Alternatively, $\mathbf{M}$ is the closed linear manifold spanned by all the eigensolutions. The projection of $x(t)$ on $\mathbf{M}$, which we call $x_1(t)$, is a stationary time-series with time-domain $\mathbf{M}$, and has the spectral representation

$$x_1(t) = x(t) P_{\mathbf{M}} = \sum_n e^{2\pi i \sigma_n t} x(0) [E(\sigma_n) - E(\sigma_n - 0)]$$

Because of equation (7.33), the spectral distribution function $F(\nu)$ of $x(t)$ has a jump (i.e. $F(\nu) - F(\nu - 0) > 0$) at and only at $\nu = \sigma_n$, i.e. a frequency in the point spectrum.

The normal $x(t) [I - P_{\mathbf{M}}]$, which we call $x_2(t) + x_3(t)$, is a stationary time-series with time-domain $\mathbf{X} - \mathbf{M}$, which is the eigenmanifold of the projection operators corresponding to the continuous spectrum of $x(t)$. Because of equation (7.33) the spectral distribution function $F(\nu)$ is continuous if and only if ν is a point of the continuous spectrum of $x(t)$.

The continuous spectrum of $x(t)$ may be decomposed into a set S_3 of frequencies where $F(\nu)$ is absolutely continuous with respect to Lebesgue measure $\mu(d\nu) = d\nu$, and into a set S_2 of frequencies where $F(\nu)$ is not absolutely continuous. Because $F(\nu)$ is a distribution function (i.e. $F(\nu)$ has bounded variation: $F(0 \cdot 5) - F(-0 \cdot 5) < \infty$), the set S_2 necessarily has Lebesgue measure zero. If we let S_1 denote the point spectrum $\sigma_1, \sigma_2, \ldots$, then we have decomposed the spectrum S of $x(t)$ into three non-overlapping sets: S_1, the point spectrum; S_2, the continuous but non-absolutely continuous spectrum; and S_3, the absolutely-continuous spectrum. That is, $S = S_1 \cup S_2 \cup S_3$, where $S_1 \cap S_2 \cap S_3 = $ null set. Both S_1 and S_2 have Lebesgue measure zero. Hence the following theorem may be readily established:

Theorem 7.50 *The Spectral Decomposition* (Cramér, 1942) Let $x(t)$, $-\infty < t < \infty$, be a stationary time-series with spectrum S and spectral distribution function $F(\nu)$. Then $x(t)$ has the unique decomposition

$$x(t) = x_1(t) + x_2(t) + x_3(t)$$

where the components $x_j(t)\,(j = 1, 2, 3)$ form a multiple stationary time-series with the properties:

(1) The $x_j(t)$ are mutually orthogonal; i.e. $((x_j(t), x_k(s))) = 0$ for $j \neq k$.

(2) The spectrum S_1 of $x_1(t)$ is the point spectrum of $x(t)$; the spectrum S_2 of $x_2(t)$ is the continuous, non-absolutely continuous spectrum of $x(t)$; and the spectrum S_3 of $x_3(t)$ is the absolutely-continuous spectrum of $x(t)$.

(3) $S_1 \cup S_2 \cup S_3 = S$; $S_1 \cap S_2 \cap S_3 = $ null set.

(4) If the spectral representation of $x_j(t)$ is

$$x_j(t) = \int_{-0\cdot5}^{0\cdot5} e^{2\pi i v t}\, dx_j(0)\, E_j(v)$$

then $dx_j(0)\,E_j(v) = dx(0)\,E(v)$ for $v \epsilon S_j$ and $= 0$ otherwise.

(5) The spectral distribution functions satisfy

$$F(v) = F_{11}(v) + F_{22}(v) + F_{33}(v)$$

$F_{11}(v)$ is a step-function and $F_{22}(v)$ and $F_{33}(v)$ are continuous functions. The derivative of $F_{11}(v)$ does not exist for v in S_1 (but $F_{11}(v) - F_{11}(v-0) = F(v) - F(v-0)$ for v in S_1), and $F_{11}'(v) = 0$ for all v, $-0\cdot5 \leqslant v \leqslant 0\cdot5$, except for the set S_1 of Lebesgue measure zero. The derivative $F_{22}'(v)$ does not exist for v in S_2, but $F_{22}'(v) = 0$ for all v except for the set S_2 of Lebesgue measure zero. The derivative $F_{33}'(v)$ exists for all v, and $F_{33}'(v) = F'(v) > 0$ for v in S_3 and $F_{33}'(v) = 0$ elsewhere.

(6) Any of the components $x_1(t)$, $x_2(t)$, and $x_3(t)$ may be the null variate. If $x_3(t)$ is non-null then the spectrum S_3 has positive Lebesgue measure. The spectra S_1 and S_2 always have zero Lebesgue measure.

The derivative $F_{33}'(v)$ is called the *spectral power function*, or *spectral density function* of $x(t)$. To verify property (6) of Theorem 7.50 we note that

$$F_{33}(0\cdot5) = \int_{S_3} F_{33}'(v)\, dv$$

will equal zero unless S_3 has positive Lebesgue measure. We also note that the component $x_j(t)$ may be constructed from $F_{jj}(t)$ by use of Theorem 7.21; i.e. form the frequency domains $\mathbf{L}^2(F_{jj})$ and then let $x_j(t) \leftrightarrow e^{2\pi i v t}$ in $\mathbf{L}^2(F_{jj})$. The time-domains $\mathbf{X}_j\ (j = 1, 2, 3)$ are mutually orthogonal and satisfy $\mathbf{X} = \mathbf{X}_1 + \mathbf{X}_2 + \mathbf{X}_3$; and the frequency-domains $\mathbf{L}^2(F_{jj})$ are mutually orthogonal and satisfy

$$\mathbf{L}(F) = \mathbf{L}(F_{11}) + \mathbf{L}(F_{22}) + \mathbf{L}(F_{33})$$

7.6 White noise

Definition 7.60   A time-series $x(t)$ is said to be a *running average* of a stationary time-series $\theta(t)$ if it is possible to choose coefficients a_s such that, as $m-n\to\infty$,

$$\sum_{s=n+1}^{m} a_s\,\theta(t-s) \Rightarrow x(t)$$

Clearly, if $x(t)$ is a running average of a stationary time-series $\theta(t)$, then $x(t)$ is stationary and subordinated to $\theta(t)$.

Definition 7.61   A stationary time-series $\theta(t)$ is called *white noise* if $\theta(t)$, $-\infty<t<\infty$, is an orthonormal sequence (i.e. if its auto-covariance $R_{\theta\theta}(s) = 1$ if $s = 0$, and $= 0$ if $s\neq 0$).

From Theorem 7.20 we see that if $\theta(t)$ is white noise, then $F_{\theta\theta} = v+0{\cdot}5$, and conversely any stationary time-series with a spectral distribution function equal to $v+0{\cdot}5$ is white noise. If $\theta(t)$ is white noise, then $\theta(t)$, $-\infty<t<\infty$, is a maximal orthonormal set in the time-domain Θ of $\theta(t)$, and hence every element in this time-domain has a unique Fourier series representation in terms of $\theta(t)$ (Theorem 4.74). Thus, for a stationary time-series $x(t)$ to be a running average of white noise $\theta(t)$ it is necessary and sufficient that $x(t)$ be subordinated to $\theta(t)$; in this case, $x(t)$ has the unique Fourier representation

$$x(t) = \sum_{s=-\infty}^{\infty} b_s\,\theta(t-s), \quad b_s = ((x(t),\theta(t-s))) \qquad (7.60)$$

Because $dF_{\theta\theta}(v) = dv$, we see that the spectrum S_θ of white noise is the whole interval $-0{\cdot}5 \leqslant v \leqslant 0{\cdot}5$, except for a set of Lebesgue measure zero. Since the Lebesgue measure of a line segment is equal to the length of the segment, S_θ has Lebesgue measure equal to $0{\cdot}5-(-0{\cdot}5) = 1$. The frequency domain $\mathbf{L}^2(F_{\theta\theta})$, which (because $F_{\theta\theta} = 0{\cdot}5+v$) we shall denote by $\mathbf{L}^2(0{\cdot}5+v)$, is the $\mathbf{L}^2$ space with Lebesgue measure dv.

Theorem 7.60 (Kolmogorov, 1941)   Given the stationary time-series $x(t)$; if $x(t)$ is a running average of white noise $\theta(t)$, then $F_{xx}(v)$ is absolutely continuous with respect to Lebesgue measure. If $F_{xx}(v)$ is absolutely continuous with respect to Lebesgue measure, and if the orthogonal complement $\mathbf{V}_\infty - \mathbf{X}$ has infinite dimension, then $x(t)$ is a running average of white noise $\theta(t)$.

Proof (1).   By hypothesis, $x(t)$ has the representation (7.60), and so $x(t)$ is subordinated to $\theta(t)$. By Theorem 7.40, we have

$$F_{xx}(v) = \int_{-0{\cdot}5}^{v} |\Phi_x^{(\theta)}(v)|^2\,dv, \quad F_{x\theta}(v) = \int_{-0{\cdot}5}^{v} \Phi_x^{(\theta)}(v)\,dv \qquad (7.61)$$

which shows that F_{xx} is absolutely continuous with respect to Lebesgue measure. We also note that, under the isomorphism

7

328 ENDERS A. ROBINSON

between Θ and $\mathbf{L}^2(0.5+\nu)$, we have $x(0)\leftrightarrow\Phi_x^{(\theta)}$ and $\theta(t)\leftrightarrow e^{2\pi i\nu t}$. Hence, by Property (4) of Theorem 7.21, the coefficient b_s in the representation (7.60) is given by

$$b_s = ((x(0), \theta(-s))) = \int_{-0.5}^{0.5} \Phi_x^{(\theta)} e^{2\pi i\nu s} \, d\nu$$

and so $\Phi_x^{(\theta)}$ has the Fourier series in $\mathbf{L}^2(0.5+\nu)$ given by

$$\Phi_x^{(\theta)} = \sum_{s=-\infty}^{\infty} b_s e^{-2\pi i\nu s} \tag{7.62}$$

with b_s as Fourier coefficients. From equation (7.61) the spectral density function is

$$F'_{xx}(\nu) = |\Phi_x^{(\theta)}(\nu)|^2 = \left| \sum_{s=-\infty}^{\infty} b_s e^{-2\pi i\nu s} \right|^2 \tag{7.63}$$

$$\text{Q.E.D. (1).}$$

Proof (2). By hypothesis, F_{xx} is absolutely continuous with respect to Lebesgue measure, and $\mathbf{V}_\infty - \mathbf{X}$ has infinite dimension. Hence, by the Radon–Nikodym theorem, there exists a non-negative function $|\Psi(\nu)|^2$ such that

$$F_{xx}(\nu) = \int_{-0.5}^{0.5} |\Psi(\nu)|^2 \, d\nu \tag{7.64}$$

This shows that $\Psi(\nu)$ (which is any square root of $|\Psi(\nu)|^2$) belongs to $\mathbf{L}^2(0.5+\nu)$. Define the functions

$$F_{11}(\nu) = F_{xx}(\nu), \quad F_{12}(\nu) = \int_{-0.5}^{\nu} \Psi(\nu) \, d\nu, \quad F_{22}(\nu) = 0.5+\nu$$

These functions satisfy $|\Delta F_{12}|^2 \leqslant \Delta F_{11} \Delta F_{22}$ and so represent the spectral functions of a multiple stationary time-series. Because $\mathbf{V}_\infty - \mathbf{X}$ is infinite, there always exists a time-series $\theta(t)$ such that $x(t)$ and $\theta(t)$ are multiple-stationary and such that

$$F_{x\theta}(\nu) = F_{12}(\nu) = \int_{-0.5}^{\nu} \Psi(\nu) \, d\nu, \quad F_{\theta\theta}(\nu) = F_{22}(\nu) = 0.5+\nu \tag{7.65}$$

Because $F_{\theta\theta} = 0.5+\nu$, $\theta(t)$ is white noise. Equations (7.64) and (7.65), by Theorem 7.41, show that $x(t)$ is subordinated to $\theta(t)$. (Note: Two cases may occur: (1) $\Psi(\nu)\neq 0$ almost everywhere with respect to Lebesgue measure. Then by Theorem 7.42, $x(t)$ is equivalent to $\theta(t)$, i.e. $\mathbf{X} = \Theta$, in which case use of the infinite dimension of $\mathbf{V}_\infty - \mathbf{X}$ is not needed. (2) $\Psi(\nu) = 0$ on a set of positive Lebesgue measure. Then $x(t)$ is not equivalent to $\theta(t)$, i.e. $\mathbf{X}$ is properly contained in Θ, and so $\Theta - \mathbf{X}$ has infinite dimension. But $\Theta - \mathbf{X}$ is contained in $\mathbf{V}_\infty - \mathbf{X}$, and so the condition that $\mathbf{V}_\infty - \mathbf{X}$ have infinite dimension is needed.) Because $x(t)$ is subordinated to $\theta(t)$, we can use equation (7.60) to express $x(t)$ as a running average. Q.E.D. (2).

Case (1) in the proof above, by noting that $F'_{xx}(\nu) = |\Psi(\nu)|^2$, becomes

Theorem 7.61   A stationary time-series $x(t)$ is a running average of white noise $\theta(t)$ with $\mathbf{X} = \mathbf{\Theta}$ if and only if $F_{xx}(\nu)$ is absolutely continuous with respect to Lebesgue measure and $F'_{xx}(\nu)$ is positive almost everywhere with respect to Lebesgue measure.

7.7   The Wold decomposition

Definition 7.70   Let $x(s)$, $-\infty < s < \infty$, be a stationary time-series. A given integer t is called the *present time*; integers $s < t$, the *past time*, and integers $s > t$, the *future time*.

Definition 7.71   The closed linear manifold $\mathbf{X} = \mathbf{X}(\infty)$ spanned by $x(s)$, $-\infty < s < \infty$, is called the *time-domain*. The closed linear manifold $\mathbf{X}(t)$ spanned by the present and past variates $x(s), s \leqslant t$, is called the *present-and-past time-domain*. The common part $\mathbf{X}(-\infty)$ of all the $\mathbf{X}(t)$, i.e. $\mathbf{X}(-\infty) \equiv \bigcap\limits_{-\infty}^{\infty} \mathbf{X}(t)$, is called the *remote-past time-domain*. The orthogonal complement $\mathbf{X}(t) - \mathbf{X}(-\infty)$ is called the *present-and-recent-past time-domain*.

The remote-past time-domain is a closed linear manifold, and so is the present-and-recent-past time-domain. Since $x(t)\,U^s = x(t+s)$, we may write $\mathbf{X}(t)\,U^s = \mathbf{X}(t+s)$. We also have

$$\mathbf{X}(-\infty)\,U^s = \mathbf{X}(-\infty)$$

Definition 7.72   A stationary time-series $x(t)$ is said to be *singular* if $\mathbf{X}(-\infty) = \mathbf{X}$. A stationary time-series that is not singular is said to be *non-singular*. A stationary time-series $x(t)$ is said to be *regular* if $x(t) \neq 0$ and $\mathbf{X}(-\infty) = 0$.

We see that a stationary time-series is singular if and only if $\mathbf{X}(t) = \mathbf{X}$ for all t. In fact, if $\mathbf{X}(t) = \mathbf{X}$ for a given t, it must be true for all t (because $x(t)$ is stationary), and

$$\mathbf{X}(-\infty) = \mathbf{X}(t) = \mathbf{X}(t+s) = \mathbf{X}$$

Thus a stationary time-series is singular if and only if its present-and-recent-past time-domain is equal to the null variate; i.e. $\mathbf{X}(t) - \mathbf{X}(-\infty) = 0$. A non-null stationary time-series is regular if and only if its remote-past time-domain is equal to the null variate; i.e. $\mathbf{X}(-\infty) = 0$.

Theorem 7.70   *The Wold decomposition* (Wold, 1938)   Let $x(t)$, $-\infty < t < \infty$, be a non-singular stationary time-series. Then $x(t)$ has the unique decomposition

$$x(t) = y(t) + z(t), \quad \text{with} \quad z(t) = b_0\,\theta(t) + b_1\,\theta(t-1) + \ldots$$

$$(7.70)$$

330   ENDERS A. ROBINSON

where the components $\theta(t), y(t), z(t)$, $-\infty < t < \infty$, form a multiple stationary time-series with the properties:

(1) Each of the variates $\theta(t), y(t), z(t)$ is in $\mathbf{X}(t)$.

(2) The stationary time-series $y(t)$ is singular and the stationary time-series $z(t)$ is regular.

(3) $\theta(t)$ is white noise.

(4) The variates $y(t)$, $-\infty < t < \infty$, are orthogonal to the variates $\theta(t)$ and $z(t)$, $-\infty < t < \infty$; i.e.

$$((y(s), \theta(t))) = ((y(s), z(t))) = 0$$

for $s = t$ as well as $s \neq t$.

(5) $b_0 > 0$ and $b_0^2 + |b_1|^2 + |b_2|^2 + \ldots < \infty$.

Proof. Application of the Projection Theorem 5.20 shows that $x(t)$ can be resolved in one and only one way into two orthogonal components

$$x(t) = x(t) P_{\mathbf{X}(t-1)} + w(t)$$

where the first component is in $\mathbf{X}(t-1)$ and the second component $w(t)$, the normal to $\mathbf{X}(t-1)$, is in $\mathbf{X}(t) - \mathbf{X}(t-1)$. Because $x(t)$ is regular, it follows that $\mathbf{X}(t) \neq \mathbf{X}(t-1)$, and so $\|w(t)\| > 0$. Define $\theta(t)$ to be $w(t)/\|w(t)\|$. Since $\theta(t)$ is in $\mathbf{X}(t) - \mathbf{X}(t-1)$, and since $\theta(t-s), s \geq 1$, is in $\mathbf{X}(t-1)$, we see that $((\theta(t), \theta(t-s))) = 0$ for $s \geq 1$, and so the $\theta(t)$, $-\infty < t < \infty$, form an orthonormal set. The $\theta(t)$ are called the (white noise) *innovations*. The closed linear manifold spanned by $\theta(t)$ is $[\theta(t)] = \mathbf{X}(t) - \mathbf{X}(t-1)$.

Let $\Theta(t)$ be the closed linear manifold spanned by the present and past innovations $\theta(s)$, $s \leq t$. Suppose $y \in \mathbf{X}(t) - \Theta(t)$; i.e. $y \in \mathbf{X}(t)$ and $y \perp \Theta(t)$. Then $y \perp [\theta(t)] = \mathbf{X}(t) - \mathbf{X}(t-1)$ and so $y \in \mathbf{X}(t-1)$. In a similar way, it follows that $y \in \mathbf{X}(t-2), y \in \mathbf{X}(t-3), \ldots$, and so by induction $y \in \mathbf{X}(-\infty)$. Hence $\mathbf{X}(t) - \Theta(t) \supset \mathbf{X}(-\infty)$. Conversely, suppose $y \in \mathbf{X}(-\infty)$. Since $\mathbf{X}(-\infty) \subset \mathbf{X}(s-1)$, it follows that $y \in \mathbf{X}(s-1)$. But $\mathbf{X}(s-1) \perp \theta(s)$, so $y \perp \theta(s)$ for arbitrary s. Thus $y \perp \Theta(t)$, and so $y \in \mathbf{X}(t) - \Theta(t)$. Hence $\mathbf{X}(-\infty) \supset \mathbf{X}(t) - \Theta(t)$. Therefore $\mathbf{X}(-\infty) = \mathbf{X}(t) - \Theta(t)$, and so

$$\mathbf{X}(t) - \mathbf{X}(-\infty) = \mathbf{X}(t) - [\mathbf{X}(t) - \Theta(t)] = \Theta(t)$$

which says that *the present-and-recent-past time-domain is equal to the closed linear manifold spanned by the present and past innovations.* In other words, the present and past innovations, $\theta(s)$ for $s \leq t$, form a maximal orthonormal set in the present-and-recent-past time-domain.

Thus we have $\mathbf{X}(t) = \mathbf{X}(-\infty) + \Theta(t)$ where $\mathbf{X}(-\infty) \perp \Theta(t)$. By the Projection Theorem 5.20, the time-series $x(t)$ has the unique decomposition

$$x(t) = x(t) P_{\mathbf{X}(-\infty)} + x(t) P_{\Theta(t)}$$

which we denote by $x(t) = y(t) + z(t)$. The time-series $y(t)$ and $z(t)$ are stationary. It can be readily shown that the present-and-past time-domain $\mathbf{Y}(t)$ of $y(t)$ is $\mathbf{X}(-\infty)$, and that the present-and-past time-domain $\mathbf{Z}(t)$ of $z(t)$ is $\mathbf{\Theta}(t)$. Since $\mathbf{Y}(t) = \mathbf{X}(-\infty)$ for all t, it follows that the time-series $y(t)$ is singular: $\mathbf{Y}(t) = \mathbf{Y}(-\infty)$.

By Theorems 4.74 and 5.20, $z(t)$ has the Fourier series representation

$$z(t) = \sum_{s=0}^{\infty} b_s \, \theta(t-s)$$

where the Fourier coefficients are given by

$$b_s = ((x(t), \theta(t-s))), \quad s \geqslant 0; \quad \sum_{s=0}^{\infty} |b_s|^2 < \infty$$

Because $\mathbf{Z}(-\infty) = \mathbf{\Theta}(-\infty) = \mathbf{X}(-\infty) - \mathbf{X}(-\infty) = 0$, we see that $z(t)$ is regular. We have

$$b_0 = ((x(t), \theta(t))) = ((x(t) P_{\mathbf{X}(t-1)} + w(t), \theta(t))) = \| w(t) \|$$

so b_0 is real and positive, and $z(t)$ is non-null.   Q.E.D.

Theorem 7.71   In the Wold decomposition (7.70), the stationary time-series $z(t)$ and $\theta(t)$ are equivalent (i.e. $\mathbf{Z} = \mathbf{\Theta}$) and their spectra S_z and S_θ coincide.

Theorem 7.72   In the Wold decomposition (7.70) the spectrum S_z of $z(t)$ is a set of Lebesgue measure one.

Proof. The spectrum S_θ of white noise has Lebesgue measure one (Section 7.6) and $S_\theta = S_z$, by Theorem 7.71.   Q.E.D.

Theorem 7.73   In the Wold decomposition (7.70) the spectrum S_y of $y(t)$ is a set of Lebesgue measure zero.

Proof. By hypothesis, $y(t)$ and $z(t)$ are orthogonal, and each subordinated to $x(t)$. From Theorem 7.40 we know that

$$F_{yy}(\nu) = \int_{-0.5}^{\nu} |\Phi_y^{(x)}(\nu)|^2 \, dF_{xx}(\nu)$$

$$F_{yx}(\nu) = \int_{-0.5}^{\nu} \Phi_y^{(x)}(\nu) \, dF_{xx}(\nu)$$

and similarly for F_{zz} and F_{zx}. But because $y(t)$ and $z(t)$ are orthogonal, $R_{yx}(s) = R_{yy}(s) + R_{yz}(s) = R_{yy}(s)$ and so $F_{yx} = F_{yy}$. Hence, by the above equations,

$$|\Phi_y^{(x)}(\nu)|^2 = \Phi_y^{(x)}(\nu), \quad |\Phi_z^{(x)}(\nu)|^2 = \Phi_z^{(x)}(\nu) \tag{7.71}$$

almost everywhere with respect to F_{xx}-measure. Moreover,

$$\Phi_y^{(x)}(\nu) + \Phi_z^{(x)}(\nu) = \frac{dF_{yy}}{dF_{xx}} + \frac{dF_{zz}}{dF_{xx}} = \frac{dF_{xx}}{dF_{xx}} = 1 \tag{7.72}$$

For equations (7.71) and (7.72) to be satisfied, we must have

$$\Phi_y^{(x)}(\nu)\,\Phi_z^{(x)}(\nu) = 0$$

almost everywhere with respect to F_{xx}-measure. Hence the spectra S_y of $y(t)$ and S_z of $z(t)$ do not overlap. That is, $S_x = S_y \cup S_z$ and $S_y \cap S_z = $ null set. But by Theorem 7.72, S_z has Lebesgue measure one. Since S_x at most can have Lebesgue measure one, S_x does have Lebesgue measure one, and S_y Lebesgue measure zero. We note that in the isomorphism between $\mathbf{X}$ and $\mathbf{L}^2(F_{xx})$

$$y(0)\leftrightarrow\Phi_y^{(x)}(\nu) = \Lambda_y(\nu), \quad z(0)\leftrightarrow\Phi_z^{(x)}(\nu) = \Lambda_z(\nu) \qquad (7.73)$$

where $\Lambda_y(\nu)$ is the characteristic function of S_y (i.e. $\Lambda_y(\nu) = 1$ for ν in S_y and $= 0$ for ν not in S_y) and $\Lambda_z(\nu)$ is the characteristic function of S_z.   Q.E.D.

Theorem 7.74   Given a non-singular stationary time-series $x(t)$ with spectral decomposition $x(t) = x_1(t)+x_2(t)+x_3(t)$ and Wold decomposition $x(t) = y(t)+z(t)$, then $y(t) = x_1(t)+x_2(t)$ and $z(t) = x_3(t)$.

Proof.   Because $F_3(\nu)$ is absolutely continuous with respect to Lebesgue measure, $x_3(t)$ is a running average of white noise, by Theorem 7.60. Because $z(t)$ is a running average of white noise, $F_{zz}(\nu)$ is absolutely continuous with respect to Lebesgue measure. Because S_1, S_2, and S_y have Lebesgue measure zero, F_{11}, F_{22}, and F_{yy} are not absolutely continuous with respect to Lebesgue measure. Since $S = S_1 \cup S_2 \cup S_3 = S_y \cup S_z$, we have $S_1 \cup S_2 = S_y$ (which has Lebesgue measure zero) and $S_3 = S_z$ (which has Lebesgue measure one).   Q.E.D.

Theorem 7.75   Given a non-singular stationary time-series $x(t)$, then its spectral power function $F'(\nu)$ exists and is strictly positive, except on a ν-set of Lebesgue measure zero.

Proof.   The spectrum $S_3 = S_z$ has Lebesgue measure one, and so by Property (5) of Theorem 7.50, $F'(\nu)>0$ for ν in S_3.   Q.E.D.

Theorem 7.76   In the Wold decomposition (7.70) we have

$$\theta(t)\leftrightarrow\frac{\Lambda_z(\nu)}{B(\nu)}\,e^{2\pi i\nu t} \qquad (7.74)$$

in the isomorphism between the time-domain $\mathbf{X}$ and the frequency-domain $\mathbf{L}^2(F_{xx})$, where

$$\Lambda_z(\nu) = \begin{cases}0, & \nu\in S_y \\ 1, & \nu\in S_z\end{cases}, \quad B(\nu) = \sum_{s=0}^{\infty} b_s\,e^{-2\pi i\nu s}$$

Proof. Let $\theta(0) \leftrightarrow \Psi(\nu)$ in the isomorphism between $\mathbf{X}$ and $\mathbf{L}^2(F_{xx})$. Then by Property (5) of Theorem 7.21 we have

$$z(0) = \sum_{s=0}^{\infty} b_s \theta(-s) \leftrightarrow \sum_{s=0}^{\infty} b_s \Psi(\nu) e^{-2\pi i \nu s}$$

in the isomorphism. But by equation (7.73), $z(0) \leftrightarrow \Lambda_z(\nu)$, and solving for $\Psi(\nu)$ we get the desired result.   Q.E.D.

A non-singular stationary time-series $x(t)$ is regular if and only if $y(t) = 0$ in its Wold decomposition. Then $x(t)$ is equivalent to $\theta(t)$, and has the one-sided (i.e. $b_s = 0$ for $s < 0$) representation

$$x(t) = \sum_{s=0}^{\infty} b_s \theta(t-s), \quad b_s = ((x(t), \theta(t-s))) \tag{7.75}$$

By equation (7.62) we have

$$x(0) \leftrightarrow \Phi_x^{(\theta)}(\nu) = \sum_{s=0}^{\infty} b_s e^{-2\pi i \nu s} \equiv B(\nu) \tag{7.76}$$

which defines $B(\nu)$. Since

$$\|x(t)\|^2 = \sum_{s=0}^{\infty} |b_s|^2 = R_{xx}(0)$$

is finite, the power series $\sum_0^{\infty} b_s \xi^s$ is analytic within the unit circle $|\xi| < 1$, and its limit value on the boundary of the unit circle is $B(\nu)$ where $\xi = e^{-2\pi i \nu}$ for $|\xi| = 1$. Since $x(t) \neq 0$, the power series is not identically zero, and hence $B(\nu) \neq 0$ almost everywhere for $-0.5 \leqslant \nu \leqslant 0.5$ (i.e. $B(\nu) \neq 0$ for ν in S_θ). Hence by Theorem 7.42, $x(t)$ is equivalent to $\theta(t)$. By equation (7.63), $F'(\nu) = |B(\nu)|^2 > 0$ almost everywhere.

Fundamental results of Kolmogorov (1941) and Wiener (1942) are

Theorem 7.77   If $x(t)$ is regular, then $\sum_{s=0}^{\infty} b_s \xi^s$ has no zeros in the unit circle $|\xi| < 1$, where the b_s are the coefficients of the Wold decomposition of $x(t)$.

Theorem 7.78 $x(t)$ is regular if and only if all the following conditions are true for $-0.5 \leqslant \nu \leqslant 0.5$:

(1) $F_{xx}(\nu)$ is absolutely continuous with respect to Lebesgue measure.

(2) $F'_{xx}(\nu) > 0$, except for a set of Lebesgue measure zero.

(3) $\int_{-0.5}^{0.5} \log F'(\nu)\, d\nu$ is finite (i.e. $> -\infty$, since it is always $< \infty$).

Because condition (1) in Theorem (7.78) is equivalent to $S_1 \cup S_2 = $ null set, we have

Theorem 7.79 $x(t)$ is non-singular if and only if both conditions (2) and (3) are true.

Example 7.70   Consider a time-series $x(t)$ for which conditions (1) and (2) hold, and (3) does not. Then we see that $x(t)$ is singular, and moreover by Theorem 7.61, $x(t)$ is a running average of white noise, with spectrum S_x of Lebesgue measure 1.

7.8   Electric filters, prediction, and filtering

The stochastic process given by the finite linear operation

$$\epsilon(t) = a_0 x(t) + a_1 x(t-1) + \dots + a_n x(t-n) \qquad (7.80)$$

was introduced by Yule (1921, 1927). Equation (7.80) states that the convolution of the *input* $x(t)$ with the constant coefficients $a_0, \dots, a_n$ yields the *output* $\epsilon(t)$. Because equation (7.80) involves only a finite number of constant coefficients, and because only inputs at or prior to the time t (and no future inputs) are required to compute the output at time t, the linear operation (7.80) can be realized physically by a linear *electric filter* with pulsed data (i.e. a pulsed filter), or by a (high-speed) digital computer. A pulsed filter is a transmission device which is supplied with input data at specified equally-spaced times and, in response, furnishes output data at the same moments. Hurewicz (1947) shows that a pulsed filter (7.80) is equivalent to the combination of a clamping device together with a linear filter of the continuous type.

The *transient-response function* of a filter is the response of the filter to a single unit input at $t = 0$ (i.e. $x(0) = 1, x(t) = 0$ for $t \neq 0$). Hence we see that the constant coefficients $a_0, a_1, \dots, a_n$ are the transient-response function of the filter (7.80). A filter is essentially an instrument for the separation of different frequency ranges. To see this, let the input $x(t)$ to the filter (7.80) be the (complex) sinusoidal wave $x(t) = e^{2\pi i \nu t}$. Then we see that the output is necessarily a sinusoidal wave of the same frequency ν but generally will differ in amplitude and phase, and so may be written $\epsilon(t) = A(\nu) e^{2\pi i \nu t}$. The function $A(\nu)$, which is a complex function expressing the amplitude and phase changes, is called the *transfer function*, or the *filter characteristics*, or the *frequency-response function*, or the *gain*, of the filter. Substituting into equation (7.80), we have $A(\nu) = \sum_0^n a_s e^{-2\pi i \nu s}$. The absolute value $|A(\nu)|$ is called the *absolute gain*, and $|A(\nu)|^2$ the *power transfer function*.

If $x(t)$ is a stationary process, then $\epsilon(t)$ given by equation (7.80) (N.B. n is finite!) is called a *Yule process*. Using the spectral representation of $x(t)$, we have

$$\epsilon(t) = \sum_0^n a_s x(t-s) = \int_{-0.5}^{0.5} e^{2\pi i \nu t} A(\nu) \, dx(0) E_x(\nu) \qquad (7.81)$$

where $A(\nu)$ is the transfer function. The Yule process $\epsilon(t)$ is stationary, and subordinated to $x(t)$. In the isomorphism between $\mathbf{X}$ and $\mathbf{L}^2(F_{xx})$, we have

$$\epsilon(0) = \sum_0^n a_s x(-s) \leftrightarrow A(\nu) = \sum_0^n a_s e^{-2\pi i \nu s}$$

From equation (7.81) we obtain the autocovariance

$$R_{\epsilon\epsilon}(s) = \int_{-0.5}^{0.5} e^{2\pi i \nu s} |A(\nu)|^2 \, dF_{xx}(\nu) \qquad (7.82)$$

so that
$$F_{\epsilon\epsilon}(\nu) = \int_{-0.5}^{\nu} |A(\nu)|^2 \, dF_{xx}(\nu) \qquad (7.83)$$

These equations show the filtering action of the transfer function. In particular, we see that the input spectral power $dF_{xx}(\nu)$ is multiplied by the power transfer function $|A(\nu)|^2$ to yield the output spectral power $dF_{\epsilon\epsilon}(\nu)$.

Now, let $\epsilon(t)$ be the limit ($\Rightarrow$) of a sequence of Yule processes $\sum_0^n a_s x(t-s)$ as $n \to \infty$ where $x(t)$ is a stationary time-series. Then $\epsilon(t)$ is stationary and subordinated to $x(t)$. By Theorem 7.40, we have $\epsilon(0) \leftrightarrow \Phi_\epsilon^{(x)}(\nu)$, which serves as the definition of the transfer function; i.e. $A(\nu) \equiv \Phi_\epsilon^{(x)}(\nu)$. Then, with this $A(\nu)$, $\epsilon(t)$, $R_{\epsilon\epsilon}(s)$, $F_{\epsilon\epsilon}(\nu)$ are equal respectively to the right-hand sides of equations (7.81), (7.82), (7.83).

Let $x(t)$ be a non-singular, stationary time-series. The problem of the linear, least-squares *prediction* of $x(t+\alpha)$ from the whole past $x(s), s \leqslant t$, is the problem of finding an element $\hat{x}(t+\alpha)$ in $\mathbf{X}(t)$ such that the distance $\|x(t+\alpha) - \hat{x}(t+\alpha)\|$ is a minimum. Clearly the solution is

$$\hat{x}(t+\alpha) = x(t+\alpha) P_{\mathbf{X}(t)}$$

which is, since $\theta(s) P_{\mathbf{X}(t)} = \theta(s)$ if $s \leqslant t$, and $= 0$ if $s > t$,

$$\hat{x}(t+\alpha) = y(t+\alpha) P_{\mathbf{X}(t)} + \sum_{s=0}^{\infty} b_s \theta(t+\alpha-s) P_{\mathbf{X}(t)}$$

$$= y(t+\alpha) + \sum_{s=\alpha}^{\infty} b_s \theta(t+\alpha-s) \qquad (7.84)$$

a result due to Wold (1938). This equation exhibits the least-squares prediction operator. We have, by equations (7.73) and (7.74),

$$\hat{x}(0+\alpha) = y(0) + \sum_{s=\alpha}^{\infty} b_s \theta(\alpha-s) \leftrightarrow \Lambda_y(\nu) + \frac{\Lambda_z(\nu)}{B(\nu)} \sum_{s=\alpha}^{\infty} b_s e^{-2\pi i \nu (s-\alpha)}$$

Hence the expression on the right is the transfer function of the optimum linear predictor, a result due to Wiener (1942) and Kolmogorov (1941). That is, the transfer function is equal to one for $\nu \in S_y$ (i.e. waves of "pure" frequencies are passed without modulation). The transfer function is equal to

$$\frac{1}{B(\nu)} \sum_{t=0}^{\infty} b_{t+\alpha}\, e^{-2\pi i \nu t} \tag{7.85}$$

for $\nu \in S_z$ (i.e. waves of "non-pure" frequencies are modulated, i.e. their spectral power density is multiplied by the factor (7.85)). The "non-pure" frequencies have their spectral power density spread over the whole spectral band,

i.e. $F'_{zz}(\nu) > 0$ for ν in $S_z = \{-0{\cdot}5 \leqslant \nu \leqslant 0{\cdot}5,$ except for $S_y\}$

The "pure" frequencies have their power so concentrated that it is crowded in the set S_y of Lebesgue measure zero. Since

$$\| y(t+\alpha) - \hat{y}(t+\alpha) \| = 0$$

the past of $y(t)$ determines its future for an infinite time (i.e. for arbitrary *prediction lead* $\alpha > 0$), so the singular component $y(t)$ is *deterministic*. On the other hand, the future of the non-singular time-series $x(t)$ is not completely determined by a linear operation on its past. That is, we have the (non-null) *prediction error*

$$x(t+\alpha) - \hat{x}(t+\alpha) = \sum_{s=0}^{\alpha-1} b_s\, \theta(t+\alpha-s) \tag{7.86}$$

and the positive *mean-square prediction error*

$$\| x(t+\alpha) - \hat{x}(t+\alpha) \| = \sum_{s=0}^{\alpha-1} |b_s|^2 > 0$$

since $b_0 > 0$.

Now suppose that $x(t)$, $m(t)$, and $n(t)$ form a multiple stationary time-series, and $x(t)$ is additively composed of the message $m(t)$ and the noise $n(t)$; i.e. $x(t) = m(t) + n(t)$. We assume that these three time-series are regular. Hence they have absolutely continuous spectral distribution functions. The problem of linear least-squares *filtering* (using the whole past $x(s), s \leqslant t$) is the problem of finding an element $\hat{m}(t+\alpha)$ in $\mathbf{X}(t)$ such that the distance $\| m(t+\alpha) - \hat{m}(t+\alpha) \|$ is a minimum. To find the solution, we use the Wold decomposition

$$x(t) = \sum_{s=0}^{\infty} b_s\, \theta(t-s)$$

Since $\theta(t)$, $-\infty < t < \infty$, is a maximal orthonormal set in $\mathbf{X}$, the projection of the message $m(t)$ on the time-domain $\mathbf{X}$ has the Fourier series representation

$$m(t)\, P_{\mathbf{X}} = \sum_{s=-\infty}^{\infty} c_s\, \theta(t-s), \quad c_s = ((m(t), \theta(t-s))) \tag{7.87}$$

Hence the least-squares approximation to $m(t+\alpha)$ is

$$\hat{m}(t+\alpha) = m(t+\alpha)P_{\mathbf{X}(t)} = m(t+\alpha)P_{\mathbf{X}}P_{\mathbf{X}(t)}$$

since projection on $\mathbf{X}$ followed by projection of $\mathbf{X}(t)$ is the same as projection on $\mathbf{X}(t)$ directly. Using equation (7.87), we have

$$\hat{m}(t+\alpha) = \sum_{s=-\infty}^{\infty} c_s\,\theta(t+\alpha-s)\,P_{\mathbf{X}(t)} = \sum_{s=\alpha}^{\infty} c_s\,\theta(t+\alpha-s) \qquad (7.88)$$

The transfer function is the expression on the right of

$$\hat{m}(0+\alpha) \leftrightarrow \frac{1}{B(v)}\sum_{t=0}^{\infty} c_{t+\alpha}\,e^{-2\pi i v t} \qquad (7.89)$$

a result due to Wiener (1942). This is the optimum transfer function for separating message from message-plus-noise by a linear operation. If α is positive, it is called the *lead* of the filter; if α is negative, then $-\alpha$ is called the *lag*. The classic papers on the intuitive content of Wiener's work as applied to electric networks are Blackman, Bode, and Shannon (1946) and Bode and Shannon (1950).

Because the transfer functions (7.85) and (7.89) in general involve an infinite number of coefficients, in this sense they are not necessarily physically realizable by an electric filter, or high-speed digital computer. Nevertheless, because $\hat{x}(t+\alpha)$ and $\hat{m}(t+\alpha)$ are elements of the present-and-past time-domain $\mathbf{X}(t)$, and since every element in $\mathbf{X}(t)$ is a limit ($\Rightarrow$) of a sequence of Yule processes (7.81) (see Section 4.6), we can approximate $\hat{x}(t+\alpha)$ and $\hat{m}(t+\alpha)$ by Yule processes which are physically realizable.

7.9 Wavelet theory

The Wold decomposition reaches its full beauty in physical applications. This application to wavelet theory was developed to solve certain problems in the interpretation of geophysical recordings of seismic wave propagation through the earth (Robinson, 1954, 1957).

Suppose that a real-valued time-series $x(t)$ is additively composed of many overlapping wavelets which arrive as time progresses. We assume that each wavelet has the same shape c_s but that each is amplified by a random factor. In keeping with physical processes, we require the wavelets to be one-sided ($c_s = 0$ for $s < 0$) and to damp out

$$\left(\sum_{0}^{\infty} c_s^2 < \infty\right)$$

Let the amplification factor of the wavelet which arrives at time t be $\theta(t)$. Then we see that the wavelet with arrival time t has zero amplitude for times prior to t, has amplitude $c_0\,\theta(t)$ at time t, has amplitude $c_1\,\theta(t)$ at time $t+1$, has amplitude $c_2\,\theta(t)$ at time $t+2$, and so forth. We assume that amplification factors $\theta(t)$, $-\infty < t < \infty$, are uncorrelated variates with the same variance, so by centring and scaling, $\theta(t)$ becomes white noise.

Consider the time instant t. The wavelet which arrives at time t contributes $c_0 \theta(t)$ to $x(t)$. The wavelet which arrived at time $t-1$ contributes $c_1 \theta(t-1)$ to $x(t)$. The wavelet with arrival time $t-2$ contributes $c_2 \theta(t-2)$ to $x(t)$, and so forth. Because $x(t)$ is the sum of all these contributions, we have

$$x(t) = c_0 \theta(t) + c_1 \theta(t-1) + c_2 \theta(t-2) + \ldots = \sum_{s=0}^{\infty} c_s \theta(t-s) \qquad (7.90)$$

In this equation we see that the basic wavelet shape c_s represents the "dynamics", and the amplification factors, or impulses, $\theta(t)$, the "random" nature, of the time-series $x(t)$.

Wavelet problem   Given a fixed realization of the time-series $x(s)$, for $s \leqslant t$, find (1) the basic wavelet shape c_s, and (2) the amplification factor $\theta(s)$ for every $s \leqslant t$.

That is, we are given the observed past values of a time-series that is the resultant of many overlapping and interfering wavelets which arrive at various times with random amplifications. We are to pick out each wavelet as a distinct entity and identify it by its arrival time.

For the wavelet problem to have a unique solution, a stability condition must be imposed on the wavelet shape. Suppose that the wavelet shape damps to zero (i.e. $c_s = 0$ for $s > n$). Form the homogeneous difference equation

$$c_0 \epsilon(t) + c_1 \epsilon(t-1) + \ldots + c_n \epsilon(t-n) = 0$$

from the wavelet shape. For an arbitrary set of initial values $\epsilon(0), \epsilon(1), \ldots, \epsilon(n-1)$, the sequence $\epsilon(n), \epsilon(n+1), \ldots$ may be generated by recursive deductions from this difference equation, and this sequence will form a general solution. A necessary and sufficient condition that $\sum_{0}^{\infty} |\epsilon(t)|^2$ converges is that the magnitude of the roots of the characteristic equation

$$c_0 + c_1 \xi + \ldots + c_n \xi^n = 0$$

be greater than one (i.e. that $\sum_{0}^{n} c_s \xi^s$ has no zeros for $|\xi| \leqslant 1$). In this case the general solution of the homogeneous difference equation is a damped oscillation, and the difference equation is said to be stable. In case there are any roots with magnitude equal to one, then the difference equation is semi-stable; whereas if there are any roots with magnitude less than one, the difference equation is unstable. Accordingly, if $c_s = 0$ for $s > n$, we shall impose the stability requirement that $\sum_{0}^{n} c_s \xi^s$ (which is analytic in the whole plane) have no zeros for $|\xi| < 1$.

In the general case, we shall impose the stability requirement that $\sum_0^\infty c_s \xi^s$ be analytic for $|\xi| < 1$ and have no zeros for $|\xi| < 1$. With this stability requirement, $x(t)$ is regular and representation (7.90) is its Wold decomposition (7.75) where $c_s = b_s$.

The solution of the wavelet problem proceeds according to the following steps.

(1) Given the observed time-series $x(s)$, $s \leqslant t$, the *first step* is to average out (i.e. destroy) the random, uncorrelated elements $\theta(s)$, $s \leqslant t$, and yet preserve as much information about the wavelet shape as possible. Phase information cannot be saved; but we can obtain the spectral energy information about b_s. To do this, we compute the autocovariance $R(u)$ from $x(t)$ by the time-average (7.34) in which $m = t - u$, and $n \to -\infty$. Because

$$R(u) = \left(\left(\sum_0^\infty b_s \, \theta(t+u-s), \, \sum_0^\infty b_s \, \theta(t-s) \right) \right) = \sum_{s=0}^\infty b_{s+u} \, b_s$$

we see that the autocovariance contains no information about individual values of $\theta(s)$. From $R(u)$ we obtain the spectral power function $F'(u)$ by Theorem 7.20. By equations (7.63) and (7.76) we have

$$|B(v)| = \left| \sum_{s=0}^\infty b_s \, \mathrm{e}^{-2\pi i v s} \right| = \sqrt{\{F'(v)\}}$$

so we may compute the absolute gain $|B(v)|$ of the transfer function $B(v)$ by taking the square root of the spectral power function. That is, the spectral energy function $|B(v)|^2$ for the wavelet shape is equal to the spectral power function $F'(v)$ of the time-series.

(2) The *second step* is to compute $B(v)$ from $|B(v)|$, and thus obtain the wavelet shape b_s. Equivalently, if we write

$$B(v) = |B(v)| \, \mathrm{e}^{i\phi \, (v)} \tag{7.91}$$

the second step is to find the phase function $\phi(v)$. We recognize this step to be the problem of the factorization of the spectral power function, which was first solved by Wold (1938) in the case of physically realizable transfer functions, and by Kolmogorov (1939, 1941) and Wiener (1942) in the transcendental case. Because $x(t)$ is regular, condition (3) of Theorem 7.78 is true. As a consequence, $\log \sqrt{\{F'(v)\}}$, which is an even real function of v, may be expanded in a real, symmetric Fourier cosine series

$$\log \sqrt{\{F'(v)\}} = \tfrac{1}{2} \log F'(v) = \sum_{-\infty}^\infty \delta_s \cos 2\pi v s = \delta_0 + 2 \sum_1^\infty \delta_s \cos 2\pi v s \tag{7.92}$$

where the Fourier coefficients δ_s are given by

$$\delta_s = \delta_{-s} = \int_0^{0 \cdot 5} \cos 2\pi v s \, \log F'(v) \, dv \tag{7.93}$$

By taking the logarithm of each side of equation (7.91), and utilizing equation (7.92), we have

$$\log B(v) = \log \sqrt{\{F'(v)\}} + i\phi(v) = \delta_0 + 2\sum_1^\infty \delta_s \cos 2\pi vs + i\phi(v) \qquad (7.94)$$

Because $x(t)$ is regular, $\sum_0^\infty b_s \xi^s$ has no singularities or zeros within the unit circle. Hence $\log \sum_0^\infty b_s \xi^s$ is analytic within the unit circle, and consequently has the power series representation

$$\log \sum_0^\infty b_s \xi^s = \beta_0 + 2\sum_1^\infty \beta_s \xi^s \qquad (7.95)$$

for $|\xi| < 1$. As $|\xi|$ approaches 1, this power series has

$$\log B(v) = \beta_0 + 2\sum_1^\infty \beta_s e^{-2\pi i vs}$$

$$= \beta_0 + 2\sum_1^\infty \beta_s \cos 2\pi vs - 2i\sum_1^\infty \beta_t \sin 2\pi vs \qquad (7.96)$$

as its limit values on the unit circle $|\xi| = 1$ where $\xi = e^{-2\pi i v}$.

Comparing equations (7.94) and (7.96), we see that $\delta_s = \beta_s$ and

$$\phi(v) = -2\sum_1^\infty \delta_s \sin 2\pi vs = -2\sum_1^\infty \sin 2\pi vs \int_0^{0\cdot5} \cos 2\pi vs \log F'(v)\, dv$$

This equation shows that the phase function $\phi(v)$ is the Hilbert transform of $\log \sqrt{\{F'(v)\}}$, and thus $\phi(v)$ is uniquely determined by $x(s)$, $s \leqslant t$. Substituting $\phi(v)$ into equation (7.91), we obtain the desired transfer function $B(v)$, from which we obtain the wavelet shape by

$$b_s = \int_{-0\cdot5}^{0\cdot5} B(v)\, e^{2\pi i vs}\, dv$$

Alternatively, we may write equation (7.95) as

$$\sum_0^\infty b_s \xi^s = \exp\left(\delta_0 + 2\sum_1^\infty \delta_s \xi^s\right)$$

which gives the wavelet shape b_s in terms of the δ_s of equation (7.93). In particular, this equation gives $b_0 = e^{\delta_0}$ which shows the reason for condition (3) in Theorem 7.78.

(3) The *last step* consists of removing the wavelet shape b_s from $x(t)$, thereby leaving, as a residual, the random components $\theta(t)$. The wavelet shape b_s uniquely determines the least-squares prediction operator, as seen by equation (7.85). If we let the prediction lead $\alpha = 1$, then equation (7.86) shows that if we predict $x(s)$ from $\mathbf{X}(s-1)$ the prediction error is $b_0\theta(s)$. We may repeat this process for all $s \leqslant t$,

thereby obtaining the amplification factors $\theta(s)$ for $s \leqslant t$. Thus we have obtained the theoretical solution to the wavelet problem.

The practical solution of the wavelet problem in which we only have a finite section $x(1), x(2), \ldots, x(N)$ of an observed time-series involves statistical estimation. The method we proposed uses first the Gauss method of least squares to obtain the prediction operator (with $n < N$)

$$\hat{x}(t+1) = c + \sum_{s=0}^{n} k_s x(t-s) \tag{7.97}$$

where the constant c appears if the section of $x(t)$ does not have a zero average. Then, provided the difference equation is stable, the wavelet shape is computed by

$$b_{t+1} = \sum_{s=0}^{n} k_s b_{t-s}, \quad (t = 0, 1, 2, \ldots) \tag{7.98}$$

where we let $b_t = 0$ for $t < 0$, and b_0^2 equal to the variance of the prediction errors $w(t) = x(t) - \hat{x}(t)$. That is, the wavelet shape b_t for $t > 0$ is determined by successive step-by-step predictions from its own past values, where its initial values are $b_t = 0$ for $t < 0$ and b_0 for $t = 0$. The amplitude factors $\theta(t)$ are the normalized prediction errors: $\theta(t) = w(t)/b_0$.

The coefficients k_s are *autoregressive* coefficients of the time-series. The operator (7.97) cannot predict from past values of the time-series the arrival of a new wavelet, and thus a prediction error $b_0 \theta(t)$ is introduced at the arrival time t of each wavelet. Nevertheless, for times subsequent to the arrival time of the wavelet, the prediction operator can perfectly predict the remaining values of the wavelet, as seen by equation (7.98). Hence the prediction error $b_0 \theta(t)$ indicates the arrival of a new wavelet at time t with initial amplitude $b_0 \theta(t)$. For this reason $\theta(t)$ is called the innovation at time t.

The wavelet method provides the estimate of the spectral power function given by

$$[\text{Estimate of } F'(\nu)] = \left| \sum_{s \geqslant 0} b_s e^{-2\pi i \nu s} \right|^2$$

where the Wold coefficients (i.e. wavelet shape) are computed from

$$b_0 = \| w(t) \|$$
$$b_1 = \| w(t) \| k_0$$
$$b_2 = \| w(t) \| (k_0^2 + k_1)$$
$$b_3 = \| w(t) \| (k_0^3 + 2k_0 k_1 + k_2), \ldots$$

and where the autoregressive coefficients $k_1, k_2, \ldots, k_n \, (n < N)$ are computed from the observed time-series $x(0), x(1), \ldots, x(N)$ by the Gauss method of least squares.

So far we have confined ourselves to a section of a time-series which we assumed to be stationary. The optimum prediction operator transformed this section into the uncorrelated prediction errors $b_0 \theta(t)$, the mean-square value of which is a minimum. As we have seen, the prediction operator cannot predict from past values of the time-series the initial arrival of a new wavelet, and thus a prediction error, in the form of an impulse $b_0 \theta(t)$, is introduced at the arrival time of each wavelet. Nevertheless, for times subsequent to the arrival time of the wavelet, the prediction operator can perfectly predict the wavelet, and thereby yield zero error of prediction for this wavelet.

Suppose now we apply the prediction operator to another section of the time-series which has dynamics different from the original section. The prediction operator, applied to the new section, will encounter wavelets of a different shape, and consequently the prediction errors will no longer be impulses, but instead will be transient time-functions. As a result the mean-square prediction error will be much greater. This procedure provides a method for the detection of various types of wavelets comprising a non-stationary time-series. In the case of multiple time-series, multiple prediction operators can be used, which take into account coherency relationships. These multiple operators yielded highly successful results in the analysis of seismic records (Wadsworth, Robinson, Bryan, and Hurley, 1953; Robinson, 1953, 1953 a). The high-speed Whirlwind digital computer at the Massachusetts Institute of Technology was used.

REFERENCES

BARTLETT, M. S. (1946). On the theoretical specification and sampling properties of autocorrelated time-series. *J. Roy. Statist. Soc.*, **B, 8**, 27.

BARTLETT, M. S. (1950). Periodogram analysis and continuous spectra. *Biometrika*, **37**, 1.

BARTLETT, M. S. (1955). *Stochastic Processes*. Cambridge Univ. Press.

BAWLY, G. M. (1936). Über einige Verallgemeinerungen der Grenzwertsätze der Wahrscheinlichkeitsrechnung. *Rec. Math. (Mat. Sbornik) N.S.*1, **43**, 917.

BESICOVITCH, A. S. (1932). *Almost Periodic Functions*. Cambridge Univ. Press.

BIRKOFF, G. and MACLANE, S. (1948). *Modern Algebra*, N.Y.

BLACKMAN, R. B., BODE, H. W., and SHANNON, C. E. (1946). *Monograph on Data Smoothing and Prediction in Fire Control Systems*. NDRC Report.

BLACKMAN, R. B. and TUKEY, J. W. (1958). The measurement of power spectra from the point of view of communications engineering. *Bell Sys. Tech. J.*, **37**, 185.

BOCHNER, S. (1955). *Harmonic Analysis and the Theory of Probability*. U. of Cal. Press, Berkeley.

BODE, H. W. and SHANNON, C. E. (1950). A simplified derivation of linear least square smoothing and prediction theory. *Proc. Inst. Rad. Eng.*, **38**, 417.

CHAMPERNOWNE, D. G. (1948). Sampling theory applied to autoregressive schemes. *J. Roy. Statist. Soc.*, **B, 10**, 204.

COOKE, R. G. (1950). *Infinite Matrices and Sequence Spaces*. Macmillan, London.

COOKE, R. G. (1953). *Linear Operators*. Macmillan, London.

CRAMÉR, H. (1940). On the theory of stationary random processes. *Ann. Math.*, **41**, 215.

CRAMÉR, H. (1942). On harmonic analysis in certain functional spaces. *Ark. Mat. Astr. Fys.*, **28B**, No. 12.

CRAMÉR, H. (1946). *Mathematical Methods of Statistics*. Princeton Univ. Press.

DOOB, J. L. (1953). *Stochastic Processes*. Wiley, N.Y.

FELLER, W. (1950). *Probability Theory*. Wiley, N.Y.

FISCHER, E. (1907). Sur la convergence en moyenne. *Comp. Rend. Acad. Sci. Paris*, **144**, 1022 and 1148.

FRÉCHET, M. (1937). *Généralités sur les Probabilités. Variables Aléatoires*. Paris.

GIBBS, W. (1902). *Elementary Principles of Statistical Mechanics*. Yale Univ. Press.

GNEDENKO, B. V. (1939). On the theory of limit theorems for sums of independent random variables (Russian). *Izvestiya Akad. Nauk. S.S.S.R., Ser. Mat.*, 181 and 643.

GNEDENKO, B. V. (1944). Limit theorems for sums of independent random variables. (Russian). *Uspekhi Mat. Nauk.*, **10**, 115. (English translation, No. 45, Am. Math. Soc., N.Y.)

GNEDENKO, B. V. and KOLMOGOROV, A. N. (1954). *Limit Distributions for Sums of Independent Random Variables* (Russian). English trans., Addison-Wesley, Cambridge, Mass.

GRENANDER, U. (1950). Stochastic processes and statistical inference. *Ark. Mat.*, **1**, 195.

GRENANDER, U. and ROSENBLATT, M. (1957). *Statistical Analysis of Stationary Time Series*. Wiley, N.Y.

HALMOS, P. R. (1950). *Measure Theory*. Van Nostrand, Princeton.

HALMOS, P. R. (1951). *Hilbert Space*. Chelsea, N.Y.

HALMOS, P. R. (1958). *Finite-Dimensional Vector Spaces*. Van Nostrand, Princeton.

HAMBURGER, H. L. and GRIMSHAW, M. E. (1956). *Linear Transformations*. Cambridge Univ. Press.

HILBERT, D. (1912). *Grundzüge einer allgemeinen Theorie der linearen Integralgleichungen*. Leipzig.

HOPF, E. (1937). *Ergodentheorie*. Springer, Berlin (also Chelsea, N.Y.).

HUREWICZ, W. (1947). Filters and servo systems with pulsed data. *M.I.T. Rad. Lab. Ser.*, **25**, 231.

KENDALL, M. G. (1945). On the analysis of oscillatory time-series. *J. Roy. Statist. Soc.*, **108**, 93.

KENDALL, M. G. (1946). *Contributions to the Study of Oscillatory Time-series*. Cambridge Univ. Press.

KENDALL, M. G. (1957). *A Course in Multivariate Analysis.* Griffin's Statistical Monographs and Courses, **2**. Charles Griffin and Co., London.

KENDALL, M. G. and STUART, A. (1958). *The Advanced Theory of Statistics*, vol. I. Vol. II (M. G. KENDALL), 1955 ; revision by Kendall and Stuart in preparation. Charles Griffin and Co., London.

KHINTCHINE, A. JA. (1933). *Asymptotische Gesetze der Wahrscheinlichkeitsrechnung.* Springer, Berlin (also Chelsea, N.Y.).

KHINTCHINE, A. JA. (1934). Korrelationstheorie der stationären stochastischen Prozesse. *Math. Ann.*, **109**, 604.

KHINTCHINE, A. JA. (1937). Zur Theorie der unbeschränkt teilbaren Verteilungsgesetze. *Rec. Math. (Mat. Sbornik) N.S.*2, **44**, 79.

KHINTCHINE, A. JA. (1948). *Mathematical Foundations of Statistical Mechanics* (Russian). English trans., Dover, N.Y.

KHINTCHINE, A. JA. (1957). *Mathematical Foundations of Information Theory* (Russian). English trans., Dover, N.Y.

KOLMOGOROV, A. N. (1932). Sulla forma generale di un processo stocastico omogeneo. *R.C. Acad. Lincei.*, (**6**), **15**, 805 and 866.

KOLMOGOROV, A. N. (1933). *Grundbegriffe der Wahrscheinlichkeitsrechnung.* Springer, Berlin. (English trans., Chelsea, N.Y.)

KOLMOGOROV, A. N. (1939). Sur L'interpolation et extrapolation des suites stationnaires. *C.R. Acad. Sci. Paris*, **208**, 2043.

KOLMOGOROV, A. N. (1941). Stationary sequences in Hilbert space. (Russian). *Bull. Moscow State Univ., Math.*, **2**. (Spanish trans., *Trab. Estad.*, **4**, 55 and 243.)

KOLMOGOROV, A. N. (1941a). Interpolation und Extrapolation von stationären zufälligen Folgen (Russian with German summary). *Bull. Acad. Sci. U.R.S.S., Ser. Math.*, **5**, 3.

KOLMOGOROV, A. N. and FOMIN, S. V. (1957). *Functional Analysis* (Russian). English trans., Graylock, Rochester, N.Y.

KOOPMAN, B. O. (1931). Hamiltonian systems and linear transformations in Hilbert space. *Proc. Nat. Acad. U.S.*, **17**, 315.

LÉVY, P. (1937). *Théorie de L'Addition des Variables Aléatoires.* Gauthier-Villars, Paris.

LÉVY, P. (1948). *Processus Stochastiques et Mouvement Brownien.* Paris.

LOÈVE, M. (1955). *Probability Theory.* Van Nostrand, Princeton, N.J.

McMILLAN, B. (1953). The basic theorems of information theory. *Ann. Math. Statist.*, **24**, 196.

MOYAL, J. E. (1949). Stochastic processes and statistical physics. *J. Roy. Statist. Soc.*, **B, 11**, 150.

MUNROE, M. E. (1953). *Measure and Integration.* Addison-Wesley, Cambridge, Mass.

QUENOUILLE, M. H. (1947). A large-sample test for the goodness-of-fit of autoregressive schemes. *J. Roy. Statist. Soc.*, **110**, 123.

QUENOUILLE, M. H. (1957). *The Analysis of Multiple Time-Series.* Griffin's Statistical Monographs and Courses, **1**. Charles Griffin and Co., London.

QUENOUILLE, M. H. (1958). *Fundamentals of Statistical Reasoning.* Griffin's Statistical Monographs and Courses, **3**. Charles Griffin and Co., London.

RIESZ, F. (1907). Sur les systèmes orthogonaux de fonctions. *C.R. Acad. Sci. Paris*, **144**, 615, 734.

RIESZ, F. and SZ.-NAGY, B. (1952). *Leçons d'Analyse Fonctionelle.* English trans., F. Ungar, N.Y.

ROBINSON, E. A. (1953). *Linear Operator Study of a Seismic Profile.* M.I.T. Geophysical Analysis Group, **4**, Cambridge, Mass.

ROBINSON, E. A. (1953a). *On the Theory and Practice of Linear Operators in Seismic Analysis.* M.I.T. Geophys. Anal. Group, **5**, Cambridge, Mass.

ROBINSON, E. A. (1954). *Predictive Decomposition of Time Series with Applications to Seismic Exploration.* M.I.T. Geophys. Anal. Group, **7**, Cambridge, Mass.

ROBINSON, E. A. (1957). Predictive decomposition of seismic traces. *Geophysics*, **22**, 767.

ROBINSON, E. A., SIMPSON, S. M., and SMITH, M. K. (1954). *Further Research on Linear Operators in Seismic Analysis.* M.I.T. Geophys. Anal. Group, **6**, Cambridge, Mass.

SAVAGE, L. J. (1954). *The Foundations of Statistics.* Wiley, N.Y.

SHANNON, C. E. (1948). A mathematical theory of communication. *Bell Sys. Tech. J.*, **27**, 379 and 623.

SHANNON, C. E. (1949). Communication in the presence of noise. *Proc. Inst. Rad. Eng.*, **37**, 10.

SIMPSON, S. M. (1955). *Linear Operators and Seismic Noise*. M.I.T. Geophys. Anal. Group, **9**, Cambridge, Mass.

SLUTSKY, E. (1927). The summation of random causes as the source of cyclic processes (Russian with English summary). *Prob. Ec. Cond.*, Inst. Ec. Conj. Moscow.

STONE, M. H. (1929 ; 1930). Linear transformations in Hilbert space. *Proc. Nat. Acad. U.S.*, **15**, 198 and 423 ; **16**, 172.

STONE, M. H. (1932). *Linear Transformations in Hilbert Space*. Am. Math. Soc. Colloquium Publ., **15**, N.Y.

TUKEY, J. W. (1949). The sampling theory of power spectrum estimates. *Sym. Appl. Autocor. Anal. Phys. Prob.*, Oceanographic Inst., Woods Hole, Mass.

TUKEY, J. W. and HAMMING, R. W. (1949). *Measuring Noise Color*. Bell Tel. Lab., Murray Hill, N.J.

VON NEUMANN, J. (1929). Eigenwerttheorie Hermitescher Funktionaloperatoren. *Math. Ann.*, **102**, 49.

VON NEUMANN, J. (1931). Uber Funktionen von Funktionaloperatoren. *Ann. Math.* (2), **32**, 191.

VON NEUMANN, J. (1932). *Mathematische Grundlagen der Quantenmechanik*. Springer, Berlin. (English trans., Princeton Univ. Press.)

VON NEUMANN, J. (1932a). Proof of the quasiergodic hypothesis. *Proc. Nat. Acad. U.S.*, **18**, 70.

VON NEUMANN, J. (1933). *Functional Operators*. Inst. Adv. Study, Princeton, N.J. (Reprinted, Princeton Univ. Press, *Ann. Math. Studies*, **21, 22**.)

WADSWORTH, G. P. and BRYAN, J. G. (1958). *Elementary Statistics*. M.I.T. Math. Dept., Cambridge, Mass.

WADSWORTH, G. P., ROBINSON, E. A., BRYAN, J. G., and HURLEY, P. M. (1953). Detection of reflections on seismic records by linear operators. *Geophysics*, **18**, 539.

WEYL, H. (1909). Über die Konvergenz von Reihen. *Math. Ann.*, **67**, 225.

WEYL, H. (1928). *Gruppentheorie und Quantenmechanik*. English trans., Dover, N.Y.

WHITTLE, P. (1951). *Hypothesis Testing in Time-series Analysis*. Almqvist and Wiksells, Uppsala.

WHITTLE, P. (1953). The analysis of multiple stationary time series. *J. Roy. Statist. Soc.*, **B, 15**, 125.

WHITTLE, P. (1954). Some recent contributions to the theory of stationary processes. Appendix 2 of 1954 edition of WOLD (1938).

WIENER, N. (1923). Differential space, *M.I.T. J. Math. Phys.*, **2**, 131.

WIENER, N. (1930). Generalized harmonic analysis. *Acta Math.*, **55**, 117.

WIENER, N. (1942). *Stationary Time Series*. M.I.T. N.D.R.C. Report, Cambridge, Mass. (Reprinted, Wiley, N.Y.)

WIENER, N. (1948). *Cybernetics*. M.I.T., Cambridge, Mass., and Wiley, N.Y.

WINTNER, A. (1929). Zur Theorie der beschränkten Bilinearformen. *Math. Zeit.*, **30**, 228.

WOLD, H. (1938). *Stationary Time Series*. Almqvist and Wiksells, Uppsala. (2nd edn, 1954.)

WOLD, H., in association with JURÉEN, L. (1953). *Demand Analysis*. Wiley, N.Y.

YULE, G. U. (1921). On the time correlation problem. *J. Roy. Statist. Soc.*, **84**, 497.

YULE, G. U. (1927). On a model of investigating periodicities in disturbed series with special reference to Wolfer's sunspot numbers. *Phil. Trans.*, **A, 226**, 267.

YULE, G. U. and KENDALL, M. G. (1937, 14th edn, 1958). *An Introduction to the Theory of Statistics*. Charles Griffin and Co., London.

CHAPTER 8

PREDICTIVE DECOMPOSITION
INTO MARKOV AND PASSIVE COMPONENTS[1]

Enders A. Robinson[2]

Let $x(t)$ be a stationary time series with a known rational spectral density that does not vanish for any real frequency. The least-squares prediction of $x(t + \epsilon)$, $\epsilon > 0$, is decomposed into two components: the Markov component and the passive component. The Markov component is a weighted average of $x(t)$ and its derivatives, all evaluated at the present time instant t . The passive component is a convolution integral involving the whole past $x(\tau)$, $\tau \le t$, of the time series. The methods of electric filter theory are used throughout the paper.

---

1. The writing of this paper was sponsored by the United States Army under Contract No. DA-11-022-ORD-2059, and was done at the Mathematics Research Center, United States Army, Madison, Wisconsin.

2. Mathematics Research Center, U. S. Army, and Mathematics Department, University of Wisconsin, Madison, Wisconsin.

§1. LINEAR FILTERS AND STATIONARY TIME SERIES

The input voltage $y(t)$ and the output voltage $x(t)$ of a four-terminal, passive, lumped-constant, linear <u>filter</u> are related by the differential equation (with real coefficients, with $a_n \neq 0$, $b_m \neq 0$, and where t denotes time)

$$a_n \frac{d^n x}{dt^n} + a_{n-1} \frac{d^{n-1} x}{dt^{n-1}} + \ldots + a_0 = b_m \frac{d^m y}{dt^m} + b_{m-1} \frac{d^{m-1} y}{dt^{m-1}} + \ldots + b_0$$

where $n \geq m$. For a derivation of this result $(n \geq m)$, see James, Nichols, and Phillips (1947, pp. 24, 25). We shall call the filter <u>strictly passive</u> if $n > m$.

The <u>transfer function</u> $W(s)$ of the filter is given by the rational function

$$W(s) = \frac{b_m s^m + b_{m-1} s^{m-1} + \ldots + b_0}{a_n s^n + a_{n-1} s^{n-1} + \ldots + a_0}$$

where $s = \rho + i\omega$ $(\rho, \omega$ real) is a complex variable, and ω denotes real angular frequency (radians per second).

It is well-known that the transfer function $W(s)$ is the Laplace transform $\mathcal{L}$ of the <u>impulse response</u> $w(t)$ of the filter:

$$W(s) = \mathcal{L}\{w(t)\} = \int_0^\infty w(t) e^{-st} dt$$

where $w(t) = 0$ for $t < 0$. The theory of the Laplace transform as applied to lumped constant systems, as well as tables of transform pairs, may be found in electrical engineering books, and reference is made in particular to Gardner and Barnes (1942) and James, Nichols, a nd Phillips (1947).

Let $A(s)$ denote the polynomial

$$A(s) = a_n s^n + a_{n-1} s^{n-1} + \ldots + a_0 = a_n (s-a_1)(s-a_2)\ldots(s-a_n) .$$

Since the coefficients of this polynomial are real, its complex roots occur

in complex conjugate pairs: a_i, $\bar{a}_i$. Let $B(s)$ denote the polynomial

$$B(s) = b_m s^m + b_{m-1} s^{m-1} + \ldots + b_0 = b_m (s-\beta_1)(s-\beta_2)\ldots(s-\beta_m)$$

where, for the same reason, the complex roots occur in complex conjugate

pairs: β_j, $\bar{\beta}_j$. We shall assume that no a_i equals any β_j , for clearly

any common factors could be cancelled out in the expression for

$W(s) = B(s)/A(s)$.

We shall call the filter <u>strictly stable</u> if the roots $a_1, a_2, \ldots, a_n$

of the polynomial $A(s)$ have negative real parts. The roots

$a_1, a_2, \ldots, a_n$ may be called the <u>eigen-frequencies</u> of the filter.

As we shall see in Sections 3 and 4, the impulse response $w(t)$ of a

strictly stable filter damps out exponentially as $t \to \infty$.

We shall call a strictly stable filter <u>strictly minimum-phase</u> if the

roots $\beta_1, \beta_2, \ldots, \beta_m$ of the polynomial $B(s)$ have negative real

parts. The concept of minimum-phase is due to Bode (1945).

Let us suppose that the filter is strictly passive and strictly minimum-

phase. If its input $y(t)$ is white noise, then its output $x(t)$ is a

stationary time-series with a rational spectral density that does not

vanish for any real angular frequency ω .

Converseley, let $x(t)$ be a stationary time-series with rational

spectral density that does not vanish for any real angular frequency ω .

Then $x(t)$ may be considered to be the output of a strictly passive, strictly minimum-phase filter with white noise input (Wold, 1938; Wiener, 1942).

§2. THE OPTIMUM PREDICTION SYSTEM

We suppose that $x(t)$ is a stationary time-series with rational spectral density that does not vanish for any real angular frequency ω. We denote the present time instant by t , and suppose that we have observed the time series from the remote past (time $\tau = -\infty$) until the present time t , which is the last point of observation. That is, we know the past values $x(\tau)$, $\tau < t$, and the present value $x(t)$, but do not know the future values $x(\tau)$, $\tau > t$.

We wish to find the filter characteristics $P(s)$ of a system whose input at time t is $x(t)$ and whose output $z(t)$ at time t approximates in the least-squares sense the future value $x(t+\epsilon)$, where $\epsilon > 0$. This optimum prediction system must be physically realizable in the sense that $z(t)$ can only depend upon the present and past values $x(\tau)$, $t \leq \tau$; that is, only on values that can be observed up to and including the present time t , and not on future values. The solution of this problem was given by Wiener (1942) and a simplified derivation by Bode and Shannon (1950).

Let $W(s)$ be the transfer function of the strictly passive, strictly minimum-phase filter with white noise input $y(t)$ and output the stationary time-series $x(t)$. Bode and Shannon (1950) show that the network of the optimum prediction system may be broken into two tandem networks, with overall transfer function

$$P(s) = P_1(s)P_2(s) \ .$$

The transfer function $P_1(s)$ of the first network is selected to be

$$P_1(s) = \frac{1}{W(s)} \ .$$

The transfer function $P_2(s)$ of the second network is the Laplace trans-

form of the impulse response given by

$$p_2(t) = \begin{cases} w(t+\epsilon) & t \geq 0 \\ 0 & t < 0 \end{cases}$$

where $w(t)$ is the impulse response of the filter $W(s)$. Hence the

transfer function $P(s)$ of the <u>optimum prediction system</u> is given by

$$P(s) = P_1(s)P_2(s) = \frac{\mathcal{L}\{p_2(t)\}}{\mathcal{L}\{w(t)\}} = \frac{A(s)\mathcal{L}\{p_2(t)\}}{B(s)} \ .$$

That is, if the stationary time-series $x(t)$ is the input to the system

$P(s)$, then the least-squares approximation $z(t)$ to the future value

$x(t+\epsilon)$ is the output.

We now wish to evaluate $P(s)$ for the cases in which (1) the eigen-

frequencies $a_1, a_2, \ldots, a_n$ are distinct and (2) the eigen-frequency

a_1 is repeated r_1 times; a_2, r_2 times; $\ldots$; a_n, r_n times.

§3. DISTINCT EIGEN-FREQUENCIES

In this section we assume that the eigen-frequencies

$a_1, a_2, \ldots, a_n$ of the filter are distinct. The transfer function $W(s)$

then has the partial fraction expansion

$$W(s) = \sum_{i=1}^{n} \frac{A_i}{s-a_i}$$

where

$$A_i = W(s)(s-a_i)]_{s=a_i} \quad .$$

The impulse response $w(t)$ is given by the inverse Laplace transform
(Pairs No. 0.11 and No. 1.102 in Gardner and Barnes, 1942),

$$w(t) = \mathcal{L}^{-1}\{W(s)\} = \begin{cases} \displaystyle\sum_{i=1}^{n} A_i e^{a_i t} & t \geq 0 \\[2em] 0 & t < 0 \end{cases}$$

which exhibits the rôle played by the eigen-frequencies a_i of the
filter. Since the filter $W(s)$ is strictly stable (i.e. the a_i have negative
real parts), the weighting function $w(t)$ damps out exponentially as
$t \to \infty$. We have

$$p_2(t) = \begin{cases} \displaystyle w(t+\epsilon) = \sum_{i=1}^{n} A_i e^{a_i(t+\epsilon)} & t \geq 0 \\[2em] 0 & t < 0 \end{cases} \quad .$$

Making use of Pair No. 1.102 in Gardner and Barnes (1942), we have

$$\mathcal{L}\{p_2(t)\} = \sum_{i=1}^{n} \frac{A_i e^{a_i \epsilon}}{s-a_i} \quad .$$

Hence the transfer function of the optimum predictor is

$$P(s) = \frac{\displaystyle\sum_{i=1}^{n} \frac{A_i}{s-a_i} e^{a_i \epsilon}}{\displaystyle\sum_{i=1}^{n} \frac{A_i}{s-a_i}} = \frac{A(s) \displaystyle\sum_{i=1}^{n} \frac{A_i}{s-a_i} e^{a_i \epsilon}}{B(s)} \quad .$$

Let $C(s)$ be the polynomial

$$C(s) = A(s) \sum_{i=1}^{n} \frac{A_i}{s-a_i} e^{a_i s} = a_n(s-a_1)(s-a_2)\ldots(s-a_n)\sum_{i=1}^{n} \frac{A_i}{s-a_i} e^{a_i \epsilon} \quad .$$

We see that $C(s)$ is of degree $n-1$ and so may be written

$$C(s) = c_{n-1} s^{n-1} + c_{n-2} s^{n-2} + \ldots + c_0 \, .$$

Letting $s = a_1$ we have

$$C(a_1) = a_n (a_1 - a_2)(a_1 - a_3) \ldots (a_1 - a_n) A_1 \, e^{a_1 \epsilon} \, .$$

But

$$A_1 = W(s)(s - a_1)]_{s = a_1} = \frac{b_m (a_1 - \beta_1)(a_1 - \beta_2) \ldots (a_1 - \beta_m)}{a_n (a_1 - a_2)(a_1 - a_3) \ldots (a_1 - a_n)}$$

so

$$C(a_1) = b_m (a_1 - \beta_1)(a_1 - \beta_2) \ldots (a_1 - \beta_m) e^{a_1 \epsilon} \, .$$

Letting $s = a_2, \ldots, a_n$ we thereby obtain the system of n simultaneous

linear equations

$$c_{n-1} a_1^{n-1} + c_{n-2} a_1^{n-2} + \ldots + c_0 = b_m (a_1 - \beta_1)(a_1 - \beta_2) \ldots (a_1 - \beta_m) e^{a_1 \epsilon}$$

$$c_{n-1} a_2^{n-1} + c_{n-2} a_2^{n-2} + \ldots + c_0 = b_m (a_2 - \beta_1)(a_2 - \beta_2) \ldots (a_2 - \beta_m) e^{a_2 \epsilon}$$

$$\ldots$$

$$c_{n-1} a_n^{n-1} + c_{n-2} a_n^{n-2} + \ldots + c_0 = b_m (a_n - \beta_1)(a_n - \beta_2) \ldots (a_n - \beta_m) e^{a_n \epsilon}$$

which may be solved for the n coefficients $c_{n-1}, c_{n-2}, \ldots, c_0$ of the

polynomial $C(s)$. In more compact notation, this system may be written

$$[c_{n-1} s^{n-1} + c_{n-2} s^{n-2} + \ldots + c_0]_{s = a_i} = [B(s) e^{s \epsilon}]_{s = a_i}$$

where $i = 1, 2, \ldots, n$.

§4. MULTIPLE EIGEN-FREQUENCIES

In this section we assume that the eigen-frequencies of the filter are multiple: i.e., the eigen-frequency a_1 is repeated r_1 times; a_2, r_2 times; ... ; a_q, r_q times, where $r_1 + r_2 + \ldots r_q = n$. The transfer function $W(s)$ has the partial fraction expansion

$$W(s) = \sum_{i=1}^{q} \sum_{j=1}^{r_i} \frac{A_{ij}}{(s-a_i)^j}$$

where

$$A_{ij} = \frac{1}{(r_i-j)!} \left[\left(\frac{d}{ds}\right)^{r_i-j} \{W(s)(s-a_i)^{r_i}\} \right]_{s=a_i}$$

for $i = 1, 2, \ldots, q$ and $j = 1, 2, \ldots, r_i$. The impulse response is

$$w(t) = \mathcal{L}^{-1}\{W(s)\} = \sum_{i=1}^{q} \sum_{j=1}^{r_i} \frac{A_{ij}}{(j-1)!} t^{j-1} e^{a_i t}, \qquad t \geq 0,$$

and where $w(t) = 0$ for $t < 0$. Here we have used Pairs No. 0.21 and No. 2.121 in Gardner and Barnes (1942). Since the filter $W(s)$ is strictly stable (i.e. the a_i have negative real parts), the weighting function $w(t)$ damps out exponentially as $t \to \infty$. We have

$$p_2(t) = \sum_{i=1}^{q} \sum_{j=1}^{r_i} \frac{A_{ij}}{(j-1)!} (t + \epsilon)^{j-1} e^{a_i(t+\epsilon)}, \qquad t \geq 0$$

and $p_2(t) = 0$ for $t < 0$. This is

$$p_2(t) = \sum_{i=1}^{q} \sum_{j=1}^{r_i} \frac{A_{ij} e^{a_i \epsilon}}{(j-1)!} \sum_{k=0}^{j-1} \frac{(j-1)!}{(j-1-k)!\,k!} \epsilon^{j-1-k} t^k e^{a_i t}, \qquad t \geq 0.$$

We therefore have

$$\mathcal{L}\{p_2(t)\} = \sum_{i=1}^{q} \sum_{j=1}^{r_i} A_{ij} e^{a_i \epsilon} \sum_{k=0}^{j-1} \frac{\epsilon^{j-1-k}}{(j-1-k)!} \mathcal{L}\{\frac{t^k e^{a_i t}}{k!}\} \ .$$

Hence the transfer function of the optimum predictor is

$$P(s) = \frac{A(s) \sum_{i=1}^{q} \sum_{j=1}^{r_i} A_{ij} e^{a_i \epsilon} \sum_{k=0}^{j-1} \frac{\epsilon^{j-1-k}}{(j-1-k)!} \frac{1}{(s-a_i)^{k+1}}}{B(s)}$$

where we have used Pair No. 2.121 in Gardner and Barnes (1942).

If we denote the numerator of $P(s)$ by $C(s)$, we see that $C(s)$ is a polynomial of degree $n-1$, and so may be written

$$C(s) = c_{n-1} s^{n-1} + c_{n-2} s^{n-2} + \ldots + c_0 \ .$$

A system of n simultaneous linear equations may be found for the n coefficients c_{n-1} , c_{n-2} , $\ldots$, c_0 by the same method as in the previous section, except now the manipulations are more involved. The final result is the system

$$(\frac{d}{ds})^{r_i - j} [c_{n-1} s^{n-1} + c_{n-2} s^{n-2} + \ldots + c_0]_{s=a_i} = (\frac{d}{ds})^{r_i - j} [B(s) e^{s \epsilon}]_{s=a_i}$$

where $i = 1, 2, \ldots , q$ and $j = 1, 2, \ldots , r_i$ $(r_1 + r_2 + \ldots + r_q = n)$.

§5. THE MARKOV AND PASSIVE COMPONENTS

In the preceding sections we have found that the optimum predictor has transfer function given by the quotient of polynomials

$$P(s) = \frac{C(s)}{B(s)} = \frac{c_{n-1} s^{n-1} + c_{n-2} s^{n-2} + \ldots + c_0}{b_m s^m + b_{m-1} s^{m-1} + \ldots + b_0} \ .$$

In the case when the polynomials $C(s)$ and $B(s)$ have common roots, the common factors should be cancelled out. Then the analysis will go through as in the case of no common factors, except that now the degree of the numerator and the degree of the denominator each will have been reduced by the number of common factors.

Let us now consider the case when $C(s)$ and $B(s)$ have no common factors. That is, the roots γ_1, γ_2, ..., γ_{n-1} of the polynomial $C(s)$ are distinct from the roots β_1, β_2, ..., β_m of the polynomial $B(s)$. The time series $x(t)$, the input, and the optimum prediction $z(t)$, the output, will formally satisfy the differential equation

$$b_m \frac{d^m z}{dt^m} + b_{m-1} \frac{d^{m-1} z}{dt^{m-1}} + \ldots + b_0 = c_{n-1} \frac{d^{n-1} x}{dt^{n-1}} + c_{n-2} \frac{d^{n-2} x}{dt^{n-2}} + \ldots + c_0$$

which describes the optimum predictor. Because the filter $W(s)$ is strictly minimum-phase (i.e. the roots β_1, β_2, ..., β_m have negative real parts), we see that the prediction transfer function $P(s)$ is strictly stable. That is, the numerator $B(s)$ of the filter transfer function $W(s)$ is the denominator of the prediction transfer function, and so the strictly minimum-phase requirement on $W(s)$ is necessary in order to make $P(s)$ strictly stable.

If $c_{n-1} \neq 0$, then the highest degree term in $P(s)$ is s^{n-m-1}.
Then by long division we may express $P(s)$ as

$$P(s) = f_{n-m-1} s^{n-m-1} + f_{n-m-2} s^{n-m-2} + \ldots + f_2 s^2 + f_1 s + f_0 + \frac{E(s)}{B(s)}$$

where the f_i $(i = 0, 1, 2, \ldots, n-m-1)$ are constants and $E(s)$ is a polynomial of at most degree $m-1$. Since $B(s)$ is of degree m, we see that

$E(s)/B(s)$ is the transfer function of a four-terminal, strictly passive, strictly stable, lumped-constant, linear filter. (We recall that for a strictly passive filter, the degree of the denominator of the transfer function must be greater than the degree of the numerator.)

Suppose that the eigen-frequency β_1 is repeated h_1 times; β_2, h_2 times; ... ; β_k, h_k times, where $h_1 + h_2 + \ldots + h_k = m$. Then we have the partial fraction expansion

$$\frac{E(s)}{B(s)} = \sum_{i=1}^{k} \sum_{j=1}^{h_i} \frac{B_{ij}}{(s-\beta_i)^j}$$

where

$$B_{ij} = \frac{1}{(h_i-j)!} \left[\left(\frac{d}{ds}\right)^{r_i-j} \left\{ \frac{E(s)}{B(s)} (s-\beta_i)^{h_i} \right\} \right]_{s=a_i}$$

for $i = 1, 2, \ldots k$ and $j = 1, 2, \ldots , h_i$. The impulse response is

$$\mathcal{L}^{-1}\left\{\frac{E(s)}{B(s)}\right\} = \begin{cases} \sum_{i=1}^{k} \sum_{j=1}^{h_i} \frac{B_{ij}}{(j-1)!} t^{j-1} e^{\beta_i t} &, \quad t \geq 0 \\ 0 &, \quad t < 0 \end{cases}.$$

If we let $\delta(t)$ denote the impulse (Dirac delta) function, then

$$\mathcal{L}\{\delta(t)\} = 1 , \qquad \mathcal{L}\{\delta^{(n)}(t)\} = s^n$$

(n a positive integer) where $\delta^{(n)}(t)$ denotes the nth derivative of the impulse function. Hence the impulse response of the optimum predictor is

$$p(t) = \mathcal{L}^{-1}\{P(s)\} = \mathcal{L}^{-1}\{f_{n-m-1}s^{n-m-1} + \ldots + f_2 s^2 + f_1 s + f_0\} + \mathcal{L}^{-1}\{\frac{E(s)}{B(s)}\}$$

$$= f_{n-m-1}\delta^{(n-m-1)}(t) + \ldots + f_2 \delta''(t) + f_1 \delta'(t) + f_0 + \sum_{i=1}^{k}\sum_{j=1}^{h_i}\frac{B_{ij}}{(j-1)!}t^{j-1}e^{\beta_i t}$$

for $t \geq 0$ and $p(t) = 0$ for $t < 0$. The optimum prediction for a lead of

ϵ time units is

$$z(t) = \int_{-\infty}^{t} p(t-\tau)x(\tau)d\tau \qquad .$$

Now we know from the properties of the impulse function that

$$\int_{-\infty}^{t} \delta^{(n)}(t-\tau)x(\tau)d\tau = [\frac{d^n x(\tau)}{d\tau^n}]_{\tau=t} \qquad .$$

Hence

$$z(t) = [\{f_{n-m-1}(\frac{d}{d\tau})^{n-m-1} + \ldots + f_2 \frac{d^2}{d\tau^2} + f_1 \frac{d}{d\tau} + f_0\}x(\tau)]_{\tau=t}$$

$$+ [\sum_{i=1}^{k}\sum_{j=1}^{h_i}\frac{B_{ij}}{(j-1)!}\int_{-\infty}^{t}(t-\tau)^{j-1}e^{\beta_i(t-\tau)}x(\tau)d\tau] \qquad .$$

The first term in brackets may properly be called the <u>Markov component</u>
because this term depends upon the value of the process $x(t)$ and its derivatives only at the last point of observation, namely the present time instant $\tau = t$. That is, the Markov component is a weighted average of
$x(t)$ and its derivatives $x^{(n-m-j)}(t)$, all evaluated at the present time
instant t . In the case when $n-m = 1$, the Markov component reduces
to $f_0 x(t)$, and the constant f_0 is the "saltus in the weighting measure at
the last point of observation" in the theorem of Chover (1958).

The second term in brackets may be called the <u>passive component</u>

because this term depends upon the whole past $x(\tau)$, $\tau \leq t$, of the

process. That is, the passive component is the output of a strictly passive

filter. Since the passive component appears as a convolution integral, no

past observation $x(\tau)$, $\tau < t$, receives more than an infinitesimal weight-

ing, i.e. there are no salti in the weighting measure for past times, or in

other words the weighting measure for past times (i.e. for $\tau < t$) is abso-

lutely continuous with respect to Lebesgue measure. An elementary pre-

sentation of the mathematical theory of measure may be found in the book by

Robinson (1959).

This predictive decomposition into Markov and passive components

is helpful in the building of prediction systems. Since the spectral density

function $\Phi(\omega)$ of the time series $x(t)$ is given by

$$\Phi(\omega) \; = \; |W(i\omega)|^2 \; = \; \frac{|B(i\omega)|^2}{|A(i\omega)|^2}$$

we see that $\Phi(\omega)$ is a rational function whose numerator is a polynomial of

degree 2m and whose denominator is a polynomial of degree 2n .

By a theorem of Lévy (1948) it is known that all the derivatives of $x(t)$

exist up to and including order n-m-1 , but not higher orders. These

are the derivatives which enter into the Markov component. Thus the

Markov component corresponds to the extrapolation based on the derivatives,

as in a Taylor series expansion with remainder, and the passive component corresponds to the remainder[1]. Thus the performance of a prediction

system will depend to a large degree on the performance of the Markov com-

ponent, and we know the Markov component does not utilize the past values

$x(\tau)$, $\tau < t$, of the time series.

---

[1] It should be noted that our predictive decomposition into Markov and

passive components is not the same as the one given by Wiener (1942,

formula 2.35, p. 72 in the 1949 edition). Wiener's formula, using our

notation, assumes knowledge of only the spectral density of the $n-m-1^{th}$

derivative of $x(t)$, whereas our formula assumes knowledge of the

spectral density of $x(t)$.

REFERENCES

Bode, H. W. (1945). Network Analysis and Feedback Amplifier Design: Van Nostrand, N.Y.

Bode, H. W., and Shannon, C. E. (1950). A simplified derivation of linear least square smoothing and prediction theory: Proceedings of the I. R. E. 38, 417-425.

Chover, J. (1958). Conditions on the realization of prediction by measures: Duke Math. Journal, 25, 305-310.

Gardner, M. F., and Barnes, J. L. (1942). Transients in Linear Systems, vol. 1: Wiley, N.Y.

James, H. M., Nichols , N. B., and Phillips, R. S. (1947). Theory of Servomechanisms : MIT Radiation Lab. Series, 25, McGraw-Hill, N.Y.

Lévy, P. (1948). Processus stochastiques et mouvement Brownien : Gauthier-Villars, Paris.

Robinson, E. A. (1959). Infinitely Many Variates : Charles Griffin & Co., London.

Wiener, N. (1942). Stationary Time Series : MIT, Cambridge, Mass. (Reprinted 1949, Wiley, N.Y.)

Wold, H. (1938). Stationary Time Series, Almqvist and Wiksells, Uppsala (Second edition, 1954).

CHAPTER 9

AUTOMATIC ALGEBRAIC REDUCTIONS FOR THE THIRD-ORDER AUTOREGRESSIVE PROCESS

Bruce A. Holum and Enders A. Robinson

§1. Introduction

Two of the most widely applied stationary stochastic processes are the Markov (or first-order autoregressive)

$$x_t + a_1 x_{t-1} = \epsilon_t \tag{1.1}$$

and the second order autoregressive

$$x_t + a_1 x_{t-1} + a_2 x_{t-2} = \epsilon_t . \tag{1.2}$$

Here x_t represents the time series realization at unit intervals of time; and the innovation ϵ_t , a sequence of uncorrelated random variables with mean zero and variance σ^2 . Although the general theory of diesrete stationary processes is known, which includes the nth order autoregressive process, a certain amount of specialization is necessary to render the general theory in a form convenient for applications. The special theories of the first and second order processes are well known and conveniently found in time series literature. This paper deals with the special theory of the third-order autoregressive process

$$x_t + a_1 x_{t-1} + a_2 x_{t-2} + a_3 x_{t-3} = \epsilon_t , \tag{1.3}$$

and considers the possibility of obtaining formulas for $E(x_t^2)$ for this and higher order autoregressive processes.

§2. Moving-average Representation

The autoregressive representation (1.3) of the x_t time series may be inverted into its moving average representation by writing

$$x_t \, (E^3 + a_1 E^2 + a_2 E + a_3) \; = \; \epsilon_{t+3} \tag{2.1}$$

where E is the displacement operator: $E\,x_t = x_{t+1}$. Now

$$E^3 + a_1 E^2 + a_2 E + a_3 \; = \; (E-r_1)\,(E-r_2)\,(E-r_3) \tag{2.2}$$

where r_1 , r_2 , and r_3 are the roots of the cubic equation in E .
Thus we may put

$$x_t \; = \; \frac{\epsilon_{t+3}}{(E-r_1)\,(E-r_2)\,(E-r_3)} \tag{2.3}$$

Expanding this expression in terms of partial fractions we have

$$x_t \; = \; \left[\frac{A}{(E-r_1)} + \frac{B}{(E-r_2)} + \frac{C}{(E-r_3)} \right] \epsilon_{t+3} \tag{2.4}$$

where

$$A \; = \; \frac{1}{(r_1-r_2)\,(r_1-r_3)}, B = \frac{1}{(r_2-r_1)\,(r_2-r_3)}, \; C = \frac{1}{(r_3-r_1)\,(r_3-r_2)} \tag{2.5}$$

Thus x_t becomes

$$x_t \; = \left[\frac{A}{E(1-\frac{r_1}{E})} + \frac{B}{E(1-\frac{r_2}{E})} + \frac{C}{E(1-\frac{r_3}{E})} \right] y_{t+3} \tag{2.6}$$

which has the expansion

$$x_t = \left[\frac{A}{E}\left(1+\frac{r_1}{E}+\frac{r_1^2}{E^2}+\ldots\right)+\frac{B}{E}\left(1+\frac{r_2}{E}+\frac{r_2^2}{E^2}+\ldots\right)+\frac{C}{E}\left(1+\frac{r_3}{E}+\frac{r_3^2}{E^2}+\ldots\right) \right] y_{t+3}$$

$$\tag{2.70}$$

provided the roots r_1, r_2, r_3 each have magnitude less than unity.

This is a necessary condition for the time series to be stationary.

The following identities may be readily verified:

$$A + B + C = 0$$

$$Ar_1 + Br_2 + Cr_3 = 0 \qquad (2.8)$$

$$Ar_1^2 + Br_2^2 + Cr_3^2 = 1$$

As a result the terms in (2.7) containing y_{t+3}, y_{t+2}, y_{t+1}

vanish (which could have been established by a general argument without

recourse to the identities (2.8)), and we have the moving-average

representation

$$x_t = \sum_{s=0}^{\infty} (Ar_1^{s+2} + Br_2^{s+2} + Cr_3^{s+2})\, y_{t-s} \qquad (2.9)$$

which explicitly is

$$x_t = \sum_{s=0}^{\infty} \left[\frac{r_1^{s+2}}{(r_1-r_2)(r_1-r_3)} + \frac{r_2^{s+2}}{(r_2-r_1)(r_2-r_3)} + \frac{r_3^{s+2}}{(r_3-r_1)(r_3-r_2)} \right] y_{t-s} \qquad (2.10)$$

That is

$$x_t = \sum_{s=0}^{\infty} b_s\, \epsilon_{t-s} \qquad (2.11)$$

with

$$b_s = \frac{1}{(r_1-r_2)(r_2-r_3)(r_3-r_1)} \left[(r_3-r_2)r_1^{s+2} + (r_1-r_3)r_2^{s+2} + (r_2-r_1)r_3^{s+2} \right]$$

§3. The Variance

Because the innovations ϵ_t form an orthonormal sequence, with

variance σ^2, the variance of the x_t is seen to be

$$E(x_t^2) = \sigma^2 \sum_{s=0}^{\infty} b_s^2 \qquad (3.1)$$

If we now square b_s we see that each term forms a geometric series

since r_i is less than 1 by hypothesis.

Therefore, we may easily form the sum

$$E(x_t^2) = \frac{\sigma^2}{D} \left[\frac{(r_3-r_2)^2 r_1^4}{1-r_1^2} + \frac{(r_1-r_3)^2 r_2^4}{1-r_2^2} + \frac{(r_2-r_1)^2 r_3^4}{1-r_2^2} \right.$$

$$+ \frac{2(r_3-r_2)(r_1-r_3)(r_1-r_2)^2}{1-r_1 r_2} + \frac{2(r_1-r_3)(r_2-r_1)(r_2-r_3)^2}{1-r_2 r_3}$$

$$\left. + \frac{2(r_2-r_1)(r_3-r_2)(r_3-r_1)^2}{1-r_3 r_1} \right] \qquad (3.2)$$

where the discriminant D of the cubic is defined by the formula

$$D = [(r_1-r_2)(r_1-r_3)(r_2-r_3)]^2 \qquad (3.3)$$

If we follow this same argument for the first and second order cases

we obtain

$$[E(x_t^2)]_{n=1} = \frac{\sigma^2 r_1^4}{1-r_1^2} \qquad (3.4)$$

$$[E(x_t^2)]_{n=2} = \frac{\sigma^2}{D} \left[\frac{r_1^4}{1-r_1^2} + \frac{r_2^4}{1-r_2^2} - \frac{2r_1^2 r_2^2}{1-r_1 r_2} \right] \qquad (3.4)$$

Now notice how rapidly the number of terms in the numerator of each

expression grows when we put everything over a common denominator.

In the first order expression there is only one term, in the second order

expression there are 12 terms and in the third order case there are 1440

terms. It is not difficult to show that the number of terms goes up much

faster than exponentially.

Clearly the amount of work involved in simplifying these expressions

becomes prodigous while remaining methodical. In view of this the question

arose, could an automatic digital computer be taught to do this methodical

work?

A program to do this on the IBM Type 704 EDPM is now nearing

completion. The algebraic expression is digitalized by using the ex-

ponents of the various factors. Then the first term in the expression

for the third-order process goes into the 704 in consecutive locations

as 0, 1, 4, 0, 0, 1, 1, 1, 0, 1, 1, 0, 0, 2. Here each number, except the first

two which are the sign and coefficient of the term, represent the exponents

of the variables of certain factors of the variables such as

$(1-r_1 r_2)$, $(r_1 - r_2)$, $(1-r_1{}^2)$, etc.

The procedure is to put all six terms over a common denominator

and then simplifying by expanding and combining terms, thus reducing

the numerator to its simplest unfactored form.

Since the exponents of known factors are known it is possible to

methodically expand the expression by "block multiplication", that is,

to multiply a block by $(r_1 - r_2)$ the exponent of this factor is reduced by

one , the block is rewritten, and the exponent of r_1 in the first block

is increased by one, the exponent of r_2 in the second block is increased

by one, and the sign of the second block is changed.

The computer systematically goes through each block until the

exponent of each factor is reduced to zero. Then the expression has

been completely expanded and the computer can begin to combine

the terms and reduce the expression to its most compact form.

In this program it is left to the programmer to factor the final

result.

In the particular case of the third-order autoregressive process, where the form of each factor is known, it is possible to do the reduction since we can digitalize the expression by using the exponents of the factors.

The logical extension of this idea is to inquire whether a digital computer could be programmed to process algebraic expressions in their symbolic form.

Would it be possible to use a language such as that of FORTRAN, the automatic program writing system for the IBM 704, to express an arbitrary algebraic expression in symbolic form, punch the expression into IBM cards and feed the cards into the computer to be processed automatically with machine precision and electronic speed?

Using magnetic tape it should be possible to handle expressions of virtually any size. Precision is assured and computers can perform about 42,000 operations per second.

In this system multiplication, division, and exponentiation are represented by $*, /,$ and $**$ respectively. Thus an expression such as

$$\frac{(2x^2 - 3y^2)}{4x}$$

would be expressed, on a single line in IBM characters as necessary for one character per column, as

$$(2*x**2 - 3*y**2)/(4*x)$$

Work is now being done to pursue this idea. Such a program should be a useful tool in obtaining results where a great deal of algebraic reduction is required.

We would like to acknowledge Midwestern Universities Research Association for providing the computer, and International Business Machines for providing the time.

CHAPTER 10

SUMS OF STATIONARY RANDOM VARIABLES[1]

ENDERS A. ROBINSON

A sequence $x(t)$ ($-\infty < t < \infty$, t an integer) of elements in Hilbert space is called stationary if the inner product $(x(t+s), x(t))$ does not depend upon t. If the Hilbert space is L^2 space with probability measure, then $x(t)$ is a random variable and the sequence $x(t)$ ($-\infty < t < \infty$) is called a second-order stationary random process. Let X be the closed linear manifold spanned by all the elements of the stationary process. Then Kolmogorov [1] has shown that the equation $x(t)U = x(t+1)$, $-\infty < t < \infty$, uniquely determines the unitary operator U with domain and range X. Using the von Neumann [2] spectral representation of U, we obtain the spectral representation of the random process

$$x(t) = \int_{-.5}^{.5} e^{2\pi i u t} dx(0) E(u), \qquad -\infty < t < \infty.$$

The von Neumann [3] ergodic theorem, in the framework of Khintchine [4], is applicable, and shows that the average $\sum_1^n x(t)/n$ converges in the mean to the random variable $x(0)[E(0+) - E(0-)]$ as $n \to \infty$. In this paper we consider sums instead of averages; that is, we consider $\sum_1^n x(t)$, and establish the following theorem.

THEOREM. *Let the random variables $x(t)$ ($-\infty < t < \infty$, t an integer) be a second-order stationary random process with spectral distribution function $F(u)$. For* variance $\{\sum_1^n x(t)\}$ *to be bounded for all positive integers n, each of the following two conditions is necessary and sufficient:*

(1)
$$\int_{-.5}^{.5} \sin^{-2} \pi u\, dF(u) < \infty.$$

(2) *There is a second-order stationary random process*

$$y(t) \ (-\infty < t < \infty) \ \textit{satisfying} \ y(t) - y(t+1) = x(t).$$

PROOF. (NECESSARY CONDITIONS). We are given that *variance* $\{\sum_1^n x(t)\} < B$ for all positive integers n. Without loss of generality we assume that the $x(t)$ are centered so that their mean values are zero. Then

Presented to the Society, February 28, 1959; received by the editors April 6, 1959.

[1] This paper was sponsored by the United States Army under Contract No. DA-11-022-ORD-2059 at the Mathematics Research Center, United States Army, Madison, Wisconsin.

$$\text{variance}\left\{\sum_1^n x(t)\right\} = \left(\sum_1^n x(t), \ \sum_1^n x(t)\right).$$

From the spectral representation we have

$$\sum_1^n x(t) = \int_{-.5}^{.5} e^{2\pi iu}\,\frac{1-e^{2\pi iun}}{1-e^{2\pi iu}}\,dx(0)E(u),$$

so

$$\text{variance}\left\{\sum_1^n x(t)\right\} = \int_{-.5}^{.5}\frac{\left|1-e^{2\pi iun}\right|^2}{\left|1-e^{2\pi iu}\right|^2}\,dF(u)$$

where $F(u) = \|x(0)E(u)\|^2$, $-.5\leqq u\leqq.5$, is the spectral distribution function. Hence we have

$$B > \text{variance}\left\{\sum_1^n x(t)\right\} = \left\{\int_{-.5}^{0-}+\int_{0+}^{.5}\right\}\frac{\sin^2\pi un}{\sin^2\pi u}\,dF(u)$$
$$+ n^2[F(0+)-F(0-)]$$

which shows that $F(0+)-F(0-)$ must vanish. Moreover, we have

$$B > \frac{1}{N}\sum_{n=1}^{N}\left\{\int_{-.5}^{0-}+\int_{0+}^{.5}\right\}\frac{\sin^2\pi un}{\sin^2\pi u}\,dF(u)$$
$$= \left\{\int_{-.5}^{0-}+\int_{0+}^{.5}\right\}\left[\frac{1}{N}\sum_{n=1}^{N}\left(\frac{1}{2}-\frac{1}{2}\cos 2\pi un\right)\right]\sin^{-2}\pi u\,dF(u).$$

Clearly the limit of the expression in brackets, as $N\to\infty$, is $1/2$, so $\int_{-.5}^{.5}\sin^{-2}\pi u\,dF(u)$ is finite. Q.E.D. (1).

The distribution function $F(u)$ defines a Lebesgue-Stieltjes measure on the real line segment $-.5\leqq u\leqq.5$. Let W denote the L^2 space of complex-valued measurable functions $\Phi(u)$ defined on $-.5\leqq u\leqq.5$ for this measure. Define a correspondence between an element x of X and an element $\Phi(u)$ of W by

$$x = \int_{-.5}^{.5}\Phi(u)dx(0)E(u)\leftrightarrow\Phi(u).$$

Then Stone [5] and Kolmogorov [1] have shown that this correspondence establishes an isomorphism between X and W that preserves inner products. The function $e^{2\pi iut}/(1-e^{2\pi iu})$ belongs to W since

$$\int_{-.5}^{.5}\left|\frac{e^{2\pi iut}}{1-e^{2\pi iu}}\right|^2 dF(u) = \frac{1}{4}\int_{-.5}^{.5}\sin^{-2}\pi u\,du < \infty.$$

If we define the element $y(t)$ of X by the correspondence $y(t) \leftrightarrow e^{2\pi iut}/(1-e^{2\pi iu})$ we see that

$$y(t) - y(t+1) \leftrightarrow \frac{e^{2\pi iut} - e^{2\pi iu(t+1)}}{1 - e^{2\pi iu}} = e^{2\pi iut}.$$

But by the spectral representation, we know that $x(t) \leftrightarrow e^{2\pi iut}$, and hence we have $y(t) - y(t+1) = x(t)$ for all integers t. Since

$$(y(t+s), y(t)) = \int_{-.5}^{.5} \frac{e^{2\pi iu(t+s)}e^{-2\pi iut}}{|1 - e^{2\pi iu}|^2} \, dF(u)$$

$$= \frac{1}{4} \int_{-.5}^{.5} e^{2\pi ius} \sin^{-2} \pi u \, dF(u)$$

depends only on s, we see that $y(t)$ is a stationary random process. Q.E.D. (2).

Proof. (Sufficient conditions). Let condition (1) of the theorem be given. Since

$$\text{variance} \left\{ \sum_1^n x(t) \right\} = \int_{-.5}^{.5} \frac{\sin^2 \pi un}{\sin^2 \pi u} \, dF(u) \leq \int_{-.5}^{.5} \sin^{-2} \pi u \, dF(u)$$

we see that the variance is bounded. Q.E.D. (1).

Let condition (2) of the theorem be given. Then $\|y(t)\|$ is a finite constant. Because $\sum_1^n x(t) = y(1) - y(n+1)$ we have $\|\sum_1^n x(t)\| \leq \|y(1)\| + \|y(n+1)\|$, and so $\textit{variance} \left\{ \sum_1^n x(t) \right\} = \|\sum_1^n x(t)\|^2$ is bounded. Q.E.D. (2).

References

1. A. N. Kolmogorov, *Stationary sequences in Hilbert space*, Bull. Moscow State Univ. Ser. Math. vol. 2 no. 6 (1941) 40 pp.

2. J. von Neumann, *Eigenwerttheorie Hermitescher Functionaloperatoren*, Math. Ann. vol. 102 (1929) pp. 49–131.

3. ———, *Proof of the quasiergodic hypothesis*, Proc. Nat. Acad. Sci. U.S.A. vol. 18 (1932) pp. 70–82.

4. A. Ja. Khintchine, *Korrelationstheorie der stationären stochastischen Prozesse*, Math. Ann. vol. 109 (1934) pp. 604–615.

5. M. H. Stone, *Linear transformations in Hilbert space*, Amer. Math. Soc. Colloquium Publications, vol. 15, 1932.

Mathematics Research Center, United States Army,
     Madison, Wisconsin

Reprinted from the
Proceedings of the American Mathematical Society
Vol. 11, No. 1, February, 1960
pp. 77-79

CHAPTER 11

Structural Properties of Stationary Stochastic Processes with Applications

Enders A. Robinson, * University of Wisconsin
and Uppsala University

1. THE STRUCTURE OF A REGULAR STATIONARY STOCHASTIC PROCESS

The purpose of this chapter is to develop the structural properties of stationary stochastic processes by use of the concept of *minimum-delay*. We shall consider processes with discrete time parameter t, so that t will be an integer. By using matrix notation,[1] we shall carry out our analysis for multiple processes.

The most straightforward method of treating stationary stochastic processes is by operational calculus. In this way we can see first of all what the mathematical operations are doing. The technical questions regarding the interpretation of the operational methods can come as a second step. Our presentation is made in this spirit.

We now wish to introduce *matrix-valued functions*. A matrix-valued function is one that assumes values that are matrices whose entries are complex numbers. Alternatively, we may say that a matrix-valued function is a matrix whose entries are *scalar-valued functions*. An example of a matrix-valued function is the square-matrix-valued time-function $\mathbf{f}(t) = [f_{jk}(t)]$, j, $k = 1, 2, \ldots, n$, where t is an integer and the $f_{jk}(t)$ are scalar-valued time-functions. The operational Fourier transform $\mathbf{F}(\omega)$ of $\mathbf{f}(t)$ is given by

$$\mathbf{F}(\omega) = \sum_{t=-\infty}^{\infty} e^{-i\omega t}\mathbf{f}(t) = \left[\sum_{t=-\infty}^{\infty} e^{-i\omega t}f_{jk}(t) \right] = [F_{jk}(\omega)], j, k = 1, 2, \cdots, n,$$

* Prepared in part in connection with research at Massachusetts Institute of Technology sponsored by Advanced Research Projects Agency (Project Vela Uniform).

[1] Boldface letters, e.g., $\boldsymbol{\alpha}$, denote matrices; a bar, e.g., $\bar{\boldsymbol{\alpha}}$, denotes the complex conjugate transpose; vertical bars indicate $|\boldsymbol{\alpha}|^2 = \boldsymbol{\alpha}\bar{\boldsymbol{\alpha}}$; $\boldsymbol{\alpha}^{-1}$ denotes the inverse of a square matrix $\boldsymbol{\alpha}$; and tr $\boldsymbol{\alpha}$ denotes the trace of a square matrix $\boldsymbol{\alpha}$, where the trace is defined as the sum of the entries on the main diagonal of $\boldsymbol{\alpha}$.

and the inverse transform is

$$\mathbf{f}(t) = \frac{1}{2\pi} \int_{-\pi}^{\pi} e^{i\omega t} \mathbf{F}(\omega)\, d\omega = \left[\frac{1}{2\pi} \int_{-\pi}^{\pi} e^{i\omega t} F_{jk}(\omega)\, d\omega \right]$$
$$= [f_{jk}(t)],\ j,\ k = 1, 2, \cdots, n.$$

We see that $\mathbf{F}(\omega)$ is a square-matrix-valued frequency-function, where ω is angular frequency $(-\pi \leqslant \omega \leqslant \pi)$. We denote Fourier transform pairs by $\mathbf{f}(t) \leftrightarrow \mathbf{F}(\omega)$.

A Fourier transform pair $\mathbf{b}(t) \leftrightarrow \mathbf{B}(\omega)$ may represent a time-invariant linear system, where $\mathbf{b}(t)$ is called the *impulse response function* and $\mathbf{B}(\omega)$ the *transfer function*. Input $\mathbf{f}(t)$ and output $\mathbf{g}(t)$ are related by the convolution

$$\sum_{s=-\infty}^{\infty} \mathbf{b}(s)\mathbf{f}(t - s) = \mathbf{g}(t),$$

which we may write more concisely as $\mathbf{b}(t) * \mathbf{f}(t) = \mathbf{g}(t)$. The impulse response, $\mathbf{b}(t)$ of a system will always be an $n \times n$ matrix but input $\mathbf{f}(t)$ and output $\mathbf{g}(t)$ may be $n \times m$ matrices. In terms of frequency functions, we have $\mathbf{B}(\omega)\, \mathbf{F}(\omega) = \mathbf{G}(\omega)$, where $\mathbf{f} \leftrightarrow \mathbf{F}$ and $\mathbf{g} \leftrightarrow \mathbf{G}$. The square-gain of the system is defined as $|\mathbf{B}(\omega)|^2$.

To make the operational concepts just introduced more precise, let us introduce the notion of an abstract space $\mathbf{L}^2$ whose elements are $n \times m$ matrices $\mathbf{f}, \mathbf{g}, \mathbf{h}, \cdots$, and constants are $n \times n$ constant matrices $\boldsymbol{\alpha}, \boldsymbol{\beta}, \cdots$. We see then that we can perform the operations of addition, e.g., $\mathbf{f} + \mathbf{g}$, and multiplication by a constant, e.g., $\boldsymbol{\alpha}\mathbf{f}$. We postulate that $\mathbf{L}^2$ is a linear space. In addition, we postulate that an inner product, denoted by $(\mathbf{f}, \mathbf{g})$, is defined for any two elements $\mathbf{f}$ and $\mathbf{g}$. For given $\mathbf{f}$ and $\mathbf{g}$, the inner product is a constant $n \times n$ matrix. We define the norm of an element $\mathbf{f}$ in space $\mathbf{L}^2$ as $\|\mathbf{f}\| = [\mathrm{tr}\ (\mathbf{f}, \mathbf{f})]^{1/2}$. We postulate that space $\mathbf{L}^2$ with norm $\|\mathbf{f}\|$ is complete. Two elements, $\mathbf{f}$ and $\mathbf{g}$, are called normal, or orthogonal, provided $(\mathbf{f}, \mathbf{g}) = \mathbf{0}$. A sequence $\mathbf{f}_N$ $(N = 0, 1, 2, \cdots)$ of elements in $\mathbf{L}^2$ is said to *converge in the mean* to an element $\mathbf{f}$ in $\mathbf{L}^2$, provided the numbers $\|\mathbf{f} - \mathbf{f}_N\|$ converge to zero as $N \to \infty$. We then write

$$\underset{N \to \infty}{\text{l.i.m.}}\ \mathbf{f}_N = \mathbf{f},$$

where l.i.m. stands for *limit in the mean*.

Definition 1. *A set $\{\mathbf{g}(r), r \in R\}$ of elements in $\mathbf{L}^2$ is said to be maximal in a subspace of $\mathbf{L}^2$ or to span a subspace of $\mathbf{L}^2$ if for every element $\mathbf{f}$ of the subspace there exist constants $\boldsymbol{\lambda}_{Nj}$ and elements $\mathbf{g}(r_{Nj})$ of the set such that*

$$\mathbf{f} = \underset{N \to \infty}{\text{l.i.m.}}\ \sum_{j=-N}^{N} \boldsymbol{\lambda}_{Nj}\mathbf{g}(r_{Nj}).$$

The following are examples of spaces that are specific realizations of abstract space $\mathbf{L}^2$:

1. Space $\mathbf{L}_t^2$ (where t is an integer denoting time).

   Elements.　All $n \times n$ matrix-valued $\mathbf{f}(t)$ such that $\operatorname{tr} \sum\limits_{t=-\infty}^{\infty} |\mathbf{f}(t)|^2 < \infty$.

   Inner product.　$(\mathbf{f}, \mathbf{g})_t = \sum\limits_{t=-\infty}^{\infty} \mathbf{f}(t)\overline{\mathbf{g}(t)}$.

2. Space $\mathbf{L}_t^2(0, \infty)$ (this space is a proper subspace of $\mathbf{L}_t^2$).
   Elements.　All $n \times n$ matrix-valued $\mathbf{w}(t)$ such that $\mathbf{w}(t) = 0$ for $t < 0$

   and $\operatorname{tr} \sum\limits_{t=0}^{\infty} |\mathbf{w}(t)|^2 < \infty$.

   Inner product.　Same as that for $\mathbf{L}_t^2$.

3. Space $\mathbf{L}_\omega^2$ (where ω is a real variable denoting angular frequency).
   Elements.　All $n \times n$ matrix-valued $\mathbf{F}(\omega)$ such that

$$\operatorname{tr} \frac{1}{2\pi} \int_{-\pi}^{\pi} |\mathbf{F}(\omega)|^2 \, d\omega < \infty.$$

   Inner product.　$(\mathbf{F}, \mathbf{G})_\omega = (1/2\pi) \int_{-\pi}^{\pi} \mathbf{F}(\omega)\overline{\mathbf{G}(\omega)} \, d\omega.$

4. Space $\mathbf{L}_P^2$ (where P is probability measure).
   Elements.　All $n \times m$ matrix-valued random variables $\mathbf{x}$ such that $\operatorname{tr} E\{|\mathbf{x}|^2\} < \infty$.
   Inner product.　$(\mathbf{x}, \mathbf{y})_P = E\{\mathbf{x}\bar{\mathbf{y}}\}.$

Let us now consider the Fourier transform pair $\mathbf{f}(t) \leftrightarrow \mathbf{F}(\omega)$. Plancherel's theorem states that $\mathbf{f}(t)$ belongs to $\mathbf{L}_t^2$ if and only if $\mathbf{F}(\omega)$ belongs to $\mathbf{L}_\omega^2$, and then $\|\mathbf{f}(t)\|_t = \|\mathbf{F}(\omega)\|_\omega$ (*Bessel's equality*). If $\mathbf{f}(t)$ and $\mathbf{g}(t)$ belong to $\mathbf{L}_t^2$ and $\mathbf{f}(t) \leftrightarrow \mathbf{F}(\omega)$, $\mathbf{g}(t) \leftrightarrow \mathbf{G}(\omega)$, then *Parseval's equality* states that $(\mathbf{f}, \mathbf{g})_t = (\mathbf{F}, \mathbf{G})_\omega.$

Nevertheless, in order to develop the structure of stationary processes, we must consider Fourier transform pairs $\mathbf{b}(t) \leftrightarrow \mathbf{B}(\omega)$ which are not necessarily members of $\mathbf{L}_t^2 \leftrightarrow \mathbf{L}_\omega^2$. To do so, let us consider the pairs $\mathbf{f}(t) \leftrightarrow \mathbf{F}(\omega)$ and $\mathbf{g}(t) \leftrightarrow \mathbf{G}(\omega)$, each of which does belong to $\mathbf{L}_t^2 \leftrightarrow \mathbf{L}_\omega^2$. Let us then define the Fourier transform pair

$$\sum_{s=-N}^{N} \mathbf{b}_N(s)\mathbf{f}(t - s) \leftrightarrow \sum_{s=-N}^{N} \mathbf{b}_N(s)e^{-i\omega s}\mathbf{F}(\omega) = \mathbf{B}_N(\omega)\mathbf{F}(\omega),$$

where the $\mathbf{b}_N(s)$, $-N \leqslant s \leqslant N$, are constants and

$$\mathbf{b}_N(s) \leftrightarrow \mathbf{B}_N(\omega) = \sum_{s=-N}^{N} \mathbf{b}_N(s)e^{-i\omega s}.$$

We are now prepared to define a generalized function.

Definition 2.　*Let $\mathbf{f}(t)$ and $\mathbf{g}(t)$ belong to $\mathbf{L}_t^2$. Then the sequence of constants* $\mathbf{b}_N(s)$, $-N \leqslant s \leqslant N$, $N = 0, 1, 2, \cdots$, *represents a generalized time-function,* *denoted by* $\mathbf{b}(s)$, *provided that in space* $\mathbf{L}_t^2$

$$\mathbf{g}(t) = \underset{N \to \infty}{\text{l.i.m.}} \sum_{s=-N}^{N} \mathbf{b}_N(s)\mathbf{f}(t - s).$$

We then write

$$\mathbf{g}(t) = \sum_{s=-\infty}^{\infty} \mathbf{b}(s)\mathbf{f}(t - s) = \mathbf{b}(t) * \mathbf{f}(t).$$

Bessel's equality yields

$$\left\| \mathbf{g}(t) - \sum_{s=-N}^{N} \mathbf{b}_N(s)\mathbf{f}(t - s) \right\|_t = \left\| \mathbf{G}(\omega) - \mathbf{B}_N(\omega)\mathbf{F}(\omega) \right\|_\omega.$$

Hence, if $\mathbf{b}_N(s)$, $-N \leqslant s \leqslant N$, $N = 0, 1, 2, \cdots$ is a generalized function, then in space $\mathbf{L}_\omega^2$

$$\underset{N \to \infty}{\text{l.i.m.}} \ \mathbf{B}_N(\omega) \ \mathbf{F}(\omega)$$

exists, which we denote by $\mathbf{B}(\omega) \ \mathbf{F}(\omega)$. We may then write the *generalized Fourier transform pair* $\mathbf{b}(t) \leftrightarrow \mathbf{B}(\omega)$.

We now want to develop properties of wavelets.

Definition 3. *An element* $\mathbf{w}(t)$ *of space* $\mathbf{L}_t^2 \ (0, \ \infty)$ *is called a wavelet.*

Definition 4. *A wavelet* $\mathbf{w}(t) \leftrightarrow \mathbf{W}(\omega)$ *is called invertible if the inverse matrix* $\mathbf{W}^{-1}(\omega)$ *exists for almost all* ω. *Then the inverse wavelet* $\mathbf{w}^{-1}(t)$ *is defined by* $\mathbf{w}^{-1}(t) \leftrightarrow \mathbf{W}^{-1}(\omega)$.

Since $\mathbf{W}^{-1}(\omega) \ \mathbf{W}(\omega) = \mathbf{I}$, where $\mathbf{I}$ is the identity matrix, we have

$$\mathbf{w}^{-1}(t) * \mathbf{w}(t) = \frac{1}{2\pi} \int_{-\pi}^{\pi} e^{i\omega t}\mathbf{I} \, d\omega = \boldsymbol{\delta}(t),$$

where $\boldsymbol{\delta}(t)$ is a matrix with the Kronecker delta function $\delta(t)$ ($= 1$ for $t = 0$ and $= 0$ for $t \neq 0$) on the main diagonal and zeros elsewhere. Because $\boldsymbol{\delta}(t)$ belongs to $\mathbf{L}_t^2$, we see from the equation

$$\boldsymbol{\delta}(t) = \sum_{s=-\infty}^{\infty} \mathbf{w}^{-1}(s)\mathbf{w}(t - s)$$

that $\mathbf{w}^{-1}(t) \leftrightarrow \mathbf{W}^{-1}(\omega)$ is a generalized Fourier transform pair; that is, $\mathbf{w}^{-1}(t)$ stands for a sequence of constants $\mathbf{w}_N^{-1}(s)$, $-N \leqslant s \leqslant N$, $N = 0, 1, 2, \cdots$, such that in space $\mathbf{L}_t^2$

$$\boldsymbol{\delta}(t) = \underset{N \to \infty}{\text{l.i.m.}} \sum_{s=-N}^{N} \mathbf{w}_N^{-1}(s)\mathbf{w}(t - s),$$

or equivalently in space $\mathbf{L}_\omega^2$

$$\mathbf{I} = \underset{N \to \infty}{\text{l.i.m.}} \sum_{s=-N}^{N} \mathbf{w}_N^{-1}(s)e^{-i\omega s}\mathbf{W}(\omega) = \underset{N \to \infty}{\text{l.i.m.}} \ \mathbf{w}_N^{-1}(\omega) \ \mathbf{W}(\omega)$$

where

$$\mathbf{W}_N^{-1}(\omega) = \sum_{s=-N}^{N} \mathbf{w}_N^{-1}(s)e^{-i\omega s}.$$

A system is called *dissipative* if its impulse response $\mathbf{b}(t)$ belongs to $\mathbf{L}_t^2$ and *realizable* if its impulse response $\mathbf{b}(t) = 0$ for $t < 0$. Thus a system is realizable and dissipative if and only if its impulse response is a wavelet. We now define *minimum-delay*.

Definition 5. *A wavelet (or realizable dissipative system) $\mathbf{w}(t)$ is called minimum-delay provided it has a realizable inverse $\mathbf{w}^{-1}(t)$ (i.e., $\mathbf{w}^{-1}(t) = 0$ for $t < 0$).*

The following minimum-delay theorem can then be established (cf. Robinson, 1962, Theorems 9.40 and 9.80):

Theorem 1. *Let $\mathbf{w}(t)$ be an invertible wavelet. Then each of the following conditions is necessary and sufficient that $\mathbf{w}(t)$ is minimum-delay:*

1. *The partial energy of the wavelet $\mathbf{w}(t)$ given by*

$$\operatorname{tr} \sum_{t=0}^{\tau} \left| \mathbf{w}(t) \right|^2$$

   is a maximum (with respect to wavelets in the class of all wavelets with the same square-gain as $\mathbf{w}(t)$) for all time $\tau \geqslant 0$.
2. *The set $\{\mathbf{w}(t - r), r \geqslant 0\}$ spans $\mathbf{L}_t^2(0, \infty)$; that is, given any wavelet $\mathbf{v}(t)$, there exists a realizable generalized function $\boldsymbol{\gamma}(t)$ such that*

$$\mathbf{v}(t) = \boldsymbol{\gamma}(t) * \mathbf{w}(t) = \sum_{s=0}^{\infty} \boldsymbol{\gamma}(s)\mathbf{w}(t - s),$$

   or, in other words, there exists a sequence of constants $\boldsymbol{\gamma}_N(s)$, $0 \leqslant s \leqslant N$, $N = 0, 1, 2, \cdots$, such that

$$\mathbf{v}(t) = \operatorname*{l.i.m.}_{N \to \infty} \sum_{s=0}^{N} \boldsymbol{\gamma}_N(s)\mathbf{w}(t - s).$$

We now define *all-pass* systems, which are also called phase-shift systems or phase-equalizers in communication engineering.

Definition 6. *A system $\mathbf{a}(t) \leftrightarrow \mathbf{A}(\omega)$ is called an all-pass system provided that it is realizable and has square-gain $\left| \mathbf{A}(\omega) \right|^2$ equal to the identity matrix.*

Because $\mathbf{A}(\omega)$ belongs to $\mathbf{L}_\omega^2$, it follows from Bessel's equality that $\mathbf{a}(t)$ belongs to $\mathbf{L}_t^2$, so $\mathbf{a}(t)$ is a wavelet. We see that an all-pass system is always invertible and has the inverse

$$\mathbf{a}^{-1}(t) \leftrightarrow \mathbf{A}^{-1}(\omega) = \overline{\mathbf{A}(\omega)}.$$

Because $\mathbf{A}^{-1}(\omega)$ belongs to $\mathbf{L}_\omega^2$, it follows that $\mathbf{a}^{-1}(t)$ belongs to $\mathbf{L}_t^2$. We shall call an all-pass system $\mathbf{a}(t)$ *nontrivial* if its inverse $\mathbf{a}^{-1}(t)$ is nonrealizable and *trivial* if its inverse is realizable. An all-pass system is trivial if and only if its transfer function $\mathbf{A}(\omega)$ is a constant, unitary matrix $\mathbf{A}_0$.

The following theorem gives a *canonical representation* for invertible wavelets (or, equivalently, invertible realizable dissipative systems; cf. Robinson, 1962, Theorem 9.61).

Theorem 2. $\mathbf{w}(t)$ *is an invertible wavelet if and only if*

$$\mathbf{w}(t) = \mathbf{w}_0(t) * \mathbf{a}(t),$$

where $\mathbf{a}(t)$ *is an all-pass wavelet and* $\mathbf{w}_0(t)$ *is a minimum-delay wavelet with the same square-gain as* $\mathbf{w}(t)$.

We now wish to show how the structural properties of stationary stochastic processes may be developed by making use of the operational calculus. The expected value, or mean, of a random variable $\mathbf{x}$ is denoted by $E\{\mathbf{x}\}$. As we have seen, the class of all random variables $\mathbf{x}$ with finite norm forms the space $\mathbf{L}_P^2$ with inner product $(\mathbf{x}, \mathbf{y})_P = E\{\mathbf{x}\bar{\mathbf{y}}\}$. A stochastic process is a family of random variables. For example, $\mathbf{x}(t)$ represents a stochastic process depending on the time-parameter t. We now have the operational definitions:

Definition 7. *A stochastic process* $\boldsymbol{\theta}(t)$ *with zero mean and autocovariance* $E\{\boldsymbol{\theta}(t)\overline{\boldsymbol{\theta}(\mathbf{s})}\} = \boldsymbol{\delta}(t - s)$, *where* $\boldsymbol{\delta}(t - s)$ *is a matrix with the Kronecker delta function* $\delta(t - s)$ *on the main diagonal and zeros elsewhere, is called white noise.*

Definition 8. *Let white noise* $\boldsymbol{\theta}(t)$ *be the input to a dissipative system* $\mathbf{b}(t) \leftrightarrow \mathbf{B}(\omega)$. *Then the output* $\mathbf{x}(t) = \mathbf{b}(t) * \boldsymbol{\theta}(t)$ *is called a moving-average multiple second-order stationary stochastic process.*

In case $\mathbf{b}(t)$ is, in addition, an invertible realizable system, then $\mathbf{x}(t)$ is called *regular*.

Definition 9. *Let white noise* $\boldsymbol{\theta}(t)$ *be the input to an invertible realizable dissipative system* $\mathbf{w}(t) \leftrightarrow \mathbf{W}(\omega)$ [*i.e.,* $\mathbf{w}(t)$ *is an invertible wavelet*]. *Then the output* $\mathbf{x}(t) = \mathbf{w}(t) * \boldsymbol{\theta}(t)$ *is called a regular multiple second-order stationary stochastic process. The process* $\mathbf{x}(t)$ *is said to have a minimum-delay structure provided that* $\mathbf{w}(t)$ *is minimum-delay.*

Two moving-average multiple stationary stochastic processes $\mathbf{x}(t)$ and $\mathbf{y}(t)$ are said to be *stationarily correlated*, provided that their *cross covariance* $\boldsymbol{\phi}_{\mathbf{xy}}(s) = E\{\mathbf{x}(t + s)\overline{\mathbf{y}(t)}\}$ does not depend on t. The cross covariance may be expressed as the Fourier transform

$$\boldsymbol{\phi}_{\mathbf{xy}}(s) = \frac{1}{2\pi} \int_{-\pi}^{\pi} \boldsymbol{\Phi}_{\mathbf{xy}}(\omega) e^{i\omega t}\, d\omega,$$

where $\boldsymbol{\Phi}_{\mathbf{xy}}(\omega)$ is called the *cross-spectral density*. We readily see that a moving-average multiple stationary process $\mathbf{x}(t) = \mathbf{b}(t) * \boldsymbol{\theta}(t)$ is stationarily correlated with itself; that is, the autocovariance $\boldsymbol{\phi}_{\mathbf{xx}}(s) = E\{\mathbf{x}(t + s)\overline{\mathbf{x}(t)}\}$ does not depend on t. We also see that the *autospectral density* $\Phi_{xx}(\omega)$ is equal to the square-gain of the system $\mathbf{B}(\omega)$; that is, $\Phi_{xx}(\omega) = |\mathbf{B}(\omega)|^2$.

Let $\mathbf{x}(t) = \mathbf{w}(t) * \boldsymbol{\theta}(t)$ be a regular stationary stochastic process. Let $\mathbf{m}(t)$ be a moving-average stationary stochastic process that is stationarily correlated with $\mathbf{x}(t)$. The process $\mathbf{m}(t)$ may be called the *message*. Let $\boldsymbol{\gamma}(t)$ be a realizable generalized function such that

$$\mathbf{v}(t) = \boldsymbol{\gamma}(t) * \mathbf{w}(t) = \underset{N \to \infty}{\text{l.i.m.}} \sum_{s=0}^{N} \boldsymbol{\gamma}_N(s) \, \mathbf{w}(t - s)$$

belongs to $\mathbf{L}_t^2(0, \infty)$. Then

$$\boldsymbol{\gamma}(t) * \mathbf{x}(t) = \underset{N \to \infty}{\text{l.i.m.}} \sum_{s=0}^{N} \boldsymbol{\gamma}_N(s) \, \mathbf{x}(t - s)$$

belongs to L_P^2 because

$$\left\| \boldsymbol{\gamma}(t) * \mathbf{x}(t) \right\|_P = \left\| \boldsymbol{\gamma}(t) * \mathbf{w}(t) * \boldsymbol{\theta}(t) \right\|_P = \left\| \mathbf{v}(t) * \boldsymbol{\theta}(t) \right\|_P$$

$$= E \left\{ \sum_{s=0}^{\infty} \mathbf{v}(s) \, \boldsymbol{\theta}(t - s) \, \overline{\sum_{r=0}^{\infty} \mathbf{v}(r) \boldsymbol{\theta}(t - r)} \right\}$$

$$= \sum_{s=0}^{\infty} \sum_{r=0}^{\infty} \mathbf{v}(s) E\{ \boldsymbol{\theta}(t - s)\overline{\boldsymbol{\theta}(t - r)} \}\overline{\mathbf{v}(r)}$$

$$= \sum_{s=0}^{\infty} \sum_{r=0}^{\infty} \mathbf{v}(s)\boldsymbol{\delta}(r - s)\overline{\mathbf{v}(r)} = \sum_{s=0}^{\infty} \mathbf{v}(s)\overline{\mathbf{v}(s)}$$

$$= \left\| \mathbf{v}(s) \right\|_t = \left\| \boldsymbol{\gamma}(s) * \mathbf{w}(s) \right\|_t.$$

Let α be a real number which may be positive, zero, or negative. We shall call

$\mathbf{m}(t + \alpha)$	: the *regressand*
$\mathbf{x}(t)$	: the *regressor*
$\boldsymbol{\gamma}(t)$	: the *regression coefficient*
$\boldsymbol{\gamma}(t) * \mathbf{x}(t)$	: the *regression*
$\mathbf{m}(t + \alpha) - \boldsymbol{\gamma}(t) * \mathbf{x}(t)$	: the *residual*, denoted by $\mathbf{e}(t)$.

Hence we may write the linear regression equation

$$\mathbf{m}(t + \alpha) = \boldsymbol{\gamma}(t) * \mathbf{x}(t) + \mathbf{e}(t),$$

which in words is

regressand = (regression coefficient) * (regressor) + residual.

If we determine the regression coefficient $\boldsymbol{\gamma}(t)$ such that the squared-norm $\|\mathbf{e}(t)\|_P^2$ of the residual is a minimum, then the regression equation is called a *least-squares linear regression equation*, the regression coefficient $\boldsymbol{\gamma}(t)$ is called the least-squares regression coefficient, and the residual $\mathbf{e}(t)$ is called the least-squares residual. We now have the following theorem (cf. Robinson, 1962, Theorem 11.50):

Theorem 3. *The equation*

$$\mathbf{m}(t + \alpha) = \boldsymbol{\gamma}(t) * \mathbf{x}(t) + \mathbf{e}(t)$$

is the least-squares linear regression equation if and only if

$$\boldsymbol{\gamma}(t) = \sum_{s=0}^{\infty} \mathbf{g}_0(s + \alpha)\mathbf{w}_0^{-1}(t - s),$$

where $\mathbf{w}_0(t)$ is a minimum-delay wavelet with square-gain $\Phi_{\mathbf{xx}}(\omega)$ and where

$$\mathbf{g}_0(t) = \boldsymbol{\phi}_{\mathbf{mx}}(t) * \overline{\mathbf{w}_0^{-1}(-t)}.$$

The minimum squared-norm of the residual is

$$\|\mathbf{e}(t)\|_P^2 = \operatorname{tr} \boldsymbol{\phi}_{\mathbf{mm}}(0) - \operatorname{tr} \sum_{s=\alpha}^{\infty} |\mathbf{g}_0(s)|^2.$$

In the special case in which $\mathbf{m}(t) = \mathbf{x}(t)$ and $\alpha > 0$, Theorem 3 becomes the following:

Theorem 4. *The equation*

$$\mathbf{x}(t + \alpha) = \mathbf{k}(t) * \mathbf{x}(t) + \mathbf{e}(t)$$

is the least-squares linear regression equation if and only if

$$\mathbf{k}(t) = \sum_{s=0}^{\infty} \mathbf{w}_0(s + \alpha)\mathbf{w}_0^{-1}(t - s),$$

where $\mathbf{w}_0(t)$ is a minimum-delay wavelet with square-gain $\Phi_{\mathbf{xx}}(\omega)$. The minimum squared-norm of the residual is given by the partial energy

$$\|\mathbf{e}(t)\|_P^2 = \operatorname{tr} \sum_{s=0}^{\alpha-1} |\mathbf{w}_0(s)|^2.$$

As a matter of terminology, we call $\mathbf{k}(t)$ the least-squares predicting system, and $\hat{\mathbf{x}}(t + \alpha) = \mathbf{k}(t) * \mathbf{x}(t)$ the least-squares prediction for predicting lead $\alpha > 0$. We see that the least-squares prediction system $\mathbf{k}(t)$ has transfer function

$$\mathbf{K}(\omega) = \left[\sum_{t=0}^{\infty} e^{-i\omega t}\mathbf{w}_0(t + \alpha) \right] \mathbf{W}_0^{-1}(\omega),$$

where $\mathbf{w}_0(t) \leftrightarrow \mathbf{W}_0(\omega)$ is minimum-delay with square-gain $\Phi_{\mathbf{xx}}(\omega)$. The following theorem brings out the connection between the concepts of minimum-delay and least-squares (cf. Robinson, 1962, Theorem 11.70).

Theorem 5. *Let a regular stationary stochastic process have the given structure $\mathbf{x}(t) = \mathbf{w}(t) * \boldsymbol{\theta}(t)$. Then the structure is minimum-delay if and only if*

$$\mathbf{x}(t + \alpha) = \sum_{s=0}^{\infty} \mathbf{w}(s + \alpha)\boldsymbol{\theta}(t - s) + \sum_{s=-\alpha}^{-1} \mathbf{w}(s + \alpha)\boldsymbol{\theta}(t - s)$$

is the least-squares linear regression equation where the first term on the right-hand side is the least-squares prediction and the second term is the least-squares residual.

A stationary process $\mathbf{x}(t)$ represents a flow of information with time. Let t be the present time. Let us define

$$\mathbf{T}(\mathbf{x}) = \text{total information} = \text{S.s.b.} \ \{\mathbf{x}(s), \ -\infty < s < \infty\},$$

$$\mathbf{P}_t(\mathbf{x}) = \text{present information} = \text{S.s.b.} \ \{\mathbf{x}(s), \ s \leqslant t\},$$

where S.s.b. stands for "subspace spanned by" with reference to space L_P^2.

Theorem 6. *Let* $\mathbf{x}(t)$ *and* $\mathbf{y}(t) = \mathbf{b}(t) * \mathbf{x}(t)$ *be moving–average stationary processes where* $\mathbf{b}(t)$ *belongs to* $\mathbf{L}_t^2$. *Then*
   1. $\mathbf{T}(\mathbf{x}) \supset \mathbf{T}(\mathbf{y})$.
   2. $\mathbf{T}(\mathbf{x}) = \mathbf{T}(\mathbf{y})$ *if and only if* $\mathbf{b}$ *is invertible.*
   3. $\mathbf{P}_t(\mathbf{x}) \supset \mathbf{P}_t(\mathbf{y})$ *if and only if* $\mathbf{b}$ *is a wavelet.*
   4. $\mathbf{P}_t(\mathbf{x}) = \mathbf{P}_t(\mathbf{y})$ *if and only if* $\mathbf{b}$ *is a minimum–delay wavelet.*

If $\mathbf{T}(\mathbf{x}) = \mathbf{T}(\mathbf{y})$, we say that the system $\mathbf{b}(t)$ does not destroy the information contained in $\mathbf{x}(t)$. If $\mathbf{P}_t(\mathbf{x}) = \mathbf{P}_t(\mathbf{y})$, we say that the system $\mathbf{b}(t)$ does not delay the information contained in $\mathbf{x}(t)$. Thus we see that a realizable dissipative system is minimum-delay if and only if, in the transmission of information from input to output, the system neither destroys nor delays the information.

2. REGULAR SINGLE STATIONARY STOCHASTIC PROCESSES FOR DISCRETE TIME

To illustrate the foregoing operational concepts in terms of the usual analytic approach, we shall now consider in detail the case of regular single stationary stochastic processes for discrete time.

We see that a wavelet w_t now becomes a sequence of complex numbers such that $w_t = 0$ for $t < 0$ and $\sum_0^\infty |w_t|^2 < \infty$. Its L^2-Fourier transform is given by

$$W(\omega) = \underset{N \to \infty}{\text{l.i.m.}} \sum_{t=0}^{N} w_t e^{-i\omega t}, \text{ a.e.}$$

The *z-transform* of a wavelet w_t is defined as

$$W^*(z) = \sum_{t=0}^{\infty} w_t z^t.$$

$W^*(z)$ is an analytic function for $|z| < 1$ and has the limit value

$$W^*(e^{-i\omega}) = \sum_{t=0}^{\infty} w_t e^{-i\omega t} = W(\omega), \text{ a.e.,}$$

on the unit circle $z = e^{-i\omega}$.

Let the square-gain of a system be given by $M^2(\omega)$, where the *gain* $M(\omega)$ satisfies

$$M(\omega) \geqslant 0, \qquad \frac{1}{2\pi} \int_{-\pi}^{\pi} M^2(\omega)\, d\omega < \infty,$$

and

$$\frac{1}{2\pi} \int_{-\pi}^{\pi} \log M(\omega)\, d\omega > -\infty.$$

Then the z-transform of *the central minimum-delay wavelet b_t* with gain $M(\omega)$ is given by

$$B^*(z) = \sum_{t=0}^{\infty} b_t z^t = \exp\left[\frac{1}{2\pi} \int_{-\pi}^{\pi} \frac{e^{i\omega} + z}{e^{i\omega} - z} \log M(\omega)\, d\omega\right], \; |z| < 1.$$

All other minimum-delay wavelets w_{0t} with the same gain $M(\omega)$ are given by $w_{0t} = cb_t$ where c is an arbitrary constant with modulus $|c| = 1$.

The z-transform of an *all-pass wavelet a_t* is given by

$$A^*(z) = \sum_{t=0}^{\infty} a_t z^t = cz^m \prod_k \frac{z_k - z}{1 - \bar{z}_k z} \frac{|z_k|}{z_k} \exp\left[-\frac{1}{2\pi} \int_{-\pi}^{\pi} \frac{e^{i\omega} + z}{e^{i\omega} - z}\, d\beta(\omega)\right],$$

where $|z| < 1$, $|c| = 1$, m is a nonnegative integer, $\{z_k\}$ is an empty or non-empty set satisfying $|z_k| < 1$, and

$$\sum_k (1 - |z_k|) < \infty;$$

$\beta(\omega)$ is a real bounded nondecreasing function whose derivative vanishes almost everywhere. If $|A^*(z)| = 1$ for $|z| < 1$, then $A^*(z)$ is *trivial;* otherwise $A^*(z)$ is *nontrivial.* An all-pass wavelet a_t has gain $|A(\omega)| = 1$, a.e.

In the scalar case that we are treating in this section we see that a wavelet is invertible if and only if it is nonnull. The *canonical representation* then becomes the following:

Theorem 7. w_t *is an invertible wavelet if and only if*

$$w_t = \sum_{s=0}^{t} b_s a_{t-s}$$

where a_t is an all-pass wavelet and b_t is the central minimum-delay wavelet with the same gain as w_t. This representation of w_t is unique.

As a corollary to Theorem 7 we have the following:

Theorem 8. $W^*(z)$ *is the z-transform of an invertible wavelet if and only if*

$$W^*(z) = B^*(z)\, A^*(z), \qquad |z| < 1,$$

where $A^*(z)$ is the z-transform of an all-pass wavelet and $B^*(z)$ is the z-transform *of the central minimum-delay wavelet with the same gain as $W^*(z)$. This representation of $W^*(z)$ is unique.*

Theorem 7 shows that the partial sums $\sum_0^N b_s a_{t-s}$ for fixed t converge to the value w_t. The following theorem shows that these partial sums also converge in the mean to the wavelet w_t.

Theorem 9. *In Hilbert space $L_t^2(0, \infty)$ we have*

$$w_t = \operatorname*{l.i.m.}_{N \to \infty} \sum_{s=0}^N b_s a_{t-s}.$$

Proof. By Bessel's equality, we have

$$\sum_{t=0}^\infty \left| w_t - \sum_{s=0}^N b_s a_{t-s} \right|^2 = \frac{1}{2\pi} \int_{-\pi}^\pi \left| W(\omega) - \sum_{s=0}^N b_s e^{-i\omega s} A(\omega) \right|^2 d\omega.$$

Because $W(\omega) = B(\omega) A(\omega)$ and $|A(\omega)| = 1$, the foregoing is equal to

$$\frac{1}{2\pi} \int_{-\pi}^\pi \left| B(\omega) - \sum_{s=0}^N b_s e^{-i\omega s} \right|^2 d\omega = \sum_{S=N+1}^\infty |b_s|^2,$$

which can be made as small as we please by choosing N large enough.   Q.E.D.

The following theorems in this section give conditions for explicit representations of the linear predictor, that is, conditions under which generalized functions reduce to wavelets.

Theorem 10. *Let x_t be a regular stationary stochastic process with spectral density $\Phi(\omega)$. A necessary and sufficient condition that the transfer function $K(\omega)$ of the least-squares linear predicting system (for any prediction lead $\alpha > 0$) be expressible as a Fourier series*

$$K(\omega) = \sum_{s=0}^\infty k_s e^{-i\omega s}$$

where k_s is a wavelet is that

$$\frac{1}{2\pi} \int_{-\pi}^\pi \frac{d\omega}{\Phi(\omega)} < \infty. \tag{1}$$

Proof (necessary condition).   By hypothesis,

$$K(\omega) = \frac{\sum_{s=0}^\infty b_{s+\alpha} e^{-i\omega s}}{\sum_{s=0}^\infty b_s e^{-i\omega s}} = \sum_{s=0}^\infty k_s e^{-i\omega s}, \qquad \sum_{s=0}^\infty |k_s|^2 < \infty,$$

where $b_0, b_1, b_2, \cdots$ is the central minimum-delay wavelet.

Let the prediction lead $\alpha = 1$.   Then, using the equation at the bottom of p. 103 in Robinson (1959), we have under the Stone-Kolmogorov isomorphism the correspondence $\hat{x}_1 \leftrightarrow K(\omega)$ (*ibid.*, p. 83).   Hence it follows that $\hat{x}_0 \leftrightarrow K(\omega)e^{-i\omega}$ (*ibid.*, p. 86).   We also have under the Stone-Kolmogorov isomorphism that $x_t \leftrightarrow e^{i\omega t}$ (*ibid.*, p. 85) and $\theta_t \leftrightarrow e^{i\omega t}/B(\omega)$ (*ibid.*, p. 100).   Therefore

$$b_0\theta_0 = x_0 - \hat{x}_0 \leftrightarrow \frac{b_0}{B(\omega)} = 1 - K(\omega)e^{-i\omega}.$$

It then follows that

$$\frac{1}{B(\omega)} = \frac{1}{b_0} - \frac{k_0}{b_0}e^{-i\omega} - \frac{k_1}{b_0}e^{-i\omega 2} - \frac{k_2}{b_0}e^{-i\omega 3} - \cdots$$

where

$$\frac{1}{2\pi}\int_{-\pi}^{\pi}\frac{d\omega}{|B(\omega)|^2} = \frac{1}{b_0^2}\left(1 + \sum_{0}^{\infty}|k_s|^2\right),$$

which converges because, by hypothesis, $\sum_{0}^{\infty}|k_s|^2 < \infty$.   Since $\Phi(\omega) = |B(\omega)|^2$, the necessary condition is established.   Q.E.D.

Proof (sufficient condition).   Now, using more explicit notation to indicate the prediction lead α, we denote the transfer function of the least-squares linear predicting system by

$$K_\alpha(\omega) = \frac{B_\alpha(\omega)}{B(\omega)} \quad\text{where}\quad B_\alpha(\omega) = \sum_{t=0}^{\infty}b_{t+\alpha}e^{-i\omega t},$$

where b_0, b_1, b_2, $\cdots$ is the central minimum-delay wavelet.   Because $B_0/B = 1$, it follows that B_0/B belong to L_ω^2.   Suppose B_α/B belongs to L_ω^2.   We note that $1/B$ belongs to L_ω^2, since $|B(w)|^2 = \Phi(\omega)$ and by hypothesis $1/\Phi(\omega)$ is integrable.   Hence the linear combination

$$e^{i\omega}\frac{B_\alpha}{B} - b_\alpha e^{i\omega}\frac{1}{B} = \frac{B_{\alpha+1}}{B}$$

belongs to L_ω^2.   Therefore by finite induction $K_\alpha = B_\alpha/B$ belongs to L_ω^2 for $\alpha = 0$, 1, 2, $\cdots$.   Since $B(\omega)$ is central minimum-delay, we have the canonical representation

$$B^*(z) = \exp\left[\frac{1}{2\pi}\int_{-\pi}^{\pi}\frac{e^{i\omega}+z}{e^{i\omega}-z}\log|B(\omega)|\,d\omega\right], \qquad |z| < 1,$$

and because b_α, $b_{\alpha+1}$, $\cdots$ is a wavelet we have the canonical representation

$$B_\alpha^*(z) = A^*(z)\exp\left[\frac{1}{2\pi}\int_{-\pi}^{\pi}\frac{e^{i\omega}+z}{e^{i\omega}-z}\log|B_\alpha(\omega)|\,d\omega\right], \qquad |z| < 1.$$

Combining the last two equations, we find

$$\frac{B_\alpha^*(z)}{B^*(z)} = A^*(z) \exp\left[\frac{1}{2\pi}\int_{-\pi}^{\pi}\frac{e^{i\omega}+z}{e^{i\omega}-z}\log\left|\frac{B_\alpha(\omega)}{B(\omega)}\right|d\omega\right], \qquad |z| < 1. \qquad (2)$$

Because B_α/B belongs to L_ω^2, the gain $|B_\alpha/B|$ is quadratically integrable. Also

$$\log\left|\frac{B_\alpha(\omega)}{B(\omega)}\right| = \log|B_\alpha(\omega)| - \log|B(\omega)|$$

is integrable. Hence (2) is the canonical representation of $K_\alpha^*(z)$, and so $k_t(\alpha)$ is a wavelet. In fact, k_t is given by

$$k_t(\alpha) = \sum_{s=0} b_{s+\alpha}b_{t-s}^{-1}, \qquad (3)$$

where b_t^{-1} is the wavelet corresponding to $1/B(\omega)$, that is,

$$B^{-1}(\omega) = \frac{1}{B(\omega)} = \sum_{t=0}^{\infty} b_t^{-1}e^{-i\omega t}.$$

We also note that since $B_\alpha = K_\alpha B$ we have

$$b_{\alpha+N} = \sum_{s=0}^{N} k_s(\alpha)b_{N-s}, \qquad N = 0, 1, 2, \cdots. \qquad (4)$$

$$\text{Q.E.D.}$$

At this point we revert to less explicit notation, now using k_s instead of $k_s(\alpha)$ and K instead of K_α. In space L_P^2 let $\Theta(t)$ denote the closed linear manifold spanned by $\{\theta_s, s \leqslant t\}$.

Theorem 11. *Let x_t be a regular stationary stochastic process with spectral density $\Phi(\omega)$. A necessary and sufficient condition that the least-squares prediction $\hat{x}_{t+\alpha}$ (for any prediction lead $\alpha > 0$) has the representation*

$$\hat{x}_{t+\alpha} = \sum_{s=0}^{\infty} k_s x_{t-s}$$

in the sense that k_s is a wavelet and that the error of approximation

$$\hat{x}_{t+\alpha} - \sum_{s=0}^{N} k_s x_{t-s}$$

belongs to the closed linear manifold $\Theta(t - N - 1)$ whose intersection

$$\bigcap_{N=0}^{\infty} \Theta(t - N - 1) = 0$$

is that

$$\frac{1}{2\pi}\int_{-\pi}^{\pi}\frac{d\omega}{\Phi(\omega)} < \infty. \qquad (1)$$

Proof. We have

$$\hat{x}_\alpha - \sum_{s=0}^{N} k_s x_{-s} = \sum_{n=0}^{\infty} b_{\alpha+n}\theta_{-n} - \sum_{s=0}^{N} k_s \left(\sum_{r=0}^{\infty} b_r \theta_{-s-r} \right)$$

$$= \sum_{n=0}^{\infty} b_{\alpha+n}\theta_{-n} - \sum_{n=0}^{\infty} \left(\sum_{s=0}^{N} k_s b_{n-s} \right) \theta_{-n}$$

$$= \sum_{n=0}^{\infty} \left(b_{\alpha+n} - \sum_{s=0}^{N} k_s b_{n-s} \right) \theta_{-n}$$

$$= \sum_{n=N+1}^{\infty} \left(b_{\alpha+n} - \sum_{s=0}^{N} k_s b_{n-s} \right) \theta_{-n} \in \Theta(-N-1)$$

if and only if

$$b_{\alpha+N} = \sum_{s=0}^{N} k_s b_{N-s}, \qquad N = 0, 1, 2, \cdots . \tag{4}$$

But, by Theorem 10, k_s is a wavelet satisfying (4) if and only if condition (1) holds.

$$\text{Q.E.D.}$$

Theorem 12. *Let x_t be a regular stationary stochastic process with spectral density $\Phi(\omega)$. Let the condition*

$$\frac{1}{2\pi} \int_{-\pi}^{\pi} \frac{d\omega}{\Phi(\omega)} < \infty \tag{1}$$

be satisfied, so that the wavelet k_t given by

$$k_t = \sum_{s=0}^{t} b_{s+\alpha} b_{t-s}^{-1} \tag{3}$$

represents the coefficients of the least-squares predicting system for prediction lead $\alpha > 0$. Then the condition

$$b_{\alpha+n} = \underset{N \to \infty}{\text{l.i.m.}} \sum_{s=0}^{N} k_s b_{n-s} \quad \text{in space} \quad L_t^2(0, \infty)$$

is necessary and sufficient that

$$\hat{x}_{t+\alpha} = \underset{N \to \infty}{\text{l.i.m.}} \sum_{s=0}^{N} k_s x_{t-s} \quad \text{in space} \quad L_P^2.$$

Proof. We have

$$E\left\{ \left| \hat{x}_\alpha - \sum_{0}^{N} k_s x_{-s} \right|^2 \right\} = \frac{1}{2\pi} \int_{-\pi}^{\pi} \left| K(\omega) - \sum_{0}^{N} k_s e^{-i\omega s} \right|^2 \Phi(\omega)\, d\omega$$

(by the Stone-Kolmogorov isomorphism), which is

$$\frac{1}{2\pi} \int_{-\pi}^{\pi} \left| B_\alpha(\omega) - \sum_0^N k_s e^{-i\omega s} B(\omega) \right|^2 d\omega = \sum_{n=0}^{\infty} \left| b_{\alpha+n} - \sum_0^N k_s b_{n-s} \right|^2$$

(by Bessel's equality). Q.E.D.

3. APPLICATION: LEAST-SQUARES APPROXIMATE INVERSE

As a corollary to Theorem 8, we have the following:

Theorem 13.   *The finite wavelet b_0, b_1, $\cdots$, b_n is minimum-delay if and only if the polynomial $B^*(z) = b_0 + b_1 z + \cdots + b_n z^n$ has no zeros lying within the unit circle.*

Now we must distinguish two cases. The first case would be that of a finite minimum-delay wavelet b_0, b_1, $\cdots$, b_n whose z-transform $B^*(z)$ has *zeros lying on the unit circle*. In this case the (realizable) inverse wavelet b_0^{-1}, b_1^{-1}, $\cdots$ would be a generalized time-function. The second case would be that of a finite minimum-delay wavelet b_0, b_1, $\cdots$, b_n whose z-transform has *no zeros lying on the unit circle*. In this case the (realizable) inverse wavelet b_0^{-1}, b_1^{-1}, $\cdots$ is a wavelet (i.e., $(b_0^{-1})^2 + (b_1^{-1})^2 + \cdots$ converges) and can be found by polynomial division, that is

$$\frac{1}{B(z)} = b_0^{-1} + b_1^{-1}z + b_2^{-1}z^2 + b_3^{-1}z^3 + \cdots .$$

In both cases the inverse wavelet is infinite, and so to facilitate the finite operations required for a digital computer[1] we shall seek a *finite* wavelet α_0, α_1, $\cdots$, α_m that is the *approximate inverse* to b_0, b_1, $\cdots$, b_n. We shall do so by the method of least-squares; that is, we wish to determine α_0, α_1, $\cdots$, α_m such that $\sum_{s=0}^{m} \alpha_s b_{t-s}$ $(t = 0, 1, \cdots, m + n)$ is the least-squares approximation to $d_0 = 1$, $d_1 = 0$, $d_2 = 0$, $\cdots$, $d_{m+n} = 0$.

The regressor $\mathbf{B}$ is the $(m + 1) \times (m + n + 1)$ matrix

$$\mathbf{B} = \begin{bmatrix} b_0 & b_1 & b_2 & \cdots & b_n & 0 & 0 & \cdots & 0 & 0 \\ 0 & b_0 & b_1 & \cdots & b_{n-1} & b_n & 0 & \cdots & 0 & 0 \\ & & \cdots & & & & & & & \cdots \\ 0 & 0 & 0 & & & \cdots & & & b_{n-1} & b_n \end{bmatrix}$$

The regression coefficient is the $1 \times (m + 1)$ row vector $\boldsymbol{\alpha} = (\alpha_0, \alpha_1, \cdots, \alpha_m)$, the regressand is the $1 \times (m + n + 1)$ row vector $\mathbf{d} = (d_0, d_1, \cdots, d_{m+n})$, and we shall let the $1 \times (m + n + 1)$ row vector $\mathbf{e} = (e_0, e_1, \cdots,$

---

[1] In this section we carry our analysis through for real-valued wavelets. A prime indicates matrix transpose.

e_{m+n}) denote the residual. The regression equation is $\mathbf{d} = \boldsymbol{\alpha}\mathbf{B} + \mathbf{e}$. The squared-norm $\mathbf{ee}'$ of the residual is a minimum if and only if the regressor $\mathbf{B}$ is *normal* to the residual $\mathbf{e}$, that is, if and only if the *normal equations* $\mathbf{eB}' = \mathbf{0}$ are satisfied. The normal equations may be written $\mathbf{eB}' = (\mathbf{d} - \boldsymbol{\alpha}\mathbf{B})\mathbf{B}' = \mathbf{0}$, which is $\boldsymbol{\alpha}\mathbf{BB}' = \mathbf{dB}'$. Define ϕ_s as follows:

$$\phi_s = \sum_{t=0}^{n-s} b_{t+s} b_t \quad \text{for} \quad s = 0, 1, 2, \cdots, n; \ \phi_s = 0 \quad \text{for} \quad s > n;$$

$$\text{and} \quad \phi_{-s} = \phi_s.$$

Clearly ϕ_s so defined is the autocovariance function of a stationary process of finite moving-average type, and so ϕ_s is a positive definite function. We see that $\mathbf{BB}'$ is the $(m + 1) \times (m + 1)$ symmetric matrix $[\phi_{t-s}]$, where $t, s = 0, 1, 2, \cdots, m$, and that $\mathbf{dB}'$ is the $1 \times (m + 1)$ row vector $(b_0, 0, 0, \cdots, 0)$. The solution of the normal equations, $\boldsymbol{\alpha} = (\mathbf{dB}')(\mathbf{BB}')^{-1}$ gives the desired *least-squares finite approximate inverse* $\boldsymbol{\alpha}$. The least-squares approximate inverse was given by Rice (1962) who gives applications to seismic record analysis. It has been further developed by Simpson (1961) and Claerbout (1961) to take into account the presence of noise. A similar development was made by Levin (1960).

We now state our main result for this section:

Theorem 14. *Let $b_0 = 1, b_1, \cdots, b_n$ be a finite minimum-delay wavelet. Then the least-squares finite approximate inverse $\alpha_0, \alpha_1, \cdots, \alpha_m$ is minimum-delay. More specifically, the polynomial $\alpha_0 + \alpha_1 z + \cdots + \alpha_m z^m$ has no zeros lying within or on the unit circle, so that the inverse of $\alpha_0, \alpha_1, \cdots, \alpha_m$ is a wavelet and can be found by polynomial division.*

Proof. Define r_s as follows: $r_s = \phi_s$ for $s = 0, 1, \cdots, m; \ r_s$ for $s > m$ given by the equation $0 = \sum_{u=0}^{m} \alpha_u r_{s-u}$, and $r_{-s} = r_s$. Define $\mathbf{A}_k$ for $k = 0, 1, 2, \cdots$ as the $(k + m) \times (k + m)$ matrix

$$\mathbf{A}_k = \begin{bmatrix} \alpha_0 & \alpha_1 & \alpha_2 & \cdots & \alpha_m & 0 & \cdots & 0 & 0 \\ 0 & \alpha_0 & \alpha_1 & \cdots & \alpha_{m-1} & \alpha_m & \cdots & 0 & 0 \\ & & \cdots & & & & & & \\ 0 & 0 & 0 & & & & & \alpha_{m-1} & \alpha_m \\ 0 & 0 & 0 & & & & & 0 & 0 \\ & & \cdots & & & & & & \\ 0 & 0 & 0 & & \cdots & & & 1 & 0 \\ 0 & 0 & 0 & & & & & 0 & 1 \end{bmatrix}$$

with all zeros below the main diagonal. The first k rows involve $\alpha_0, \alpha_1, \cdots, \alpha_m$ and the bottom right-hand corner is the $m \times m$ unit matrix. Define

$\mathbf{R}_{k+m}$ as the $(k+m) \times (k+m)$ matrix $[r_{t-s}]$ where $t, s = 1, 2, \cdots, k+m$. Then, performing the operations, we find that $\det (\mathbf{A}_k \mathbf{R}_{k+m}) = \det \mathbf{R}_m$. But $\mathbf{R}_m = [r_{t-s}] = [\phi_{t-s}]$ (where $t, s = 1, 2, \cdots, m$), and $\det [\phi_{t-s}] > 0$ because ϕ_s is the autocovariance of a stationary process of finite moving averages. Now $\alpha_0 > 0$, so $\det \mathbf{A}_k = \alpha_0^k > 0$. Hence $\det \mathbf{R}_{k+m} > 0$ showing that the function r_s $(-\infty < s < \infty)$ is a positive definite function. Thus the r_s form the autocovariance function of a stationary process, say y_t. Because $\sum\limits_{u=0}^{m} \alpha_u r_{s-u}$ $= 0$ for $s > 0$, the process y_t is autoregressive with coefficients $\alpha_0, \alpha_1, \cdots, \alpha_m$, and the coefficients of an autoregressive process satisfies the conclusions of the theorem. Q.E.D.

Let us note that the sequence of least-squares finite approximate inverses for $m = 0, 1, 2, \cdots$, is one representation of the generalized function b_s^{-1}, the inverse of the wavelet $b_0, b_1, \cdots, b_n$. As we have seen, if $B^*(z)$ has no zeros on the unit circle, the generalized function b_s^{-1} reduces to a wavelet.

Suppose now that the finite wavelet $b_0, b_1, \cdots, b_n$ (with $b_n \neq 0$) is not minimum-delay. The z-transform may be written as

$$B^*(z) = b_0 + b_1 z + \cdots + b_n z^n = b_n(z - \beta_1)(z - \beta_2) \cdots (z - \beta_n).$$

If any of the zeros $\beta_1, \beta_2, \ldots, \beta_n$ lie on the unit circle, the (nonrealizable) inverse b_s^{-1} will be a generalized time-function. If none of the zeros lies on the unit circle, the (nonrealizable) inverse b_s^{-1} will belong to L_t^2. Let us now consider this second case. Relabeling the zeros if necessary, we may suppose that $\beta_1, \beta_2, \cdots, \beta_h$ lie outside the unit circle and $\beta_{h+1}, \beta_{h+2}, \cdots, \beta_n$ lie inside the unit circle. Supposing for the moment that the zeros are distinct, the z-transform of the inverse wavelet has the partial fraction expansion (where the B_j are constants)

$$[B^*(z)]^{-1} = \sum_{j=1}^{m} \frac{B_j}{z - \beta_j} = -\sum_{j=1}^{h} B_j \sum_{s=0}^{\infty} \beta_j^{-s-1} z^s + \sum_{j=h+1}^{n} B_j \sum_{s=-\infty}^{-1} \beta_j^{-s-1} z^s,$$

so that the inverse wavelet is

$$b_s^{-1} = \begin{cases} -\sum\limits_{j=1}^{h} B_j \beta_j^{-s-1} & \text{for} \quad s \geq 0 \\[2em] \sum\limits_{j=h+1}^{n} B_j \beta_j^{-s-1} & \text{for} \quad s < 0. \end{cases}$$

If the zeros are not distinct, the inverse wavelet may be found by the same method.

Let us now return to the general case of an arbitrary nonminimum delay wavelet $b_0, b_1, \cdots, b_n$. Its nonrealizable inverse, which may or may not be a function in L_t^2, is infinite, and so we seek a finite approximation. We now let the regressand be the $1 \times (m + n + 1)$ row vector $\mathbf{d} = (d_0, d_1, \cdots, d_{m+n})$ where $d_k = 1$ for some positive integer k and $d_t = 0$ otherwise. Then let us carry through the analogous least-squares argument, as we did above, and the finite operator $\alpha_0, \alpha_1, \cdots, \alpha_m$ given by the solution of the normal equations is called the *least-squares finite approximate delayed inverse* of b_0, $b_1, \cdots, b_n$, where the delay is equal to k. We note that the sequence of least-squares finite approximate delayed inverses for $m = 0, 1, 2, \cdots$, with delay $k = [m/2]$ is one representation of the generalized function b_s^{-1}, the inverse of the wavelet $b_0, b_1, \cdots, b_n$. As we have seen, if $B^*(z)$ has no zeros on the unit circle, the generalized function b_s^{-1} reduces to a function in L_t^2.

4. APPLICATION: POSITIVE-TYPE SPECTRAL ESTIMATES

Let the finite time-series $\mathbf{x}_0, \mathbf{x}_1, \mathbf{x}_2, \cdots, \mathbf{x}_N$ be $(N + 1)$ consecutive observations from a multiple (i.e., $n \times m$ matrix-valued) stationary stochastic process. We wish to consider the type of spectral estimate obtained by the following tandem operations:

1. Let the infinitely long time series

$$\cdots, 0, 0, \mathbf{x}_0, \mathbf{x}_1, \mathbf{x}_2, \cdots, \mathbf{x}_N, 0, 0, 0, \cdots$$

formed by letting $\mathbf{x}_t = \mathbf{0}$ for $t < 0$ and for $t > N$ be the input to a nonrealizable system $\mathbf{b}_t \leftrightarrow \mathbf{B}(\omega)$, called the *filter*. Let $\mathbf{b}_t = b_t \mathbf{I}$ and $\mathbf{B}(\omega) = B(\omega)\mathbf{I}$, where $\mathbf{I}$ is the $n \times n$ unit matrix and $b_t \leftrightarrow B(\omega)$ are scalar-valued functions. The impulse response b_t may be infinitely long in both directions (i.e., b_t may be different from zero for both $t \to -\infty$ and $t \to \infty$). Denote the output time-series, which, in general, will be infinitely long in both directions, by

$$\cdots, \mathbf{y}_{-2}, \mathbf{y}_{-1}, \mathbf{y}_0, \mathbf{y}_1, \mathbf{y}_2, \cdots, \mathbf{y}_N, \mathbf{y}_{N+1}, \mathbf{y}_{N+2}, \cdots.$$

2. Let $\mathbf{y}_t$ ($-\infty < t < \infty$) be the input to a *squared-magnitude device*. The output will therefore be the infinitely long $n \times n$ matrix-valued time-series

$$\cdots, \left|\mathbf{y}_{-2}\right|^2, \left|\mathbf{y}_{-1}\right|^2, \left|\mathbf{y}_0\right|^2, \left|\mathbf{y}_1\right|^2, \left|\mathbf{y}_2\right|^2, \cdots, \left|\mathbf{y}_N\right|^2, \left|\mathbf{y}_{N+1}\right|^2, \left|\mathbf{y}_{N+2}\right|^2, \cdots.$$

3. Let $\left|\mathbf{y}_t\right|^2$ ($-\infty < t < \infty$) be the input to a device that sums the input from $t = -\infty$ to $t = \infty$ and then divides by $(N + 1)$. We call the output

$$\hat{\Psi} = \frac{1}{N + 1} \sum_{t=-\infty}^{\infty} \left|\mathbf{y}_t\right|^2,$$

which is a $n \times n$ matrix, the *positive-type spectral estimate* associated with the filter $B(\omega)$.

We summarize our notation as follows:

$$b_t(-\infty < t < \infty) \leftrightarrow B(\omega) = \sum_{t=-\infty}^{\infty} b_t e^{-i\omega t}.$$

$$u_t = \begin{cases} 1 & \text{for } t = 0, 1, \cdots, N \\ 0 & \text{otherwise} \end{cases} \leftrightarrow U(\omega) = \sum_{t=0}^{N} e^{-i\omega t},$$

where $U(\omega)$ is called the *Dirichlet kernel*.

$$u_t * u_t = \begin{cases} N+1-|t| & \text{for } |t| \leqslant N \\ 0 & \text{otherwise} \end{cases} \leftrightarrow S(\omega) = |U(\omega)|^2$$

$$= \left\{ \frac{\sin[(N+1)\omega/2]}{\sin \omega/2} \right\}^2$$

where $S(\omega)$ is called the *Fejér kernel*.

$$\mathbf{x}_t u_t = \begin{cases} \mathbf{x}_t & \text{for } t = 0, 1, \cdots, N \\ \mathbf{0} & \text{otherwise} \end{cases} \leftrightarrow \mathbf{X}_N(\omega) = \sum_{t=0}^{N} \mathbf{x}_t e^{-i\omega t},$$

$$\mathbf{y}_t = \sum_{s=-\infty}^{\infty} b_{t-s}\mathbf{x}_s u_s = \sum_{s=0}^{N} b_{t-s}\mathbf{x}_s (-\infty < t < \infty) \leftrightarrow \mathbf{Y}(\omega) = B(\omega)\mathbf{X}_N(\omega).$$

$$\mathbf{r}_t = \begin{cases} \displaystyle\sum_{s=0}^{N-t} \mathbf{x}_{s+t}\overline{\mathbf{x}_s} & \text{for } t = 0, 1, \cdots, N \\ \mathbf{0} & \text{for } t > N \\ \tilde{\mathbf{r}}_{-t} & \text{for } t < 0 \end{cases} \leftrightarrow \mathbf{R}(\omega) = \sum_{t=-N}^{N} \mathbf{r}_t e^{-i\omega t}$$

$$= |\mathbf{X}_N(\omega)|^2 = \overline{\mathbf{R}(\omega)},$$

where $\mathbf{r}_t/(N+1)$ is the empirical autocovariance and $\mathbf{R}(\omega)/(N+1)$ is the periodogram.

$$f_t = \sum_{s=-\infty}^{\infty} b_{s+t}\overline{b_s}(-\infty < t < \infty) \leftrightarrow F(\omega) = |B(\omega)|^2 = \sum_{t=-\infty}^{\infty} f_t e^{-i\omega t} = \overline{F(\omega)},$$

where the square-gain $F(\omega) = |B(\omega)|^2$ is called the *filter window*.

$$\boldsymbol{\phi}_t = E\{\mathbf{x}_{s+t}\mathbf{x}_s\}(-\infty < t < \infty) \leftrightarrow \Phi(\omega) = \sum_{t=-\infty}^{\infty} \boldsymbol{\phi}_t e^{-i\omega t} = \overline{\Phi(\omega)},$$

where $\boldsymbol{\phi}_t$ is the autocovariance and $\Phi(\omega)$ is the spectral density.

$$E\{\mathbf{r}_t\} = \begin{cases} (N+1-|t|)\boldsymbol{\phi}_t & \text{for } |t| \leqslant N \\ \mathbf{0} & \text{otherwise} \end{cases} \leftrightarrow \frac{1}{2\pi} \int_{-\pi}^{\pi} S(\mu)\, \Phi(\omega - \mu)\, d\mu.$$

$$g_t = \begin{cases} f_t(N+1-|t|) & \text{for } |t| \leqslant N \\ 0 & \text{otherwise} \end{cases} \leftrightarrow G(\omega) = \frac{1}{2\pi} \int_{-\pi}^{\pi} F(\mu)\, S(\omega - \mu)\, d\mu,$$

where $G(\omega)$ is called the *spectral window*. We note that $G(\omega) \geqslant 0$. Using Bessel's equality we have

$$\hat{\Psi} = \frac{1}{N+1} \sum_{t=-\infty}^{\infty} |\mathbf{y}_t|^2 = \frac{1}{2\pi(N+1)} \int_{-\pi}^{\pi} |\mathbf{Y}(\omega)|^2 \, d\omega$$

$$= \frac{1}{2\pi(N+1)} \int_{-\pi}^{\pi} |B(\omega)|^2 |\mathbf{X}_N(\omega)|^2 \, d\omega = \frac{1}{2\pi(N+1)} \int_{-\pi}^{\pi} F(\omega)\overline{\mathbf{R}(\omega)} \, d\omega,$$

which shows the effect of the filter window $F(\omega)$. Using Parseval's equality, we see that the *positive-type spectral estimate* reduces to the *finite sum*

$$\hat{\Psi} = \frac{1}{2\pi(N+1)} \int_{-\pi}^{\pi} F(\omega)\overline{\mathbf{R}(\omega)} \, d\omega = \frac{1}{N+1} \sum_{t=-N}^{N} f_t \overline{\mathbf{r}_t}.$$

The *expected value* of the positive-type spectral estimate $\hat{\Psi}$ is

$$E\{\hat{\Psi}\} = \frac{1}{N+1} \sum_{t=-N}^{N} f_t E\{\overline{\mathbf{r}_t}\} = \frac{1}{N+1} \sum_{t=-N}^{N} f_t \overline{E\{\mathbf{r}_t\}}$$

$$= \frac{1}{N+1} \sum_{t=-N}^{N} f_t \overline{(N+1-|t|)\boldsymbol{\phi}_t} = \frac{1}{N+1} \sum_{t=-N}^{N} f_t (N+1-|t|)\overline{\boldsymbol{\phi}_t}$$

$$= \frac{1}{N+1} \sum_{t=-N}^{N} g_t \overline{\boldsymbol{\phi}_t} = \frac{1}{2\pi(N+1)} \int_{-\pi}^{\pi} G(\omega)\overline{\Phi(\omega)} \, d\omega,$$

which shows the effect of the spectral window $G(\omega)$. Let us now consider the asymptotic properties of the spectral estimate $\hat{\Psi}$. If $N \to \infty$, then $S(\omega)/2\pi(N+1) \to \delta(\omega)$, where $\delta(\omega)$ is the Dirac delta function. Therefore

$$\frac{G(\omega)}{N+1} = \int_{-\pi}^{\pi} F(\mu) \frac{S(\omega-\mu)}{2\pi(N+1)} \, d\mu \to \int_{-\pi}^{\pi} F(\mu)\, \delta(\omega-\mu)\, d\mu = F(\omega)$$

as $N \to \infty$. Hence, since $N \to \infty$, $E\{\hat{\Psi}\}$ tends to

$$\Psi = \frac{1}{2\pi} \int_{-\pi}^{\pi} F(\omega)\, \overline{\Phi(\omega)} \, d\omega,$$

which shows that $\hat{\Psi}$ is an asymptotically unbiased estimate of Ψ. Moreover, it may be shown that $\hat{\Psi}$ is also a consistent estimate of Ψ.

The Daniell spectral estimate $\hat{\Psi}_{\mathrm{D}}$ (Hannan, 1960, p. 61) is the positive-type spectral estimate obtained by requiring the filter to be the bandpass filter

$$b_t = \frac{\sin ht}{2\pi ht}\, e^{i\omega_0 t} \leftrightarrow B(\omega) = \begin{cases} \dfrac{1}{2h} & \text{for} \quad \omega_0 - h \leqslant \omega \leqslant \omega_0 + h \\ 0 & \text{otherwise} \end{cases}.$$

The filter window is thus the box-shaped function

$$F(\omega) = |B(\omega)|^2 = \begin{cases} \dfrac{1}{4h^2} & \text{for} \quad \omega_0 - h \leqslant \omega \leqslant \omega_0 + h \\ 0 & \text{otherwise} \end{cases},$$

so that

$$f_t = \frac{1}{2\pi} \int_{-\pi}^{\pi} |B(\omega)|^2 e^{i\omega t}\, d\omega = \frac{1}{2\pi} \int_{\omega_0 - h}^{\omega_0 + h} \frac{1}{4h^2}\, e^{i\omega t}\, d\omega = \frac{1}{2h}\, b_t = \frac{\sin ht}{4\pi h^2 t}\, e^{i\omega_0 t}.$$

Hence the *Daniell spectral estimate* may be written

$$\hat{\mathbf{\Psi}}_D = \frac{1}{N+1} \sum_{t=-N}^{N} \frac{\sin ht}{4\pi h^2 t}\, e^{i\omega_0 t}\overline{\mathbf{r}_t}.$$

S. M. Simpson (1961b) has developed a high-speed autocovariance program for the IBM 709 and IBM 7090, utilizing grouping techniques, which computes the empirical autocovariance r_t for $t = 0, 1, \cdots, M$ for a time-series x_t for $t = 0, 1, \cdots, N$ (where x_t is expressed in integers such that $|x_t| \leqslant S$). The machine computing time in seconds is given by the formula

$$[31N + 70S + M(7.5S + 0.53N)]t_0,$$

where t_0 is the *add time* of the machine. For example, if we suppose $S = 100$ and $N = 5000$, then on the IBM 7090, which has an add time $t_0 = 4.36 \times 10^{-6}$ sec, it would take 75 sec to compute r_t for $t = 0, 1, \cdots, 5000$ (and only 8 sec to compute r_t for $t = 0, 1, \cdots, 500$). A high-speed spectral program developed by Simpson (1961) would take 12 sec on the IBM 7090 to compute the Daniell spectral estimates from r_t (for $t = 0, 1, \cdots, 5000$) for 500 different values of ω_0. Hence the entire computing time required on an IBM 7090 to compute the Daniell spectral estimates for 500 different values of ω_0 from a time-series with $N = 5000$ observations is 75 plus 12, or 87 sec.

In those cases in which we wish to reduce computing time, we may instead compute a modified Daniell estimate. Break up the time-series of N observations into p successive sections of N' observations each, where $pN' = N$. Then, for a fixed value of ω_0 and a fixed value of h, compute the Daniell estimate based on the N' observations in each of the p sections. The arithmetic average of these p Daniell estimates gives the desired modified Daniell estimate (for the given value of ω_0) for the time series of N observations. For example, suppose the time series has $N = 500,000$ observations. Let us divide this time series into $p = 100$ nonoverlapping sections of $N' = 5000$ observations each. As we have already seen, it takes 87 sec to compute the Daniell estimates (for 500 different values of ω_0) for each of the p sections. Hence the modified Daniell estimates (for 500 different values of ω_0) for the given time series of $N = 500,000$ observations would take 100 times 87 sec, or 8700 sec, or 145 min, of computing time.

I would like to thank J. Claerbout, E. J. Hannan, E. Lyttkens, M. Rosenblatt, S. M. Simpson, S. Treitel, J. W. Tukey, and H. Wold for helpful comments and discussions.

REFERENCES

Bartlett, M. S. *Stochastic Processes.* Cambridge: Cambridge University Press, 1955.

Blackman, R. B., and Tukey, J. W. *The Measurement of Power Spectra*, New York: Dover, 1959.

Bode, H. W., and Shannon, C. E. A simplified derivation of linear least-square smoothing and prediction theory. *Proc. IRE*, **38**, 417–425 (1950).

Claerbout, J. Least-squares inverse in the presence of noise (1961). Paper in Simpson (1961b).

Davenport, W. B., and Root, W. L. *An Introduction to the Theory of Random Signals and Noise.* New York: McGraw-Hill, 1958.

Doob, J. L. *Stochastic Processes.* New York: Wiley, 1953.

Grenander, U., and Rosenblatt, M. *Statistical Analysis of Stationary Time-Series.* New York: Wiley, 1957.

Hannan, E. J. *Time Series Analysis.* London: Methuen, 1960.

Helson, H., and Lowdenslager, D. Prediction theory and Fourier series in several variables. *Acta Math.*, **99**, 165–202 (1958).

Karhunen, K. Über die Struktur stationärer zufälliger Funktionen. *Ark. Mat.*, **1**, 141–160 (1949).

Levin, M. J. Optimum estimation of impulse response in the presence of noise. *IRE Trans. Circuit Theory* (1960).

Parzen, E. *Mathematical considerations in the estimation of spectra.* Tech. Report 3, Stanford University Statistics Lab. (1960).

Quenouille, M. H. *The analysis of multiple time-series.* London: Charles Griffin and Co., New York: Stechert-Hafner, 1957.

Rice, R. B. Inverse convolution filters. *Geophysics*, **27**, 4–18 (1962).

Riesz, F. Über die Randwerte einer analytischen Function. *Math. Z.*, **18**, 87–95 (1923).

Robinson, E. A. *An introduction to infinitely many variates.* London: Charles Griffin and Co. New York: Stechert-Hafner, 1959.

Robinson, E. A. *Random wavelets and cybernetic systems.* London: Charles Griffin and Co. New York: Stechert-Hafner, 1962.

Robinson, E. A., and H. Wold. Minimum-delay structure of least-squares/eo-ipso predicting systems for stationary stochastic processes. *Brown University Symposium on Time Series.* New York: Wiley, 1962.

Simpson, S. M. *Initial studies on underground nuclear detection with seismic data prepared by a novel digitization system.* Air Force Cambridge Research Laboratories, Scientific Report No. 1 of Contract No. AF 19(604)7378(1961a).

Simpson, S. M. *Time series techniques applied to underground nuclear detection and further digitalized seismic data.* Air Force Cambridge Research Laboratories 62–262, Scientific Report No. 2 of Contract AF 19(604)7378 (1961b).

Smirnov, V. I. Sur les valeurs limites des fonctions régulières à l'intérieur d'un cercle. *Leningrad Phys-Math 2* (1928).

Szego, G. Über die Randwerte einer analytischen Function. *Math. Ann.*, **84**, 232–244 (1921).

Wiener, N., and Masani, P. The prediction theory of multivariate stochastic processes, II. *Acta Math. (Uppsala)*, **99**, 93–137 (1958).

Wold, H.　*A study in the analysis of stationary time-series:* Thesis, University of Stockholm. 1938. Second Edition. Almqvist & Wiksell, Uppsala, 1954.

Wold, H. Forecasting by the chain principle. *Brown University Symposium on Time Series,* New York: Wiley, 1962.

Yaglom, A. M.　*Introduction to the theory of stationary random functions* (Russian). 1952. English translation by R. A. Silverman. Englewood Cliffs, N.J.: Prentice-Hall, 1962.

APPENDIX. MINIMUM-DELAY STRUCTURE OF LEAST-SQUARES AND EO-IPSO PREDICTING SYSTEMS FOR STATIONARY STOCHASTIC PROCESSES

Enders A. Robinson[1] *and Herman Wold*[2]

This appendix establishes a relation between the concepts of minimum-delay, least-squares, and *eo-ipso* predicting systems which are treated separately by the writers in their individual chapters.

Definition 1.　*The finite set of real constants $a_0, a_1, \cdots, a_m$ (with $a_m \neq 0$) is called strongly minimum-delay provided that the zeros $z_1, z_2, \cdots, z_m$ of the polynomial*

$$A(z) = a_0 + a_1 z + \cdots + a_m z^m$$

lie outside the unit circle (i.e., $|z_j| > 1, j = 1, 2, \cdots, m$).

If $a_0, a_1, \cdots, a_m$ is strongly minimum-delay, then

$$\frac{1}{A(z)} = B(z) = b_0 + b_1 z + b_2 z^2 + \cdots$$

where

$$b_0^2 + b_1^2 + b_2^2 + \cdots < \infty;$$

that is, the wavelets $a_0, a_1, \cdots, a_m$ and $b_0, b_1, b_2, \cdots$ are the inverse of each other:

$$\sum_{s=0}^{m} a_s b_{t-s} = \delta_t = \begin{cases} 1 & \text{if} \quad t = 0 \\ 0 & \text{if} \quad t \neq 0. \end{cases}$$

Let x_t $(-\infty < t < \infty)$ be a second-order stationary, nondeterministic (real-valued) stochastic process for discrete time with zero mean and autocovariances

$$\phi_s = E\{x_{t+s} x_t\}.$$

Definition 2.　*Let x_t allow the representation*

$$x_t + a_1 x_{t-1} + \cdots + a_m x_{t-m} = \epsilon_t \tag{1}$$

with

$$E\{x_{t-s}\epsilon_t\} = 0 \quad \text{for} \quad s = 1, 2, \cdots, m. \tag{2}$$

Then (1) with (2) is called a linear least-squares predicting system of order m for x_t.

[1] University of Wisconsin and Uppsala University.

[2] Uppsala University.

Definition 3. *Let x_t allow the representation*

$$x_t + a_1 x_{t-1} + \cdots + a_m x_{t-m} = \epsilon_t \tag{3}$$

with

$$E\{x_t | x_{t-1}, \cdots, x_{t-m}\} = -a_1 x_{t-1} - \cdots - a_m x_{t-m}. \tag{4}$$

Then (3) with (4) is called an eo ipso predicting system of order m for x_t.

We note that an *eo ipso* predicting system is a least-squares predicting system but that the converse is not necessarily true.

We are now in a position to state the following:

Theorem 1.[3] *Let*

$$x_t + a_1 x_{t-1} + \cdots + a_m x_{t-m} = \epsilon_t$$

be a least-squares or an eo ipso predicting system.   Then

$$a_0 = 1, a_1, a_2, \cdots, a_m$$

is strongly minimum-delay.

First Proof.   The minimum mean-square error (denoted by σ^2) is

$$E\{\epsilon_t^2\} = E\{x_t \epsilon_t\} = \phi_0 + a_1\phi_1 + \cdots + a_m\phi_m = \sigma^2 > 0.$$

The normal equations are

$$
\begin{aligned}
E\{x_{t-1}\epsilon_t\} &= \phi_1 + a_1\phi_0 \quad + \cdots + a_m\phi_{m-1} = 0 \\
E\{x_{t-2}\epsilon_t\} &= \phi_2 + a_1\phi_1 \quad + \cdots + a_m\phi_{m-2} = 0 \\
&\cdots \cdots \cdots \cdots \cdots \cdots \cdots \cdots \cdots \\
E\{x_{t-m}\epsilon_t\} &= \phi_m + a_1\phi_{m-1} + \cdots + a_m\phi_0 \quad = 0.
\end{aligned}
$$

Define r_t as follows:

$$
\begin{aligned}
r_t &= r_{-t} \\
r_t &= \phi_t \quad \text{for} \quad t = 0, 1, 2, \cdots, m;
\end{aligned}
$$

and r_t for $t > m$ by the equation

$$r_t + a_1 r_{t-1} + \cdots + a_m r_{t-m} = 0.$$

Thus r_t satisfies, by definition,

$$
\begin{aligned}
r_0 \quad + a_1 r_1 \quad &+ \cdots + a_m r_m \quad = \sigma^2 \\
r_1 \quad + a_1 r_0 \quad &+ \cdots + a_m r_{m-1} = 0 \\
r_2 \quad + a_1 r_1 \quad &+ \cdots + a_m r_{m-2} = 0 \\
\cdots \cdots \cdots &\cdots \cdots \cdots \cdots \cdots \\
r_m \quad + a_1 r_{m-1} &+ \cdots + a_m r_0 \quad = 0 \\
r_{m+1} + a_1 r_m \quad &+ \cdots + a_m r_1 \quad = 0 \\
r_{m+2} + a_1 r_{m+1} &+ \cdots + a_m r_2 \quad = 0 \\
\cdots \cdots \cdots &\cdots \cdots \cdots \cdots \cdots
\end{aligned} \tag{5}
$$

[3] This theorem was stated by E. R. in a seminar at Uppsala.   It was first proved by H. W. (see second proof) and subsequently proved independently by E. R. (see first proof).

Define $\mathbf{R}_{k+m}$ for any $k = 0, 1, 2, \cdots$ to be the $(k + m) \times (k + m)$ auto-covariance matrix:

$$\mathbf{R}_{k+m} =$$

$$\begin{bmatrix}
r_0 & r_1 & \cdots & r_{k-2} & r_{k-1} & r_k & r_{k+1} & \cdots & r_{k+m-2} & r_{k+m-1} \\
r_1 & r_0 & \cdots & r_{k-3} & r_{k-2} & r_{k-1} & r_k & \cdots & r_{k+m-3} & r_{k+m-2} \\
\cdots & & & & & & & & & \\
r_{k-2} & r_{k-3} & \cdots & r_0 & r_1 & r_2 & r_3 & \cdots & r_m & r_{m+1} \\
r_{k-1} & r_{k-2} & \cdots & r_1 & r_0 & r_1 & r_2 & \cdots & r_{m-1} & r_m \\
r_k & r_{k-1} & \cdots & r_2 & r_1 & r_0 & r_1 & \cdots & r_{m-2} & r_{m-1} \\
r_{k+1} & r_k & \cdots & r_3 & r_2 & r_1 & r_0 & \cdots & r_{m-3} & r_{m-2} \\
\cdots & & & & & & & & & \\
r_{k+m-2} & r_{k+m-3} & \cdots & r_m & r_{m-1} & r_{m-2} & r_{m-3} & \cdots & r_0 & r_1 \\
r_{k+m-1} & r_{k+m-2} & \cdots & r_{m+1} & r_m & r_{m-1} & r_{m-2} & \cdots & r_1 & r_0
\end{bmatrix}.$$

Define $\mathbf{A}_k$ for any $k = 0, 1, 2, \cdots$ to be the $(k + m) \times (k + m)$ matrix:

$$\mathbf{A}_k =
\begin{bmatrix}
1 & 0 & \cdots & 0 & 0 & 0 & 0 & \cdots & 0 & 0 \\
a_1 & 1 & \cdots & 0 & 0 & 0 & 0 & \cdots & 0 & 0 \\
\cdots & & & & & & & & & \\
a_{m-1} & a_{m-2} & \cdots & 1 & 0 & 0 & 0 & \cdots & 0 & 0 \\
a_m & a_{m-1} & \cdots & a_1 & 1 & 0 & 0 & \cdots & 0 & 0 \\
0 & a_m & \cdots & a_2 & a_1 & 1 & 0 & \cdots & 0 & 0 \\
0 & 0 & \cdots & a_3 & a_2 & 0 & 1 & \cdots & 0 & 0 \\
\cdots & & & & & & & & & \\
0 & 0 & \cdots & a_m & a_{m-1} & 0 & 0 & \cdots & 1 & 0 \\
0 & 0 & \cdots & 0 & a_m & 0 & 0 & \cdots & 0 & 1
\end{bmatrix},$$

where the first k columns involve the entries $1, a_1, \cdots, a_m$ and the bottom right-hand corner is the $m \times m$ unit matrix. Then we see that

$$\mathbf{R}_{k+m}\mathbf{A}_k =
\begin{bmatrix}
\sigma^2 & * & * & \cdots & * & * & * & * & \cdots & * & * \\
0 & \sigma^2 & * & \cdots & * & * & * & * & \cdots & * & * \\
\cdots & & & & & & & & & & \\
0 & 0 & 0 & \cdots & \sigma^2 & * & * & * & \cdots & * & * \\
0 & 0 & 0 & \cdots & 0 & \sigma^2 & * & * & \cdots & * & * \\
0 & 0 & 0 & \cdots & 0 & 0 & r_0 & r_1 & \cdots & r_{m-2} & r_{m-1} \\
0 & 0 & 0 & \cdots & 0 & 0 & r_1 & r_0 & \cdots & r_{m-3} & r_{m-2} \\
\cdots & & & & & & & & & & \\
0 & 0 & 0 & \cdots & 0 & 0 & r_{m-2} & r_{m-3} & \cdots & r_0 & r_1 \\
0 & 0 & 0 & \cdots & 0 & 0 & r_{m-1} & r_{m-2} & \cdots & r_1 & r_0
\end{bmatrix},$$

where the stars * denote unspecified entries. Taking determinants, we have

$$\det \mathbf{R}_{k+m} \cdot \det \mathbf{A}_k = \sigma^{2k} \cdot \det \mathbf{R}_m.$$

But

$$\det \mathbf{R}_m =
\begin{vmatrix}
\phi_0 & \phi_1 & \cdots & \phi_{m-1} \\
\phi_1 & \phi_0 & \cdots & \phi_{m-2} \\
\cdots & & & \\
\phi_{m-1} & \phi_{m-2} & \cdots & \phi_0
\end{vmatrix} > 0,$$

since x_t is nondeterministic. Also we see that det $\mathbf{A}_k = 1$. Hence

$$\det \mathbf{R}_{m+k} > 0,$$

showing that $\mathbf{R}_{m+k}$ is positive definite. It follows that the r_t form the auto-covariance function of a stationary stochastic process, say y_t. Moreover, because of equations (5) the process y_t is autoregressive, with coefficients $a_0 = 1, a_1, \cdots, a_m$; that is, θ_t defined by

$$y_t + a_1 y_{t-1} + \cdots + a_m y_{t-m} = \theta_t$$

satisfies $E\{\theta_t y_{t-v}\} = 0$ for $v > 0$, $E\{\theta_t y_t\} = E\{\theta_t \theta_t\} = \sigma^2$ and $E\{\theta_t \theta_s\} = 0$ for $t \neq s$. For

$$\sum_{u=0}^{m} a_u r_{v-u} = \sum_{u=0}^{m} a_u E\{y_{t-u} y_{t-v}\} = E\left\{\left(\sum_{u=0}^{m} a_u y_{t-u}\right) y_{t-v}\right\} = E\{\theta_t y_{t-v}\},$$

and so equations (5) may be written

$$E\{\theta_t y_{t-v}\} = \begin{cases} \sigma^2 & \text{for} \quad v = 0 \\ 0 & \text{for} \quad v = 1, 2, 3, \cdots, \end{cases}$$

which shows that θ_t is orthogonal to the closed linear manifold spanned by the set $\{y_{t-1}, y_{t-2}, y_{t-3}, \cdots\}$. But θ_s for all $s < t$ belongs to this closed linear manifold, from which it follows that $E\{\theta_t \theta_s\} = 0$ for $s < t$, hence $E\{\theta_t \theta_s\} = 0$ for $t \neq s$. Now it is well-known that the set of coefficients of an autoregressive process is strongly minimum-delay. Therefore the set $1, a_1, a_2, \cdots, a_m$ is strongly minimum-delay. Q.E.D.

Second Proof: The argument is to show by actual construction that there exists a stationary stochastic process y_t with autocovariances

$$r_0, r_1, \cdots, r_m, r_{m+1}, \cdots,$$

where $r_0, r_1, \cdots, r_{m-1}$ are the same as for the given process x_t, whereas $r_m, r_{m+1}, \cdots$ are obtained by the same relations as in the first proof:

$$r_{m+k} + a_1 r_{m+k-1} + \cdots + a_m r_k = 0; \qquad k = 0, 1, 2, \cdots.$$

We shall indicate the construction briefly. Let

$$\epsilon_1, \epsilon_2, \cdots, \epsilon_{m-1}, \epsilon_m, \epsilon_{m+1}, \cdots$$

be a sequence of random variables, mutually independent and normally distributed with zero mean and unit variance. With reference to the formulas for Gramm-Schmidt's orthogonalization, we construct the sequence of random variables

$$y_1, y_2, \cdots, y_{m-1}, y_m, y_{m+1}, \cdots$$

in such manner that the orthogonalization of $y_1, y_2, \cdots, y_n$ leads to $\epsilon_1, \epsilon_2, \cdots, \epsilon_n$. This construction determines the variables y_n one by one, y_n being determined in terms of the variables $y_1, y_2, \cdots, y_{n-1}$ and ϵ_n and the autocovariances $r_0, r_1, \cdots, r_{n-1}, r_n$.   Considering the matrices

$$\mathbf{R}_{m+k} = \begin{bmatrix} r_0 & r_1 & \cdots & r_{m+k-1} \\ r_1 & r_0 & \cdots & r_{m+k-2} \\ \cdots\cdots\cdots\cdots\cdots\cdots\cdots \\ r_{m+k-1} & r_{m+k-2} & \cdots & r_0 \end{bmatrix}; \qquad k = 0, 1, 2, \cdots,$$

it is formally verified that $\mathbf{R}_{m+k}$ is the covariance matrix of $y_1, y_2, \cdots,$ y_{m+k+1}. It remains to show that the matrices $\mathbf{R}_{m+k}$ are positive definite. Now $\mathbf{R}_m$ is positive definite, since it is the autocovariance process of m variables $x_t, x_{t+1}, \cdots, x_{t+m-1}$ in the given process, and that $\mathbf{R}_{m+1}, \mathbf{R}_{m+2}, \cdots$ are positive definite can be shown by induction, making use of Jacobi's determinant theorem.

Comments.

1. If we form the least-squares regression

$$y_{m+k} = c_1 y_{m+k-1} + c_2 y_{m+k-2} + \cdots + c_{m+k-2} y_2 + c_{m+k-1} y_1,$$

it can be shown, again making use of Jacobi's theorem, that

$$c_{m+1} = c_{m+2} = \cdots = c_{m+k-1} = 0.$$

Hence the foregoing construction is such that

$$r_{0(m+k)\cdot 12\cdots(m-1)} = 0; \qquad k = 0, 1, 2, \cdots,$$

where r is the partial correlation coefficient of y_{t+m} and y_{t-k} for given $y_{t+m-1},$ $\cdots, y_t$. In retrospect, the construction can be interpreted by saying that we have formed the variables $y_{t+m}, y_{t+m-1}, y_{t+m-2}, \cdots, y_t$ in accordance with the given matrix $\mathbf{R}_m$ and so that the past variables $y_{t-1}, y_{t-2}, \cdots$ will influence y_{t+m} only via $y_{t+m-1}, \cdots, y_t$.

2. Minimum-delay, least-squares, and *eo-ipso* are three optimal properties, conceptually different, minimum-delay belonging to information theory and least-squares and *eo-ipso* to prediction theory. Theorem 1 establishes the close relationship between the three notions, since least-squares and *eo-ipso* for stationary structures implies minimum-delay.

3. The theorem and the two proofs extend to the case of vector variables X_t; that is, the theorem then refers to the representation of X_t in terms of a recursive system (which reduces to the autogression (1) when the vector involves just one component), with the vector components taken in a specified ordering, and within each period t the statement of minimum-delay refers to this ordering.

4. The theorem extends to complex-valued processes.

Communicated 22 February 1961 by H. Cramér and H. Wold

Extremal representation of stationary stochastic processes

By Enders A. Robinson

1. We let the real variables t and ω represent time and angular frequency respectively. A time-function $b\,(t)$ may be considered to be the *impulse response* function of a time-invariant linear system. If $b\,(t)$ is one-sided (i. e. if $b\,(t)=0$ for $t<0$ so that no response occurs prior to its stimulus) then the system is said to be *realizable*. The *transfer function* of a system with impulse response $b\,(t)$ is defined to be the formal Laplace transform

$$B\,(p) = \int_{-\infty}^{\infty} b\,(t)\, e^{-pt}\, d\,t, \quad p = \sigma + i\,\omega \quad (\sigma,\,\omega \text{ real}).$$

Letting $\sigma = 0$ we obtain the formal Fourier transform

$$B\,(i\,\omega) = \left|\, B\,(i\,\omega)\,\right| e^{iP\,(i\omega)} = \int_{-\infty}^{\infty} b\,(t)\, e^{-i\omega t}\,d\,t,$$

where $\left|\,B\,(i\,\omega)\,\right|$ is called the *gain*, and $P\,(i\,\omega)$ the *phase-shift*, of the system. The *group-delay* of the system is defined to be

$$\tau_g = -\frac{d\,P\,(i\,\omega)}{d\,\omega}.$$

Input $f\,(t)$ and output $g\,(t)$ of the system are related by the convolution: $f\,(t)*b\,(t) = g\,(t)$. The Laplace transforms $F\,(p)$ and $G\,(p)$ of input and output respectively are related by the multiplication: $F\,(p)\,B\,(p) = G\,(p)$.

We shall call a function $w\,(t)$ a *wavelet* if it is one-sided (i. e. $w\,(t)=0$ for $t<0$) and L^2 (i. e. $\int_0^{\infty} \left|\,w\,(t)\,\right|^2 d\,t < \infty$). The function space of all wavelets $w\,(t)$ (with measure $d\,t$) will be denoted by $L^2\,(0,\,\infty)$.

2. A purely non-deterministic, second-order, stationary stochastic process $x\,(t)$ (for continuous time parameter t) has a one-sided *moving-average representation*

$$x\,(t) = \int_{s=-\infty}^{t} w\,(t-s)\, d\,y\,(s), \tag{1}$$

where $w\,(t)$ is a wavelet and $y\,(s)$ is a process with orthogonal increments for which $E\,\{\left|d\,y\,(s)\right|^2\} = d\,s$. The corresponding *spectral representation* [Cramér, 1] is

397

$$x(t) = \frac{1}{2\pi} \int_{\omega=-\infty}^{\infty} e^{i\omega t} W(i\omega) \, dY(i\omega), \tag{2}$$

where $W(i\omega)$ is the L^2-Fourier transform of $w(t)$ and the formal derivative $dY(i\omega)/d\omega$ is the formal Fourier transform of the formal derivative $dy(s)/ds$. The $Y(i\omega)$ process has orthogonal increments for which

$$E\{|dY(i\omega)|^2\} = 2\pi \, d\omega.$$

For any given stationary stochastic process $x(t)$ there are infinitely many such representations; more precisely, there is a one-sided moving average representation, and a corresponding spectral representation, for every wavelet $w(t)$ that satisfies $\int_0^{\infty} w(t+s)\,w(s)\,ds = \phi(t)$, where $\phi(t)$ is the autocovariance of the process. In this paper, we shall give necessary and sufficient conditions that a representation possess extremal properties.

3. A constant A_0 of modulus 1 is called a trivial all-pass transfer function. The function

$$A_1(p) = \prod_k \frac{p_k - p}{\overline{p}_k + p} \frac{|p_k - 1|}{p_k - 1} \frac{|p_k + 1|}{p_k + 1},$$

where $\{p_k\}$ is a non-empty set satisfying $\mathrm{Re}\ p_k > 0$ and

$$\sum_k \frac{\mathrm{Re}\ p_k}{1 + |p_k|^2} < \infty$$

is called a Type 1 all-pass transfer function. The function

$$A_2(p) = e^{-\alpha p} \quad (\alpha > 0)$$

is called a Type 2 all-pass transfer function. The function

$$A_3(p) = \exp\left[-\frac{1}{\pi} \int_{-\infty}^{\infty} \frac{1 - i\lambda p}{p - i\lambda} \, d\beta(\lambda) \right],$$

where $\beta(\lambda)$ is a non-decreasing function whose derivative vanishes almost everywhere and $0 < \beta(\infty) - \beta(-\infty) < \infty$ is called a Type 3 all-pass transfer function. The function

$$A(p) = A_0 A_1(p) A_2(p) A_3(p)$$

(where any or several of the factors on the right may be absent) is called an *all-pass transfer function*. If all the factors except A_0 are absent, then $A(p)$ is called *trivial*; otherwise $A(p)$ is called *non-trivial*.

4. If $M(i\omega) \geqslant 0$,

$$\int_{-\infty}^{\infty} |M(i\omega)|^2 \, d\omega < \infty, \qquad \int_{-\infty}^{\infty} \frac{\log M(i\omega)}{1 + \omega^2} \, d\omega > -\infty,$$

then
$$W_0(p) = \exp\left[\frac{1}{\pi}\int_{-\infty}^{\infty}\frac{1-i\lambda p}{p-i\lambda}\,\frac{\log M(i\lambda)}{1+\lambda^2}\,d\lambda\right]$$

is called the *minimum-delay* transfer function for the gain $M(i\lambda)$. The wavelet $w_0(t)$ whose Laplace fransform is $W_0(p)$ is called the minimum-delay wavelet for the gain $M(i\lambda)$.

Krylov [4] and Karhunen [3] have established the following *canonical representation* for the transfer function of a system whose impulse response is a wavelet.

Lemma 1. *$W(p)$ is the L^2-Laplace transform of a wavelet if and only if, in the right half p-plane (i. e. Re $p > 0$),*

$$W(p) = A(p)\,W_0(p),$$

where $A(p)$ is an all-pass transfer function and $W_0(p)$ is the minimum-delay transfer function with the same gain as $W(p)$. This representation of $W(p)$ is unique.

In addition Karhunen [3] has established the following result.

Lemma 2. *Let $w(t)$ and $v(t)$ be wavelets and let*

$$\int_0^{\infty} w(t-r)\,\overline{v(t)}\,dt = 0 \quad \text{for} \quad r > 0.$$

Then $v(t) \equiv 0$ if and only if
$$w(t) = A_0\,w_0(t)$$

where $|A_0| = 1$ and $w_0(t)$ is the minimum-delay wavelet with the same gain as $w(t)$.

5. From the canonical representation (Lemma 1) it can be shown that a system $A(p)$ is realizable and has gain $|A(i\omega)| = 1$ for almost all frequencies ω if and only if the system is an all-pass system. Moreover it is seen that in the right half p-plane (i. e. Re $p > 0$) an all-pass transfer function $A(p)$ is analytic and has zeros $\{p_k\}$ (where the set $\{p_k\}$ may be empty).

The following three lemmas can be established directly from the definition of the all-pass transfer function.

Lemma 3. *The group-delay τ_g of an all-pass transfer function is positive (resp. zero) for $-\infty < \omega < \infty$ if and only if the all-pass transfer function is non-trivial (resp. trivial).*

Lemma 4. *The modulus $|A(p)|$ of an all-pass transfer function satisfies $|A(p)| < 1$ (resp. $|A(p)| = 1$) in the right half p-plane if and only if the all-pass transfer function is non-trivial (resp. trivial).*

It follows from the canonical representation (Lemma 1) that if a wavelet $w(t)$ is the input to an all-pass system $A(p)$, then the output $v(t)$ is also a wavelet. Because $|A(i\omega)| = 1$, we have

$$|V(i\omega)| = |A(i\omega)W(i\omega)| = |W(i\omega)|$$

so the output wavelet $v(t)$ has the same gain as the input wavelet $w(t)$. Consequently, from Bessel's equality (for quadratic integrals of L^2-Fourier transforms) it follows that

$$\int_0^\infty |w(t)|^2\,dt = \int_0^\infty |v(t)|^2\,dt$$

which says that input and output wavelets have the same *total energy*. The following lemma, however, tells us that the *partial energy* of the output wavelet is delayed with respect to the partial energy of the input wavelet.

Lemma 5. *Let a wavelet $w(t)$ be the input to an all-pass system $A(p)$ and let the wavelet $v(t)$ be the resulting output. If the all-pass system is non-trivial, the partial energy of the input exceeds the partial energy of the output for some $\alpha > 0$:*

$$\int_0^\alpha |w(t)|^2\,dt > \int_0^\alpha |v(t)|^2\,dt.$$

If the all-pass system is trivial, the partial energy of the input equals the partial energy of the output for all $\alpha > 0$.

6. We now have the minimum-delay wavelet theorem.

Theorem 1. *Let $w(t)$ be a wavelet in the class of all wavelets with gain $M(i\omega)$. Then each of the following conditions is necessary and sufficient that*

$$w(t) = A_0\,w_0(t),$$

where $w_0(t)$ is the minimum-delay wavelet with gain $M(i\omega)$ and $|A_0| = 1$:

(a)   *The set $\{w(t-r),\, r \geqslant 0\}$ is closed in $L^2(0, \infty)$.*
(b)   *The group-delay of $w(t)$ is a minimum for $-\infty < \omega < \infty$.*
(c)   *The modulus $|W(p)|$ is a maximum in the right half p-plane.*

(d)   *The partial energy $\int_0^\alpha |w(t)|^2\,dt$ is a maximum for all $\alpha \geqslant 0$.*

(e)   *For a purely non-deterministic, stationary stochastic process $x(t)$ with spectral density $\Phi(i\omega) = M^2(i\omega)$ and with moving-average representation (1), the least-squares linear prediction $z(t)$ of $x(t+\alpha)$, $\alpha > 0$, from the whole past $x(s)$, $s \leqslant t$, is*

$$z(t) = \int_{-\infty}^t w(t+\alpha-s)\,dy(s),$$

the minimum prediction error is

$$x\,(t+\alpha) - z\,(t) = \int_t^{t+\alpha} w\,(t+\alpha-s)\,d\,y\,(s)$$

and the minimum mean-square prediction error is given by the partial energy

$$E\,\{|x\,(t+\alpha)-z\,(t)|^2\} = \int_t^{t+\alpha} |w\,(t+\alpha-s)|^2\,d\,s = \int_0^\alpha |w\,(\tau)|^2\,d\,\tau.$$

(f) *For a purely non-deterministic, stationary stochastic process $x\,(t)$ with spectral density $\Phi\,(i\,\omega) = M^2\,(i\,\omega)$ and with moving-average representation (1), the closed linear manifold spanned by $x\,(s)$, $s \leqslant t$, is the same as the closed linear manifold spanned by $y\,(s)$, $s \leqslant t$, for all t.*

(g) *The function $\Gamma\,(i\,\omega)$ of the form*

$$\Gamma\,(i\,\omega) = \underset{N \to \infty}{\mathrm{l.\,i.\,m.}} \sum_{j=1}^N c_{Nj}\,e^{-i\,\omega\,r_{Nj}}$$

(where c_{Nj} are complex constants and $r_{Nj} \geqslant 0$) that is determined by the method of least squares to approximate $e^{i\,\omega\,\alpha}\,(\alpha > 0)$ with respect to measure $M^2\,(i\,\omega)\,d\,\omega/2\,\pi$ is

$$\Gamma\,(i\,\omega) = \frac{1}{W\,(i\,\omega)} \int_0^\infty w\,(\alpha+t)\,e^{-i\,\omega t}\,d\,t$$

and the minimum mean-square error is given by the partial energy

$$\frac{1}{2\,\pi} \int_{-\infty}^\infty |e^{i\,\omega\,\alpha} - \Gamma\,(i\,\omega)|^2\,M^2\,(i\,\omega)\,d\,\omega = \int_0^\alpha |w\,(\tau)|^2\,d\,\tau.$$

Proof. Condition (a) follows immediately from Lemma 2 by noting that a set in Hilbert space is closed if and only if any element orthogonal to each member of the set identically vanishes. Conditions (b), (c), and (d) follow from Lemmas 3, 4, and 5, respectively, and from the canonical representation (Lemma 1). Conditions (e) and (f) follow from the work on prediction theory by Hanner [2], Karhunen [3], and Wiener [6]. Condition (g) follows from condition (e) by utilizing the isomorphism of the closed linear manifold spanned by $x\,(s)$, $s \leqslant t$ (probability measure) and the closed linear manifold spanned by $e^{i\,\omega s}$, $s \leqslant t$ (measure $M^2\,(i\,\omega)\,d\,\omega/2\,\pi$) such that $x\,(t)$ and $e^{i\,\omega t}\,(-\infty < t < \infty)$ are corresponding elements (see, e. g., Robinson [5], p. 83). Q.E.D.

7. Summing up, the spectral density $\Phi\,(i\,\omega)$ of a purely non-deterministic stationary stochastic process being given, we see that there exists a class of different spectral representations (2), such that the transfer function $W\,(p)$ satisfies $W\,(i\,\omega) = A\,(i\,\omega)\sqrt{\Phi\,(i\,\omega)}$, where $A\,(p)$ is an arbitrary all-pass transfer function. Alternatively, the autocovariance function $\phi\,(t)$ (which is the L^1-Fourier transform of $\Phi\,(i\,\omega)$) being given, there exists a corresponding class of moving-average representations (1), such that the wavelet $w\,(t)$ (which is the L^2-Fourier transform of $W\,(i\,\omega)$) satisfies $\int_0^\infty w\,(t+s)\,\overline{w\,(s)}\,d\,s = \phi\,(t)$. Among these representations there is one called the predictive (or Woldian) decomposition given by

$$x(t) = A_0 \int_{s=-\infty}^{t} w_0(t-s)\, d\, y_0(s) = \frac{A_0}{2\pi} \int_{\omega=-\infty}^{\infty} e^{i\omega t}\, W_0(i\omega)\, d\, Y_0(i\omega)$$

(where $|A_0| = 1$, and $w_0 \leftrightarrow W_0$ is minimum-delay for the gain $\sqrt{\Phi(i\omega)}$) which has extremal properties as given by Theorem 1.

The realizable system with input $x(t)$ and output $z(t)$ where $z(t)$ is the least-squares linear prediction of $x(t+\alpha)$, $\alpha > 0$, has transfer function

$$\Gamma(i\omega) = \underset{N \to \infty}{\text{l.i.m.}} \sum_{j=1}^{N} c_{Nj}\, e^{-i\omega r_{Nj}} = \frac{1}{W_0(i\omega)} \int_0^{\infty} w_0(\alpha+t)\, e^{-i\omega t}\, d\, t$$

(where c_{Nj} are complex constants and $r_{Nj} \geqslant 0$). The minimum mean-square prediction error has the decomposition

$$E\left\{ \left| x(t+\alpha) - z(t) \right|^2 \right\} = \frac{1}{2\pi} \int_{-\infty}^{\infty} \left| e^{i\omega\alpha}\, W(i\omega) - \Gamma(i\omega)\, W(i\omega) \right|^2 d\, \omega$$

$$= \int_{-\infty}^{\infty} \left| w(t+\alpha) - \underset{N \to \infty}{\text{l.i.m.}} \sum_{j=1}^{N} c_{Nj}\, w(t-r_{Nj}) \right|^2 d\, t$$

$$= \int_{-\alpha}^{0} \left| w(t+\alpha) \right|^2 d\, t + \int_{0}^{\infty} \left| w(t+\alpha) - \underset{N \to \infty}{\text{l.i.m.}} \sum_{j=1}^{N} c_{Nj}\, w(t-r_{Nj}) \right|^2 d\, t.$$

We note that the first term is the partial energy of the wavelet $w(t)$, and the second term vanishes for arbitrary α if and only if the set $\{w(t-r),\ r \geqslant 0\}$ is closed, i.e. if and only if $w(t) = A_0\, w_0(t)$.

Uppsala, January, 1961.

REFERENCES

1. Cramér, H., On harmonic analysis in certain functional spaces. *Ann. Math. 41*, pp. 215–230 (1942).
2. Hanner, O., Deterministic and non-deterministic stationary random processes. *Ark. Matematik 1*, pp. 161–177 (1949).
3. Karhunen, K., Über die Struktur stationärer zufälliger Funktionen. *Ark. Matematik 1*, pp. 141–160 (1949).
4. Krylov, V. I., On functions regular in a half-plane (Russian). *Mat. Sbornik* N.S. *6* (48), pp. 95–138 (1939).
5. Robinson, E. A., *Infinitely many variates*. Griffin, London, 1959.
6. Wiener, N., *Stationary time series*. Wiley, New York, 1949.
7. Wold, H., *A study in the analysis of stationary time series*. Thesis, University of Stockholm. Uppsala, 1938 (second edition, 1954).

Tryckt den 30 januari 1962

Uppsala 1962. Almqvist & Wiksells Boktryckeri AB

CHAPTER 13

Extremal Properties of the Wold Decomposition

ENDERS A. ROBINSON

Institute of Statistics, Uppsala University
Uppsala, Sweden

Submitted by Lofti Zadeh

I. WAVELETS AND SYSTEMS

We let the integers t represent equally-spaced time points, and the real variable ω (where $-\pi \leq \omega \leq \pi$) represent angular frequency. We call

$$\delta_t = \begin{cases} 1 \text{ for } t = 0 \\ 0 \text{ for other } t \end{cases}$$

the *unit impulse* function. If we let the unit impulse δ_t be the input to a time-invariant, linear, digital system, then the resulting output g_t is called the *impulse response* of the system. A system is called *stable* if

$$\sum_{t=-\infty}^{\infty} |g_t|^2 < \infty.$$

A system is called *realizable* if g_t is one-sided (i.e., if $g_t = 0$ for $t < 0$). Clearly, a realizable system is a system for which no response (i.e., output) can occur prior to its stimulus (i.e., input).

If f_t is the input to a system with impulse response g_t, then the output h_t is given (formally) by the convolution

$$h_t = \sum_{s=-\infty}^{\infty} g_s f_{t-s}.$$

If we define the (formal) z-transforms

$$H^*(z) = \sum_{s=-\infty}^{\infty} h_s z^s, \qquad G^*(z) = \sum_{s=-\infty}^{\infty} g_s z^s, \qquad F^*(z) = \sum_{s=-\infty}^{\infty} f_s z^s$$

then we have (formally)

$$H^*(z) = G^*(z)\, F^*(z).$$

The impulse response of a realizable, stable system will be called a *wavelet* w_t, i.e., $w_t = 0$ for $t < 0$ and $\sum_0^\infty |w_t|^2 < \infty$. If we define the inner product between two wavelets w_t and v_t by

$$(w_t, v_t) \equiv \sum_{t=0}^\infty w_t \bar{v}_t,$$

then the space of all wavelets is a Hilbert space, which we shall denote by $L^2(0, \infty)$. The z-transform of the realizable stable system with impulse response w_t is

$$W^*(z) = \sum_{t=0}^\infty w_t z^t.$$

$W^*(z)$ is an analytic function for $|z| < 1$, and has the limit value

$$W(\omega) \equiv W^*(e^{-i\omega}) = \sum_{t=0}^\infty w_t e^{-i\omega t}, \qquad \text{a.e.,}^1$$

on the unit circle $z = e^{-i\omega}$. The wavelet w_t may be recovered from $W(\omega)$ or $W^*(z)$ by

$$w_t = \frac{1}{2\pi} \int_{-\pi}^{\pi} W(\omega)\, e^{i\omega t} d\omega = \frac{1}{2\pi i} \int_{|z|=1} W^*(z)\, z^{-t-1} dz.$$

The function $W(\omega)$ is called the transfer function of the system. It may be written in polar form as

$$W(\omega) = |W(\omega)|\, e^{i\psi(\omega)}, \qquad \text{a.e.,}$$

where $|W(\omega)|$ is called the *gain*, and $\psi(\omega)$ the *phase-shift* of the system. The *group-delay* of the system is defined to be

$$\tau_g = -\frac{d\psi(\omega)}{d\omega}, \qquad \text{a.e.}$$

II. ALL-PASS SYSTEMS AND MINIMUM-DELAY SYSTEMS

A constant A_0 of modulus 1 is called the z-transform of a trivial all-pass system. The function

$$A_1^*(z) = \prod_k \frac{z_k - z}{1 - \bar{z}_k z} \frac{|z_k|}{z_k}$$

---

[1] The notation "a.e." stands for *almost everywhere*, i.e., for all ω in the set $-\pi \leqslant \omega \leqslant \pi$ except possibly for an ω-set of Lebesgue measure zero.

where $\{z_k\}$ is a nonempty set satisfying $|z_k| < 1$ and

$$\sum_k (1 - |z_k|) < \infty$$

is called the z-transform of a Type 1 all-pass system. The function

$$A_2^*(z) = z^m$$

where m is a positive integer is called the z-transform of a Type 2 all-pass system. The function

$$A_3^*(z) = \exp\left[-\frac{1}{2\pi} \int_{-\pi}^{\pi} \frac{e^{i\lambda} + z}{e^{i\lambda} - z} \, d\beta(\lambda)\right]$$

where $\beta(\lambda)$ is a nondecreasing function whose derivative vanishes almost everywhere and $0 < \beta(\pi) - \beta(-\pi) < \infty^2$ is called the z-transform of a Type 3 all-pass system. The function

$$A^*(z) = A_0 A_1^*(z) \, A_2^*(z) \, A_3^*(z)$$

(where any or several of the factors on the right may be absent) is called the z-transform of an *all-pass system*. If all the factors except A_0 are absent, then $A^*(z)$ is called *trivial*; otherwise $A^*(z)$ is called *nontrivial*. We see that, for $|z| < 1$, $A^*(z)$ is analytic and has zeros $\{z_k\}$, (where the set $\{z_k\}$ would be empty if the factor $A_1(z)$ is missing).

If $M(\omega) \geq 0$,

$$\int_{-\pi}^{\pi} |M(\omega)|^2 \, d\omega < \infty, \qquad \int_{-\pi}^{\pi} \log M(\omega) \, d\omega > -\infty$$

then

$$B^*(z) = \exp\left[\frac{1}{2\pi} \int_{-\pi}^{\pi} \frac{e^{i\lambda} + z}{e^{i\lambda} - z} \log M(\lambda) \, d\lambda\right]$$

is called the z-transform of the *minimum-delay system with gain $M(\omega)$*.

The following *canonical representation* may be derived from the work of Szegö [1] and Riesz [2]:

THEOREM 1. $W^*(z)$ *is the z-transform of a realizable, stable system if and only if*

$$W^*(z) = A^*(z) \, B^*(z), \qquad |z| < 1,$$

---

[2] That is, on the interval $-\pi \leqslant \lambda \leqslant \pi$, $\beta(\lambda)$ is a bounded distribution function with no absolutely continuous component, so that $\beta(\lambda)$ has the decomposition $\beta(\lambda) = \beta_2(\lambda) + \beta_3(\lambda)$ where $\beta_2(\lambda)$ is a step-function and $\beta_3(\lambda)$ is a "singular" function.

where $A^*(z)$ is an all-pass z-transform and $B^*(z)$ is the minimum-delay z-transform with the same gain as $W^*(z)$. This representation of $W^*(z)$ is unique.

We now derive:

THEOREM 2. *A system $A^*(z)$ is realizable, stable, and has gain $|A^*(e^{-i\omega})| = 1$ a.e. if and only if $A^*(z)$ is an all-pass system.*

PROOF ("if" statement): By hypothesis, $A^*(z)$ is an all-pass z-transform. The minimum-delay z-transform for the gain 1 is

$$B^*(z) = \exp \frac{1}{2\pi} \int_{-\pi}^{\pi} \frac{e^{i\lambda} + z}{e^{i\lambda} - z} \log 1 \, d\lambda = \exp 0 = 1.$$

Hence, by Theorem 1,

$$A^*(z) \, B^*(z) = A^*(z) \cdot 1 = A^*(z)$$

is the z-transform of a wavelet with gain 1, a.e. Q.E.D.

PROOF ("only if" statement): By hypothesis, $A(z)$ is the z-transform of a wavelet with gain 1, a.e. But the minimum-delay z-transform with gain 1 is 1. Hence, by Theorem 1, $A^*(z) = A^*(z) \cdot 1$ is the canonical representation of $A^*(z)$, from which it follows that $A^*(z)$ is an all-pass system. Q.E.D.

Letting $A^*(z) = 1$ in Theorem 1, it follows that $B^*(z)$ is a realizable, stable system and hence

$$\beta_t = \frac{1}{2\pi i} \int_{|z|=1} B^*(z) \, z^{-t-1} \, dz = \frac{1}{2\pi} \int_{-\pi}^{\pi} B(\omega) \, e^{i\omega t} \, d\omega$$

is a wavelet, which we shall call the *minimum-delay wavelet for the gain* $|W(\omega)| = |B(\omega)|$. From Theorem 2, it follows that for any all-pass $A^*(z)$

$$a_t = \frac{1}{2\pi i} \int_{|z|=1} A^*(z) \, z^{-t-1} \, dz = \frac{1}{2\pi} \int_{-\pi}^{\pi} A(\omega) \, e^{i\omega t} \, d\omega$$

is a wavelet, which we shall call an *all-pass* wavelet. We therefore have the *canonical representation* :

THEOREM 3. *w_t is a wavelet if and only if*

$$w_t = \sum_{s=0}^{t} a_s \beta_{t-s}$$

where a_t is an all-pass wavelet and β_t is the minimum-delay wavelet with the same gain as w_t. This representation of w_t is unique.

The first examples of all-pass wavelets were given by Wold [3, p. 130].

The following three theorems can be established directly from the definition of the all-pass transfer function.

THEOREM 4. *The group-delay τ_g of an all-pass system is positive (resp. zero) for $-\pi \leq \omega \leq \pi$ if and only if the all-pass system is nontrivial (resp. trivial).*

THEOREM 5. *The modulus $|A^*(z)|$ of an all-pass z-transform satisfies $|A^*(z)| < 1$ (resp. $|A^*(z)| = 1$) for $|z| < 1$ if and only if the all-pass z-transform is nontrivial (resp. trivial).*

It follows from the canonical representation (Theorem 1) that, if a wavelet w_t is the input to an all-pass system $A^*(z)$, then the output v_t is also a wavelet. Because $|A(\omega)| = 1$, a.e., we have

$$|V(\omega)| = |A(\omega)||W(\omega)| = |W(\omega)|, \qquad \text{a.e.,}$$

which says that the output wavelet v_t has the same gain as the input wavelet w_t. Bessel's equality states that

$$\sum_{t=0}^{\infty} |w_t|^2 = \frac{1}{2\pi} \int_{-\pi}^{\pi} |W(\omega)|^2 \, d\omega$$

and likewise that

$$\sum_{t=0}^{\infty} |v_t|^2 = \frac{1}{2\pi} \int_{-\pi}^{\pi} |V(\omega)|^2 \, d\omega.$$

Consequently, it follows that

$$\sum_{t=0}^{\infty} |w_t|^2 = \sum_{t=0}^{\infty} |v_t|^2$$

which says that input and output wavelets have the same *total energy*. The following theorem, however, tells us that the *partial energy* of the output wavelet is delayed with respect to the partial energy of the input wavelet.

THEOREM 6. *Let a wavelet w_t be the input to an all-pass system $A^*(z)$ and let the wavelet v_t be the resulting output. Then the partial energy of the output never exceeds the partial energy of the input; i.e.,*

$$\sum_{t=0}^{\alpha} |v_t|^2 \leq \sum_{t=0}^{\alpha} |w_t|^2$$

for all nonnegative integers α. *More particularly,* $|v_0|^2 < |w_0|^2$ *(resp.* $|v_0|^2 = |w_0|^2$*) if and only if the all-pass system is nontrivial (resp. trivial).*

PROOF. The last statement of this theorem follows from the observation that since

$$V^*(z) = A^*(z)\, W^*(z), \qquad v_0 = V^*(0), \qquad w_0 = W^*(0)$$

we have

$$|v_0|^2 = |A^*(0)|^2 |w_0|^2$$

From the definition of an all-pass z-transform, it can be easily verified that $|A^*(0)| < 1$ if and only if $A^*(z)$ is nontrivial.   Q.E.D.

III. THE TIME-SERIES IDENTIFICATION PROBLEM

Let x_t be a purely, nondeterministic (second-order) stationary time-series with zero mean and autocovariance

$$E\{x_{t+s}\bar{x}_t\} = \phi(s).$$

By the Khintchine-Wold theorem, the spectral density $\Phi(\omega)$ satisfies

$$\phi(s) = \frac{1}{2\pi} \int_{-\pi}^{\pi} e^{i\omega s}\, \Phi(\omega)\, d\omega.$$

Let w_t be any wavelet that satisfies

$$\sum_{t=0}^{\infty} w_{t+s}\bar{w}_t = \phi(s).$$

Then there is a orthonormal sequence of random variables ϵ_t (i.e., $E\{\epsilon_{t+s}\bar{\epsilon}_t\} = \delta_s$) with zero mean such that x_t has the *one-sided moving-average representation*

$$x_t = \sum_{s=0}^{\infty} w_s \epsilon_{t-s}. \tag{1}$$

This equation has the interpretation: White noise ϵ_t is the input to a realizable, stable system w_t, and the purely nondeterministic stationary time series x_t is the resulting output.

The *time-series identification problem* is the problem of determining which wavelet w_t (and hence the corresponding orthonormal sequence ϵ_t) from the class of all admissible wavelets (and their corresponding orthonormal sequences) is to be used in the representation (1).

The fundamental result in the theory of stationary stochastic processes is the *Wold decomposition* [3]. The purpose of this paper is to show the extremal properties of the Wold decomposition in relationship to the time-series identification problem.

Specializing the Wold decomposition to the case of a single, purely, non-deterministic stationary time series, we have:

THEOREM 7. *Let x_t be a purely, nondeterministic stationary time series. Then x_t has the unique decomposition*

$$x_t = \sum_{s=0}^{\infty} b_s \vartheta_{t-s} \tag{2}$$

where

(a) *ϑ_t is an orthonormal sequence of random variables.*

(b) *ϑ_t lies in the closed linear manifold spanned by the set $\{x_s,\, s \le t\}$.*

(c) *$b_0 > 0$.*

(d) $\sum_{t=0}^{\infty} |\, b_t \,|^2 < \infty.$

We also have the following two theorems:

THEOREM 8. *A stationary time series x_t is purely nondeterministic if and only if all of the following three conditions are satisfied:*

(a) *The spectral distribution function is absolutely continuous for $-\pi \le \omega \le \pi$.*

(b) *$\Phi(\omega) > 0$, a.e.*

(c) $\int_{-\pi}^{\pi} \log \Phi(\omega)\, d\omega$ *is finite.*

THEOREM 9. *The coefficients b_0, b_1, b_2, $\cdots$ of the Wold decomposition satisfy and are uniquely determined by the conditions*

$$\sum_{t=0}^{\infty} |\, b_t \,|^2 < \infty$$

$$\Phi(\omega) = \left|\, \sum_{t=0}^{\infty} b_t e^{-i\omega t} \,\right|^2, \qquad \text{a.e.,}$$

$$b_0 = \exp\left[\frac{1}{4\pi} \int_{-\pi}^{\pi} \log \Phi(\omega)\, d\omega\right].$$

For proofs of the above three theorems, the reader is referred to Doob [4, Chapter 12, § 4] and Robinson [5, Section 7.7].

THEOREM 10. *Let x_t be a purely nondeterministic stationary time series with spectral density function $\Phi(\omega)$. Then the coefficients b_0, b_1, b_2, $\cdots$ of the Wold decomposition and the minimum-delay wavelet β_0, β_1, β_2, $\cdots$ for the gain $\sqrt{\Phi(\omega)}$ are identical; i.e., $b_t = \beta_t$ for all (nonnegative) integers t.*

PROOF. Because the wavelets b_t and β_t have the same gain $\sqrt{\Phi(\omega)}$, and because by definition β_t is minimum-delay, we may apply Theorem 1 to obtain

$$\sum_{t=0}^{\infty} b_t z^t = A^*(z) \sum_{t=0}^{\infty} \beta_t z^t, \qquad |z| < 1.$$

From Theorem 9, it follows that $b_0 = \beta_0$, so $A^*(0) = 1$. Hence by Theorem 6, $A^*(z)$ is trivial, and $A^*(z) = A^*(0) = 1$. Q.E.D.

IV. EXTREMAL PROPERTIES OF THE WOLD DECOMPOSITION

We now come to our main theorem:

THEOREM 11. *Let w_t be a wavelet in the class of all wavelets with gain $M(\omega)$. Then each of the following conditions is necessary and sufficient that*

$$w_t = A_0 \beta_t, \qquad t = 0, 1, 2, \cdots,$$

where β_t is the minimum-delay wavelet for the gain $M(\omega)$ and $|A_0| = 1$:

(a) *The group-delay of w_t is a minimum for $-\pi \leq \omega \leq \pi$.*

(b) *The modulus $|W(z)|$ is a maximum for all z satisfying $|z| < 1$.*

(c) *The partial energy $\sum_0^{\alpha} |w_t|^2$ is a maximum for $\alpha = 0, 1, 2, \cdots$,*

(d) *For the stationary time series x_t with spectral density $\Phi(\omega) = M^2(\omega)$ and with one-sided, moving-average representation (1), the least-squares linear prediction $\hat{x}_{t+\alpha}$ of $x_{t+\alpha}$, $\alpha > 0$, from the whole past x_s, $s \leq t$, is*

$$\hat{x}_{t+\alpha} = \sum_{s=\alpha}^{\infty} w_s \epsilon_{t+\alpha-s},$$

the prediction error is

$$x_{t+\alpha} - \hat{x}_{t+\alpha} = \sum_{s=0}^{\alpha-1} w_s \epsilon_{t-\alpha-s},$$

and the minimum mean-square prediction error is given by the partial energy

$$E \{| x_{t+\alpha} - \hat{x}_{t+\alpha} |^2\} = \sum_{s=0}^{\alpha-1} | w_s |^2.$$

(e) *For the stationary time series x_t with spectral density function $\Phi(\omega) = M^2(\omega)$ and with one-sided, moving-average representation* (1), *the closed linear manifold spanned by x_s, $s \leq t$ is the same as the closed linear manifold spanned by ϵ_s, $s \leq t$, for all integers t.*

(f) *The stationary time series x_t with spectral density function $\Phi(\omega) = M^2(\omega)$ and with one-sided, moving-average representation* (1) *has an autoregressive representation in the sense that there exists c_{Ns} such that*

$$\epsilon_t = \underset{N \to \infty}{\text{l.i.m.}} \sum_{s=0}^{N} c_{Ns} x_{t-s}$$

where

$$\sum_{s=0}^{N} c_{Ns} z^s \neq 0 \qquad \text{for} \qquad | z | < 1.$$

(g) *For the stationary time series x_t with spectral density $\Phi(\omega) = M^2(\omega)$ and with one-sided, moving-average representation* (1), *the realizable system with input x_t and output $\hat{x}_{t+\alpha}$ (where $\hat{x}_{t+\alpha}$ is the least-squares linear prediction of $x_{t+\alpha}$) has transfer function*

$$\Gamma(\omega) = \frac{1}{W(\omega)} \sum_{t=0}^{\infty} w_{t+\alpha} e^{-i\omega t}.$$

(h) *The set $\{w_{t+s}, s = 0, -1, -2, \cdots\}$ is closed in $L^2(0, \infty)$.*

PROOF. Conditions (a), (b), and (c) follow from Theorems 4, 5, and 6, respectively, and from the canonical representation (Theorem 1). Condition (d) follows from Theorem 10 and from the solution of the prediction problem (see [5, section 7.8, equations (7.84) and (7.86)]). Conditions (e) and (f) follow from Theorem 10 and the Wold decomposition (Theorem 7). Condition (g) follows from condition (d) by utilizing the isomorphism of the closed linear manifold spanned by $x(s)$, $s \leq t$ (probability measure) and the closed linear manifold spanned by $e^{i\omega s}$, $s \leq t$ (measure $M^2(\omega) \, d\omega/2\pi$) such that $x(t)$ and $e^{i\omega t}$ are corresponding elements (see [5, Sections 7.2 and 7.8, especially equation (7.85)]). The closed linear manifold spanned by x_s, $s \leq 0$, is the same as the closed linear manifold spanned by ϵ_s, $s \leq 0$ if and only if every

$$z = \sum_{t=0}^{\infty} c_t \epsilon_{-t} \qquad \left(\text{where } \sum_{t=0}^{\infty} | c_t |^2 < \infty\right)$$

such that

$$E\{z\bar{x}_s\} = 0 \qquad \text{for all} \qquad s \leq 0$$

satisfies $c_t \equiv 0$ (for all integers t). But

$$E\{z\bar{x}_s\} = E\left\{\sum_{t=0}^{\infty} c_t \epsilon_{-t} \sum_{r=0}^{\infty} \bar{w}_r \bar{\epsilon}_{s-r}\right\}$$

$$= \sum_{t=0}^{\infty} \sum_{r=0}^{\infty} c_t \bar{w}_r \delta_{-t-s+r} = \sum_{t=0}^{\infty} c_t \bar{w}_{t+s}.$$

Condition (h) now follows by noting that the set $\{w_{t+s}, s \leq 0\}$ is closed in $L^2(0, \infty)$ if and only if any wavelet c_t orthogonal to each member of the set identically vanishes.   Q.E.D.

Let a purely nondeterministic stationary time-series x_t with autocovariance $\phi(s)$ and spectral density $\Phi(\omega)$ be given. Then there exists a class of different one-sided, moving-average representations

$$x_t = \sum_{s=0}^{\infty} w_s \epsilon_{t-s} \tag{1}$$

in which w_t is determined by

$$w_t = \sum_{s=0}^{\infty} a_s b_{t-s}$$

where

$$A^*(z) = \sum_{s=0}^{\infty} a_s z^s$$

is an *arbitrary* all-pass system and

$$B^*(z) = \sum_{s=0}^{\infty} b_s z^s = \exp\left\{\frac{1}{4\pi} \int_{-\pi}^{\pi} \frac{e^{i\lambda} + z}{e^{i\lambda} - z} \log \Phi(\omega)\, d\omega\right\}$$

is the minimum-delay system with gain $\sqrt{\Phi(\omega)}$. Among these representations, there is one called the Wold (or predictive) decomposition given by

$$x_t = \sum_{s=0}^{\infty} b_s \vartheta_{t-s}, \tag{2}$$

which is the representation obtained by choosing the all-pass system

$A^*(z) = 1$. If A_0 is any constant of modulus one (i.e., if A_0 is a trivial all-pass system), then the representation (1) in which

$$w_s = A_0 b_s, \qquad \epsilon_t = \frac{1}{A_0} \vartheta_t$$

(for all integers s, t) has the extremal properties given by Theorem 11. In particular, if we have a real-valued time-series x_t and specify that x_t can be simulated by an autoregressive representation, then we necessarily obtain either the Wold decomposition (2) or the representation (1) with $w_s = -b_s$ and $\epsilon_s = -\vartheta_s$.

ACKNOWLEDGMENTS

I should like to express my sincere thanks to Professor Herman Wold for his help and encouragement.

REFERENCES

1. SZEGÖ, G. Über die Randwerte einer analytischen Funktion. *Math. Ann.* 84, 232-244 (1921).
2. RIESZ, F. Über die Randwerte einer analytischen Funktion. *Math. Z.* 18, 87-95 (1923).
3. WOLD, H. A study in the analysis of stationary time-series. Thesis, University of Stockholm, 1938 (2nd ed. 1954).
4. DOOB, J. L. "Stochastic Processes." Wiley, New York, 1953.
5. ROBINSON, E. A. "Infinitely Many Variates." Griffin, London, 1959.

АКАДЕМИЯ НАУК СССР

ТЕОРИЯ ВЕРОЯТНОСТЕЙ
И ЕЕ ПРИМЕНЕНИЯ

Том VIII

2

(Отдельный оттиск)

МОСКВА · 1963

414

CHAPTER 14

НЕКОТОРЫЕ СВОЙСТВА РАЗЛОЖЕНИЯ ВОЛЬДА СТАЦИОНАРНЫХ СЛУЧАЙНЫХ ПРОЦЕССОВ

Е. А. РОБИНСОН (УПСАЛА, ШВЕЦИЯ)

(*Резюме*)

Основные результаты работы (теоремы 11—13) касаются вопроса о представимости величин $\hat{x}_{t+a}$ (являющихся наилучшим прогнозом значений x_{t+a} стационарного в широком смысле процесса по значениям x_s, $s \leqslant t$) в. виде ряда

$$\hat{x}_{t+a} \sim \sum_{s=0}^{\infty} k_s \dot{x}_{t-s},$$

где коэффициенты k_s удовлетворяют условию: $\sum |k_s|^2 < \infty$. Предварительно приводятся некоторые свойства последовательностей $\{w_t\}$, $\sum |w_t|^2 < \infty$.

PROPERTIES OF THE WOLD DECOMPOSITION OF STATIONARY STOCHASTIC PROCESSES

ENDERS A. ROBINSON

A sequence w_t of complex numbers (where t is an integer) is called a *wavelet* if $w_t = 0$ for $t < 0$ and $\sum_0^{\infty} |w_t|^2 < \infty$. If we define the inner product between two wavelets w_t and v_t by

$$(w_t, v_t) = \sum_{t=0}^{\infty} w_t \overline{v_t}$$

then the space of all wavelets is a Hilbert space, which we shall denote by $L^2(0, \infty)$. The *z-transform* of a wavelet w_t is defined to be

$$W^*(z) = \sum_{t=0}^{\infty} w_t z^t.$$

$W^*(z)$ is an analytic function for $|z| < 1$, and has the limit value

$$W^*(e^{-2\pi i \nu}) = \sum_{t=0}^{\infty} w_t e^{-2\pi i \nu t}, \text{ a. e.,}$$

on the unit circle $z = e^{-2\pi i \nu}$.

416   E. A. Robinson

Let us consider all complex-valued functions $F(v)$ on the interval $-0.5 \leqslant v \leqslant 0.5$ such that

$$\int_{-0.5}^{0.5} |F(v)|^2 dv < \infty.$$

If we define the inner product between two such functions $F(v)$ and $G(v)$ by

$$(F(v), G(v)) = \int_{-0.5}^{0.5} F(v)\, \overline{G(v)}\, dv$$

then the space of all such functions is a Hilbert space, which we shall denote by $L^2(dv)$

It is well-known that if w_t is a wavelet, then there is a function $W(v)$ in space $L^2(dv)$ given by

$$W(v) = \underset{N\to\infty}{\text{l. i. m.}} \sum_{t=0}^{N} w_t e^{-2\pi i v t} = W^*(e^{-2\pi i v}), \text{ a. e.} \tag{1}$$

Moreover, if the wavelets w_t and v_t yield $W(v)$ and $V(v)$, respectively, then

$$\sum_{0}^{\infty} w_t \overline{v_t} = \int_{-0.5}^{0.5} W(v)\, \overline{V(v)}\, dv.$$

This equation is known as *Bessel's equality* if $w_t = v_t$ (and consequently $W(v) = V(v)$); otherwise it is known as *Parseval's equality.*

The function $W(v)$, given by equation (1), may be written in polar form as

$$W(v) = |W(v)|\, e^{i \psi(v)}$$

where $|Y(v)|$ is called the *gain*, and $\psi(v)$ the *phase-shift*. The *group-delay* is defined to be

$$\tau_g = - \frac{d\psi(v)}{dv}, \text{ a. e.}$$

The function

$$P^*(z) = cz^m \prod_k \frac{z_k - z}{1 - \bar{z}_k z} \frac{|z_k|}{z_k} \exp\left\{ -\int_{-0.5}^{0.5} \frac{e^{2\pi i v} + z}{e^{2\pi i v} - z}\, d\beta(v) \right\}$$

(where: $|z| < 1$; $|c| = 1$; m is a non-negative integer; $\{z_k\}$ is an empty or non-empty set satisfying $|z_k| < 1$ and $\sum_k (1 - |z_k|) < \infty$; and $\beta(v)$ is a real, bounded, non-decreasing function whose derivative vanishes almost everywhere) is called an *all-pass* z-transform. If $|P^*(z)| = 1$ for $|z| < 1$ then $P^*(z)$ is called *trivial*: otherwise $P^*(z)$ is called *non-trivial.*

The function

$$B^*(z) = \exp\left\{ \int_{-0.5}^{0.5} \frac{e^{2\pi i v} + z}{e^{2\pi i v} - z} \log M(v)\, dv \right\}$$

(where $|z| < 1$; $M(v) \geqslant 0$; $\int_{-0.5}^{0.5} |M(v)|^2 dv < \infty$; and $\int_{-0.5}^{0.5} \log M(v)\, dv > -\infty$) is called the *minimum-delay* z-transform with gain $M(v)$.

The following *canonical representation* follows from Theorems 6.2 and 11.4 in Privalov (1956, pp. 78 and 110). See also Szegö (1921), Riesz (1923), Smirnov (1928), Krylov

(1939), Kolmogorov (1941), Karhunen] (1949), Yaglom (1955), and Grenander and Rosenblatt (1957).

Theorem 1. *$W^*(z)$ is the z-transform of a wavelet if and only if*

$$W^*(z) = P^*(z)B^*(z), \quad |z| < 1,$$

where $P^*(z)$ is an all-pass z-transform and $B^*(z)$ is the minimum-delay z-transform with the same gain as $W^*(z)$. This representation of $W^*(z)$ is unique.

We now derive:

Theorem 2. *$P^*(z)$ is an all-pass z-transform if and only if $P^*(z)$ is the z-transform of a wavelet with gain equal to 1, a. e.*

P r o o f. The gain $M(v) = 1$ has the minimum-delay z-transform $B^*(z) = \exp 0 = 1$. If $P^*(z)$ is all-pass, then $P^*(z) = P^*(z) \cdot 1$ is a canonical representation. Hence $P^*(z)$ is the z-transform of a wavelet with gain 1. Conversely. if $P^*(z)$ is the z-transform of a wavelet with gain 1, then its canonical representation must be $P^*(z) = P^*(z) \cdot 1$. Hence $P^*(z)$ is an all-pass z-transform. Q. E. D.

By this theorem, an all-pass z-transform $P^*(z)$ is the z-transform of a wavelet p_t which we call an *all-pass wavelet*. The first examples of all-pass wavelets were given by Wold (1938, p. 130). An all-pass wavelet has gain $|P(v)| = 1$, a. e. This unit gain property is the reason for using the term «all-pass». The minimum-delay wavelet with gain equal to one is the *unit-impulse* δ_t ($= 1$ if $t = 0$ and $= 0$ if $t \neq 0$), which has z-transform $P^*(z) = 1$. Using this z-transform in Theorem 1, we see that $B^*(z)$ is the z-transform of a wavelet β_t, which we call the *minimum-delay wavelet* with gain $M(v)$.

The *canonical representation* in terms of wavelets is

Theorem 3. *w_t is a wavelet if and only if*

$$w_t = \sum_{s=0}^{t} \beta_s p_{t-s}$$

where p_t is an all-pass wavelet and β_t is the minimum-delay wavelet with the same gain as w_t. This representation of w_t is unique. Theorem 3 shows that the partial sums $\sum_0^N \beta_s p_{t-s}$ for fixed t converge to the value w_t. The following theorem shows that these partial sums also converge in the mean to the wavelet w_t.

Theorem 4. *In Hilbert space $L^2(0, \infty)$, we have*

$$w_t = \underset{N \to \infty}{\text{l. i. m.}} \sum_{s=0}^{N} \beta_s p_{t-s}.$$

P r o o f. By Bessel's equality, we have

$$\sum_{t=0}^{\infty} \left| w_t - \sum_{s=0}^{N} \beta_s p_{t-s} \right|^2 = \int_{-0.5}^{0.5} \left| W(v) - \sum_{s=0}^{N} \beta_s e^{-2\pi i vs} P(v) \right|^2 dv.$$

Because $W(v) = B(v)P(v)$ and $|P(v)| = 1$, the above is equal to

$$\int_{-0.5}^{0.5} |B(v) - \sum_{s=0}^{N} \beta_s e^{-2\pi i vs}|^2 dv = \sum_{s=N+1}^{\infty} |\beta_s|^2$$

which can be made as small as we please by choosing N large enough. Q. E. D.

The following three theorems can be established directly from the definition of the all-pass z-transform.

Theorem 5. *The group-delay* τ_g *of an all-pass z-transform is positive (resp. zero) for* $-0.5 \leqslant v \leqslant 0.5$ *if and only if the z-transform is non-trivial (resp. trivial).*

Theorem 6. *The modulus* $|P^*(z)|$ *of an all-pass z-transform satisfies* $|P^*(z)| < 1$ *(resp.* $|P^*(z)| = 1$) *for* $|z| < 1$ *if and only if the all-pass z-transform is non-trivial (resp. trivial).*

It follows from Theorem 3 that if w_t is a wavelet, and p_t is an all-pass wavelet, then

$$v_t = \sum_0^t w_s p_{t-s} \quad \text{is also a wavelet. By a proof similar to that of Theorem 4, it may}$$

be shown that in Hilbert space $L^2(0,\infty)$

$$v_t = \underset{N \to \infty}{\text{l. i. m.}} \sum_{s=0}^{N} w_s p_{t-s}.$$

Because $|P(v)| = 1$, a. e., application of Bessel's equality gives

$$\sum_0^\infty |v_t|^2 = \sum_0^\infty |w_t|^2$$

which says that the two wavelets v_t and w_t have the same *total energy*. The following theorem, however, says that the all-pass wavelet p_t delays the partial energy of v_t with respect to the partial energy of w_t.

Theorem 7. *Let* w_t *be a wavelet, and* p_t *be an all-pass wavelet. Then the partial energy of the wavelet* $v_t = \sum_0^t w_s p_{t-s}$ *never exceeds the partial energy of* w_t; *i. e.*

$$\sum_0^\alpha |v_t|^2 \leqslant \sum_0^\alpha |w_t|^2$$

for $\alpha = 0,1,2, \ldots$ *. More particularly,* $|v_0|^2 < |w_0|^2$ *(resp.* $|v_0|^2 = |w_0|^2$) *if and only if the all-pass wavelet* p_t *is non-trivial (resp. trivial).*

Fundamental to the theory of stationary stochastic processes is the Wold decomposition (Wold, 1938, Theorem 7; Kolmogorov, 1941, Theorem 18). The following version can be derived from Theorem 7.70 in Robinson (1959, p. 97).

Theorem 8. x_t *is a regular stationary stochastic process if and only if*

$$x_t = \sum_{s=0}^\infty b_s \vartheta_{t-s}$$

where

    (a) ϑ_t *is an orthonormal sequence of variates,*

    (b) ϑ_t *lies in the closed linear manifold spanned by* x_s, $s \leqslant t$,

    (c) $b_0 > 0$,

    (d) $\sum_0^\infty |b_t|^2 < \infty$.

This representation of x_t, *called the Wold decomposition, is unique.*

Theorem 9. *Let* x_t *be a regular stationary stochastic process with spectral density function* $\Phi(v)$. *Then the coefficients* b_0, b_1, b_2, $\ldots$ *of the Wold decomposition and the coefficients* β_0, β_1, β_2, $\ldots$ *of the minimum-delay wavelet with gain* $\sqrt{\Phi(v)}$ *are identical; i. e.* $b_t = \beta_t$ *for* $t = 0, 1, 2, \ldots$

Proof. Because the wavelets b_t and β_t have the same gain $\sqrt{\Phi(v)}$, and because, by definition, β_t is minimum-delay, we may apply Theorem 1 to obtain

$$\sum_{t=0}^{\infty} b_t z^t = P^*(z) \sum_{t=0}^{\infty} \beta_t z^t, \quad |z| < 1.$$

From Theorem 4.3* in Doob (1953, p. 577), we have

$$b_0 = \exp\left[\int_{-0.5}^{0.5} \log \sqrt{\Phi(v)}\, dv \right].$$

Hence $b_0 = \beta_0$. It then follows from Theorem 7 that $P^*(z)$ is trivial, and in fact $P^*(z) = P^*(0) = 1$. Q. E. D.

The following theorem exhibits the extremal properties of the Wold decomposition, or in other words, of the minimum-delay wavelet.

Theorem 10. *Let w_t be a wavelet in the class of all wavelets with gain $M(v)$. Then each of the following conditions is necessary and sufficient that*

$$w_t = c\beta_t, \quad t = 0, 1, 2, \ldots$$

where β_t is the minimum-delay wavelet for the gain $M(v)$ and c is a complex constant of modulus $|c| = 1$.

(a) *The group-delay of w_t is minimum for $-0.5 \leqslant v \leqslant 0.5$.*

(b) *The modulus $|W^*(z)|$ is a maximum for all z satisfying $|z| < 1$.*

(c) *The partial energy $\sum_{0}^{\alpha} |w_t|^2$ is a maximum for $\alpha = 0, 1, 2, \ldots$*

(d) *For the regular stationary stochastic process $x_t = \sum_{0}^{\infty} w_s \varepsilon_{t-s}$ (here and everywhere in present theorem ε_t is an orthonormal sequence of variates) the least-squares linear prediction $\hat{x}_{t+\alpha}$ of $x_{t+\alpha}$, $\alpha > 0$, from the whole past x_s, $s \leqslant t$, is*

$$\hat{x}_{t+\alpha} = \sum_{s=\alpha}^{\infty} w_s \varepsilon_{t+\alpha-s},$$

the prediction error is

$$\hat{x}_{t+\alpha} - x_{t+\alpha} = \sum_{s=0}^{\alpha-1} w_s \varepsilon_{t+\alpha-s}$$

and the minimum mean-square prediction error is the partial energy

$$E\{|\hat{x}_{t+\alpha} - x_{t+\alpha}|^2\} = \sum_{s=0}^{\alpha-1} |w_s|^2.$$

(e) *For the regular stationary stochastic process $x_t = \sum_{0}^{\infty} w_s \varepsilon_{t-s}$, the closed linear manifold spanned by x_s, $s \leqslant t$, is the same as the closed linear manifold spanned by ε_s, $s \leqslant t$, for all integers t.*

---

* The condition $\sum_{0}^{\infty} b_t z^t \neq 0$ for $|z| < 1$ given in Doob's Theorem 4.3 is redundant.

(f) *The regular stationary stochastic process* $x_t = \sum\limits_{0}^{\infty} w_s \varepsilon_{t-s}$ *has an autoregressive representation in the sense that there exists complex numbers* a_{Ns} *such that*

$$\varepsilon_t = \underset{N\to\infty}{\text{l.i.m.}} \sum_{s=0}^{N} a_{Ns} x_{t-s}$$

where

$$\sum_{s=0}^{N} a_{Ns} z^s \neq 0 \text{ for } |z| < 1.$$

(g) *For the regular stationary stochastic process* $x_t = \sum\limits_{0}^{\infty} w_s \varepsilon_{t-s}$ *the transfer function* $K(v)$ *of the least-squares linear predictor of* $x_{t+\alpha}$, $\alpha > 0$, *from the whole past* x_s, $s \leqslant t$, *is*

$$K(v) = \frac{1}{W(v)} \sum_{t=0}^{\infty} w_{t+\alpha} e^{-2\pi i v t}.$$

(h) *The set* $\{w_{t+s},\ s = 0, -1, -2, \ldots\}$ *is closed in* $L^2(0, \infty)$.

P r o o f. Conditions (a), (b), and (c) follow from Theorems 5, 6, and 7, respectively and Theorem 1. Condition (d) follows from Theorem 9 and from the Kolmogorov solution of the prediction problem (see Robinson, 1959, Section 7.8, equations (7.84) and (7.86)) Conditions (e) and (f) follow from Theorems 8 and 9. Condition (g) follows from condition (d) by utilizing the Stone — Kolmogorov isomorphism (see Robinson, 1959, Sections 7.2 and 7.8, especially equation (7.85)). The closed linear manifold spanned by x_s, $s \leqslant 0$, is the same as the closed linear manifold spanned by ε_s, $s \leqslant 0$ if and only if every

$$z = \sum_{t=0}^{\infty} c_t \varepsilon_{-t} \quad \left(\text{where } \sum_{t=0}^{\infty} |ct|^2 < \infty\right)$$

such that $\mathbf{E}\{z\bar{x}_s\} = 0$ for all $s \leqslant 0$ satisfies $c_t \equiv 0$ (for all integers t). But

$$\mathbf{E}\{z\bar{x}_s\} = \mathbf{E}\left\{\sum_{t=0}^{\infty} c_t \varepsilon_{-t} \sum_{r=0}^{\infty} \bar{w}_r \bar{\varepsilon}_{s-r}\right\} = \sum_{t=0}^{\infty} \sum_{r=0}^{\infty} c_t \bar{w}_r \delta_{-t-s+r} = \sum_{t=0}^{\infty} c_t \bar{w}_{t+s}.$$

Condition (h) now follows by noting that the set $\{w_{t+s},\ s \leqslant 0\}$ is closed in $L^2(0, \infty)$ if and only if any wavelet c_t orthogonal to each member of the set identically vanishes. Q.E.D.

The following theorems given conditions for explicit representations of the linear pred ictor.

Theorem 11. *Let* x_t *be a regular stationary stochastic process with spectral density* $\Phi(v)$. *A necessary and sufficient condition that the transfer function* $K(v)$ *of the least squares linear predictor (for any prediction lead* $\alpha > 0$) *be expressible as a Fourier series*

$$K(v) = \sum_{s=0}^{\infty} k_s e^{-2\pi i v s}$$

where k_s *is a wavelet is that*

$$\int_{-0.5}^{0.5} \frac{dv}{\Phi(v)} < \infty. \tag{2}$$

P r o o f. N e c e s s a r y condition. By hypothesis,

$$K(v) = \frac{\sum\limits_0^\infty b_{\alpha+s} e^{-2\pi i v s}}{\sum\limits_0^\infty b_s e^{-2\pi i v s}} = \sum_0^\infty k_s e^{-2\pi i v s}, \quad \sum_0^\infty |k_s|^2 < \infty.$$

Let $\alpha = 1$. Under the Stone—Kolmogorov isomorphism, $\hat{x}_1 \leftrightarrow K(v)$ (see equation at bottom of p 103 in Robinson, 1959). Hence $\hat{x}_0 = \hat{x}_1 U^{-1} \leftrightarrow K(v) e^{-2\pi i v}$ (see p. 86 in Robinson, 1959). Also $x_t \leftrightarrow e^{2\pi i v t}$ and $\vartheta_t \leftrightarrow e^{2\pi i v t}/B(v)$ (see equation (7.74) in Robinson, 1959, p. 100). Therefore

$$b_0 \vartheta_0 = x_0 - \hat{x}_0 \leftrightarrow \frac{b_0}{B(v)} = 1 - K(v) e^{-2\pi i v}.$$

It then follows that

$$\frac{1}{B(v)} = \frac{1}{b_0} - \frac{k_0}{b_0} e^{-2\pi i v} - \frac{k_1}{b_0} e^{-2\pi i v 2} - \ \ldots$$

where

$$\int_{-0.5}^{0.5} \frac{dv}{|B(v)|^2} = \frac{1}{b_0^2} \left[1 + \sum_0^\infty |k_s|^2 \right]$$

which converges because, by hypothesis, $\sum\limits_0^\infty |k_s|^2 < \infty$. Since $\Phi(v) = |B(v)|^2$, the necessary condition is established. Q.E.D.

S u f f i c i e n t condition. Now using more explicit notation to indicate the prediction lead α, we denote the transfer function of the optimum linear predictor by

$$K_\alpha(v) = \frac{B_\alpha(v)}{B(v)} \quad \text{where} \quad B_\alpha(v) = \sum_{t=0}^\infty b_{t+\alpha} e^{-2\pi i v t}.$$

Because $B_0/B = 1$, it follows that B_0/B belongs to $L^2(dv)$. Suppose B_α/B belongs to $L^2(dv)$. We note that $1/B$ belongs to $L^2(dv)$ since $|B(v)|^2 = \Phi(v)$ and, by hypothesis, $1/\Phi(v)$ is integrable. Hence the linear combination

$$e^{2\pi i v} \frac{B_\alpha}{B} - b_\alpha e^{2\pi i v} \frac{1}{B} = \frac{B_{\alpha+1}}{B}$$

belongs to $L^2(dv)$. Therefore, by finite induction $K_\alpha = B_\alpha/B$ belongs to $L^2(dv)$ for $\alpha = 0, 1, 2, \ldots$. Since, by Theorem 9, $B(v)$ is minimum-delay, we have the canonical representation

$$B^*(z) = \exp\left[\int_{-0.5}^{0.5} \frac{e^{2\pi i v} + z}{e^{2\pi i v} - z} \log|B(v)| \, dv \right].$$

Since $b_\alpha, b_{\alpha+1}, \ldots$ is a wavelet, we have the canonical representation

$$B_\alpha^*(z) = P^*(z) \exp\left[\int_{-0.5}^{0.5} \frac{e^{2\pi i v} + z}{e^{2\pi i v} - z} \log|B_\alpha(v)| \, dv \right].$$

Combining the above two equations, we have

$$\frac{B_\alpha^*(z)}{B^*(z)} = P^*(z) \exp\left[\int_{-0.5}^{0.5} \frac{e^{2\pi i v} + z}{e^{2\pi i v} - z} \log\left| \frac{B_\alpha(v)}{B(v)} \right| \, dv \right]. \tag{3}$$

Because B_α/B belongs to $L^2(dv)$, the gain $|B_\alpha/B|$ is quadratically integrable. Also

$$\log \frac{B_\alpha(v)}{B(v)} = \log|B_\alpha(v)| - \log|B(v)|$$

is integrable. Hence equation (3) is the canonical representation of $K^*(z)$, and so k_t is a wavelet. In fact, k_t is given by

$$k_t = \sum_{s=0}^{t} b_{\alpha+s} a_{t-s} \tag{4}$$

where a_t is the wavelet corresponding to $1/B(v)$. Q.E.D.

Theorem 12. *Let x_t be a regular stationary stochastic process with spectral density $\Phi(v)$. A necessary and sufficient condition that the least-squares linear prediction $\hat{x}_{t+\alpha}$ (for any prediction lead $\alpha > 0$) has the representation*

$$\hat{x}_{t+\alpha} = \sum_{s=0}^{\infty} k_s x_{t-s}$$

in the sense that k_s is a wavelet and that the error of approximation

$$\hat{x}_{t+\alpha} - \sum_{s=0}^{N} k_s x_{t-s}$$

belongs to the linear manifold $\Theta(t-N-1)$ whose intersection

$$\bigcap_{N=0}^{\infty} \Theta(t-N-l) = 0$$

is that (2).

Proof. We let $\Theta(t)$ denote the closed linear manifold spanned by $\vartheta(s)$, $s \leqslant t$. We have

$$\hat{x}_\alpha - \sum_{s=0}^{N} k_s x_{-s} = \sum_{n=0}^{\infty} b_{\alpha+n}\vartheta_{-n} - \sum_{s=0}^{N} k_s \left[\sum_{r=0}^{\infty} b_r \vartheta_{-s-r} \right] =$$

$$= \sum_{n=0}^{\infty} b_{\alpha+n}\vartheta_{-n} - \sum_{n=0}^{\infty} \left[\sum_{s=0}^{N} k_s b_{n-s} \right] \vartheta_{-n} = \sum_{n=0}^{\infty} \left[b_{\alpha+n} - \sum_{s=0}^{N} k_s b_{n-s} \right] \vartheta_{-n} =$$

$$= \sum_{n=N+1}^{\infty} \left[b_{\alpha+n} - \sum_{s=0}^{N} k_s b_{n-s} \right] \vartheta_{-n} \quad \varepsilon\Theta(-N-1)$$

if and only if

$$b_{\alpha+N} = \sum_{s=0}^{N} k_s b_{N-s}, \quad N = 0, 1, 2, \ldots \tag{5}$$

But, by Theorem 11, k_s is a wavelet satisfying equation (5) if and only if condition (2) holds. Q.E.D.

If we define the inner product between two variates y and z with finite second absolute moments by $(y, z) = \mathbf{E}\{y\bar{z}\}$, then the space of all such variates is a Hilbert space, denoted by $L^2(dP)$. Consider all complex-valued functions $F(v)$ for which

$$\int_{-0.5}^{0.5} |F(v)|^2 \Phi(v)\, dv < \infty$$

where $\Phi(v)$ is a spectral density function. If we define the inner product between two such functions to be

$$\int_{-0.5}^{0.5} F(v)\overline{G(v)}\, \Phi(v)\, dv$$

then the space of all such functions is a Hilbert space, denoted by $L^2[\Phi(v)\, dv]$.

Theorem 13. *Let x_t be a regular stationary stochastic process with spectral density $\Phi(v)$. Let condition (2) be satisfied, so that the wavelet k_t given by equation (4) represents the coefficients of the least-squares linear predictor for prediction lead $\alpha > 0$. Then each of the following conditions is necessary and sufficient that*

$$\hat{x}_{t+\alpha} = \underset{N \to 0}{\text{l.i.m.}} \sum_{s=0}^{N} k_s x_{t-s} \qquad in\ space\ \ L^2(dP).$$

(a) $\quad K(v) = \underset{N \to \infty}{\text{l.i.m.}} \sum_{s=0}^{N} k_s e^{-2\pi i v s} \qquad in\ space\ \ L^2(\Phi dv).$

(b) $\quad b_{\alpha+n} = \underset{N \to \infty}{\text{l.i.m.}} \sum_{s=0}^{N} k_s b_{n-s} \qquad in\ space\ \ L^2(0,\ \infty).$

P r o o f. We have

$$\mathbf{E}\left\{ \left| \hat{x}_\alpha - \sum_0^N k_s x_{-s} \right|^2 \right\} = \int_{-0.5}^{0.5} \left| K(v) - \sum_0^N k_s e^{-2\pi i v s} \right|^2 \Phi(v)\, dv$$

by the Stone — Kolmogorov isomorphism), which is

$$\int_{-0.5}^{0.5} \left| B_\alpha(v) - \sum_0^N k_s e^{-2\pi i v s} B(v) \right|^2 dv = \sum_{n=0}^{\infty} \left| b_{\alpha+n} - \sum_0^N k_s b_{n-s} \right|^2$$

(by Bessel's equality). Q.E.D.

Sufficient conditions that $\sum_0^N k_s x_{t-s}$ converge in the mean (in space $L^2(dP)$) to $\hat{x}_{t+\alpha}$ have been given by Wold (1938), Doob (1953), and Yaglom (1955). These sufficient conditions involve the uniform convergence of the Fourier series $\sum_0^{\infty} k_s e^{-2\pi i v s}$. Another sufficient condition given by Wiener and Masani (1958) is that $0 < C_1 < \Phi(v) < C_2 < \infty$. Condition (c) in the theorem below improves on the result of Wiener and Masani.

Theorem 14. *Let x_t be a regular stationary stochastic process with spectral density $\Phi(v)$. Let $K(v)$ be the transfer function of the least-squares linear predictor for prediction lead $\alpha > 0$. Then each of the following conditions is sufficient that there is a wavelet k_t such that*

$$\hat{x}_{t+\alpha} = \underset{N \to \infty}{\text{l.i.m.}} \sum_0^N k_s x_{t-s} \ \ in\ space\ \ L^2(dP).$$

(a) $\quad \sum_0^N k_s e^{-2\pi i v s}$ *converges uniformly to* $K(v)$.

(b) $\quad K(v) = \sum_0^{\infty} k_s e^{-2\pi i v s}$ *where* $\sum_0^{\infty} |k_s| < \infty$.

(c) $\quad \displaystyle\int_{-0.5}^{0.5} \frac{dv}{\Phi(v)} < \infty$ *and* $\Phi(v) < C$ *where C is a positive constant.*

P r o o f. (a). Uniform convergence implies convergence (in the mean) in space $L^2(dv)$ so that k_s is a wavelet. From Theorem 11 it follows that condition (2) holds, and that k_s is given by equation (4). Uniform convergence implies convergence (in the mean) in space $L^2(\Phi\, dv)$. Hence by condition (a) of Theorem 13, the desired result follows. Q.E.D.

(b) Because

$$\left| K(v) - \sum_{0}^{N} k_s e^{-2\pi i v s} \right| = \left| \sum_{N+1}^{\infty} k_s e^{-2\pi i v s} \right| \leqslant \sum_{N+1}^{\infty} |k_s|,$$

it follows that the partial sums converge uniformly to $K(v)$, so that condition (a) is satisfied. Q.E.D.

(c) Theorem 11 applies and gives the wavelet k_s of equation (4). Because

$$\int_{-0.5}^{0.5} \left| K(v) - \sum_{0}^{N} k_s e^{-2\pi i v s} \right|^2 \Phi(v)\, dv < C \int_{-0.5}^{0.5} \left| K(v) - \sum_{0}^{N} k_s e^{-2\pi i v s} \right|^2 dv,$$

the partial sums converge (in the mean) to $K(v)$ in space $L^2(\Phi\, dv)$. The desired result follows from condition (a) of Theorem 13. Q.E.D.

In closing, we recall that Kolmogorov (1941, Theorem 24) proved that condition (2) is necessary and sufficient that x_t have least-square interpolation error η_t such that $\mathbf{E}\{|\eta_t|^2\} > 0$.

I should like to express my sincere thanks to Professor Herman Wold for his help and encouragement.

Institute of Statistics
Uppsala University

Поступила в редакцию
6.5.61

REFERENCES

[1] J. L. D o o b, Stochastic processes, Wiley, New York, 1953.

[2] U. G r e n a n d e r and M. R o s e n b l a t t, Statistical analysis of stationary time-series, Wiley, New York, and Almqvist and Wicksell, Uppsala, 1957.

[3] K. K a r h u n e n, Über die Struktur stationärer zufälliger Funktionen, Ark. Mathematik, 1 (1949), 141—160.

[4] V. I. K r y l o v, On functions regular in a half-plane (Russian), Mat. Sbornik, 6 (48), (1939), 95—138. (English translation by D. V. Thampuran, University of Wisconsin).

[5] A. N K o l m o g o r o v, Stationary Sequences in Hilbert space (Russian), Bull. Moscow State University. Math, 2 (1941). (Spanish translation in Trab. Estad, 4, 55 and 243).

[6] I. I. P r i v a l o v, Randeigenschaften analytischer Functionen (German translation fro 1950 Russian edition) VEB Deutscher Verlag der Wissenschaften, Berlin, 1956.

[7] F. R i e s z, Über die Randwerte einer analytischen Function, Math. Zeit, 18 (1923), 87—95.

[8] E. A. R o b i n s o n, Infinitely many variates. Griffin, London, 1959.

[9] V. I. S m i r n o v, Sur les valuers limites des fonctions régulières a l'interieur d'un cercle, Leningrad Phys-Math, 2 (1928).

[10] G. S z e g ö, Über die Randwerte einer analytischen Function, Math. Ann., 84 (1921), 232—244.

[11] N. W i e n e r and P. M a s a n i, The prediction theory of multivariate stochastic processes, II. Acta Math. (Uppsala), 99 (1958), 93—137.

[12] H. W o l d, A study in the analysis of stationary time-series, Thesis, University of Stockholm (Second Edition, Almqvist and Wiksell, Uppsala, 1954), 1938.

[13] A. M. Y a g l o m, Introduction to the theory of stationary random functions (Russian), Uspehi Mat. Nauk, 7 (1952), 3—168. (English translation by D. V. Thampuran, University of Wisconsin).

[14] A. M. Y a g l o m, Correlation theory of processes with random stationary n-th increments (Russian), Math. Sbornik, 37 (79), (1955), 141—196; (English translation in A. M. S. Translations, 8 (Series 2), 1958).

CHAPTER 15

JOURNAL OF GEOPHYSICAL RESEARCH   VOL. 68, No. 19   OCTOBER 1, 1963

Mathematical Development of Discrete Filters for the Detection of Nuclear Explosions

ENDERS A. ROBINSON

University Institute of Statistics
Uppsala, Sweden

Abstract. In this paper the mathematical theory of the following discrete filters to be used on digitized data from microbarographic and seismic records for the detection of nuclear explosions is developed: (1) matched filter, (2) modified matched filter, (3) modified matched filter for a multiparameter model, (4) filter for the estimation of the trend component, (5) wave-shaping filter, (6) wave-shaping filter in the noiseless case, (7) spike filter. F'nally a simplified computational method is given for the case of stationary noise.

INTRODUCTION

Microbarographic records and seismic records may be analyzed in order to detect signals caused by nuclear explosions. Signals of high-energy content are easily detected visually, but as the signal energy decreases, visual detection becomes progressively more difficult until the point is reached where the signal is completely masked by noise. As a result increased attention is being given to various types of filtering operations for the detection of signals in the presence of noise. Moreover, much work is being done toward the development of digitization systems which make microbarographic and seismic recordings available in the form of time series, that is, discrete numerical readings at equally spaced time instants. See, for example, *Simpson et al.* [1961a]. The various filtering operations can then be performed on the discrete data by means of a digital computer. The flexibility of digital-computer processing of data makes this approach extremely useful in the research stages of the design of filters. See, for example, *Simpson et al.* [1961b, 1962]. The purpose of this paper is to give in one place the mathematical assumptions and the derivations of various discrete filters which can be used for the detection of nuclear explosions. Several important discrete filters, for example *Wiener* [1949] filters, are not included here, but will be treated in a separate paper.

MATCHED FILTER

Assumptions. (*a*) The signal s_t has a known fixed shape. (*b*) The noise n_t is a normal ran-dom process. (*c*) The mean value $E\{n_t\}$ is known to be equal to zero. (*d*) The covariance $\phi_{t\tau} = E\{n_t n_\tau\}$ of the noise is known. (*e*) The time t is a discrete, integer-valued parameter. (An alternative symbol for time is τ.) (*f*) The observed random process is x_t. (*g*) The random process x_t is observed for $t = 1, 2, \cdots , N$.

Problem. We wish to test whether hypothesis H_0 or hypothesis H_1 is true, where H_0 is $x_t = n_t$ and H_1 is $x_t = s_t + n_t$.

Matrix notation.

$$x = [x_1 \; x_2 \; \cdots \; x_N]$$

$$n = [n_1 \; n_2 \; \cdots \; n_N]$$

$$s = [s_1 \; s_2 \; \cdots \; s_N]$$

are $1 \times N$ row vectors,

$$\phi = \begin{bmatrix} \phi_{11} & \phi_{12} & \cdots & \phi_{1N} \\ \phi_{21} & \phi_{22} & \cdots & \phi_{2N} \\ & \cdots & & \\ \phi_{N1} & \phi_{N2} & \cdots & \phi_{NN} \end{bmatrix} = [\phi_{t\tau}]$$

is the covariance matrix of the noise (assumed to be nonsingular),

$$\mu = [\mu_{t\tau}] = \phi^{-1}$$

Probability density. Under hypothesis H_0, x is normally distributed with mean zero and covariance matrix ϕ; that is, the probability density $f_0(x)$ is

$$f_0(x) = [(2\pi)^N \det \phi]^{-1/2} \; \exp \; (-\tfrac{1}{2} x\mu x^T)$$

Under hypothesis H_1, x is normally distributed

with mean s and covariance matrix ϕ; that is, the probability density $f_1(x)$ is

$$f_1(x) = [(2\pi)^N \det \phi]^{-1/2}$$

$$\exp \left[-\tfrac{1}{2}(x - s)\,\mu\,(x - s)^T\right]$$

Likelihood ratio. The likelihood ratio $\Lambda(x)$ is formed by taking the quotient of these two density functions; that is,

$$\Lambda(x) = f_1(x)/f_0(x)$$
$$= \exp\left\{-\tfrac{1}{2}[(x - s)\mu(x - s)^T - x\mu x^T]\right\}$$
$$= \exp\left[-\tfrac{1}{2}(x\mu x^T - x\mu s^T - s\mu x^T\right.$$
$$\left. + s\mu s^T - x\mu x^T)\right]$$
$$= \exp\left[-\tfrac{1}{2}(-2x\mu s^T + s\mu s^T)\right]$$

(since $x\mu s^T = s\mu x^T$). The observer will choose hypothesis H_0 when $\Lambda(x) < \Lambda_0$, where Λ_0 is a constant determined by the decision criterion used. Setting $\Lambda(x) = \Lambda_0$ and taking logarithms, we obtain $-\tfrac{1}{2}(-2x\mu s^T + s\mu s^T) = \log \Lambda_0$. Thus the inequality $\Lambda(x) < \Lambda_0$ is the same as the inequality $-\tfrac{1}{2}(-2x\mu s^T + s\mu s^T) < \log \Lambda_0$, or, what is the same thing, $x\mu s^T < \log \Lambda_0 + \tfrac{1}{2} s\mu s^T$.

Decision rule. Setting $G = \log \Lambda_0 + \tfrac{1}{2} s\mu s^T$, we have the following decision rule: Choose H_0 (that is, say that the signal is not present) if $x\mu s^T < G$; choose H_1 (that is, say that the signal is present) if $x\mu s^T > G$. That is, the decision is based on the *test statistic*

$$x\mu s^T = \sum_{t,\tau=1}^{N} x_t \mu_{t\tau} s_\tau$$

computed from the observed process $x = [x_1 \; x_2 \; \cdots \; x_N]$ and the known signal $s = [s_1 \; s_2 \; \cdots \; s_N]$. Hence we may say that the observed process x is compared with, or matched to, the signal s. For this reason, the filter which is used in computing $x\mu s^T$ is called a *matched filter*.

Distribution of the test statistic. Since the test statistic $x\mu s^T$ is a linear combination of normal random variables x, it follows that $x\mu s^T$ itself is normally distributed, with mean and variance as follows.

Under H_0 (i.e., $x = n$)

$$E\{x\mu s^T\} = E\{x\}\mu s^T = 0$$

since $E\{x\} = 0$,

$$\mathrm{var}\,\{x\mu s^T\} = E\{(x\mu s^T)^2\} = E\{(x\mu s^T)^T x\mu s^T\}$$
$$= E\{s\mu^T x^T x\mu s^T\}$$
$$= s\mu^T E\{x^T x\}\mu s^T = s\mu^T \phi \mu s^T$$
$$= s\mu s^T$$

since $E\{x^T x\} = \phi = E\{n^T n\}$, $\phi\mu = I$ (identity matrix), and $\mu^T = \mu$.

Under H_1 (i.e., $x = s + n$)

$$E\{x\mu s^T\} = E\{x\}\mu s^T = s\mu s^T$$

since $E\{x\} = s$,

$$\mathrm{var}\,\{x\mu s^T\} = s\mu s^T$$

(the same as the variance under H_0, because the variance of a random variable does not depend upon its mean).

Relaxation of the normality assumption. Let us now drop assumption (b), that is, we no longer assume that n_t is a normal random process, but instead we assume (b_1) n_t is a random process with an unknown distribution. We now consider a filter with coefficients given by $a = [a_1 \; a_2 \; \cdots \; a_N]^T$, an $N \times 1$ column vector. The input to this filter is the observed random process x, and the output is xa.

The mean-square output is defined as

$$E\{(xa)^2\} = E\{(xa)^T xa\} = E\{a^T x^T x a\}$$
$$= a^T E\{x^T x\} a$$

Under hypothesis H_0 ($x = n$, or no signal present), the mean-square output is

$$E\{(xa)^2\} = a^T E\{n^T n\} a = a^T \phi a$$

Under hypothesis H_1 ($x = s + n$, or signal present), the mean-square output is

$$E\{(xa)^2\}$$
$$= a^T E\{(s + n)^T(s + n)\} a$$
$$= a^T E\{s^T s\} a + a^T E\{s^T n\} a + a^T E\{n^T s\} a$$
$$+ a^T E\{n^T n\} a$$
$$= a^T s^T s a + a^T \phi a$$

since $E\{n\} = 0$, $E\{n^T\} = 0$, and $E\{n^T n\} = \phi$. We wish to determine a so that the ratio

$$\frac{\text{Mean-square output under } H_1}{\text{Mean-square output under } H_0}$$
$$= \frac{a^T s^T s a + a^T \phi a}{a^T \phi a} = 1 + \frac{a^T s^T s a}{a^T \phi a}$$

is a maximum, or equivalently so that the ratio

$$\lambda = a^T s^T s a / a^T \phi a = (sa)^2 / a^T \phi a$$

is a maximum. Now ϕ and μ can be expressed as $\phi = U \Lambda U^T$ and $\mu = U \Lambda^{-1} U^T$, where U is an orthogonal matrix (that is, $U^T = U^{-1}$) and Λ is a positive diagonal matrix. Define σ and α by $\sigma = sU\Lambda^{-1/2}$ and $\alpha = \Lambda^{1/2} U^T a$, so that $\sigma\alpha = sa$, $\alpha^T \alpha = a^T \phi a$, and $\sigma\sigma^T = s\mu s^T$. The Schwarz inequality $(\sigma\alpha)^2 \leq (\alpha^T \alpha)(\sigma\sigma^T)$ then becomes $(sa)^2 \leq (a^T \phi a)(s\mu s^T)$. Thus the ratio λ satisfies

$$\lambda = (sa)^2 / a^T \phi a \leq s\mu s^T$$

so $s\mu s^T$ is the upper bound for λ. Because the filter $a = \mu s^T$ yields this upper bound for λ, this filter is the optimal one. The output of the filter is $xa = x\mu s^T$, which is the same as the test statistic obtained under the assumption that n was normally distributed.

Under H_0

$$E\{xa\} = 0 \qquad E\{(xa)^2\} = s\mu s^T$$

Under H_1

$$E\{xa\} = s\mu s^T \qquad E\{(xa)^2\} = (s\mu s^T)^2 + s\mu s^T$$

MODIFIED MATCHED FILTER

Assumptions. We now modify assumption (a), given in the preceding section, to (a_1): The signal s_t is given by $s_t = cf_t$, where c is an unknown constant and f_t is a known fixed function.

Problem. We wish to test whether hypothesis H_0 or hypothesis H_1 is true, where H_0 is $x_t = n_t$ ($c = 0$) and H_1 is $x_t = cf_t + n_t$ ($c \neq 0$).

Matrix notation.

$$f = [f_1\ f_2\ \cdots\ f_N]$$

$$s = cf$$

Likelihood ratio. The likelihood ratio is

$$\Lambda(x) = \exp\{-\tfrac{1}{2}[-2x\mu(cf)^T + cf\mu(cf)^T]\}$$

$$= \exp[cx\mu f^T - \tfrac{1}{2}c^2 f\mu f^T]$$

The observer will choose H_0 when the likelihood ratio Λ satisfies $\Lambda < \Lambda_0$ for some fixed threshold Λ_0. Because c is unknown, we consider $\max \Lambda(x) < \Lambda_0$, where $-\infty < c < \infty$, instead of $\Lambda(x) < \Lambda_0$. The maximum occurs when the negative of the exponent of $\Lambda(x)$ is a minimum, that is, when $J = -cx\mu f^T + \tfrac{1}{2}c^2 f\mu f^T = $ minimum. We then have

$$\frac{\partial J}{\partial c} = 0 \qquad -x\mu f^T + c^* f\mu f^T = 0$$

or $c^* = x\mu f^T / f\mu f^T$.

Hence

$$\max(\Lambda x) = \exp[c^* x\mu f^T - \tfrac{1}{2}(c^*)^2 f\mu f^T]$$

$$-\infty < c < \infty$$

$$= \exp\left[\frac{x\mu f^T}{f\mu f^T} x\mu f^T - \frac{(x\mu f^T)^2}{2(f\mu f^T)^2} f\mu f^T\right]$$

$$= \exp[\tfrac{1}{2}(x\mu f^T)^2 / f\mu f^T]$$

Taking logarithms, we have $\tfrac{1}{2}(x\mu f^T)^2 / f\mu f^T < \log \Lambda_0$.

Decision rule. Letting $G = 2\log\Lambda_0$, we obtain the decision rule: Choose H_0 (that is, say that the signal is not present) if $(x\mu f^T)^2 / f\mu f^T < G$; choose H_1 (that is, say that the signal is present) if $(x\mu f^T)^2 / f\mu f^T > G$. That is, the decision is based on the *test statistic*

$$(x\mu f^T)^2 / f\mu f^T = \left(\sum_{t,\tau=1}^{N} x_t \mu_{t\tau} f_\tau\right)^2 \bigg/ \sum_{t,\tau=1}^{N} f_t \mu_{t\tau} f_\tau$$

computed from the observed process $x = [x_1\ x_2\ \cdots\ x_N]$ and the known $f = [f_1\ f_2\ \cdots\ f_N]$. We see that the test statistic is a quadratic function of the observations $x = [x_1\ x_2\ \cdots\ x_N]$.

Maximum-likelihood estimates. Let us now find the maximum-likelihood estimates of c and s. We recall that the density function of x, under the normality assumption, is

$$f(x) = [(2\pi)^N \det \phi]^{-1/2}$$

$$\cdot \exp[-\tfrac{1}{2}(x - s)\mu(x - s)^T]$$

where $s = cf$. The maximum-likelihood estimate of s, denoted by s^*, is that value of $s = cf$ for which $f(x)$ is a maximum. The maximum occurs when the negative of the exponent of $f(x)$ is a minimum, that is, when $P = (x - s)\mu(x - s)^T = $ minimum, subject to the constraint that $s = cf$. We shall now use the method of Lagrange multipliers to solve this constrained minimization problem. We therefore introduce the undetermined multipliers $\lambda = [\lambda_1\ \lambda_2\ \cdots\ \lambda_N]^T$. We now wish to minimize $J = (x - s)\mu(x - s)^T + (s - cf)\lambda + \lambda^T(s - cf)^T$ with respect to s, λ, and c. We thus obtain

$$\partial J / \partial s = 0 \qquad -(x - s^*)\mu + \lambda^{*T} = 0$$

$$\partial J / \partial \lambda = 0 \qquad s^* - c^* f = 0$$

$$\partial J / \partial c = 0 \qquad \lambda^{*T} f^T = 0$$

Solving these equations, we have $x\mu f^T - c^* f\mu f^T = 0$. Hence $c^* = x\mu f^T / f\mu f^T$ and $s^* = c^* f = (x\mu f^T / f\mu f^T) f$.

Let us now find the maximum-likelihood estimate of c. By the above reasoning, it is that value of c obtained by requiring $P = (x - s)\mu (x - s)^T = (x - cf)\mu (x - cf)^T = $ minimum. We thus have

$$\partial P / \partial c = 0 \qquad -(x - c^* f)\mu f^T = 0$$

or $c^* = x\mu f^T / f\mu f^T$, which is the same c^* as found above, as we would expect.

Miscellaneous notes. Now it may be shown that the quadratic form $x\mu f^T$ is

$$x\mu f^T = - \begin{vmatrix} \phi_{11} & \phi_{12} & \cdots & \phi_{1N} & f_1 \\ \phi_{21} & \phi_{22} & \cdots & \phi_{2N} & f_2 \\ \cdots \\ \phi_{N1} & \phi_{N2} & \cdots & \phi_{NN} & f_N \\ x_1 & x_2 & \cdots & x_N & 0 \end{vmatrix}$$

$$\div \begin{vmatrix} \phi_{11} & \phi_{12} & \cdots & \phi_{1N} \\ \phi_{21} & \phi_{22} & \cdots & \phi_{2N} \\ \cdots \\ \phi_{N1} & \phi_{N2} & \cdots & \phi_{NN} \end{vmatrix}$$

Similarly,

$$f\mu f^T = - \begin{vmatrix} \phi_{11} & \phi_{12} & \cdots & \phi_{1N} & f_1 \\ \phi_{21} & \phi_{22} & \cdots & \phi_{2N} & f_2 \\ \cdots \\ \phi_{N1} & \phi_{N2} & \cdots & \phi_{NN} & f_N \\ f_1 & f_2 & \cdots & f_N & 0 \end{vmatrix}$$

$$\div \begin{vmatrix} \phi_{11} & \phi_{12} & \cdots & \phi_{1N} \\ \phi_{21} & \phi_{22} & \cdots & \phi_{2N} \\ \cdots \\ \phi_{N1} & \phi_{N2} & \cdots & \phi_{NN} \end{vmatrix}$$

Hence

$$c^* = \begin{vmatrix} \phi_{11} & \phi_{12} & \cdots & \phi_{1N} & f_1 \\ \cdots \\ \phi_{N1} & \phi_{N2} & \cdots & \phi_{NN} & f_N \\ x_1 & x_2 & \cdots & x_N & 0 \end{vmatrix}$$

$$\div \begin{vmatrix} \phi_{11} & \phi_{12} & \cdots & \phi_{1N} & f_1 \\ \cdots \\ \phi_{N1} & \phi_{N2} & \cdots & \phi_{NN} & f_N \\ f_1 & f_2 & \cdots & f_N & 0 \end{vmatrix}$$

MODIFIED MATCHED FILTER FOR A MULTIPARAMETER MODEL

Assumptions. Instead of assumption (a_1) given at the beginning of the preceding section, let us introduce assumption (a_2): The signal s_t is given by $s_t = \Sigma_{i=1}^{p} c_i f_{it}$, where $c_1, c_2, \cdots, c_p$ are unknown constants and $f_{1t}, f_{2t}, \cdots, f_{pt}$ are known fixed functions.

Problem. We wish to test $H_0 : x_t = n_t$ ($c_1 = 0, c_2 = 0, \cdots, c_p = 0$) and $H_1 : x_t = \Sigma_{i=1}^{p} c_i f_{it} + n_t$ (some or all of the $c_1, c_2, \cdots, c_p \neq 0$).

Matrix notation.

$$c = (c_1 \; c_2 \; \cdots \; c_p)$$

$$f = \begin{bmatrix} f_{11} & f_{12} & \cdots & f_{1N} \\ f_{21} & f_{22} & \cdots & f_{2N} \\ \cdots \\ f_{p1} & f_{p2} & \cdots & f_{pN} \end{bmatrix}$$

$$s = cf$$

Likelihood ratio. The likelihood ratio is $\Lambda(x) = \exp(x\mu f^T c^T - \frac{1}{2} cf\mu f^T c^T)$. The maximum of $\Lambda(x)$ with respect to all values of c occurs when $x\mu f^T - c^* f\mu f^T = 0$, or $c^* = x\mu f^T (f\mu f^T)^{-1}$. This maximum is

$$\max \Lambda(x) = \exp(x\mu f^T c^{*T} - \tfrac{1}{2} c^* f\mu f^T c^{*T})$$

$$-\infty < c < \infty$$

$$= \exp[\tfrac{1}{2} x\mu f^T (f\mu f^T)^{-1} f\mu x^T]$$

where we have used $[(f\mu f^T)^{-1}]^T = (f\mu f^T)^{-1}$ and $\mu^T = \mu$.

Decision rule. As before, we let $G = 2 \log \Lambda_0$. We thus obtain the decision rule: Choose H_0 if $(x\mu f^T)(f\mu f^T)^{-1} f\mu x^T < G$; choose H_1 if $(x\mu f^T)(f\mu f^T)^{-1} f\mu x^T > G$.

Maximum-likelihood estimates. As before, we find that the *maximum-likelihood estimates* of c and s are $c^* = (x\mu f^T)(f\mu f^T)^{-1}$ and $s^* = c^* f = (x\mu f^T)(f\mu f^T)^{-1} f$.

Let us now find the expected value of c^*. It is $E\{c^*\} = E\{x\mu f^T (f\mu f^T)^{-1}\} = c$, so c^* is an unbiased estimate of c. Also, $E\{s^*\} = E\{c^* f\} = cf = s$, so s^* is an unbiased estimate of s.

The covariance matrix of c^* is

$$\Psi = E\{(c^* - c)^T (c^* - c)\}$$

$$= E\{[(x - cf)\mu f^T (f\mu f^T)^{-1}]^T$$

$$[(x - cf)\mu f^T (f\mu f^T)^{-1}]\}$$

But under both hypotheses we have $x - cf = n$. Hence

$$\Psi = E\{[n\mu f^T(f\mu f^T)^{-1}]^T[n\mu f^T(f\mu f^T)^{-1}]\}$$

$$= E\{(f\mu f^T)^{-1}f\mu n^T n\mu f^T(f\mu f^T)^{-1}\}$$

$$= (f\mu f^T)^{-1}f\mu E\{n^T n\}\mu f^T(f\mu f^T)^{-1}$$

$$= (f\mu f^T)^{-1}f\mu\phi\mu f^T(f\mu f^T)^{-1}$$

$$= (f\mu f^T)^{-1}f\mu f^T(f\mu f^T)^{-1} \qquad (\text{since} \quad \phi\mu = I)$$

$$= (f\mu f^T)^{-1}$$

Hence c^* has mean c and covariance matrix $(f\mu f^T)^{-1}$.

The covariance matrix of s^* is

$$E\{(s^* - s)^T(s^* - s)\}$$

$$= E\{[(x - cf)\mu f^T(f\mu f^T)^{-1}f]^T$$

$$\cdot [(x - cf)\mu f^T(f\mu f^T)^{-1}f]\}$$

$$= f^T(f\mu f^T)^{-1}f\mu E\{n^T n\}\mu f^T(f\mu f^T)^{-1}f$$

$$= f^T(f\mu f^T)^{-1}f\mu\phi\mu f^T(f\mu f^T)^{-1}f$$

$$= f^T(f\mu f^T)^{-1}f$$

Hence $s^* = c^*f$ has mean $s = cf$ and covariance matrix $f^T(f\mu f^T)^{-1}f$.

Miscellaneous notes. We note that the maximum-likelihood estimate c^* and s^* are independent of the scale of the covariance matrix. Let us suppose that $\phi = \sigma^2\psi$ and $\mu = \phi^{-1} = \sigma^{-2}\psi^{-1}$, where ψ is known but the scale factor σ^2 is unknown. Then $c^* = x\psi^{-1}f^T(f\psi^{-1}f^T)^{-1}$ and $s^* = c^*f$, independently of σ^2, and assumption (d) may be so relaxed.

FILTER FOR THE ESTIMATION OF THE TREND COMPONENT

Assumptions. As a preliminary step in the analysis of random processes, we might try various methods of estimating the trend component as well as possible so that it may be separated from the trend-free component. One approach to this problem is to use the models discussed in the foregoing sections, with the following changes in terminology: (1) s_t, instead of being called the signal, is now called the *trend* component; (2) n_t, instead of being called the noise, is now called the *trend-free* component; (3) as before, the observed random process is x_t. We shall make assumptions (a_2), (b), (c), (d), (e), (f), and (g).

Problem. Given that $x_t = s_t + n_t$, filter x_t so as to estimate the trend component s_t.

Maximum-likelihood estimates. As a solution to the problem, the maximum-likelihood estimates of c and s may be used. Also, we recall that we may relax assumption (d) to (d_1): The covariance $\phi_{t\tau} = E\{n_t n_\tau\}$ is equal to $\phi_{t\tau} = \sigma^2\psi_{t\tau}$, where $\psi_{t\tau}$ is known and σ^2 is an unknown scale factor.

We let $\phi = \sigma^2\psi$, and hence $\mu = \sigma^{-2}\psi^{-1}$. Substituting $\mu = \sigma^{-2}\psi^{-1}$ into the expressions for c^* and s^* we obtain $c^* = x\psi^{-1}f^T(f\psi^{-1}f^T)^{-1}$ and $s^* = x\psi^{-1}f^T(f\psi^{-1}f^T)^{-1}f$, which shows that the maximum-likelihood estimates are independent of the unknown scale factor σ^2.

Relaxation of the normality assumption. Instead of (b) we now assume (b_1): n_t is a random process with an unknown distribution.

Least-squares estimate. We have $x = cf + n$ and we wish to estimate c. The least-squares estimate $c^\dagger$ is that value of c such that the sum of squared errors $J = (x - cf)(x - cf)^T = $ minimum. We have

$$\frac{\partial J}{\partial c} = 0 \qquad (x - c^\dagger f)(-f^T) = 0$$

or $c^\dagger ff^T = xf^T$. Hence the least-squares estimate is $c^\dagger = x f^T(ff^T)^{-1}$.

To compute $c^\dagger$ we see that we do not need assumption (d); that is, we do not need to know the covariance matrix ϕ of n_t. This fact makes the least-squares estimate extremely useful in practice. The least-squares estimate is unbiased since $E\{c^\dagger\} = E\{xf^T(ff^T)^{-1}\} = E\{x\} f^T(ff^T)^{-1}$ and (using $E\{x\} = cf$) $E\{c^\dagger\} = cff^T(ff^T)^{-1} = c$. The covariance matrix of $c^\dagger$ is defined as $E\{(c^\dagger - c)^T(c^\dagger - c)\}$.

Since

$$c^\dagger = xf^T(ff^T)^{-1} = (n + s)f^T(ff^T)^{-1}$$

$$= (n + cf)f^T(ff^T)^{-1}$$

$$= nf^T(ff^T)^{-1} + cff^T(ff^T)^{-1}$$

$$= nf^T(ff^T)^{-1} + c$$

we have $c^\dagger - c = nf^T(ff^T)^{-1}$. Hence the covariance matrix of $c^\dagger$ is

$$E\{(c^\dagger - c)^T(c^\dagger - c)\}$$

$$= E\{[nf^T(ff^T)^{-1}]^T[nf^T(ff^T)^{-1}]\}$$

$$= E\{(ff^T)^{-1}fn^T nf^T(ff^T)^{-1}\}$$

$$= (ff^T)^{-1}fE\{n^T n\}f^T(ff^T)^{-1}$$

which is $E\{(c^\dagger - c)^T(c^\dagger - c)\} = (ff^T)^{-1}$ $f\phi f^T(ff^T)^{-1}$.

Best unbiased linear estimate. We wish to find an estimate c^* which is (1) linear in the observations, i.e. $c^* = xb$, where

$$b = \begin{bmatrix} b_{11} & b_{12} & \cdots & b_{1p} \\ b_{21} & b_{22} & \cdots & b_{2p} \\ \cdots & & & \cdots \\ b_{N1} & b_{N2} & \cdots & b_{Np} \end{bmatrix}$$

is (2) an unbiased estimate, i.e. $E\{c^*\} = c$, and is (3) best in the sense that the covariance matrix of c^* is 'less than' the covariance matrix of any other linear unbiased estimate $c^\ddagger$, i.e. $E\{(c^* - c)^T(c^* - c)\} < E\{(c\ddagger - c)^T(c\ddagger - c)\}$. (More precisely, the sign $<$ as used here means that $E\{(c\ddagger - c)^T(c\ddagger - c)\} - E\{(c^* - c)^T$ $(c^* - c)\}$ is a positive definite matrix.)

Using (1) we have $c^* = xb = (s + n)b = cfb + nb$. Using (2) we have $c = E\{c^*\} = cfb + E\{n\}b = cfb$, since $E\{n\} = 0$. Thus we have the constraint $fb = I = p \times p$. Also we have $c^* - c = nb$, so the covariance matrix of c^* is

$$E\{(c^* - c)^T(c^* - c)\} = E\{(nb)^T(nb)\}$$
$$= E\{b^T n^T nb\} = b^T E\{n^T n\}b = b^T \phi b$$

Hence b may be determined by minimizing the $p \times p$ matrix $b^T \phi b$ subject to the constraint $fb = I$.

We may use the method of Lagrange multipliers. We introduce the $p \times p$ matrix λ as an undetermined multiplier. We then wish to minimize $J = b^T \phi b + 2\lambda(fb - I)$ with respect to b and λ. We have

$$\frac{\partial J}{\partial b} = 0 \qquad b^T \phi + \lambda f = 0$$

$$\frac{\partial J}{\partial \lambda} = 0 \qquad fb = I$$

Solving these equations for b and λ we have $b^T = -\lambda f\phi^{-1}$; thus $b = -\phi^{-1}f^T\lambda^T$. Also $fb = I$ gives $-f\phi^{-1}f^T\lambda^T = I$, or $\lambda^T = -(f\phi^{-1}f^T)^{-1}$;

hence $b = -\phi^{-1}f^T\lambda^T = \phi^{-1}f^T(f\phi^{-1}f^T)^{-1}$. We recall that the notation for ϕ^{-1} was $\mu = \phi^{-1}$; hence $b = \mu f^T(f\mu f^T)^{-1}$.

Thus the best unbiased linear estimate c^* of c is $c^* = xb = x\mu f^T(f\mu f^T)^{-1}$, which, we see, is the same as the maximum-likelihood estimate obtained under the normality assumption (b). The covariance matrix of c^* is

$$E\{(c^* - c)^T(c^* - c)\} = b^T\phi b$$
$$= (f\mu f^T)^{-1}f\mu\phi\mu f^T(f\mu f^T)^{-1}$$

which is $E\{(c^* - c)^T(c^* - c)\} = (f\mu f^T)^{-1}$.

WAVE-SHAPING FILTER

Assumptions. (a) The signal s_t has a known fixed shape. (b) The noise n_t is a random process with an unknown distribution. (c) The mean value $E\{n_t\}$ of the noise is known to be equal to zero. (d) The covariance $\phi_{t\tau} = E\{n_t n_\tau\}$ of the noise is known. (e) The time t is a discrete, integer-valued parameter. (f) The observed random process is $x_t = s_t + n_t$. (g) The random process x_t is observed for $t = 1, 2, \cdots,$ N. (h) The observed random process $x_1, x_2,$ $\cdots, x_N$ is passed into a time-invariant linear filter with coefficients $\beta_0, \beta_1, \cdots, \beta_M$ (to be determined). (i) The actual output of the filter is given by the equation

$$y_t = \sum_{s=0}^{M} \beta_s x_{t-s} \quad (t = 0, 1, 2, \cdots, M + N)$$

where values of x_t outside the range of t are taken to be zero. (j) The desired output of the filter is z_t, which is a known fixed function.

Problem. We wish to determine such values of the coefficients $\beta_0, \beta_1, \cdots, \beta_M$ that the mean of the sum of squared errors between the desired output z_t and the actual output y_t is a minimum; that is,

$$E\left\{\sum_{t=0}^{M+N}(z_t - y_t)^2\right\} = \text{minimum}$$

Schematically, we have:

Actual input, $\longrightarrow$ Filter, $\longrightarrow$ Actual output,

$x_t = s_t + n_t$ $\quad \beta_0, \beta_1, \cdots, \beta_M$ $\quad y_t$

$\longrightarrow$ Desired output, z_t

Matrix notation. We define the $(N + M + 1) \times (M + 1)$ matrices

$$X = \begin{bmatrix} x_0 & 0 & \cdots & 0 \\ x_1 & x_0 & & 0 \\ \cdots & & & \\ x_N & x_{N-1} & \cdots & \\ 0 & x_N & \cdots & \\ \cdots & & & \\ 0 & 0 & & x_{N-1} \\ 0 & 0 & \cdots & x_N \end{bmatrix}$$

$$S = \begin{bmatrix} s_0 & 0 & \cdots & 0 \\ s_1 & s_0 & & 0 \\ \cdots & & & \\ s_N & s_{N-1} & \cdots & \\ 0 & s_N & \cdots & \\ \cdots & & & \\ 0 & 0 & & s_{N-1} \\ 0 & 0 & \cdots & s_N \end{bmatrix}$$

and

$$N = \begin{bmatrix} n_0 & 0 & \cdots & 0 \\ n_1 & n_0 & & 0 \\ \cdots & & & \\ n_N & n_{N-1} & \cdots & \\ 0 & n_N & \cdots & \\ \cdots & & & \\ 0 & 0 & & n_{N-1} \\ 0 & 0 & \cdots & n_N \end{bmatrix}$$

Thus, $X = S + N$.

Let y, z, and e be the column vectors

$$y = [y_0 \ y_1 \ \cdots \ y_{M+N}]^T$$

$$z = [z_0 \ z_1 \ \cdots \ z_{M+N}]^T$$

$$e = z - y = [e_1 \ e_2 \ \cdots \ e_{M+N}]^T$$

and let β be the column vector

$$\beta = [\beta_0 \ \beta_1 \ \cdots \ \beta_M]^T$$

Thus, $y = X\beta$.

Determination of β. We wish to find that value $\beta^\dagger$ of β that minimizes

$$J = E\left\{ \sum_{t=0}^{M+N} (z_t - y_t)^2 \right\}$$

$$= E\{(z - y)^T(z - y)\}$$

$$= E\{(z - X\beta)^T(z - X\beta)\}$$

Now $J = E\{(z - X\beta)^T(z - X\beta)\} = E\{z^T z - z^T X\beta - \beta^T X^T z + \beta^T X^T X\beta\}$. Because $X = S + N$, $E\{N\} = 0$, and $E\{N^T\} = 0$, we have

$$J = E\{z^T z - 2z^T X\beta + \beta^T X^T X\beta\}$$

$$= E\{z^T z - 2z^T(S + N)\beta + \beta^T(S + N)^T(S + N)\beta\}$$

$$= E\{z^T z - 2z^T S\beta + \beta^T S^T S\beta + \beta^T N^T N\beta\}$$

$$= z^T z - 2z^T S\beta + \beta^T S^T S\beta + \beta^T E\{N^T N\}\beta$$

Hence

$$\partial J/\partial \beta = 0 \qquad -2z^T S + 2\beta^{\dagger T} S^T S$$

$$+ 2\beta^{\dagger T} E\{N^T N\} = 0$$

or $S^T z = S^T S\beta^\dagger + E\{N^T N\}\ \beta^\dagger$, or $\beta^\dagger = [S^T S + E\{N^T N\}]^{-1} S^T z$.

WAVE-SHAPING FILTER IN THE NOISELESS CASE

Assumptions. The assumptions are the same as those given in the preceding section, except that it is now assumed that there is no noise (i.e., $n_t \equiv 0$). In other words, assumptions (*b*), (*c*), and (*d*) are omitted and (*f*) becomes (*f*₁): The observed process x_t is the fixed function s_t, i.e., $x_t \equiv s_t$.

Problem. We wish to find that value $\beta^\dagger$ of β for which the sum of squared errors between the desired output z and the actual output y is a minimum, that is

$$\sum_{t=0}^{M+N} (z_t - y_t)^2 = (z - y)^T(z - y)$$

$$= \text{minimum}$$

Determination of β. From the result of the last section the desired $\beta^\dagger$ is $\beta^\dagger = [S^T S]^{-1} S^T z$.

SPIKE FILTER

Problem. The spike filter is a specialization of the wave-shaping filter. The spike filter is designed so that with the signal as input it will produce little or no output while the signal is entering the filter, a large positive spike when the signal has fully entered the filter, and little or no output thereafter. The spike filter is also designed to have little output when noise is its only input.

Suppose we want the spike to occur at time $t = t_0$. We let the desired output be

$$z_t = \begin{array}{ll} 1 & \text{when} \quad t = t_0 \\ 0 & \text{when} \quad t \neq t_0 \end{array}$$

or in matrix notation, $z = [0\ 0\ \cdots\ 1\ \cdots\ 0]^T$, where the one occurs in the position of z_{t_0}.

Solution. With this value of z, the desired spike filter is $\beta^\dagger = [S^T S + E\{N^T N\}]^{-1} S^T z$.

Simplification for the Case of Stationary Noise

The covariance matrix ϕ of the noise as used thus far does not involve any assumptions about the stationarity of the noise. If the noise is non-stationary, its covariance matrix would have to be estimated from an ensemble of realizations of the noise process (in those applications for which such an ensemble is available) and/or from theoretical considerations. On the other hand, if the noise is stationary and ergodic, its covariance matrix can be estimated from one realization of the noise process. In this case, the covariance $\phi_{t\tau}$ depends only on the difference $t - \tau$, so that the covariance matrix ϕ is simplified in that all the elements on any diagonal (main, sub-, or super-) are equal. This equidiagonal property can in turn simplify the computations.

As an illustration, suppose we wish to solve the set of equations

$$\sum_{\tau=1}^{N} f_\tau r_{t-\tau+1} = g_t \quad (t = 0, 1, 2, \cdots, N-1)$$

where the filter coefficients $f_1, f_2, \cdots, f_N$ are the unknowns and the autocovariances r_{-N+1}, $r_{-N+2}, \cdots, r_{N-1}$ and the right-hand side g_0, g_1, $\cdots, g_{N-1}$ are the knowns. Because the autocovariance matrix on the left-hand side is an equidiagonal matrix, we may use the following iterative method whereby the equations are solved for $N = 1$, then $N = 2$, then $N = 3$, then $N = 4$, and so on, until the desired value of N is reached. This method is like one given by N. Levinson in the appendix of *Wiener* [1949], but we have extended it so that it also works for matrix-valued stochastic processes (as developed by *Robinson* [1962]). Let us now

illustrate the iteration from $N = 2$ to $N = 3$.

From $N = 2$, we retain in machine storage the numerical values of the coefficients which satisfy the equations

$$(3) \qquad b_1 r_{-1} + r_{-2} = \mu$$

$$(2) \qquad b_1 r_0 + r_{-1} = 0$$

$$(1) \qquad b_1 r_1 + r_0 = \beta$$

$$(\mathrm{I}) \qquad r_0 + a_1 r_{-1} = \alpha$$

$$(\mathrm{II}) \qquad r_1 + a_1 r_0 = 0$$

$$(\mathrm{III}) \qquad r_2 + a_1 r_1 = \lambda$$

$$(\mathrm{i}) \qquad f_1 r_0 + f_2 r_{-1} = g_0$$

$$(\mathrm{ii}) \qquad f_1 r_1 + f_2 r_0 = g_1$$

$$(\mathrm{iii}) \qquad f_1 r_2 + f_2 r_1 = \epsilon$$

For $N = 3$ we wish to find the numerical values of the (primed) coefficients which satisfy the (primed) equations (where a prime now indicates a new numerical value or a new equation)

$$(4') \qquad b_2' r_{-1} + b_1' r_{-2} + r_{-3} = \mu'$$

$$(3') \qquad b_2' r_0 + b_1' r_{-1} + r_{-2} = 0$$

$$(2') \qquad b_2' r_1 + b_1' r_0 + r_{-1} = 0$$

$$(1') \qquad b_2' r_2 + b_1' r_1 + r_0 = \beta'$$

the corresponding (I'), (II'), (III'), (IV'), and

$$(\mathrm{i}') \qquad f_1' r_0 + f_2' r_{-1} + f_3' r_{-2} = g_0$$

$$(\mathrm{ii}') \qquad f_1' r_1 + f_2' r_0 + f_3' r_{-1} = g_1$$

$$(\mathrm{iii}') \qquad f_1' r_2 + f_2' r_1 + f_3' r_0 = g_2$$

$$(\mathrm{iv}') \qquad f_1' r_3 + f_2' r_2 + f_3' r_1 = \epsilon'$$

In the iteration process we form $(3')$ and $(2')$ by

$$(3') \equiv (3) + w \cdot (\mathrm{I})$$

$$(2') \equiv (2) + w \cdot (\mathrm{II})$$

where w is a yet unknown parameter. That is, we form

$$(3') \qquad w r_0 + (b_1 + w a_1) r_{-1} + r_{-2} = \mu + w\alpha$$

$$(2') \qquad w r_1 + (b_1 + w a_1) r_0 + r_{-1} = 0$$

Because the right-hand side of $(3')$ is zero, it

follows that w must satisfy $\mu + w\alpha = 0$ or $w = -\mu\alpha^{-1}$. Thus we may compute the desired $b_1' = b_1 + wa_1$ and $b_2' = w$, and then compute μ' and β' by means of (4′) and (1′), respectively. Likewise

$$(\text{II}') \equiv (\text{II}) + u \cdot (2)$$

$$(\text{III}') \equiv (\text{III}) + u \cdot (1)$$

so that $\lambda + u\beta = 0$ or $u = -\lambda\beta^{-1}$. Thus we may compute $a_1' = a_1 + ub_1$ and $a_2' = u$, and then α' and λ'. (Note: This step may be omitted for real scalar-valued processes, because then $a_t = b_t$, $\alpha = \beta$, and $\lambda = \mu$.) Finally, we form equations

$$(\text{i}') \equiv (\text{i}) + \theta \cdot (3')$$

$$(\text{ii}') \equiv (\text{ii}) + \theta \cdot (2')$$

$$(\text{iii}') \equiv (\text{iii}) + \theta \cdot (1')$$

so that $\epsilon + \theta\beta' = g_2$ or $\theta = (g_2 - \epsilon)(\beta')^{-1}$. Thus we may compute $f_1' = f_1 + \theta b_2'$, $f_2' = f_2 + \theta b_1'$, $f_3' = \theta$, and then ϵ', which completes the case for $N = 3$.

Acknowledgment. I would like to thank Professor Richard A. Haubrich for his helpful comments.

REFERENCES

Robinson, E. A., *Random Wavelets and Cybernetic Systems,* 135 pp., Charles Griffin and Company, London, and Stechert-Hafner, New York, 1962.

Simpson, S. M., E. A. Robinson, J. F. Claerbout, J. N. Galbraith, and J. Clark, Initial studies on underground nuclear detection with seismic data prepared by a novel digitization system, *Rept. 1, Contract AF 19(604)7378,* prepared at Massachusetts Institute of Technology for Air Force Cambridge Research Center, Bedford, Mass., 1961a.

Simpson, S. M., E. A. Robinson, J. F. Claerbout, J. N. Galbraith, W. P. Ross, and J. Clark, Time series techniques applied to underground nuclear detection and further digitized seismic data, *Rept. 2* of *ibid.,* 1961b.

Simpson, S. M., E. A. Robinson, J. F. Claerbout, J. N. Galbraith, R. Wiggins, and C. Pan, Continued numerical studies on underground nuclear detection and further digitized seismic data, *Rept. 3* of *ibid.,* 1962.

Wiener, N., *The Extrapolation, Interpolation, and Smoothing of Stationary Time Series,* 163 pp., John Wiley & Sons, New York, 1949.

(Manuscript received May 23, 1963;
revised July 11, 1963.)

CHAPTER 16

JOURNAL OF GEOPHYSICAL RESEARCH   VOL. 70, No. 8   APRIL 15, 1965

Reprinted by special permission of the JOURNAL OF GEOPHYSICAL RESEARCH.

Recursive Solution to the Multichannel Filtering Problem

RALPH A. WIGGINS

Department of Geology and Geophysics
Massachusetts Institute of Technology, Cambridge

ENDERS A. ROBINSON

Seismological Institute, Uppsala University
Uppsala, Sweden

Abstract. Systems made up of arrays of sensors (seismographs, hydrophones, antennas, meteorological stations) receive geophysical data on a number of channels. Such multichannel data can be processed by means of a multichannel least-squares (Wiener) digital filter; the filter is determined by requiring that the sum of the mean-square errors between the desired outputs and the actual outputs for each channel be minimized. This requirement leads to a set of simultaneous linear equations, called the normal equations, where each element is itself a square matrix. The coefficients of the digital filter are given by the solution of these normal equations. An exact recursive procedure is derived for the numerical solution of the normal equations. In this recursive procedure advantage is taken of the fact that the matrix of the normal equations is an autocorrelation matrix, and hence is in the form of a Toeplitz matrix, namely one with equal elements along any diagonal. A set of auxiliary coefficients that are generated in the recursive procedure are the coefficients of matrix-valued polynomials which are the counterparts of the classical (scalar-valued) polynomials orthogonal on the unit circle. This set of auxiliary coefficients may also be used for a sideward recursion that corresponds to shifting the time index of the right-hand side of the normal equations. This sideward recursion is valuable in applications because it represents a relatively inexpensive way of determining the filter that incorporates the optimum time delay between the multichannel input signal and the multichannel output signal. The machine time required to solve the normal equations for a filter with m (matrix-valued) coefficients is proportional to m^2 for the proposed recursive procedure, as compared with m^3 for conventional techniques for the solution of simultaneous equations. Another advantage of using the recursive procedure is that it only requires computer storage space proportional to m, rather than to m^2 as in the case of the conventional techniques.

Introduction. In geophysics it is almost always necessary to work with systems whose states are described by several time variables. The system that we have specifically in mind is made up of an array of seismographs used in a nuclear surveillance program. Other examples of importance include hydrophone array systems used to map the ocean floor, systems of ocean level sensing stations, systems of multi-component electrical and magnetic probes, antenna array systems used in radar and radio communications, and systems of meteorological stations at various geographical locations. The sensors used in these array systems will not always be in a line or in a regular grid, but will often have some arbitrary spatial distribution. In an array system, one or more sensors can be connected to a channel; each of these channels provides one of the signals to be processed. In-stead of processing the signals received on each channel separately, it is possible to process all the received signals collectively by means of a multichannel filter that takes into account the interrelations among the individual signals.

In recent work in the processing of digital records of seismic array data use is made of multichannel least-squares (Wiener) filters [*Backus et al.*, 1964; *Burg*, 1964; *Claerbout*, 1964; *Schneider et al.*, 1964; *Simpson et al.*, 1963]. The numerical coefficients of such filters are found by the solution of a set of linear equations, called the normal equations. Since a large number of coefficients are usually required in geophysical applications, this solution requires a great deal of computation when one of the conventional computational techniques for solving a set of linear equations is used. However, because the matrix of this set of equations is in the form of a Toeplitz matrix (i.e., a matrix with equal elements on any diagonal) a recursive technique may be used which greatly decreases the amount of computation necessary. Such a recursive technique was developed by *Levinson* [1949]. Levinson's method applies to the single-channel case. The present interest in multichannel filters led to an extension of this method [*Robinson*, 1963]. In this paper we wish to improve on this multichannel recursion so that it will speed up computations and thereby save computational costs. At the same time we are able to give the method a more mathematical justification by relating the procedure to the theory of polynomials orthogonal on the unit circle [*Geronimus*, 1960; *Szego*, 1959.]

Multichannel least-squares filters. Multichannel least-squares (Wiener) filtering is a data processing method that makes use of the signals from each of k channels. It is convenient to represent this multichannel data by x_t, where x_t is a column vector whose elements are the signals from each channel. That is, if we let x_{it} represent the signal coming from the ith channel (where $i = 1, 2, \cdots, k$), the multichannel signal can be written as

$$x_t = (x_{1t}, x_{2t}, \cdots, x_{kt})^T$$

where the superscript T indicates matrix transpose. Since we are interested in digital processing methods, we suppose that the signals are sampled at discrete, equally spaced time points which are represented by the time index t. Without loss of generality, we may require that t take on successive integer values.

The filter is represented by the coefficients $f_1, f_2, \cdots, f_m$, where each coefficient f_s ($s = 1, 2, \cdots, m$) is an $l \times k$ matrix. The multichannel signal x_t received by the array system represents the input to the filter; the resulting output of the filter is a multichannel signal, which we denote by the column vector $y_t = (y_{1t}, y_{2t}, \cdots, y_{lt})^T$. The relationship between input x_t and output y_t is given by the convolution formula

$$y_t = f_1 x_t + f_2 x_{t-1} + \cdots + f_m x_{t-m+1}$$

The determination of the filter coefficients is based on the concept of a desired or preferred output. The desired output is a multichannel signal, which we denote by the column vector

$z_t = (z_{1t}, z_{2t}, \cdots, z_{lt})^T$. On each channel ($i = 1, 2, \cdots, l$) there will be an error between the desired output Z_{it} and the actual output y_{it}; the mean-square value of this error is $E\{(z_{it} - y_{it})^2\}$, where the operator E denotes the ensemble average. The sum of the mean-square errors for all the channels is

$$\sum_{i=1}^{l} E\{(z_{it} - y_{it})^2\}$$

The least-squares determination of the filter coefficients requires that this sum be a minimum. This minimization leads to a set of simultaneous linear equations, called the normal equations. When the signals represent a multivariate stationary stochastic process, the normal equations obtained for the least-squares determination of a digital filter take the form

$$[f_1, f_2, \cdots, f_m] \begin{bmatrix} r_0 & r_1 & \cdots & r_{m-1} \\ r_{-1} & r_0 & \cdots & r_{m-2} \\ \vdots & \vdots & \ddots & \vdots \\ r_{-m+1} & r_{-m+2} & \cdots & r_0 \end{bmatrix}$$
$$= [g_1, g_2, \cdots, g_m]$$

Schneider et al. [1964] and *Simpson et al.* [1963] include derivations of this formula.

In this set of normal equations, the $l \times k$ matrix elements $f_1, f_2, \cdots, f_m$ are the filter coefficients, and represent the unknowns. The square matrix is in the form of a Toeplitz matrix; that is, the elements along any diagonal are the same. The $k \times k$ matrix elements of this matrix, namely $r_{-m+1}, r_{-m+2}, \cdots, r_{m-1}$, are the autocorrelation coefficients of the input signal x_t; that is

$$r_\tau = E\{x_t x_{t-\tau}{}^T\}$$
$$\tau = 0, \pm 1, \cdots, \pm(m - 1)$$

These autocorrelation coefficients are assumed to be known. The $l \times k$ matrix elements on the right-hand side, namely $g_1, g_2, \cdots, g_m$, are the cross-correlation coefficients between the desired output z_t and the input signal x_t; that is

$$g_\tau = E\{z_t x_{t-\tau+1}{}^T\} \qquad \tau = 1, 2, \cdots, n$$

These cross-correlation coefficients are also assumed to be known. The autocorrelation matrix elements have the property that $r_t = r_{-t}{}^T$.

Our purpose in this paper is to provide a solution to these equations.

Auxiliary coefficients. Corresponding to each $(n + 2) \times (n + 2)$ element autocorrelation matrix there exist two auxiliary sequences that satisfy the equations

$$[a_{0n}, a_{1n}, \cdots, a_{nn}, 0]\begin{bmatrix} r_0 & r_1 & \cdots & r_{n+1} \\ r_{-1} & r_0 & \cdots & r_n \\ \vdots & \vdots & \ddots & \vdots \\ r_{-n-1} & r_{-n} & \cdots & r_0 \end{bmatrix}$$

$$= [\alpha_n, 0, \cdots, 0, \alpha_n']$$

$$[0, b_{nn}, \cdots, b_{1n}, b_{0n}]\begin{bmatrix} r_0 & r_1 & \cdots & r_{n+1} \\ r_{-1} & r_0 & \cdots & r_n \\ \vdots & \vdots & \ddots & \vdots \\ r_{-n-1} & r_{-n} & \cdots & r_0 \end{bmatrix}$$

$$= [\beta_n', 0, \cdots, 0, \beta_n]$$

where $a_{on} = b_{on} = I$, the unit matrix.

It is convenient to write these sequences as the coefficients of polynomials:

$$a_n(z) = a_{0n} + a_{1n}z + \cdots + a_{nn}z^n$$

$$b_n(z) = b_{0n} + b_{1n}z + \cdots + b_{nn}z^n$$

To extend the length of these auxiliary sequences we form a linear combination of $a_n(z)$ and $b_n(z)$ to find

$$[a_{0n}, a_{1n} + k_{an}b_{nn}, \cdots, k_{an}b_{0n}]$$

$$\cdot \begin{bmatrix} r_0 & r_1 & \cdots & r_{n+1} \\ r_{-1} & r_0 & \cdots & r_n \\ \vdots & \vdots & & \vdots \\ r_{-n-1} & r_{-n} & \cdots & r_0 \end{bmatrix}$$

$$= [\alpha_n + k_{an}\beta_n', 0, \cdots, 0, \alpha_n' + k_{an}\beta_n]$$

and

$$[k_{bn}a_{0n}, b_{nn} + k_{bn}a_{1n}, \cdots, b_{0n}]$$

$$\cdot \begin{bmatrix} r_0 & r_1 & \cdots & r_{n+1} \\ r_{-1} & r_0 & \cdots & r_n \\ \vdots & \vdots & & \vdots \\ r_{-n-1} & r_{-n} & \cdots & r_0 \end{bmatrix}$$

$$= [\beta_n' + k_{bn}\alpha_n, 0, \cdots, 0, \beta_n + k_{bn}\alpha_n']$$

where k_{an} and k_{bn} are so chosen that

$$\alpha_n' + k_{an}\beta_n = 0 \qquad \beta_n' + k_{bn}\alpha_n = 0$$

The new filters of length $n + 2$ are

$$[a_{0,n+1}, a_{1,n+1}, \cdots, a_{n+1,n+1}]$$

$$= [a_{0n}, \cdots, a_{nn}, 0]$$

$$+ k_{an}[0, b_{nn}, \cdots, b_{0n}]$$

and

$$[b_{n+1,n+1}, \cdots, b_{1,n+1}, b_{0,n+1}]$$

$$= [0, b_{nn}, \cdots, b_{0n}]$$

$$+ k_{bn}[a_{0n}, \cdots, a_{nn}, 0]$$

where

$$k_{an} = -\alpha_n'\beta_n^{-1} = a_{n+1,n+1}$$

$$k_{bn} = -\beta_n'\alpha_n^{-1} = b_{n+1,n+1}$$

Thus the polynomials satisfy the following recursion formulas:

$$a_{n+1}(z) = a_n(z) + k_{an}z^{n+1}b_n(1/z)$$

$$b_{n+1}(1/z) = b_n(1/z) + k_{bn}(1/z^{n+1})a_n(z)$$

The new α_{n+1} and β_{n+1} are given by

$$\alpha_{n+1} = \alpha_n + k_{an}\beta_n'$$

and

$$\beta_{n+1} = \beta_n + k_{bn}\alpha_n'$$

J. P. Burg (personal communication) has shown that $\alpha_n' = \beta_n'^T$ for each recursion. Thus only one coefficient need be evaluated.

The polynomials $a_n(z)$ and $b_n(z)$ are the multivariable counterparts of the polynomials orthogonal on the unit circle treated by *Geronimus* [1960] and *Szego* [1959].

The filter recursion. We now make use of the auxiliary coefficients $b_n(z)$ to extend the general filter. The general filter polynomial is

$$f_n(z) = f_{1n}z + \cdots + f_{nn}z^n$$

where

$$[f_{1n}, f_{2n}, \cdots f_{nn}, 0] \begin{bmatrix} r_0 & r_1 & \cdots & r_n \\ r_{-1} & r_0 & \cdots & r_{n-1} \\ \vdots & \vdots & \cdot & \vdots \\ r_{-n} & r_{-n+1} & \cdots & r_0 \end{bmatrix}$$

$$= [g_1, g_2, \cdots, g_n, \gamma_n]$$

The linear combination of $f_n(z)$ and $b_n(z)$ gives

$$[f_{1n} + k_{fn}b_{nn}, \cdots, f_{nn} + k_{fn}b_{1n}, k_{fn}b_{0n}]$$

$$\cdot \begin{bmatrix} r_0 & r_1 & \cdots & r_n \\ r_{-1} & r_0 & \cdots & r_{n-1} \\ \vdots & \vdots & \cdot & \vdots \\ r_{-n} & r_{-n+1} & \cdots & r_0 \end{bmatrix}$$

$$= [g_1, \cdots, g_n, \gamma_n + k_{fn}\beta_n]$$

Thus the new general filter is obtained by setting

$$\gamma_n + k_{fn}\beta_n = g_{n+1}$$

so that

$$[f_{1,n+1}, \cdots, f_{n,n+1}, f_{n+1,n+1}]$$

$$= [f_{1,n}, \cdots, f_{nn}, 0] + k_{fn}[b_{nn}, \cdots, b_{1n}, b_{0n}]$$

The general filter polynomial satisfies the recursion formula

$$f_{n+1}(z) = f_n(z) + k_{fn}z^{n+1}b_n(1/z)$$

The derivation of the normal equations for a stochastic process also yields an expression for the expected error associated with a given least-squares filter.

$$\epsilon = d - (f_{1n}g_1{}^T + \cdots + f_{nn}g_n{}^T)$$

where d is the power matrix of *desired* output of the filter. Thus, if desired, the recursions can be monitored according to the value of ϵ and stopped when it reaches a certain limit.

The details of this system are illustrated in Figure 1.

Sideward recursion. Another recursion has been proposed by *Simpson et al.* [1963] to move the output origin. This corresponds to shifting the time index of the right-hand side of the normal equations. Thus, given the filter $f_n(z)$

and the auxiliary coefficients $a_{n-1}(z)$ and $b_{n-1}(z)$, we will generate a new filter $f_n'(z)$ such that

$$[f_{1n}', f_{2n}', \cdots, f_{nn}'] \begin{bmatrix} r_0 & r_1 & \cdots & r_{n-1} \\ r_{-1} & r_0 & \cdots & r_{n-2} \\ \vdots & \vdots & & \vdots \\ r_{-n+1} & r_{-n+2} & \cdots & r_0 \end{bmatrix}$$

$$= [g_0, g_1, \cdots, g_{n-1}]$$

First, note that

$$[0, f_{1n} - f_{nn}b_{n-1,n-1}, \cdots, f_{n-1,n} - f_{nn}b_{1,n-1}]$$

$$\cdot \begin{bmatrix} r_0 & r_1 & \cdots & r_{n-1} \\ r_{-1} & r_0 & \cdots & r_{n-2} \\ \vdots & \vdots & & \vdots \\ r_{-n+1} & r_{-n+2} & \cdots & r_0 \end{bmatrix}$$

$$= [\gamma_n', g_1, \cdots, g_n]$$

and that

$$[k_{f'n}a_{0,n-1}, f_{1n} - f_{nn}b_{n-1,n-1} + k_{f'n}a_{1,n-1},$$

$$\cdots, f_{n-1,n} - f_{nn}b_{1,n-1} + k_{f'n}a_{n-1,n-1}]$$

$$\cdot \begin{bmatrix} r_0 & r_1 & \cdots & r_{n-1} \\ r_{-1} & r_0 & \cdots & r_{n-2} \\ \vdots & \vdots & & \vdots \\ r_{-n+1} & r_{-n+2} & \cdots & r_0 \end{bmatrix}$$

$$= [\gamma_n' + k_{f'n}\alpha_n, g_1, \cdots, g_n]$$

Consequently, we obtain a new filter by (1) shifting $f_n(z)$ one unit, (2) adding a weighted version of the shifted operator $b_n(z)$, and (3) adding a weighted version of $a_n(z)$. Thus if

$$\gamma_n' - k_{f'n}\alpha_{n-1} = g_0$$

we find that

$$[f_{1n}', f_{2n}', \cdots, f_{nn}'] = [0, f_{0n}, \cdots, f_{0,n-1}]$$

$$- f_{nn}[0, b_{n-1,n-1}, \cdots, b_{1,n-1}]$$

$$+ k_{f'n}[a_{0,n-1}, a_{1,n-1}, \cdots, a_{n-1,n-1}]$$

The general filter polynomial satisfies the recursion formula

$$f_n'(z) = zf_n(z) - f_{nn}z^n b_{n-1}(1/z) + k_{f'n}a_n(z)$$

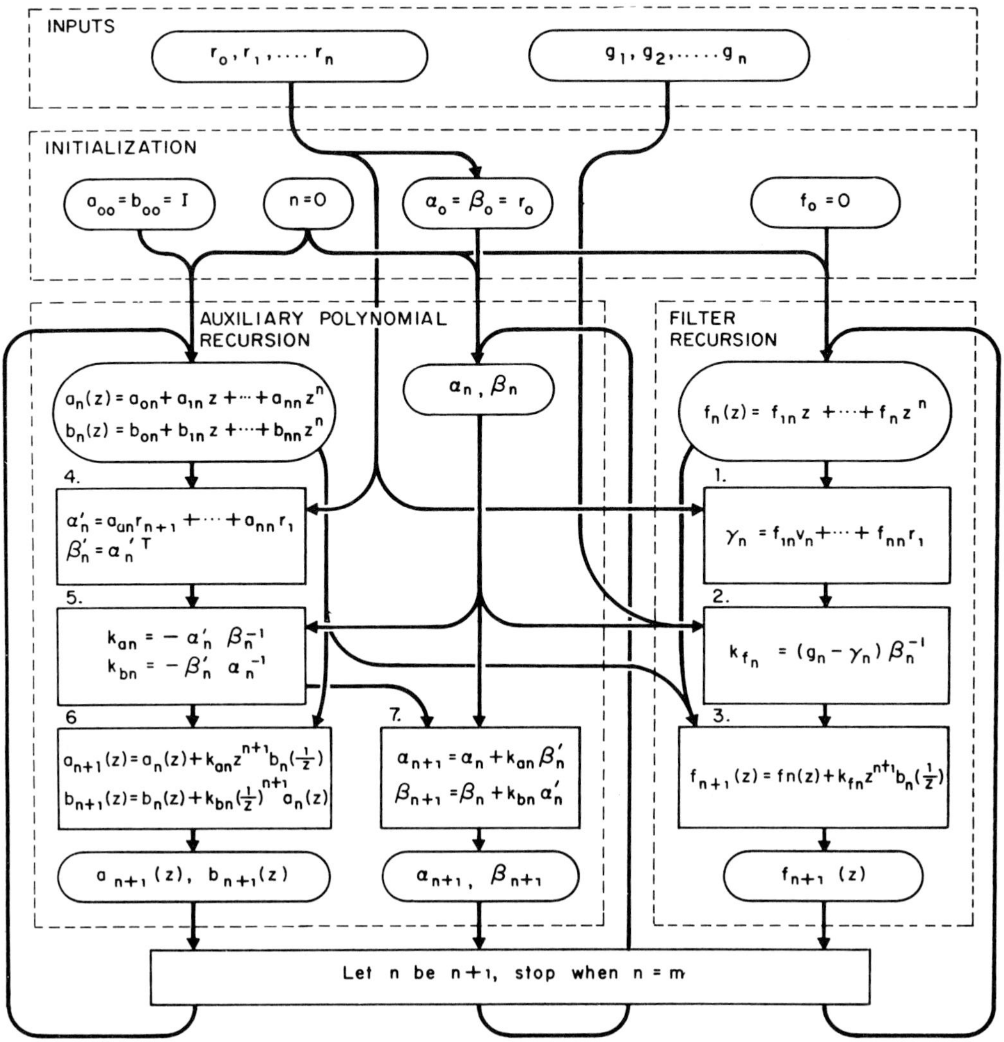

Fig. 1. Flow diagram of the recursive process. Numbers on boxes illustrate a possible sequence of computations.

We could equally well have chosen to shift the output origin in the opposite direction. In this case the recursion formula is

$$f_n''(z) = (1/z)[f_n(z) - f_{1n} a_{n-1}(z) + k_{f''n} z^{n+1} b_{n-1}(1/z)]$$

where $k_{f''n}$ is determined by

$$\gamma_n'' - k_{f''n} \beta_{n-1} = g_{n+1}$$

and

$$\gamma_n'' = (f_{2n} - f_{1n} a_{1,n-1}) r_{-1} + \cdots$$
$$+ (f_{nn} - f_{1n} a_{n-1,n-1}) r_{-n}$$

This recursion formula is extremely valuable in seismic applications because it provides a relatively inexpensive way to find the filter of a fixed length that has the optimum lag between the input data and the output data. This may be accomplished by monitoring the expected error for all possible lags.

Timing. The recursions outlined above have

been programmed in Fortran II and FAP for the IBM 7094. If the r_i matrices are $k \times k$ and the f_i and g_i matrices are $1 \times k$, the approximate time for computing a filter of length m is $(36 \, k^3 + 30 \, k^2 + 20 \, k + 400) \, m^2 \, \mu\text{sec}$ for the auxiliary coefficients recursion and $(36 \, h^2 + 40 \, k + 580) \, m^2 \, \mu\text{sec}$ for the filter recursion. The time to make one sideward recursion is $(108k^2 + 140k + 830) \, m \, \mu\text{sec}$. These times should be compared with the time taken for the normal simultaneous equations reduction for a $1 \times k$ filter: $26(km)^3 + 140(km)^2 \, \mu\text{sec}$. Thus the time taken to compute a 20-term, 5-channel filter by recursion is about 2.3 seconds. Simultaneous equations would take about 27 seconds.

Another advantage of using the recursive techniques is that much larger operators are allowed by the computer storage space limitations. The space required for recursion is $3k^2m + 2km$ registers. The simultaneous equations approach takes $k^2m^2 + km$ registers.

Numerical example. Let us consider the following 2×2 autocorrelation and 1×2 cross correlation:

$$r_0 = \begin{bmatrix} 7 & -1 \\ -1 & 3 \end{bmatrix}, \qquad r_1 = \begin{bmatrix} -2 & -2 \\ 1 & 1 \end{bmatrix},$$

$$g_1 = [1, 0], \qquad g_2 = [0, 1]$$

The first step in the recursion is to set the initial values

$$a_{00} = b_{00} = I \qquad f_0(z) = 0$$

from which we see that

$$\alpha_0 = \beta_0 = r_0 \qquad \gamma_1 = 0$$

We extend the length of the filter to 1.

$$k_{f0} = g_0 \beta^{-1} = [0.15, 0.05]$$

$$f_1(z) = f_0(z) + k_{f0} b_0(z) = [0.15, 0.05]z$$

We now extend the auxiliary sequences $a_0(z)$ and $b_0(z)$

$$\alpha_0' = a_{00} r_1 = \begin{bmatrix} -2 & -2 \\ 1 & 1 \end{bmatrix}$$

$$\beta_0' = \alpha_0'^T = \begin{bmatrix} -2 & 1 \\ -2 & 1 \end{bmatrix}$$

$$k_{a0} = -\alpha_0' \beta_0^{-1} = \begin{bmatrix} 0.4 & 0.8 \\ -0.2 & -0.4 \end{bmatrix}$$

$$k_{b0} = -\beta_0' \alpha_0^{-1} = \begin{bmatrix} 0.25 & -0.25 \\ 0.25 & -0.25 \end{bmatrix}$$

$$a_1(z) = a_0(z) + k_{a0} z b_0(1/z)$$

$$= I + \begin{bmatrix} 0.4 & 0.8 \\ -0.2 & -0.4 \end{bmatrix} z$$

$$b_1(z) = b_0(z) + k_{b0}(1/z) a_0(z)$$

$$= I + \begin{bmatrix} 0.25 & -0.25 \\ 0.25 & -0.25 \end{bmatrix} z$$

and determine the new multipliers α_1 and β_1:

$$\alpha_1 = \alpha_0 + k_{a0} \beta_0' = \begin{bmatrix} 4.6 & 0.2 \\ 0.2 & 2.4 \end{bmatrix}$$

$$\beta_1 = \beta_0 + k_{b0} \alpha_0' = \begin{bmatrix} 6.25 & -0.25 \\ -0.25 & 2.25 \end{bmatrix}$$

The first recursion is complete; therefore we set $n = 1$. We can now use the values generated for determining the next filter coefficients:

$$\gamma_1 = f_{11} r_2 = 0$$

$$k_{f'1} = g_2 \beta_1^{-1} = [-0.01786, 0.44683]$$

$$f_2(z) = f_1(z) + k_{f1} z b_1(1/z)$$

$$= [0.25719, -0.05719]z$$

$$+ [-0.01786, 0.44643]z^2$$

This process can be continued as indicated to obtain a filter of any length.

Acknowledgments. We wish to thank Dr. S. M. Simpson, Jr., for his help and ideas for this development.

This research was supported through the Air Force Cambridge Research Laboratories under contract AF 19(604)7378 as part of the Advanced Research Projects Agency's project Vela-Uniform.

REFERENCES

Backus, M., J. Burg, D. Baldwin, and E. Bryan, Wide band extraction of mantle P waves from ambient noise, *Geophysics, 24*(5), 672–692, 1964.

Burg, J. P., Three-dimensional filtering with an array of seismometers, *Geophysics, 29*(5), 693–713, 1964.

Claerbout, J. F., Detection of P waves from weak sources at great distances, *Geophysics, 29*(2), 197–211, 1964.

Geronimus, Ya. L., *Polynomials Orthogonal on a Circle and Interval*, Pergamon Press, London, 1960.

Levinson, N., The Wiener RMS (root mean square) error criterion in filter design and prediction, Appendix B of Wiener, N., *Extrapolation, Interpolation, and Smoothing of Stationary Time Series with Engineering Applications*, pp. 129–148, John Wiley & Sons, New York, 1949.

Robinson, Enders A., Mathematical development of discrete filters for detection of nuclear explosions, *J. Geophys. Res., 68*, 5559–5567, 1963.

Schneider, W. A., K. L. Larner, J. P. Burg, and M. M. Backus, A new data processing technique for the elimination of ghost arrivals on reflection seismograms, *Geophysics, 29*(5), 783–805, 1964.

Simpson, S. M., E. A. Robinson, R. A. Wiggins, and C. I. Wunsch, Studies in optimum filtering of single and multiple stochastic processes, *Sci. Rept. 7, Contract AF 19(604)7378*, Massachusetts Institute of Technology, Cambridge, 1963.

Szego, G., *Orthogonal Polynomials*, American Mathematical Society, New York, 1959.

(Manuscript received November 9, 1964; revised January 5, 1965.)

CHAPTER 17

DECONVOLUTION OF TIME SERIES
AS APPLIED TO SPEECH

ENDERS A. ROBINSON
Lincoln, Massachusetts

The vocal tract shape and the excitation function that drives the vocal
tract can be obtained from the human speech signal by deconvolution. The speech-
producing acoustic tube model of the vocal tract is a deep-source model. If the
corresponding surface-source model is constructed, then the sound waves in this
model are the physical embodiments of the recursive method of solving the normal
equations for the deconvolution operator.

1. INTRODUCTION

Human speech is produced as the result of passing air from the lungs
through the vocal tract. The vocal tract is a non-uniform acoustic tube that extends from the glottis to the lips. Each sound produced at the lips is the resultant of two factors, namely (1) the type of wave motion used to excite the vocal
tract and (2) the shape of the vocal tract. For example, in order to produce a
voiced sound, the excitation function is a series of periodic pulses generated by
the vocal cords. In order to produce an unvoiced sound, the excitation function
is white random noise generated by turbulent air. These are the two main types
of excitation functions that are inputs into the vocal tract at the glottis. The
various sounds that are outputs from the vocal tract at the lips are produced by
changing the shape of the vocal tract by moving the lips, jaw, tongue, and velum.

A particular sound at the lips represents a segment of a stationary time
series that is produced by a given type of excitation function and a given shape
of the vocal tract. More specifically, a particular sound is equal to the convolution of its excitation function with the impulse response function of the vocal
tract shape. We can make this statement because the vocal tract acts like a linear filter in producing speech.

The problem of deconvolution is one of decomposing the observed output
sound into its excitation function and its vocal-tract impulse response. As it
turns out, the actual shape of the excitation function is not important, so only
the gross features of the excitation function need to be saved, such as (1) its
r.m.s. amplitude, (2) knowledge of whether it is a pulse train (in case of a
voiced sound) or whether it is random (in case of an unvoiced sound) and (3) the
period of the pulses (in case of a voiced sound). That is, the excitation function can be characterized by three parameters, namely (1) r.m.s. value, (2) a

443

binary voiced-unvoiced parameter, and (3) in case of a voiced sound, the pitch
period. It also turns out that the vocal-tract impulse response can be characterized by as few as 12 numbers. Thus the result of the deconvolution of a particular sound can be as few as 15 parameters, namely 3 excitation parameters and 12
vocal-tract parameters. Since the sound in question would be represented by a
time series of several hundreds or thousands of observations, and since deconvolution would replace these observations by 15 parameters characterizing that sound,
this approach results in a considerable reduction of the data. In fact, a reduction of 50 to 1 seems to be a goal in speech analysis that is within reach. Another goal is the identification of the parameters representing a sound with the
phonetic symbol for that sound, so that an automatic typewriter could be constructed that would type out printed text from the spoken word.

Finally the results of the deconvolution of speech can be evaluated by reconvolution. For each sound an excitation function is generated from its parameters. This excitation function is convolved with the vocal-tract impulse
response function to produce the output time-series segment for that sound. This
convolution process is called reconvolution because we are not using the original
excitation function but a reconstructed one which agrees with the original in its
gross features only. The result of this process is the reconvolved sound. By
attaching together all these reconvolved sound segments, we obtain the reconvolved
speech signal which can be compared to the original speech signal. Experimental
listening tests with various auditors establish that the quality of the reconvolved speech is very close to that of the original speech for a wide range of
spoken messages and for many different speakers both male and female.

X-Ray techniques have been used as a direct method of determining the vocal
tract shape. Wakita (1973) gave a mathematical method for extracting the vocal-
tract shape by showing the relationship of the acoustic tube model to the recursive method given in Robinson (1967) for solving the normal equations. In this
paper we show that the vocal tract shape and the excitation function can be obtained from the speech signal by deconvolution. We make use of a constructive
proof that relates the speech-producing acoustic tube model, which is a deep-
source model, to the corresponding surface-source acoustic tube model. In this
way we can physically construct the recursive method for solving the normal
equations from the acoustic properties of the system, and can physically obtain
the autocorrelation, the prediction error operators, and the reflection coefficients in terms of the model.

Because of the limitation of space, we have been forced to make some of our
derivations in this paper quite brief. For background material on the model and
the polynomial manipulations, the reader is referred to Robinson (1967, Chapter 3).

2. ACOUSTIC TUBE MODEL

The acoustic tube model is an approximation to the vocal tract consisting of a set of interconnected cylindrical sections piled on top of each other. See Figure 1. Each section may have a different cross-sectional area but has the same height. The cross-sectional areas vary according to the sound being uttered. For any given sound let the cross-sectional area of the nth section be denoted by S_n, where n runs from 1 to N. We let section 1 be the closest to the lips, and let section N be the one closest to the glottis. For the purposes of description, we speak of the lips as being at the top, and the glottis as being at the bottom, of the vertically standing tube made up of the N sections.

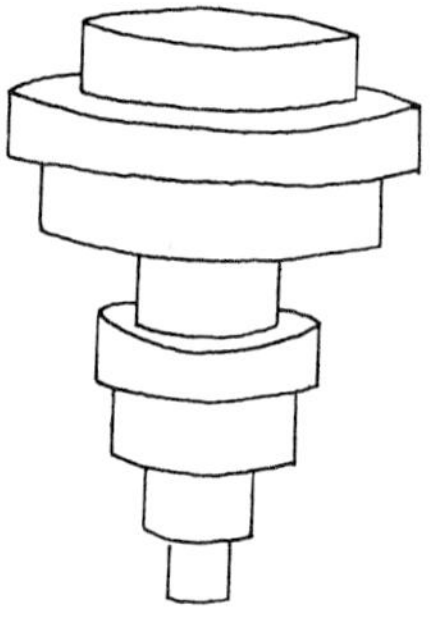

Fig. 1

The acoustic tube can support travelling wave-motion from the glottis to the lips, which we call upgoing waves, and also travelling wave-motion from the lips to the glottis, which we call downgoing waves. At each interface between two adjacent sections of the acoustic tube, a travelling wave will be partially reflected and partially transmitted, the division of energy between the reflected and transmitted portions being governed by the reflection coefficient associated with that interface. Let n denote the interface between sections n and n+1, and let c_n denote the refelction coefficient of this interface. The interpretation of this reflection coefficient is as follows. Let us use the convention that all wave motion is measured in physical units that are proportional to the square-root of energy. Thus the square of the amplitude of a wave is in terms of energy. If a downgoing impulse of unit amplitude is incident on the given interface, then a reflected impulse and a transmitted impulse are produced. See Figure 2.

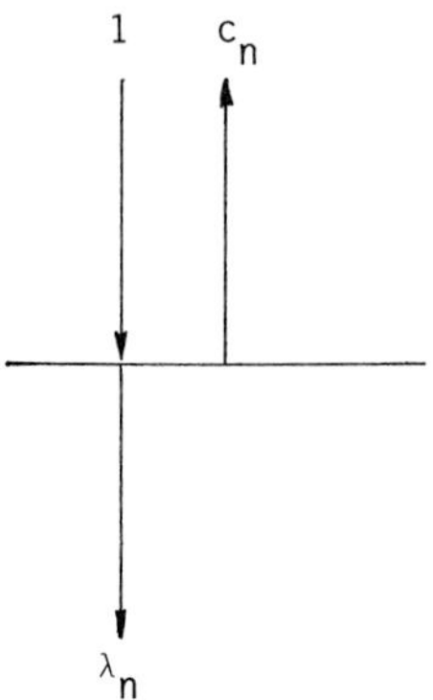

Fig. 2

The reflected impulse has implitude c_n and the transmitted impulse has amplitude
λ_n. By the law of the conservation of energy, the input energy is equal to the
output energy, namely the sum of the reflected energy and the transmitted energy,
that is

$$1^2 = c_n^2 + \lambda_n^2 .$$

Solving for λ_n we have

$$\lambda_n = \sqrt{1 - c_n^2}$$

The constant λ_n is called the (square-root energy) transmission coefficient. In
the above example we considered a downgoing impulse. If instead we consider an
upgoing impulse striking the same interface, the reflection coefficient becomes
$-c_n$ and the transmission coefficient remains the same. See Figure 3.

Let us now consider the operation of the acoustic tube model in producing
speech. The source is the excitation function which represents an upgoing waveform that enters at the lower interface N, namely at the glottis. As a result,
speech production represents a deep-source acoustic tube model. This deep source
gives rise to upgoing and downgoing wave motion in each section, due to the reflections and transmissions at each interface. In other words, the excited vocal
tract represents a reverberating system. When we reach interface 0 (namely, the
lips) we meet a *boundary condition*, because the lips are connected to the open
air which has infinite cross-sectional area. As a result there is perfect reflection at the lips. If we assume that our wave motion is related to pressure, then

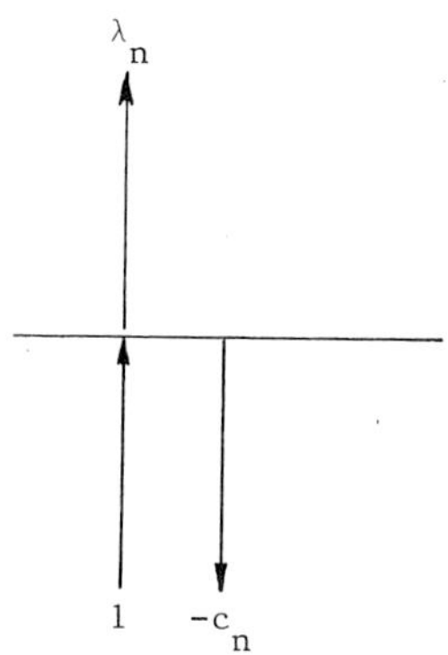

Fig. 3

the pressure must vanish at the lips, so at the lips the downgoing wave in section 1 is the negative of the upgoing wave in section 1.

The velocity of sound waves is constant, and it is convenient to choose the common height of each section such that it takes exactly one-half discrete time unit for a wave to travel from one face of the section to the other. We introduce the unit delay operator z. The delay operator z is defined as that operator which delays a time variable by one time unit. It follows that $z^{\frac{1}{2}}$ is the operator that delays a time variable by one-half time unit, namely the one-way travel time through a section.

For the deep-source model let $x_n(t)$ and $y_n(t)$ be the downgoing and upgoing waves respectively at the top of section n, and let $x_n'(t)$ and $y_n'(t)$ be the corresponding waves at the bottom of section n. See Figure 4.

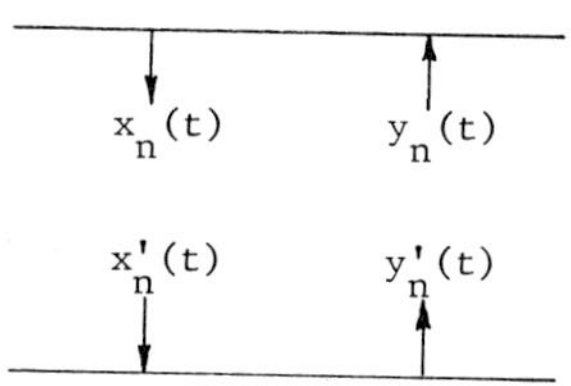

Fig. 4

Then we have

$$
\begin{bmatrix} x_n'(t) \\ y_n'(t) \end{bmatrix} = \begin{bmatrix} z^{\frac{1}{2}} & 0 \\ 0 & z^{-\frac{1}{2}} \end{bmatrix} \begin{bmatrix} x_n(t) \\ y_n(t) \end{bmatrix}
$$

Here and throughout, the variable t is the discrete time variable.

The waves at interface n are shown in Figure 5. They can be related as follows. See Figure 6. The wave $y_n'(t)$ is made up of a part due to the reflection of $x_n'(t)$ and a part due to the transmission of $y_{n+1}(t)$, that is,

$$
y_n'(t) = c_n\, x_n'(t) + \lambda_n\, y_{n+1}(t).
$$

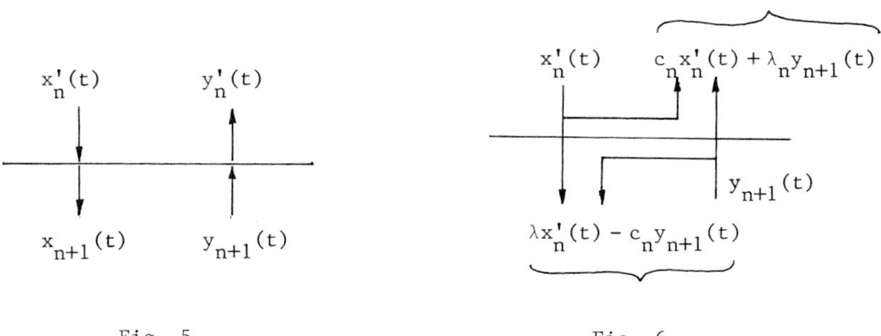

Fig. 5   Fig. 6

The wave $x_{n+1}(t)$ is made up of a part due to the reflection of $y_{n+1}(t)$ and a part due to the transmission of $x_n'(t)$, that is

$$
x_{n+1}(t) = -\, c_n\, y_{n+1}(t) + \lambda_n\, x_n'(t)\ .
$$

If we solve these two equations for the unprimed quantities we obtain

$$
\begin{bmatrix} x_{n+1}(t) \\ y_{n+1}(t) \end{bmatrix} = \frac{1}{\lambda_n} \begin{bmatrix} 1 & -c_n \\ -c_n & 1 \end{bmatrix} \begin{bmatrix} x_n'(t) \\ y_n'(t) \end{bmatrix}
$$

Combining this matrix equation with the previous matrix equation, we obtain the *scattering equation:*

$$\begin{bmatrix} x_{n+1}(t) \\ \\ y_{n+1}(t) \end{bmatrix} = \frac{z^{-\frac{1}{2}}}{\lambda_n} \begin{bmatrix} z & -c_n \\ \\ -c_n z & 1 \end{bmatrix} \begin{bmatrix} x_n(t) \\ \\ y_n(t) \end{bmatrix}$$

The z-transform of the scattering equation is

$$\begin{bmatrix} X_{n+1}(z) \\ \\ Y_{n+1}(z) \end{bmatrix} = \frac{z^{-\frac{1}{2}}}{\lambda_n} \begin{bmatrix} z & -c_n \\ \\ -c_n z & 1 \end{bmatrix} \begin{bmatrix} X_n(z) \\ \\ Y_n(z) \end{bmatrix}$$

The z-transform is defined as

$$X_n(z) = \Sigma \, x_n(t) z^t$$

where the summation is over all discrete t values. That is, the z-transform of a
time sequence denoted by a lower case letter is denoted by the corresponding capital letter. For simplicity, we often drop the argument z, so $X_n(z)$ would be
simply X_n. A bar over a z-transform indicates that each z in the z-transform has
been replaced by z^{-1}, that is $\bar{X}_n = X_n(z^{-1})$.

   If we assume that the time sequence $x_n(t)$ has finite energy, then the quantity $X_n \bar{X}_n$ represents the energy spectrum of the time sequence $x_n(t)$; that is,
$X_n \bar{X}_n$ is the z-transform of the autocorrelation of $x_n(t)$. From the scattering
equation we obtain

$$X_{n+1} \, \bar{X}_{n+1} - Y_{n+1} \, \bar{Y}_{n+1} = X_n \bar{X}_n - Y_n \bar{Y}_n$$

which says that the net downgoing energy in section n+1 is equal to the net downgoing energy in section n. A recursive application of this result shows that the
net downgoing energy in any two sections is the same, provided of course that
there are no energy sources or sinks between the two sections in question.

   The polynomials P_n and Q_n are defined as

$$\begin{bmatrix} P_n^R & Q_n^R \\ \\ Q_n & P_n \end{bmatrix} = \begin{bmatrix} z & -c_n \\ \\ -c_n z & 1 \end{bmatrix} \begin{bmatrix} z & -c_{n-1} \\ \\ -c_{n-1} z & 1 \end{bmatrix} \cdots \begin{bmatrix} z & -c_1 \\ \\ -c_1 z & 1 \end{bmatrix}$$

where the polynomial P^R is the reverse of P_n and Q_n^R is the reverse of Q_n, that is

$$P_n^R = z^n \bar{P}_n \quad , \qquad Q_n^R = z^n Q_n$$

We also define $\sigma_n = \lambda_n \lambda_{n-1} \cdots \lambda_1$ and $A_n = P_n - Q_n$. We note that in the polynomial A_n the coefficient of z^0 is one and the coefficient of z^n is c_n. The
recursive use of the scattering equation gives

$$\begin{bmatrix} X_{n+1} \\ \\ Y_{n+1} \end{bmatrix} = \frac{z^{-n/2}}{\sigma_n} \begin{bmatrix} P_n^R & Q_n^R \\ \\ Q_n & P_n \end{bmatrix} \begin{bmatrix} X_1 \\ \\ Y_1 \end{bmatrix}$$

3. DEEP SOURCE MODEL

We are interested in the impulse response of the acoustic tube in the case
of speech, that is, in the case of a deep source. If we let the upgoing wave at
the glottis be a unit impulse, that is, if we let $Y_{N+1} = 1$, and if we impose the
boundary condition of perfect reflection $Y_1 = -X_1$ at the lips (see Figure 7), we
obtain

$$\begin{bmatrix} X_{N+1} \\ \\ 1 \end{bmatrix} = \frac{z^{-N/2}}{\sigma_N} \begin{bmatrix} P_N^R & Q_N^R \\ \\ Q_N & P_N \end{bmatrix} \begin{bmatrix} X_1 \\ \\ -X_1 \end{bmatrix}$$

which gives

$$1 = - \frac{z^{-N/2}}{\sigma_N} (P_N - Q_N) X_1 \quad .$$

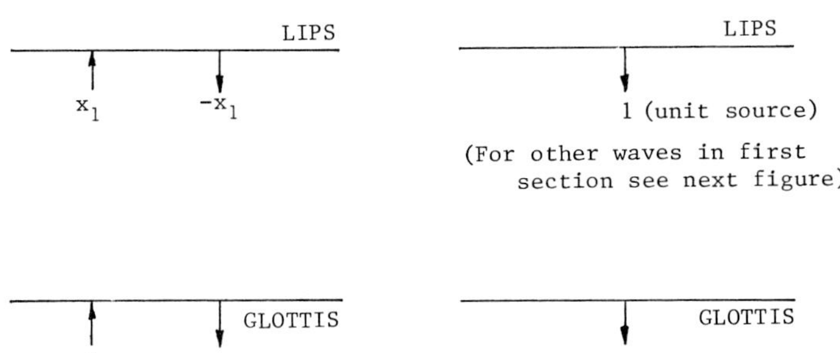

Fig. 7   Fig. 8

Thus the impulse response of the vocal tract is

$$X_1 = - \frac{\sigma_N z^{N/2}}{P_N - Q_N} = - \frac{\sigma_N z^{N/2}}{A_N}$$

4. SURFACE-SOURCE MODEL

Speech is produced by an acoustic tube subject to a deep source. Let us now consider the same acoustic tube but subject to a surface source. More specifically, we want to find the downgoing response D_{N+1} at the glottis subject to a downgoing unit impulsive source at the lips. See Figure 8. In the previous section we have found the vocal-tract impulse response, namely the upgoing response X_1 at the lips subject to an upgoing unit impulse source at the glottis. The physical principle of reciprocity states that if we interchange source and receiver we observe the same waveform. Thus the above two responses are the same, i.e. $D_{N+1} = X_1$. We note that no energy is reflected back from the glottis in this surface-source model, so the upgoing waveform at the glottis is zero, i.e. $U_{N+1} = 0$. The net downgoing energy in the glottis is therefore

$$D_{N+1} \bar{D}_{N+1} - U_{N+1} \bar{U}_{N+1} = D_{N+1} \bar{D}_{N+1} = X_1 \bar{X}_1 .$$

Let us now look at the first section of the surface-source model (i.e. the section at the lips). The downgoing impulsive source can be represented by the Kronecker delta function δ_t at the top of the first section. This source gives rise to an upgoing wave in the first section as a result of internal reflections from interfaces below; we denote this upcoming wave by $-r_1$, $-r_2$, $-r_3$, $\ldots$ where the subscript represents the time of occurrence. Because the top interface (i.e. the lips) is a perfect reflector, this upgoing wave is reflected at the lips back into the downgoing wave r_1, r_2, r_3, $\ldots$ Thus the entire downgoing wave in section 1 is made up of the downgoing impulse δ_t and the reflected wave r_1, r_2, r_3, $\ldots$; that is $\delta_t + r_t$. The z-transform of this downgoing wave is

$$D_1 = 1 + R$$

The upgoing wave in section 1 is $-r_t$ with z-transform

$$U_1 = - R$$

See Figure 9.

The net downgoing energy in section 1 is

$$D_1 \bar{D}_1 - U_1 \bar{U}_1 = (1+R)(1+\bar{R}) - R\bar{R} = 1 + R + \bar{R}$$

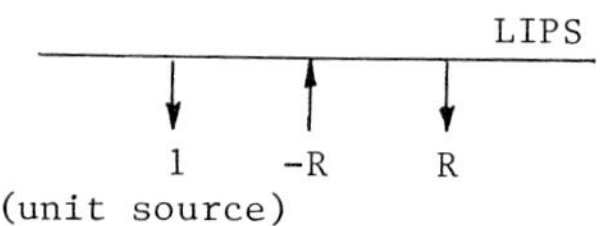

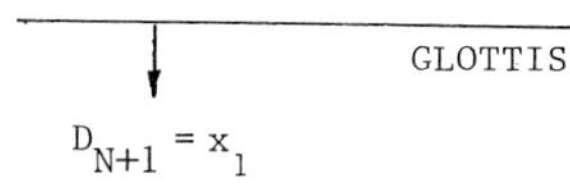

$$D_{N+1} = x_1$$

Fig. 9

which is equal to the net downgoing energy in the glottis, that is

$$X_1 \bar{X}_1 = 1 + R + \bar{R}$$

Thus we see that $1 + R + \bar{R}$ is the z-transform of the autocorrelation of $x_1(t)$. That is, the autocorrelation of the impulse response function $x_1(t)$ of the vocal tract for speech is

$$\ldots, r_2, r_1, 1, r_1, r_2, \ldots$$

where 1 is the autocorrelation coefficient for lag 0, r_1 for lag −1 and also for lag 1, and so on.

5. NORMAL EQUATIONS

In the surface source model let $d_{n+1}(t)$ and $u_{n+1}(t)$ denote the downgoing and upgoing waves respectively at the top of section n+1, and let their respective z-transforms be D_{n+1} and U_{n+1}. Then the scattering equation gives

$$
\begin{bmatrix} D_{n+1} \\ \\ U_{n+1} \end{bmatrix}
= \frac{z^{-n/2}}{\sigma_n}
\begin{bmatrix} P_n^R & Q_n^R \\ \\ Q_n & P_n \end{bmatrix}
\begin{bmatrix} 1 + R \\ \\ - R \end{bmatrix}
$$

In this equation we first solve for D_{n+1} and U_{n+1}, from which we find that

$$\bar{D}_{n+1} - U_{n+1} = \frac{z^{-n/2}}{\sigma_n} A_n (1 + R + \bar{R})$$

which is

$$A_n (1 + R + \bar{R}) = z^{n/2} \sigma_n (\bar{D}_{n+1} - U_{n+1}) .$$

The function D_{n+1} is the z-transform of the downgoing wave at the top of section n+1 in the surface source model. This downgoing wave is made up of the direct pulse $d_{n+1}(\frac{n}{2})$ together with succeeding pulses $d_{n+1}(\frac{n}{2}+1)$, $d_{n+1}(\frac{n}{2}+2)$, ... Because the direct pulse is the result of transmissions through the first n interfaces, the direct pulse is equal to the product of the first n transmission coefficients, i.e.

$$d_{n+1} (\frac{n}{2}) = \lambda_1 \lambda_2 \cdots \lambda_n = \sigma_n$$

and the time of the direct pulse is n/2. Thus

$$D_n = \sigma_n z^{n/2} + \text{terms in higher powers of } z.$$

The function U_{n+1} is the z-transform of the upgoing wave at the top of section n+1. The first pulse in the upgoing wave is the reflection of the direct downgoing pulse at interface n+1. Thus the magnitude of this pulse is $\sigma_n c_{n+1}$; that is, the magnitude is equal to the magnitude of the direct downgoing pulse times the reflection coefficient. Because one time unit elapses for the round trip in section n+1, the first pulse of the upgoing wave occurs at one time unit later than the first pulse of the downgoing wave in section n+1; that is, the first upgoing pulse at the top of section n+1 occurs at $\frac{n}{2} + 1$. Thus

$$U_{n+1} = \sigma_n c_{n+1} z^{\frac{n}{2} + 1} + \text{terms in higher powers of } z.$$

We therefore have

$$A_n (1 + R + \bar{R}) = \sigma_n (\text{terms in negative powers of } z + \sigma_n - \sigma_n c_{n+1} z^{n+1}$$
$$- \text{terms in higher powers of } z)$$

We note that the powers of z from 1 to n are missing on the right-hand side of the above equation; it is this fact that allows us to extract the normal equations. More specifically, let us now equate coefficients on each side of this equation for the powers of z from 0 to n+1. We obtain

$$a_{no} r_o + a_{n1} r_1 + \ldots + a_{nn} r_n = \sigma_n^2$$
$$a_{no} r_1 + a_{n1} r_o + \ldots + a_{nn} r_{n-1} = 0$$
$$\cdots$$
$$a_{no} r_n + a_{n1} r_{n-1} + \ldots + a_{nn} r_o = 0$$
$$a_{no} r_{n+1} + a_{n1} r_n + \ldots + a_{nn} r_1 = -c_{n+1} \sigma_n^2$$

where a_{no}, a_{n1}, ..., a_{nn} are the coefficients of A_n.

We note that $r_o \equiv 1$ and $a_{no} \equiv 1$.

Let us now look at these equations. The time function r_t is the autocorrelation of the vocal-tract impulse response function $x_1(t)$ for speech. Thus the first $n+1$ of these equations are the normal equations for the prediction-error operator a_{no}, a_{n1}, ..., a_{nn}. The last of these equations allows us to find c_{n+1}, and we shall make use of this fact in the next section.

By the definition of P_n and Q_n we have

$$
\begin{bmatrix} P_{n+1}^R & Q_{n+1}^R \\ \\ Q_{n+1} & P_{n+1} \end{bmatrix}
=
\begin{bmatrix} z & -c_{n+1} \\ \\ -c_{n+1} z & 1 \end{bmatrix}
\begin{bmatrix} P_n^R & Q_n^R \\ \\ Q_n & P_n \end{bmatrix}
$$

which gives the recursions

$$P_{n+1} = P_n - c_{n+1}\, z\, Q_n^R$$

$$Q_{n+1} = Q_n - c_{n+1}\, z\, P_n^R$$

Because $A_n = P_n - Q_n$ we can subtract the above two equations to obtain the *polynomial recursion*

$$A_{n+1} = A_n + c_{n+1}\, z\, A_n^R$$

where $A_n^R = P_n^R - Q_n^R = z^n \bar{A}$. Thus if we know A_n and c_{n+1} we can compute the polynomial A_{n+1}.

6. DECONVOLUTION OF SPEECH

As we have seen, each sound in speech is produced by the convolution of excitation function with vocal-tract impulse response function. In case the excitation function is white noise, it follows that the autocorrelation of the vocal-tract impulse response function can be approximated by the autocorrelation of the sound signal. In case the excitation function is a periodic pulse sequence, the autocorrelation of the vocal-tract impulse response function also can be derived approximately from knowledge of the autocorrelation of the sound signal. Thus the first step in the deconvolution of a sound signal measured at the lips is to compute its autocorrelation which is then used to approximate the autocorrelation

$$\ldots, \; r_2, \; r_1, \; 1, \; r_1, \; r_2, \; \ldots \qquad \text{(where } r_o \equiv 1)$$

of the impulse- response function of the vocal tract.

The next step is to compute the sequence of prediction error operators and reflection coefficients for n=1 to N. Initially, we have

$$a_{oo} = 1$$

$$\sigma_o^2 = a_{oo} r_o = 1 \; .$$

Then given the prediction error operator $a_{no}, \; \ldots, \; a_{nn}$ and σ_n^2 we compute the reflection coefficient

$$c_{n+1} = - \; (a_{no} r_{n+1} + a_{n1} \, r_n + \ldots + a_{nn} \, r_1) / \sigma_n^2 \; ,$$

the variance

$$\sigma_{n+1}^2 = \lambda_{n+1}^2 \, \sigma_n^2 = (1 - c_{n+1}^2) \; \sigma_n^2$$

and the new prediction error operator by the polynomial recursion

$$a_{n+1,o} = a_{no} = 1$$

$$a_{n+1,1} = a_{n1} + c_{n+1} a_{nn}$$

$$\ldots$$

$$a_{n+1,n} = a_{nn} + c_{n+1} a_{n1}$$

$$a_{n+1,n+1} = \qquad c_{n+1} a_{no} = c_{n+1}$$

Using this recursive algorithm we can compute the reflection coefficients $c_1, \; c_2, \; \ldots, \; c_N$ of the vocal tract. These reflection coefficients are related to the cross-sectional areas $S_1, \; S_2, \; \ldots, \; S_N$ of the vocal tract by the equation

$$c_n = \frac{S_n - S_{n+1}}{S_n + S_{n+1}}$$

Solving this equation for S_{n+1} we have

$$S_{n+1} = \frac{1 - c_n}{1 + c_n} \, S_n$$

Thus given the cross-sectional area S_1 at the lips we can compute in a stepwise fashion the areas of all the sections from the lips to the glottis. Thus the deconvolution process has given us the shape of the vocal tract.

The final prediction error operator $a_{N0}, a_{N1}, \ldots, a_{NN}$ is the required deconvolution operator. This statement follows from the fact that the impulse response function of the vocal tract, as we have seen, is

$$X_1 = - \frac{\sigma_N z^{N/2}}{A_N}$$

If E is the z-transform of the excitation function then the z-transform of the sound signal at the lips is $X_1 E$. The z-transform of the deconvolution operator is A_N. The z-transform of the deconvolved sound signal is

$$A_N X_1 E = - \sigma_N z^{N/2} E$$

Since σ_N represents a scale factor and $z^{N/2}$ represents a delay (both of which we know), the deconvolution has yielded E. That is, the deconvolution of the sound signal yields the excitation function.

7. CONCLUSION

The deconvolution of a sound produced in speech yields the reflection coefficients, or equivalently the cross-sectional areas, which make up the numerical parameters necessary to characterize the vocal tract shape for that sound. The deconvolution also yields the excitation function which drives the vocal tract for that sound. In order to establish the results, we have made use of an acoustic tube model with a source at the lips, as this model physically gives us the autocorrelation of the vocal tract impulse response, the normal equations for the prediction error operators, and the reflection coefficients.

REFERENCES

[1] ROBINSON, ENDERS A. (1967). *Multichannel Time Series Analysis with Digital Computer Programs*. Holden-Day Publishing Co., San Francisco.

[2] WAKITA, HISASHI (1973). Direct estimation of the vocal tract shape by inverse filtering of acoustic speech waveforms. *IEEE Transactions on Audio and Electroacoustics*, AU-21, 417-427.

CHAPTER 18

WAVES PROPAGATING IN RANDOM MEDIA

AS STATISTICAL TIME SERIES

Enders A. Robinson
Consultant for Amoco Production Company
Tulsa, Oklahoma

Abstract.

Wave propagation in an inhomogeneous medium may be
studied by the familiar device of replacing a generally continuous function by a step function. In this way, the
medium is divided into a set of strata with constant parameters in each stratum. The amplitudes of waves propagating
in the stratified medium are normalized so that the square
of wave amplitude gives energy. Alternatively, one can
consider wave motion as a particle process where square of
wave amplitude gives the probability of the particle. The
total energy (or probability) injected in the system is unity,
and the flow of this energy through time and space is followed as reactive energy is transformed into reflected and
transmitted energy. In those cases in which the reflection
coefficients of the interfaces are small, the reflected wave
is equal to the convolution of a minimum-delay reverberation
wavelet with the reflection coefficient sequence. In many
physical problems in which large amounts of data are handled,
the reflection coefficient sequence may be considered as a
realization of a random process, in which case the reflected
wave is the corresponding realization of a time series
generated by a statistical minimum-delay model. In the
seismic exploration for petroleum and natural gas, the reflection coefficient series represents the reflectivity of
the earth's crust and so can be used to make contour maps
of the earth's structure at depth. Empirically, it is known
that sections of the earth's crust have uncorrelated reflection coefficient sequences, and as a result the reflectivity

457

of the deep earth's crust can be obtained by deconvolution of
reflection seismograms recorded on the surface.

The Diffusion Equation and the Wave Equation.

In many problems in physics and engineering, traveling
waves are studied because it is important to pinpoint the
time for a disturbance to propagate from one place to
another. Traveling wave phenomena appear in signal and power
transmission, mechanics, acoustics, seismology, thermal
conduction, diffusion, and in many other areas of science and
engineering.

The first-order differential equations for electrical
transmission lines are called the telegrapher's equations,
which are

$$\frac{\partial v}{\partial x} = - Ri - L \frac{\partial i}{\partial t} , \quad \frac{\partial i}{\partial x} = - Gv - C \frac{\partial v}{\partial t}$$

where v is voltage across the transmission line, i is current
in the transmission line, x is distance on the line, R is
series resistance per unit length, L is series inductance per
unit length, G is shunt conductance per unit length, and C
is shunt capacitance per unit length.

Both of these equations contain loss terms, namely Ri
and Gv, and energy storage terms, namely L $\partial i/\partial t$ and
C $\partial v/\partial t$.   For different circumstances, these equations
can be specialized. There are two particularly important
cases, namely (1) the diffusion case in which the series
inductance and shunt conductance of the electric transmission line are neglected (i.e., L = 0 and G = 0) and (2) the
lossless case in which loss on the transmission line is
neglected (i.e., R = 0 and G = 0). In the first case, the
two telegrapher's equations may be combined to give the
second-order partial differential equation, called the
diffusion equation,

$$\frac{\partial^2 v}{\partial x^2} = RC \frac{\partial v}{\partial t}$$

where 1/RC is the diffusion constant. This equation represents the case of diffusion waves. In the second case, the
two telegrapher's equations may be combined to give the
second-order partial differential equation, called the
<u>wave</u> <u>equation</u>,

$$\frac{\partial^2 v}{\partial x^2} = LC \frac{\partial^2 v}{\partial t^2}$$

where $1/\sqrt{LC}$ is the velocity of propagation. This equation
represents the case of lossless waves. The characteristic
impedance $Z = \sqrt{L/C}$ also is defined.

Equations for waves in other branches of science can be
obtained by use of analogies. That is, the telegrapher's
equations have analogs in nearly all the other wave phenomena
commonly studied. For example, one of the telegrapher's
equations for heat conduction is Fourier's law of heat conduction, and the other is a continuity equation for heat
based upon the principle of storage of heat and heat capacity.
These equations for heat conduction correspond to the
electric transmission line in the diffusion case, so the
resulting equation is the diffusion equation. On the other
hand, the telegrapher's equations for various types of
acoustic waves are based on Newton's second law and
D'Alembert's principle, and on the principle of conservation
of mass which results in a continuity equation. These
acoustic equations correspond to the electric transmission
line in the lossless case, so the resulting equation is the
wave equation.

As it turns out, many of the wave phenomena commonly
studied fall into one of these two cases, i.e., diffusion
(with the diffusion equation) or lossless (with the wave
equation). For example, by using the analogy of longitudinal
particle velocity and longitudinal stress for voltage and
current, we obtain the wave equation for sound waves in
solids. Because of such analogies, we will describe our
variables in electric terms, and the interested reader may
then use the corresponding analogy to convert the variables
into the terms of the particular phenomena at hand.

Furthermore, for the rest of this paper, we will be concerned
only with lossless waves as described by the wave equation.

<u>Stratified Medium.</u>

In setting up the mathematical model of a physical
system, we use the familiar method of replacing an arbitrary
function by a step function. Consider the lower half-space
where $x = 0$ denotes the surface and positive x is measured
vertically down. A function $c(x)$ is specified for each value
of x. Instead of the continuous dependence of c on x, we
think of the half-space as being divided into a set of
horizontal strata, with $c(x)$ having a constant value in each
stratum.

We write the wave equation as

$$\frac{\partial^2 v}{\partial x^2} = \frac{1}{c^2} \frac{\partial^2 v}{\partial t^2}$$

where $c = c(x)$ is a positive constant for the stratum in
question. The wave equation is one of the few partial differential equations whose general solution can be found
explicitly. To do so, we introduce new coordinates of
$\eta = x - ct$ and $\xi = x + ct$. Then, by the chain rule, we have

$$\frac{\partial^2 v}{\partial x^2} = \frac{\partial^2 v}{\partial \xi^2} + 2 \frac{\partial^2 v}{\partial \xi \partial \eta} + \frac{\partial^2 v}{\partial \eta^2}$$

$$\frac{\partial^2 v}{\partial t^2} = c^2 \left[\frac{\partial^2 v}{\partial \xi^2} - 2 \frac{\partial^2 v}{\partial \xi \partial \eta} + \frac{\partial^2 v}{\partial \eta^2} \right]$$

so the wave equation becomes

$$\frac{\partial^2 v}{\partial \xi \partial \eta} = 0 \ .$$

This equation is satisfied if and only if

$$v = v_d(\eta) + v_u(\xi)$$

where v_d and v_u are arbitrary waveforms. That is, the solution of the wave equation is of the form

$$v(x, t) = v_d(x - ct) + v_u(x + ct)$$

which is due to D'Alembert in 1747. For constant propagation
velocity c, the voltage and current each propagate along the
x-axis as a wave. The waveform $v_d(x - ct)$ propagates in the
plus x-direction and represents a downgoing wave. The
waveform $v_u(x + ct)$ propagates in the minus x-direction and
represents an upgoing wave. From the telegrapher's equations,
it follows that the voltage for each waveform is proportional
to the current of that waveform and the constant of pro-
portionality is just the characteristic impedance Z. Calling
i_d and i_u the downgoing and upgoing current waveforms
respectively, we have

$$v_d = Z i_d \, , \quad v_u = - Z i_u \, .$$

Thus, voltage and current may be expressed in terms of downgoing and upgoing waves as

$$v = v_d + v_u = Z(i_d - i_u)$$

$$i = i_d + i_u = \frac{1}{Z} (v_d - v_u) \, .$$

Scattering Variables.

    The choice of an appropriate formalism is particularly
important in studies of time series generated by a stratified
system. The scattering quantities represent a set that is
particularly appropriate in dealing with a stratified system.
While the conventional quantities are voltage and current,
the quantities employed in the scattering formalism are
linear combinations of voltage and current. More particularly,
the scattering quantities are the downgoing and upgoing waves.

However, instead of choosing the scattering quantities in
terms of either voltage or current, it is more convenient
to make the unit (voltage · current)$^{1/2}$. Thus, we write

$$i\sqrt{Z} = i_d\sqrt{Z} + i_u\sqrt{Z}$$

$$\frac{v}{\sqrt{Z}} = i_d\sqrt{Z} - i_u\sqrt{Z} \quad .$$

If we define normalized voltage $\tilde{v}$ and normalized current
$\tilde{i}$ as

$$\tilde{v} = \frac{v}{\sqrt{Z}} \, , \quad \tilde{i} = i\sqrt{Z}$$

and the normalized downgoing waveform d and normalized upgoing waveform u as

$$d = i_d\sqrt{Z} \, , \quad u = i_u\sqrt{Z}$$

we have

$$\tilde{i} = d + u$$

$$\tilde{v} = d - u \quad .$$

The energy is voltage times current, which is

$$\text{Energy} = vi = \tilde{v}\tilde{i} \quad .$$

In case of complex quantities, the energy should be written
as

$$\text{Energy} = \text{Re}(vi^*) = \text{Re}(v^*i)$$

where Re stands for "real part" and the asterisk denotes
complex conjugate.

In certain applications, the wave motion may be considered as representing a particle process where energy may
be considered as the probability of the particle.

Reflection and Transmission Coefficients.

     The subject of this section is the reflection and trans-
mission of traveling waves at interfaces. We want to emphasize that the amplitude of a reflection is not just a
property of the material in the stratum where the reflection
occurs, but it is an interface property depending on the
material on both sides of the interface. In fact, a thin
layer of extraneous material on the interface beween two
materials will change the reflection. As a result, in
constructing the mathematical model of our physical system,
we must make the thicknesses of the discrete strata sufficiently fine in order to account for all significant thin
layers.

     We depict distance x as increasing downward on a verti-
cal axis and time t as increasing to the right on a horizontal scale. An interface (x = constant) is then represented by a horizontal line. All our wave motion is at
normal incidence, that is, all waves move up and down
vertically, but in order to include the dimension of time, a
wave component at a particular place and time will be indicated by a vector in the x,t plane. The vector points
downward and to the right for a downgoing wave and upward and
to the right for an upgoing wave.

     Let us now consider the case of an incident unit down-
going pulse on an interface. See Figure 1. Throughout this
paper, we use the word "pulse" to indicate a spike (i.e., a
sharply peaked waveform of very short time duration) which
contains energy equal to the square of its amplitude. Thus,
a unit pulse contains energy equal to unity squared which is
one. We assume the incident pulse is at the interface
(x = constant) at a given instant of time that defines the
timing line (t = constant). The vector shows the direction
of the incident pulse, but the circled ⊗ shows the location.
More specifically, we assume that the incident pulse is
slightly above the interface and slightly to the left of the
timing line, that is, the incident pulse is in the second
quadrant defined by the axes of the interface and timing line.
This incident pulse of unit amplitude is converted by the

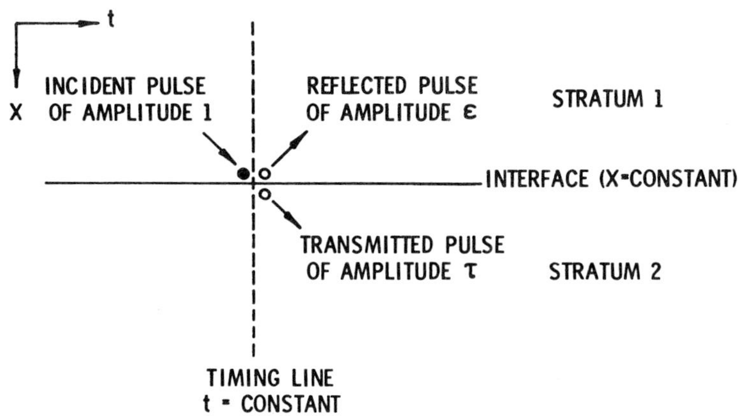

Fig. 1. Reflected and transmitted pulses due to an incident unit pulse.

interface into a reflected pulse of amplitude ε and a transmitted pulse of amplitude τ. Both the reflected and transmitted pulses occur at a split second (or split nanosecond or whatever) later than the incident pulse. Moreover, the reflected pulse is just above the interface while the transmitted pulse is just below the interface. In other words, the reflected pulse is in the first quadrant and the transmitted pulse is in the fourth quadrant. In Figure 1, the vectors show the space-time direction of these pulses and the small circles show their location. In summary, a downgoing incident pulse and the resulting reflected and transmitted pulses occur at a given interface at a given instant of time, but in actuality we must regard them as being within the proper quadrants formed by the interface and timing line. This convention is important because the energy in the incident pulse is divided into two parts, one part going into the reflected pulse and the other part going into the transmitted pulse. That is, the incident pulse appears at the interface slightly before the resulting reflected and

transmitted pulses, even though the mathematical notation
says that all three pulses appear at the interface at time t.
More specifically, we should say that the incident pulse
arrives at the interface at time $t - t_o$ and the resulting
reflected and transmitted pulses emanate from the interface
at time $t + t_o$, where t_o is a very small positive number.

All we have done so far is to describe the three pulses;
our problem now is to work out the amplitudes of the reflected
and transmitted pulses. The three pulses we have described
satisfy the wave equation (or, alternatively, the telegrapher's equations) within their respective strata, but the
telegrapher's equations must also be satisfied at the interface between the two strata. Thus, we must look at what
happens right at the boundary. It will turn out that the
telegrapher's equations demand that the three pulses fit
together in a certain way.

From a purely physical argument, we can say that
voltage must be the same on each side of an interface, which
we describe by saying that voltage is continuous across the
interface. Similarly, current must be the same on each side
of the interface, or in other words, current must be continuous across the interface. In fact, continuity of voltage
across an interface as well as continuity of current across
the interface can be derived from the two telegrapher's
equations, as we will now show.

Each telegrapher's equation holds in stratum 1 (above
the interface) and in stratum 2 (below the interface). In
addition, these equations also must hold in the interface,
which we call region 1 - 2. Although we usually think of the
interface as being sharply discontinuous, in reality it is
not. The physical properties change very rapidly but not
infinitely fast. In any case, we can imagine that there is
a very rapid but continuous transformation of the characteristic impedance between stratum 1 and stratum 2 within the
short distance we call region 1 - 2. Also, voltage and
current will make a similar kind of transition in intermediate region 1 - 2. In this intermediate region, the
telegrapher's equations will still be satisfied, and it is by
following these equations in this region that we can arrive

at the needed interface conditions.

Let us look at the first telegrapher's equation,

$$\frac{\partial v}{\partial x} = -L\,\frac{\partial i}{\partial t}\ .$$

The left-hand side involves the (partial) derivative of
voltage with respect to distance. Suppose that the voltage
jumped by some smooth but rapid transition from stratum 1 to
stratum 2. (Our purpose here is to show that this supposition
is wrong.) As a result of this supposition, the derivative
$\partial v/\partial x$ will have some very large values in the intermediate
region 1 - 2, because of the tremendous slope of v. In
other words, the derivative $\partial v/\partial x$ will have a sharp peak at
the interface. If we imagine squeezing the interface to even
a thinner layer, the peak would get much highter. If the
boundary is really sharp for the waves we are interested in,
the magnitude of $\partial v/\partial x$ in the intermediate region will
approach a Dirac delta function. As a result, the magnitude
of $\partial v/\partial x$ in the intermediate region would be much greater
than any contribution we might have from any other variation,
so we ignore any variations other than those due to the
boundary. We remember that voltage and current themselves
do not get especially large at the interface; only the
derivatives with respect to x can become so huge that they
dominate the telegrapher's equations. But now let us look at
the right-hand side of the above telegrapher's equation.
The x-derivative is on the left-hand side. We have supposed
that this x-derivative is huge. But there is nothing on the
right-hand side to match the huge left-hand side. Therefore,
our supposition is wrong; voltage cannot have any jump in
going from stratum 1 to stratum 2, for if it did there would
be a Dirac delta function on the left of the first telegrapher's equation but none on the right, and the equation
would be false. As a result, we have the condition that
voltages in stratum 1 and stratum 2 are equal at the interface: $v_1 = v_2$ at interface 1 - 2. By the same argument,
the second telegrapher's equation,

$$\frac{\partial i}{\partial x} = - C \frac{\partial v}{\partial t} \quad ,$$

shows that currents in stratum 1 and stratum 2 are equal at
the interface: $i_1 = i_2$ at interface $1 - 2$.

Let us now return to our incident, reflected, and transmitted pulses as shown in Figure 1. All three pulses occur
at the interface, but in reality the unit incident pulse and
the reflected pulse ε are just above the interface and so are
in stratum 1; whereas, the transmitted pulse τ is just below
the interface and so is in stratum 2. These pulses are all
in terms of normalized units of (voltage times current)$^{1/2}$
so in order to apply our interface conditions, we must transform them to both voltage and current. In stratum 1, the
incident unit pulse is the downgoing wave d_1 and the
reflected pulse ε is the upgoing wave u_1, so the voltage is

$$v_1 = \sqrt{Z_1}\, \tilde{v}_1 = \sqrt{Z_1}(d_1 - u_1) = \sqrt{Z_1}(1 - \varepsilon)$$

and the current is

$$i_1 = \frac{\tilde{i}_1}{\sqrt{Z_1}} = \frac{1}{\sqrt{Z_1}}\,(d_1 + u_1) = \frac{1}{\sqrt{Z_1}}\,(1 + \varepsilon) \ .$$

In stratum 2, the transmitted pulse τ is the downgoing wave
d_2, and the upgoing wave u_2 is zero, so the voltage is

$$v_2 = \sqrt{Z_2}\, \tilde{v}_2 = \sqrt{Z_2}\, d_2 = \sqrt{Z_2}\, \tau$$

and the current is

$$i_2 = \frac{\tilde{i}_2}{\sqrt{Z_2}} = \frac{d_2}{\sqrt{Z_2}} = \frac{\tau}{\sqrt{Z_2}} \ .$$

Here Z_1 and Z_2 are the characteristic impedances of strata 1
and 2, respectively. By continuity, we have $v_1 = v_2$ and
$i_1 = i_2$ so

$$\sqrt{Z_1}\,(1 - \epsilon) = \sqrt{Z_2}\,\tau$$

$$\frac{1}{\sqrt{Z_1}}\,(1 + \epsilon) = \frac{\tau}{\sqrt{Z_2}}\ .$$

Solving for ϵ and τ, we find that

$$\epsilon = \frac{Z_1 - Z_2}{Z_1 + Z_2}\ ,\qquad \tau = \frac{2\sqrt{Z_1 Z_2}}{Z_1 + Z_2}\ .$$

Because ϵ is the amplitude of the upgoing reflected pulse due to a downgoing incident pulse of unit amplitude, we call ϵ the (down-to-up) reflection coefficient of the interface between two strata with characteristic impedance Z_1 above and Z_2 below. Because τ is the amplitude of the downgoing transmitted pulse due to a downgoing incident unit pulse, we call τ the (down-to-down) transmission coefficient between the two strata.

Let us now consider an upgoing incident pulse of unit amplitude on the same interface, that is, with characteristic impedance Z_1 above and Z_2 below. Everything is the same as before, except now the roles down and up (and consequently Z_1 and Z_2 are interchanged) so the up-to-down reflection coefficient is $(Z_2 - Z_1)/(Z_2 + Z_1)$ which is $-\epsilon$, and the (up-to-up) transmission coefficient is the same as before, namely τ.

Let us now consider energy. Again, we refer to Figure 1. We recall that amplitude is measured in units of (voltage $\cdot$ current)$^{1/2}$, so the square of the amplitude is in units of energy. The incident pulse of amplitude 1 is converted into a reflected pulse of amplitude ϵ and a transmitted pulse of amplitude τ. Because no energy is lost, the energy of the incident pulse must equal the sum of the energies of the reflected and transmitted pulses; that is, we have the law of conservation of energy given by

$$1 = \epsilon^2 + \tau^2\ .$$

Solving this equation for τ, we have

$$\tau = \sqrt{1 - \varepsilon^2} \, .$$

Hence, all the coefficients can be expressed as follows:

    Down-to-up reflection coefficient $= \varepsilon$;

    Down-to-down transmission coefficient $= \tau = \sqrt{1 - \varepsilon^2}$;

    Up-to-down reflection coefficient $= -\varepsilon$;

    Up-to-up transmission coefficient $= \tau = \sqrt{1 - \varepsilon^2} \, .$

Wave Propagation.

We now want to consider wave propagation in a stratified
medium. Each wave is in fact a collection of pulses in space
and/or time. Figure 2 shows the type of stratified model
assumed, where the distance variable x varies in a positive
direction downward and the time t variable varies in a
positive direction to the right. Now let us divide the x
coordinate into increments which define strata, and let us
make the subdivision in such a way that all time events
are synchronized. More specifically, we divide the x axis
by means of discrete points x_n (where n is an integer) in
the following way. The point x_n is called interface n, the
point x_{n+1} is interface n+1, and the material between these
two interfaces is called stratum n, with characteristic
impedance Z_n and velocity c_n. Each stratum is chosen with
such a thickness Δx_n so that the two-way travel time in the
stratum is a constant Δt; that is,

$$\frac{2\Delta x_n}{c_n} = \Delta t = \text{constant} \quad (\text{where } \Delta x_n = x_{n+1} - x_n) \, .$$

In other words, given a fixed time increment Δt and the
depth x_n of layer n, then the depth of layer n+1 is determined by the formula

$$x_{n+1} = x_n + \frac{c_n \, \Delta t}{2} \, .$$

It is convenient to choose a time scale so that our time
interval is unity; that is, we choose $\Delta t = 1$.

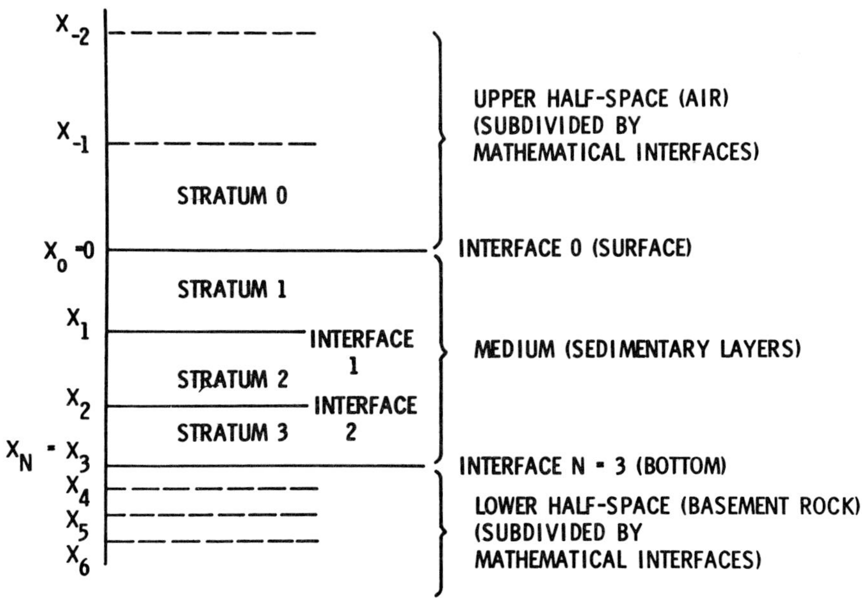

Fig. 2.   Medium with three strata.

We want to consider a medium of finite thickness X which
is sandwiched between two half-spaces. As we have just
described, we divide the medium into N strata, where $x_0 = 0$
is the surface and $x_N = X$ is the bottom. The upper half-
space has a certain characteristic impedance Z_0 and velocity
c_0. With the velocity c_0, we can subdivide the upper half-
space into strata. All of these upper half-plane strata
have the same characteristic impedance Z_0 so the reflection

coefficients of all the upper half-space interfaces are zero and transmission coefficients are one. Likewise, we can subdivide the lower half-space into such strata. We thus have the picture as shown in Figure 2.

As an input signal into the medium, we inject a unit downgoing pulse at zero time at the bottom of stratum 0. More precisely, the input pulse occurs just above interface 1 and just before time 0. The energy of this input spike is 1^2 or unity. This energy must be conserved. The energy in this input spike now propagates through time and space. Because it takes one-half time unit to travel through a layer, the energy must follow the ray paths as shown in Figure 3.

Upgoing waves are denoted by u and downgoing waves by d. Waves just above an interface are indicated by a prime; whereas, waves just below an interface are not tagged by a prime. A subscript indicates the stratum in which a wave occurs. Thus, the waves associated with interfaces n-1 and n are shown in Figure 4.

Let us now look at interface n at time t. Waves $d_n'(t)$ and $u_{n+1}(t)$ are inputs, and waves $u_n'(t)$ and $d_{n+1}(t)$ are outputs. Each output is the sum of transmitted and reflected parts of the inputs; that is,

$$d_{n+1}(t) = \tau_n\, d_n'(t) - \varepsilon_n\, u_{n+1}(t)$$

$$(1)$$

$$u_n'(t) = \varepsilon_n\, d_n'(t) + \tau_n\, u_{n+1}(t)$$

where ε_n is the reflection coefficient and τ_n is the transmission coefficient.

Flow of Energy.

Let us now consider the time-flow of energy at interface n at time t as shown in Figure 4. Just before time t, the energy at the space-time point (n, t) is

$$d_n'^{\,2}(t) + u_{n+1}^{\,2}(t) \; ;$$

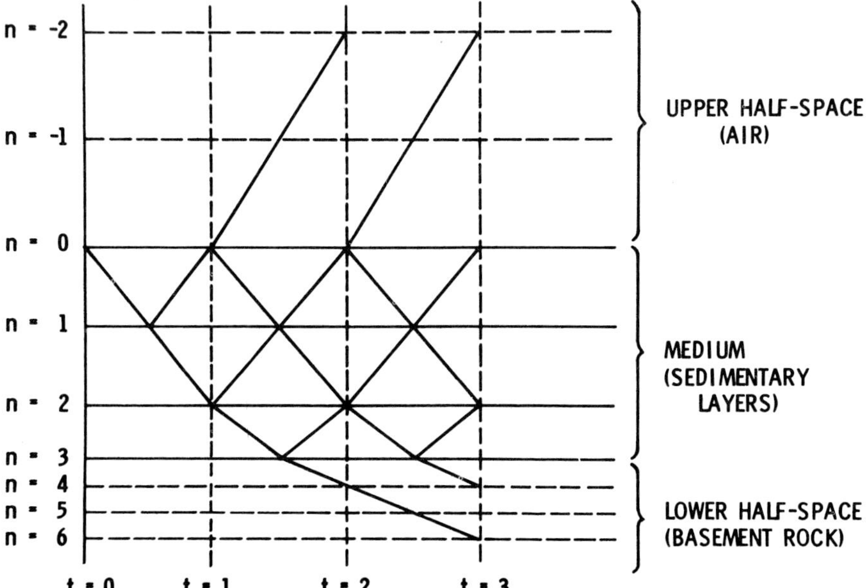

Fig. 3. Ray paths. Note that reflections can occur at real interfaces (shown by solid horizontal lines). The rays are synchronized at the interfaces, the rays crossing at integer times on even-numbered interfaces and at integer + 0.5 times on odd-numbered interfaces.

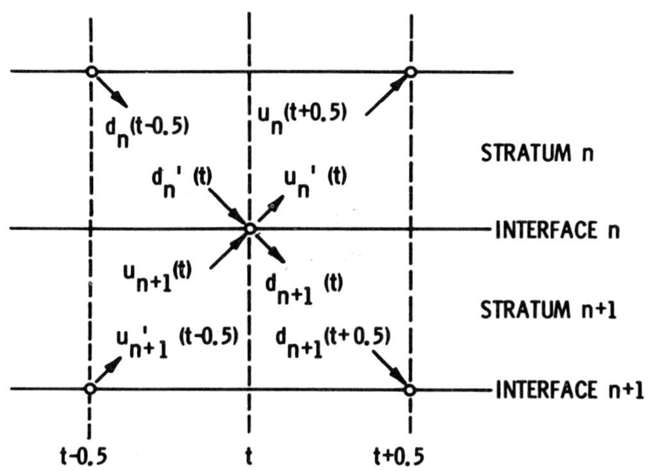

Fig. 4. Designation of waves. Note that t is an integer if n is even and t is an integer plus one-half if n is odd. The vectors of all the waves shown are in their correct quadrants with respect to interfaces and timing lines.

whereas, just after time t, the energy at space-time point (n, t) is

$$u_n'^2(t) + d_{n+1}^2(t) \; .$$

By the law of conservation of energy, the above two expressions should be equal, and indeed this is so, as can be directly verified by use of Eq. (1). The pulse $u_n'(t)$ travels up in stratum n without charge of form to become $u_n(t + 0.5)$; whereas, $d_{n+1}(t)$ travels down in stratum n+1, also without change of form to become $d_{n+1}'(t + 0.5)$. As a result, the above expressions for energy may be written as

$$d_n'^2(t) + u_{n+1}^2(t) = d_{n+1}'^2(t + 0.5) + u_n^2(t + 0.5) \; .$$

The left-hand side gives the total energy in strata n and n+1

just an instant before time t, and the right-hand side gives
the total energy in strata n and n+1 just before time
t + 0.5. If we sum each side of the above equation over all
pairs of strata (including hypothetical strata in the upper
and lower half-spaces), we get an equation that states that
the total energy in all the strata just before time t is
equal to the total energy in all the strata just before time
t + 0.5. We have thus established the law of conservation
of energy as energy flows with time. By recursive application of the law of conservation of energy, the energy at any
time must be equal to the energy originally introduced into
the system, namely one.

For ease of presentation, let us fix time t and the
number N of strata, so we can get across the basic concept
involved. Let t = 2 and N = 3, as shown in Figure 5.

The energy in the space waveform at time 2 is

$$u'_{-4}{}^2(2) + u'_{-2}{}^2(2) + u'_0{}^2(2) + d_1{}^2(2) + u'_2{}^2(2) +$$

$$+ d_3{}^2(2) + d_5{}^2(2) = 1$$

which is

$$[u'_0{}^2(0) + u'_0{}^2(1) + u'_0{}^2(2)] + [d_1{}^2(2) + u'_2{}^2(2) +$$

$$+ d_3{}^2(2)] + [d_4{}^2(1.5)] = 1 \ .$$

The three brackets in this expression show the division of the
energy into three parts. The first bracket is the energy
that has escaped by being reflected up and out of the medium
as of time t. The second bracket is the energy that is still
within the medium as of time t, so it is reactive energy.
The third bracket is energy that has escaped by being transmitted down and out of the medium as of time t. The first
bracket is the partial energy of the reflected time-waveform

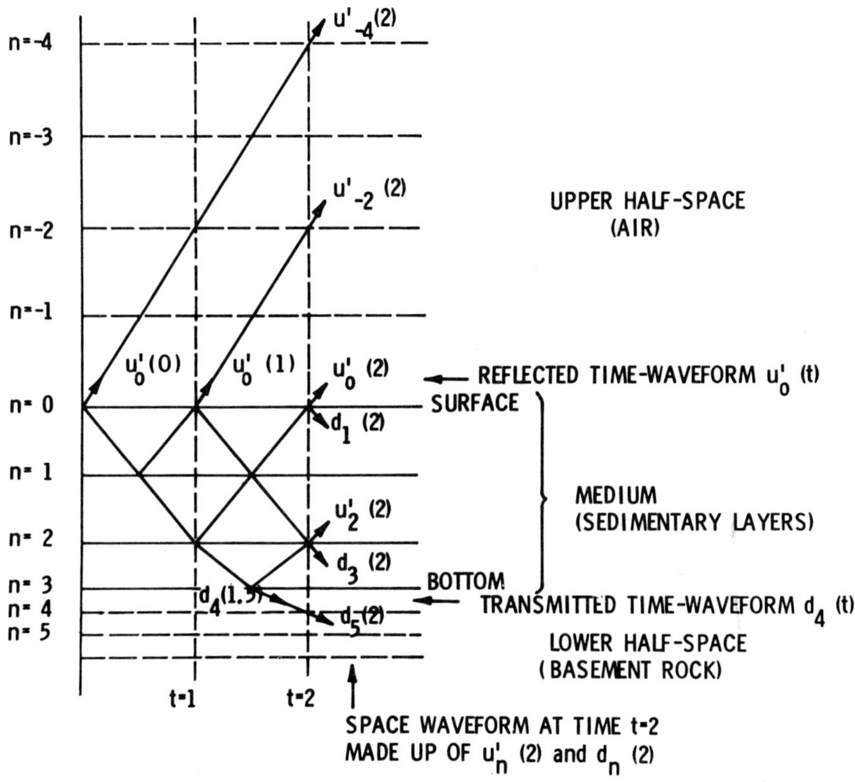

Fig. 5. The energy at $t = 2$ is the sum of squares of the pulses in the space-waveform. However, the space-waveform pulses above the surface are the same as the pulses in the reflected time-waveform up to and including time t, and the space-waveform pulses below the bottom are the same as the pulses in the transmitted time-waveform up to and including time t.

$u_0'(t)$ up to and including time t. The third bracket is the
partial energy of the transmitted time-waveform $d_{N+1}(t)$ up
to and including time t. As t tends to infinity, all the
energy escapes and the second bracket tends to zero, the
first bracket becomes the total energy in the reflected
time-waveform, and the third bracket becomes the total
energy in the transmitted time-waveform, i.e., for $t \to \infty$ we
have

$$[u_0'^2(0) + u_0'^2(1) + u_0'^2(2) + u_0'^2(3) + \cdots] + [d_4^2(1.5) +$$

$$+ d_4^2(2.5) + \cdots] = 1 .$$

In the following sections, we further discuss the
reflected wave $u_0'(t)$ and the transmitted wave $d_{N+1}(t)$ and
introduce the special notation for them given by

$$u_0'(t) = y_t \qquad\qquad t = 0,1,2,\ldots$$

$$d_{N+1}(t + \frac{N}{2}) = \tau\, b_t \qquad t = 0,1,2,\ldots$$

where $\tau = \tau_0\tau_1\ldots\tau_N$ and $b_0 = 1$. Then the above equation for
$t \to \infty$ is, in this new notation,

$$[y_0^2 + y_1^2 + y_2^2 + \cdots] + [\tau^2 + \tau^2 b_1^2 +$$

$$+ \tau^2 b_2^2 + \cdots] = 1 .$$

It can be established (Robinson and Treitel, 1976) that
there is a constant net downgoing flow of energy in each
layer which results in the fact that the downgoing wave in
each layer is minimum delay.

Wave Propagation Across Interfaces.

A wave such as $d_{n+1}(t)$ considered as a function of time
is actually a series of pulses of different amplitudes, each
pulse in the time-wave occurring one time unit apart. A wave

such as $d_{n+1}(t)$ considered a function of space is a series of pulses of different amplitudes, each pulse in the space-wave occurring at a distance of two strata apart (since a downgoing pulse in $d_{n+1}(t)$ travels through two strata per time unit).

The z-transform of the downgoing wave $d_{n+1}(t)$ is defined as

$$D_{n+1}(z) = \sum d_{n+1}(t) \, z^t$$

where the summation is for

$$t = \frac{n}{2}, \; \frac{n}{2} + 1, \; \frac{n}{2} + 2, \; \ldots$$

Likewise, the z-transform of the upgoing wave $u_{n+1}(t)$ is

$$U_{n+1}(z) = \sum u_{n+1}(t) \, z^t$$

where the summation is for

$$t = \frac{n}{2} + 1, \; \frac{n}{2} + 2, \; \ldots$$

(since $u_{n+1}(n/2) = 0$).   See Figure 6.

The wave $u_n'(t - 1/2)$ travels up in stratum n without change of form to become the wave $u_n(t)$.   Since z is a unit delay operator (i.e., $zf(t) = f(t - 1)$ where $f(t)$ is any time function), we have

$$u_n(t) = u_n'(t - \frac{1}{2}) = z^{1/2} \, u_n'(t).$$

Thus, the corresponding z-transforms satisfy

$$U_n(z) = z^{1/2} \, U_n'(z) . \tag{2}$$

The wave $d_n(t)$ travels down in stratum n without change of form to become the wave $d_n'(t + 1/2)$.   Thus,

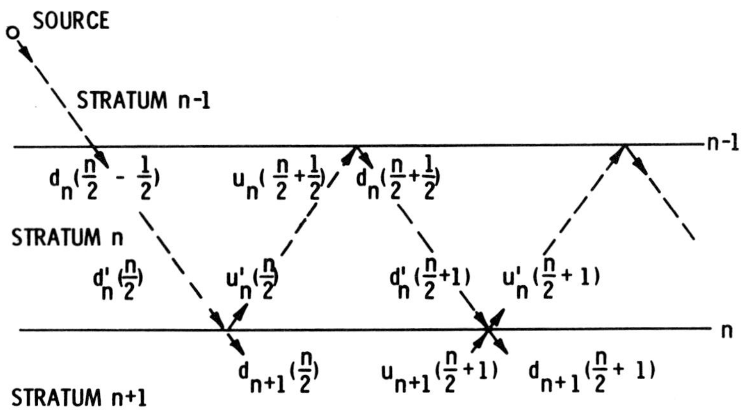

Fig. 6. Pulses associated with strata n and n+1.

$$d_n(t) = d_n'(t + \frac{1}{2}) = z^{-1/2}\, d_n'(t)$$

with the resulting equation

$$D_n(t) = z^{-1/2}\, D_n'(z) \ . \tag{3}$$

Let us now take the z-transforms of Eq. (1). We have

$$D_{n+1}(z) = \tau_n\, D_n'(z) - \varepsilon_n\, U_{n+1}(z)$$

$$\tag{4}$$

$$U_n'(z) = \varepsilon_n\, D_n'(z) + \tau_n\, U_{n+1}(z).$$

Using Eqs. (2) and (3) in Eq. (4), we obtain after some algebra the <u>propagation</u> <u>equation,</u>

$$
\begin{bmatrix} D_{n+1} \\ \\ U_{n+1} \end{bmatrix} = \frac{z^{-1/2}}{\tau_n} \begin{bmatrix} z & -\varepsilon_n \\ \\ -\varepsilon_n z & 1 \end{bmatrix} \begin{bmatrix} D_n \\ \\ U_n \end{bmatrix} \ . \tag{5}
$$

At the surface (i.e., interface 0), we have

$$
\begin{bmatrix} D_1 \\ \\ U_1 \end{bmatrix} = \frac{1}{\tau_0} \begin{bmatrix} 1 & -\varepsilon_0 \\ \\ -\varepsilon_0 & 1 \end{bmatrix} \begin{bmatrix} D_0' \\ \\ U_0' \end{bmatrix} \ . \tag{6}
$$

Use of Eq. (6) and recursive use of Eq. (5) allow us to relate the waves at the bottom and top of the medium by the equation

$$
\begin{bmatrix} D_{N+1} \\ \\ U_{N+1} \end{bmatrix} = \frac{z^{-N/2}}{\tau_N \ \tau_{N-1} \ \cdots \ \tau_1 \ \tau_0} \begin{bmatrix} A^R & -E^R \\ \\ -E & A \end{bmatrix} \begin{bmatrix} D_0' \\ \\ U_0' \end{bmatrix} \tag{7}
$$

where we define

$$
\begin{bmatrix} A^R & -E^R \\ \\ -E & A \end{bmatrix} = \begin{bmatrix} z & -\varepsilon_N \\ \\ -\varepsilon_N z & 1 \end{bmatrix} \begin{bmatrix} z & -\varepsilon_{N-1} \\ \\ -\varepsilon_{N-1} z & 1 \end{bmatrix} \cdots
$$

$$
\cdots \begin{bmatrix} z & -\varepsilon_1 \\ \\ -\varepsilon_1 z & 1 \end{bmatrix} \begin{bmatrix} 1 & -\varepsilon_0 \\ \\ -\varepsilon_0 & 1 \end{bmatrix} \ . \tag{8}
$$

We see that A and E are polynomials of degree N and A^R and
E^R are their respective reverse polynomials given by

$$A^R(z) = z^N A(z^{-1}) \ , \quad E^R(z) = z^N E(z^{-1}) \ .$$

That is, the coefficients of A^R are the same as those of A,
but in reverse order. The coefficients of E^R are the same
as those of E, but in reverse order. It is seen that the
coefficients of A and of E depend only upon the reflection
coefficients $\varepsilon_0, \varepsilon_1, \ \ldots \ , \varepsilon_N$. Explicit expressions may be
obtained for the coefficients of A and E in terms of the
reflection coefficients, but for our purposes we will make
use of the following approximation. We assume that the
reflection coefficients are small in magnitude, so that the
product of three or more reflection coefficients may be
neglected. Then Eq. (8) yields the following approximations
for A and E:

$$A \doteq 1 + \psi_1 z + \psi_2 z^2 + \cdots + \psi_N z^N$$

$$(9)$$

$$E \doteq \varepsilon_0 + \varepsilon_1 z + \varepsilon_2 z^2 + \cdots + \varepsilon_N z^N$$

where $\varepsilon_0, \varepsilon_1, \varepsilon_2, \ldots, \varepsilon_N$ is the reflection coefficient sequence
and where $\psi_1, \psi_2, \ldots, \psi_N$ is the autocorrelation

$$\psi_t = \varepsilon_t \varepsilon_0 + \varepsilon_{t+1} \varepsilon_1 + \cdots + \varepsilon_N \varepsilon_{N-t} \quad (\text{for } t = 1, 2, \ldots, N)$$

of the reflection coefficient sequence. Thus, the polynomial
E(z) according to this approximation is the z-transform of
the reflection coefficient sequence.

The Transmitted and Reflected Waves.

We recall our only input to the stratified medium is a
downgoing unit surface pulse at time zero, so $D_0' = 1$ and
$U_{N+1} = 0$. The resulting upgoing surface wave U_0' is designated
by Y and is called the reflection response. The resulting

downgoing bottom wave D_{N+1} is designated by T and is called the transmission response. Also, the product $\tau_N \tau_{N-1} \cdots \tau_1 \tau_0$ of transmission coefficients is simply denoted by τ. Thus, Eq. (7) becomes

$$\begin{bmatrix} T \\ \\ O \end{bmatrix} = \frac{z^{-N/2}}{\tau} \begin{bmatrix} A^R & -E^R \\ \\ -E & A \end{bmatrix} \begin{bmatrix} 1 \\ \\ Y \end{bmatrix} .$$

This matrix equation gives

$$T = \frac{z^{-N/2}}{\tau} (A^R - E^R Y)$$

$$O = -E + AY$$

so the reflection response is

$$Y = \frac{E}{A}$$

and the transmission response is

$$T = \frac{z^{-N/2}}{\tau A} (A^R A - E^R E) .$$

But $A^R A - E^R E$ is the determinant of the matrix given in Eq. (8). Taking the determinant of each side of Eq. (8), we see that

$$A^R A - E^R E = (1 - \varepsilon_N^2) \cdots (1 - \varepsilon_0^2) z^N$$

$$= \tau_N^2 \cdots \tau_0^2 z^N = \tau^2 z^N .$$

As a result, the transmission response is

$$T = \frac{\tau \, z^{N/2}}{A} \, .$$

Let us look at this transmission response. The direct pulse travels from the surface source to the bottom (i.e., interface N). Since the first coefficient a_o of the polynomial A is unity, the direct pulse has amplitude τ, namely the product of the transmission coefficients of the interfaces through which the direct pulse travels. The travel time of the direct pulse in going through N strata is N/2; this travel time represents the time delay of the transmitted wave; this time delay is indicated by the factor $z^{N/2}$. Because the source contained unit energy, the transmitted wave must have finite energy and hence be stable, which means that the polynomial A(z) in the denominator of T must be minimum delay (i.e., A(z) must have no zeros within the unit circle). In fact, let us define B(z) as the inverse of A(z), i.e.,

$$B(z) = \sum_{t=0}^{\infty} b_t \, z^t = \frac{1}{A(z)}$$

and we call the one-sided waveform $b_0 = 1, b_1, b_2, \dots$ the _reverberation_ _wavelet_, or simply the reverberation. Because the inverse of a minimum-delay function also is minimum delay, the reverberation is minimum delay. In terms of the reverberation, the transmitted response is

$$T = \tau \, z^{N/2} \, B$$

and the reflection response is

$$Y = BE \, .$$

Let $y_0, y_1, y_2, \dots$ denote the pulses of the reflection response. For example, in exploration seismology, $y_0, y_1, y_2, \dots$ is the reflection seismogram. If we use the

approximation for E given by Eq. (9), the above equation for Y
is

$$\sum_{t=0}^{\infty} y_t z^t = (1 + b_1 z + b_2 z^2 + \cdots)(\varepsilon_0 + \varepsilon_1 z + \cdots + \varepsilon_N z^N).$$

As a result, the reflected waveform y_t is the convolution of
the minimun-delay reverberation wavelet b_t with the reflection
coefficient sequence ε_t; that is,

$$y_t = \sum_{s=0}^{\infty} b_s \varepsilon_{t-s} \qquad (\text{for } t = 0,1,2,\ldots) . \qquad (10)$$

Equation (10) for a stratified medium is the basic mathematical
result of this paper. The purpose of this paper is the
statistical interpretation which we give to Eq. (10) in the
next section.

The Statistical Minimum-Delay Model and Predictive
  Deconvolution.

In order to fix ideas, let us consider a specific
physical situation, namely the problem of seismic exploration
for oil in the earth's sedimentary strata. The source is an
explosion or another form of energy which is introduced into
the ground at the surface. The reflection response y_t is
the seismic reflection record (or trace or time series) which
is digitally recorded at the surface. The reflection coefficient sequence ε_t is a digitized representation of the
reflectivity of the earth as a function of depth. As a
result, knowledge of the ε_t sequences for various geographic
locations on the surface allows the seismic interpreter to
make contour maps of the earth's sedimentary structure at
depth. In other words, the reflection seismic method is a
method of obtaining a picture of what the earth looks like
thousands of feet below the surface from an experiment (i.e.,
a surface source of energy and a surface recording of the
resulting reflection response to that source) carried out on
the surface.

The nature of the geophysicist's problem can be imagined

by pretending that the earth's crust is a stack of Plexiglas
sheets, all opaque with different properties and thicknesses,
each having irregular surfaces, some of the plates being
tilted, bent or deformed, and some containing pockets or
traps of varying shapes and sizes. Some of these traps may
contain accumulations of oil and natural gas. Now, the
geophysicist taps the surface of this medium and he analyzes
the resulting response at the surface to determine the thickness and physical properties of each plate, details of the
interfaces between plates, and the locations and shapes of
the traps in the plates. With this analogy in mind, the
earth's crust may be described as an acoustically trans-
lucent, heterogeneous, multilayered medium. The traps for
oil and natural gas may have an areal extent of only a half
of a square mile, and be as much as four miles deep within
the earth, with no geologic evidence at the surface that
there is a trap deep below.

Let us now consider the magnitude of this problem. The
United States consumes nearly 20 million barrels of oil each
day and a correspondingly great amount of natural gas. An
oil field with total reserves of 100 million barrels would
be considered a good discovery, yet the United States would
consume that much oil in five days. The only great oil
discovery in the United States in the lifetime of the
majority of Americans is the North Slope of Alaska, and if
the United States had to rely entirely on the North Slope's
known reserves, they would be consumed in two years. What
makes matters really difficult is that not only in the United
States but worldwide, the consumption of oil and natural gas
is increasing at a geometric (i.e., exponential) rate.

In order to discover more oil and gas to replace the
amount consumed, the oil exploration industry records about
two million reflected seismic traces, y_t, each and every day,
year after year. Each of these two million time series y_t
has about 1000 data points (i.e., t = 0,1,2,...,1000). These
reflected traces, y_t, must be analyzed to yield a picture
of the subsurface structure.

Before we describe the geophysical methods of today,
we should first look at the history of exploration geophysics.

Before 1920, most oil was discovered by chance or by looking
at surface indicators such as oil seeps or surface geological
manifestations of subsurface traps. In the 1920s, great
advances were made in geophysical methods of discovery, and
since 1930 the reflection seismic method has been the chief
method for discovering oil. Up until the digital revolution
in seismology during the early 1960s, seismic interpretation
was carried out on the basis of the reflected seismic trace,
y_t. That is, up to about 1960, the recorded trace y_t
(subject to only some electric filtering) was analyzed by
eye in order to detect reflected events from the subsurface
interfaces. From these observations, the seismic interpreter
was able to draw subsurface contour maps. The amazing thing
at the time was that the method worked at all; the inter-
pretations on the whole were correct, and great oil fields
were discovered. The question is why did the method work?
But the question "why" is not important in comparison to the
fact that the method did work; the seismic interpreter was
able to detect primary reflected events by eye. The primary
reflected event from interface t is defined as the pulse
that travels directly down from the source to interface t and
then directly back to the surface; its z-transform is

$$z^t \; \tau_0 \; \tau_1 \; \cdots \; \tau_{t-1} \; \epsilon_t \; \tau_{t-1} \; \cdots \; \tau_1 \; \tau_0 \; .$$

However, for each primary reflected event, there are many
multiple reflected events, and from a strictly mathematical
point of view, these multiple events should overwhelm the
primary events on the recorded trace y_t.   That is,
from a strictly mathematical point of view, the reverberations
of this complex stratified system we call the earth should
overwhelm our ability to pick distinct primary reflections
on the seismic records.

By 1950, the search for oil extended into new areas all
over the world, and it was realized that there were many
areas where seismic record quality was poor, or in other
words, the multiple reflections were overwhelming the
primary reflections. In particular, in water-covered areas
such as Lake Maracaibo in Venezuela and in the Persian Gulf,

strong multiple waves in the water layer gave the observed
seismic trace y_t a ringing character which made it impossible
to detect the primary reflections coming up from the deep
interfaces. Strong reverberations from surface layers
masked the desired primary events in many land areas of
exploration. In fact, as exploration pushed into more difficult areas, all types of multiple reflected energy began
to be recognized on the traces and made conventional visual
interpretation difficult to impossible. Moreover, conventional methods of electric filtering and field techniques
could not overcome these difficulties.

From a theoretical point of view, it is important to
develop models of the earth. In classical science, a great
many models are known in which the behavior of the system is
fully determined by a precisely fixed mathematical structure
subject to idealized boundary-value or initial-value conditions. Such is the case for the earth model first treated
in a classic paper of Lamb (1904) on the propagation of
seismic waves in an elastic solid. Classical geophysics is
concerned with the development of such deterministic models.
However, because of the mathematical complexity of these
models, only a very few simple geometric configurations
within the earth are susceptible to analytic solutions.
Because the field geophysicist historically was faced with a
large number of different physical configurations, each of a
complicated nature, the methods of classical geophysics were
of no direct use and could only serve as a guide. The situation in exploration seismology in the early 1950s was summarized by one of the leading field geophysicists in this
statement: "We have sharpened our tools as much as we can;
what we need now are some new tools."

As it turned out, the new tool was the digital computer
in conjunction with digital handling of seismic data from
beginning to end. Use of digital computers to analyze geophysical data began in 1950 with an MIT research project which
developed into the MIT Geophysical Analysis Group, an
industry-supported project which ran from 1953 to 1957.
Theoretical geophysical calculations made a significant
advance when Norman Haskell, the geophysical advisor to this

MIT project, published his results on how seismic surface
waves could be computed for a stratified earth model
(Haskell, 1953). After extensive digital analysis of field
records, Robinson (1954) proposed the statistical minimum-
delay model of the earth together with the idea of deconvolution to separate the statistical component from the minimum-
delay component. The present paper may be described as a
synthesis of the stratified earth model and the statistical
minimum-delay model. The stratified earth model for reflection seismograms has been extensively developed by Peterson,
Fillippone and Coker (1955), Wuenschel (1960), Goupillaud
(1961), Kunetz (1964) as well as many others.

Let us now explain the statistical minimum-delay model.
The earth's crust is made out of a large number of strata in
complex configurations. In the analysis of seismic waves
from a physical point of view, we try to understand why the
various waves that propagate through these strata behave the
way they do. It is apparent that this is a difficult subject,
and so we have to deal with it differently than with simpler
situations. One way to analyze the properties of the strata
is to start from a physical point of view, namely the point
of view acquired over the years by the field geophysicists
who analyze and interpret the seismic records. These geophysicists have a rough idea where they are going and they
begin by making the right kind of approximations, knowing
what is big and what is small in a given complicated situation. These interpretation problems are so complex that
even an elementary understanding, although inaccurate and
incomplete, is worth having. One of the things learned by
these field geophysicists is that the analysis requires an
understanding of the theory of probability. We do not want
to know where every wave is actually moving, but rather how
many move here and there on the average and the mutual
buildup or cancelling of effects. Large numbers of seismic
reflection records are needed to carry out an exploration
program over a geographic area. The quantity of data
necessarily requires the consideration of each record as a
member of a larger group or ensemble of records. Thus, the
reliability of a single record is considerably less than the

reliability of the ensemble average in connection with the description of the geological conditions existing in that area. Also, from an economic standpoint, the amount of control in such an exploration program must be kept at a bare minimum consistent with worthwhile results. As a result, the working geophysicist must proceed to fit his empirical data into the larger overall framework from a statistical point of view. A major justification for using the statistical approach in seismology is due to the fact that large amounts of data must be processed; any data in large enough quantities takes on a statistical character, even though each individual piece of data is of a deterministic nature. Thus, the field geophysicist in working with the physical properties of the earth is faced with a situation which is essentially statistical. For example, a reflection which may be followed from trace to trace, record to record, usually has more value to the seismic interpreter, and hence is statistically more significant, than a reflection which appears only on a few traces.

Because the wavelet b_t in Eq. (10) is minimum-delay, the model represented by Eq.(10) is a minimum-delay model. Let us now give reasons why Eq.(10) may be regarded as a statistical model. According to the statistical point of view, the reflection coefficient sequence ε_t may be considered as a realization of a random (or stochastic) process. Equation (10), therefore, represents the reflection seismic trace as the composition of a minimum-delay wavelet and a random sequence. Equation (10) gives the entire reflected trace y_t from time zero to infinity. In practice, it is usually not feasible to analyze the whole trace in its entirety, and so instead the trace y_t is cut up into time sections (i.e., time gates) and then is analyzed section by section. Often these gates overlap to some extent, and then the results from the several sections are blended together so as to give a time-varying character to the overall analysis.

The part of the seismic trace within a given time gate contains the primary reflections from the interfaces within a certain section of the earth's crust. If we assume that the reflection coefficients ε_t within this earth section are mutually uncorrelated (i.e., that these ε_t represent a realization from a white-noise process), then the classical method of predictive deconvolution becomes applicable. At this point, let us note that the terms "minimum-delay" and "minimum-phase" are the same so they may be used interchangeably. Also, the terms "predictive deconvolution" and "predictive decomposition" may be used interchangeably.

The method of predictive deconvolution may be described as follows. Within the time gate, the ε_t are white; hence, all the spectral shape of the trace y_t within the time gate is due to b_t. As a result, computation of the autocorrelation of y_t within the gate averages out the white-noise elements ε_t and, therefore, gives us the autocorrelation of b_t. Since b_t is minimum-delay, its inverse a_t can be computed directly from its autocorrelation by solving the normal equations. The inverse operator a_t is the required deconvolution operator. Deconvolution is the process of convolving the inverse operator a_t with the trace y_t. The result of the deconvolution is the reflection coefficient sequence ε_t which is called the deconvolved trace. The deconvolved trace ε_t represents the earth's reflectivity as a function of depth within the earth section being analyzed. In practice, of course, many variations of the basic deconvolution method described here can be used. For example, the inverse operator a_t represents an autoregressive operator, that is, a prediction error operator with unit prediction distance; prediction error operators for other prediction distances also can be used. Adaptive and time-varying deconvolution methods as well as multitrace methods may be used.

By the early 1960s, the petroleum exploration industry had undergone what was called a digital revolution. Digital methods were used for recording, processing, and display of seismic data. This digital approach was based on the statistical point of view. Since 1965, it has been routine practice to deconvolve each seismic trace.

During these years, oil companies and geophysical contractors have made careful studies which shed light on the statistical minimum-delay model. For example, White and O'Brien (1974) of British Petroleum consider the assumption

that the reflection coefficient sequence ε_t is white, that
is, the assumption that the power spectrum of ε_t is flat and
that the shape of the power spectrum of the reflected trace
y_t is due solely to the wavelet b_t. They say: "This is a
crucial assumption. An investigation of some reflection spike
sequences estimated from well-log data has shown that it is
usually valid. The only definite exceptions found occurred
beyond the seismic frequency range, although there also may
be a departure from a flat spectrum at low frequencies." In
their study of sonic logs from oil wells, Schoenberger and
Levin (1974) of Exxon find the assumption of a white reflection sequence ε_t is a good one for depth intervals of an oil
well, but breaks down for certain intervals where a few
strong reflections dominate.

Let us now look at the minimum-phase assumption. The
fact that reverberating seismic energy is minimum-phase is
well established (Kunetz, 1964; Treitel and Robinson, 1966).
We must also consider the source waveform which is relatively
narrow although it cannot be considered as narrow as a spike.
Many studies have been made on source pulses. As stated by
White and O'Brien (1974): "The source measurements already
mentioned have shown that the pulse radiated by an explosion
is minimum-phase or very close to it. It is possible for
underwater sources to emit non-minimum-phase signals but
operational conditions usually avoid this. Absorption in
the earth is generally held to be minimum-phase, and so too
is the response of most recording systems. Minimum-phase
then is likely to be satisfied in practice." As a result,
all these minimum-phase effects can be included in the
minimum-delay wavelet b_t in the model as given by Eq. (10).

What then is the final test for the statistical minimum-
delay model and the associated method of predictive decomposition or deconvolution? The answer lies in the fact that
today as well as during the past 15 years, virtually every
single seismogram taken in the exploration of oil and
natural gas has been processed according to this model.
These 10 billion seismic traces from every part of the world
must speak for themselves, and they say the statistical
minimum-delay model and the method of predictive deconvolution

are physically valid.

Conclusions.

Let us now answer the question we asked in the last
section, namely why did the seismic method work in so many
areas before the advent of deconvolution and other forms of
digital processing? The answer is that in these good seismic
areas, the reflection coefficient sequence ε_t is white or
nearly white for the entire crustal section (Treitel, 1976).
Thus, the autocorrelations ψ_t of the reflection coefficient
sequence are approximately zero, so Eq. (9) becomes $A \doteq 1$.
As a result, B is also approximately equal to 1, so b_t becomes
a unit pulse and Eq. (10) becomes simply $y_t \doteq \varepsilon_t$. In other
words, in those areas where the reflection coefficients are
small and uncorrelated, the reflection seismogram reduces
simply to the sequence of reflection coefficients ε_t. The
net effect of the white nature of the subsurface reflection
coefficients is to cancel out all the multiplies and to
amplify all the primaries so they have amplitude ε_t instead
of amplitude $(\tau_0 \tau_1 \cdots \tau_{t-1})^2 \varepsilon_t$. Here is a case in nature
where randomness has taken a very complex physical situation
and made it extremely simple and beautiful. As we have seen
in the application of deconvolution, we make use of those
crustal sections in which we can consider the sequence ε_t
uncorrelated.

The digital revolution in seismology as well as in other
sciences is based on statistical (or data-oriented) models
which make possible the unraveling (or deconvolution) of the
observed time series. The essence of this statistical
approach is that, in addition to its physical structure, a
system has a structure of behavior within which the system
operates. This structure of behavior can be deduced from
field observations. Experiments with the statistical
approach have confirmed the existence of this behavior and
have brought us closer to understanding the workings of
nature. The basis of this approach lies within the behavior
of the physical media and in the interaction of their
components. The underlying permanence is in the rules by

which energy is exchanged between components, and in this
respect the concept of the minimum-delay of energy becomes
a central concept. We see that there is a subtle shift of
emphasis here between the deterministic and the statistical
approach. The statistical approach is concerned about the
transactions rather than the things; that is it deals with
the dynamics of what happens rather than with characteristics of the system as an object. The field data in the
form of time series are the keys to this approach, for the
analysis of the time series tells us what happens and in this
way depicts the structure of the system. For example, in the
statistical approach, we are interested in the correlation
among reflection coefficients in a sequence of strata, and
with the flow of energy within this structure. In other
words, the statistical approach looks at the relationship
between things rather than the things themselves.

References.

Goupillaud, P. (1961). An· approach to inverse filtering of
     near surface layer effects from seismic records.
     Geophysics 26, 754-760.

Haskell, N. A. (1953). The dispersion of surface waves on
     multilayered media. *Bull. Seismological Soc. Amer. 43*,
     17-34.

Kunetz, G. (1964). Généralisation des opérateurs d'anti-
     résonance à un nombre quelconque de réflecteurs.
     Geophys. Prospecting 12, 283-289.

Peterson, R. A., W. R. Fillippone, and F. B. Coker (1955).
     The synthesis of seismograms from well log data.
     Geophysics 20, 516-538.

Robinson, E. A. (1954). Predictive decomposition of time
     series with application to seismic exploration. MIT
     Geophysical Analysis Group, Report No. 7, Cambridge,
     Massachusetts. Reprinted in *Geophysics 32*, 418-484
     (1967).

Robinson, E. A., and S. Treitel (1976). Net downgoing energy
     and the resulting minimum-phase property of downgoing
     waves. *Geophysics 41*, 1394-1396.

Schoenberger, M., and F. K. Levin (1974). Apparent attenuation due to intrabed multiples. *Geophysics* *39*, 278-291.

Treitel, S. (1976). Personal communication.

Treitel, S., and E. A. Robinson (1966). Seismic wave propagation in layered media in terms of communication theory. *Geophysics* *31*, 17-32.

White, R. E., and P. N. S. O'Brien (1974). Estimation of the primary seismic pulse. *Geophys. Prospecting* *22*, 627-651.

Wuenschel, P. E. (1960). Seismogram synthesis including multiples and transmission coefficients. *Geophysics* *25*, 106-129.

CHAPTER 19

Geoexploration, 16 (1978) 55—73
© Elsevier Scientific Publishing Company, Amsterdam — Printed in The Netherlands

USE OF THE KEPSTRUM IN SIGNAL ANALYSIS

MANUEL T. SILVIA[1] and ENDERS A. ROBINSON[2]

[1] *Naval Underwater Systems Center, Newport, R.I. 02840 (U.S.A.)*
[2] *Northeastern University, Boston, Mass. 02115 (U.S.A.)*

(Received September 3, 1977)

ABSTRACT

Silvia, M.T. and Robinson, E.A., 1978. Use of the kepstrum in signal analysis. In: C.H. Chen (Editor), Computer-Aided Seismic Analysis and Discrimination. Geoexploration, 16: 55—73.

The kepstrum has evolved as an important mathematical construct in the area of digital signal processing. In particular, kepstral analysis has been used for the deconvolution of reflection seismograms generated by the exploration for oil and natural gas. Here, we review the development of the normal-incidence reflection seismogram from a physical point of view and show how the kepstrum relates to the fine-grain structure of the physical model. Moreover, we develop symmetry properties of the kepstrum, introduce the concept of minimum-advance, and discuss the relationship between spectral factorization and the kepstrum. In the context of deconvolution, we discuss the methods of predictive deconvolution and homomorphic deconvolution and their dependence on the physical model of the reflection seismic process.

INTRODUCTION

The idea of the kepstrum appears in the classical work of Poisson (1823), Schwarz (1872), Szegö (1915), and Kolmogorov (1939), and has been applied to geophysical problems by Robinson (1954), Bogert et al. (1963), Schafer (1969), Oppenheim and Schafer (1975), Tribolet (1977), and others. In this paper we bring all this work together and introduce additional properties of the kepstrum so as to bring out its essential features and symmetries. As a result, the kepstrum evolves as an important construct in the theory of signal analysis and time series, and takes the place among such established functions as the autocorrelation and spectrum. For further discussion see Båth (1968, 1974) and Kulhánek (1975).

In their classical works on potential theory, Poisson (1823) and Schwarz (1872) addressed the problem of determining a potential function whose real part was an assigned value on the unit circle. As is well known in potential theory (see Bateman, 1944), Schwarz's classical expression is:

$$H(z) = \frac{1}{2\pi} \int_{-\pi}^{\pi} \left(\frac{1 + ze^{i\omega'}}{1 - ze^{i\omega'}}\right) \mathrm{Re}\,[H(\omega')]\,d\omega', \, |z| < 1 \tag{1}$$

with ω' as the integration variable and $z = re^{-i\omega}$. This equation gives the potential function $H(z)$ whose real part on the unit circle is $\mathrm{Re}\,[H(\omega)]$. The Schwarz expression follows from the famous result of Poisson on potential theory. An interesting result arises when we consider a logarithmic potential $\log H(z)$, which can be written:

$$\log H(z) = \log |H(z)| + i \arg [H(z)]$$

The real part of the logarithmic potential evaluated on the unit circle $z = e^{-i\omega}$ is then $\log |H(z = e^{-i\omega})| = \log |H(\omega)|$. In terms of the logarithmic potential, Schwarz's expression becomes:

$$\log H(z) = \frac{1}{2\pi} \int_{-\pi}^{\pi} \left(\frac{1 + ze^{i\omega'}}{1 - ze^{i\omega'}}\right) \log |H(\omega')|\,d\omega', \, |z| < 1$$

where $\log |H(\omega)|$ is the assigned or known value on the unit circle and the function $H(z)$ has no zeroes or poles inside or on the unit circle, i.e., $H(z)$ is minimum-delay. (This definition of minimum-delay is based on the Laplace definition of the z-transform, i.e., the z-transform of a sequence h_n is defined by $H(z) = \sum_{n=-\infty}^{\infty} h_n z^n$.) As a result, $\log H(z)$ has a Taylor-Maclaurin series expansion in the unit circle:

$$\log H(z) = \sum_{n=0}^{\infty} f_n z^n, \, |z| < 1 \tag{2}$$

The coefficients f_n of this Taylor-Maclaurin series are the coefficients of a stable causal system. We call these coefficients the *kepstrum* of the minimum-delay system $H(z)$. However, as we shall see, the concept of the kepstrum is not limited to minimum-delay systems.

In addition to its appearance in the logarithmic potential problem, the kepstrum was used by Szegö (1915) and Kolmogorov (1939) to solve the problem of factoring the power spectrum $\Phi(\omega)$ of a random process in order to extract a stable causal system $H(z)$. A power spectral density function $\Phi(\omega)$ must satisfy the following conditions in order to be factored so as to yield a causal system:

(1) $\Phi(\omega)$ is positive on the interval $-\pi \leqslant \omega \leqslant \pi$.
(2) $\Phi(\omega)$ has finite area on the interval $-\pi \leqslant \omega \leqslant \pi$.
(3) $\log \Phi(\omega)$ has finite area on the interval $-\pi \leqslant \omega \leqslant \pi$.

Now the problem of spectral factorization is the problem of extracting a causal system $H(z)$ whose magnitude spectrum is the square root of the power spectrum. In symbols:

$$|H(\omega)| = [\Phi(\omega)]^{\frac{1}{2}}$$

Szegö (1915) and Kolmogorov (1939) recognized that a solution to this problem could be obtained by making use of the Schwarz expression (1). Now $\log|H(\omega)|$ is the real part of the function $\log H(\omega)$. Further:

$$\text{Re}\,[\log H(\omega)] = \log|H(\omega)| = \tfrac{1}{2}\log \Phi(\omega)$$

Thus, the required solution is found by merely substituting $\tfrac{1}{2}\log \Phi(\omega)$ for $\text{Re}\,[H(\omega)]$ and substituting $\log H(z)$ for $H(z)$ in Schwarz's expression to obtain:

$$\log H(z) = \frac{1}{2\pi} \int_{-\pi}^{\pi} \left(\frac{1 + ze^{i\omega'}}{1 - ze^{i\omega'}}\right)\frac{1}{2}\log \Phi(\omega')d\omega', \; |z| < 1$$

Hence, the Szegö-Kolmogorov solution to the spectral factorization problem is:

$$H(z) = \exp\,[\log H(z)] = \exp \left[\frac{1}{4\pi} \int_{-\pi}^{\pi} \left(\frac{1 + ze^{i\omega'}}{1 - ze^{i\omega'}}\right) \log \Phi(\omega')d\omega'\right], \; |z| < 1$$

The desired sequence $h_0, h_1, h_2, \ldots$ can be written in terms of the kepstrum $f_0, f_1, f_2, \ldots$ as:

$$h_0 + h_1 z + h_2 z^2 + \ldots = \exp\,[f_0 + f_1 z + f_2 z^2 + \ldots] \tag{3}$$

We call this equation *Kolmogorov's equation*. Putting $z = 0$ in eq.3 we observe that:

$$h_0 = \exp(f_0) = \exp \left(\frac{1}{4\pi} \int_{-\pi}^{\pi} \log \Phi(\omega')d\omega'\right)$$

Thus, the leading kepstral coefficient f_0 is the logarithm of the leading impulse response coefficient h_0, i.e.,

$$f_0 = \log h_0$$

As a result, it is often convenient to normalize an impulse response function h_n so that its leading coefficient is unity; thus, the leading kepstral coefficient is zero, i.e., $f_0 = \log 1 = 0$. With this normalization, Kolmogorov's equation (eq.3) is rewritten as:

$$1 + \frac{h_1}{h_0} z + \frac{h_2}{h_0} z^2 + \ldots = \exp\,(f_1 z + f_2 z^2 + f_3 z^3 + \ldots) \tag{4}$$

We shall define the power series expansion $f_1 z + f_2 z^2 + f_3 z^3 + \ldots$ appearing in the exponent of Kolmogorov's equation (4) as the Kolmogorov Equation Power Series (KEPS). Further, since KEPS is the z-transform of the sequence $0, f_1, f_2, f_3, \ldots$, we can think of f_n as being a time response. Thus, we shall define the sequence $0, f_1, f_2, f_3, \ldots$ as the Kolmogorov Equation Power Series Time Response and call this sequence the KEPSTRUM, where we have added the Latin singular ending "um" to denote one kepstrum and shall use the Latin plural ending "a" to denote more than one kepstrum, i.e., kepstra.

Whereas Szegö and Kolmogorov were only concerned with the real part of the function $\log H(z)$ on the unit circle, Robinson (1954) was concerned with the imaginary part as well; namely, both the real and the imaginary parts have physical meaning. Thus, it is important to consider the logarithm of the spectrum $H(\omega)$. Let us elaborate. For a minimum-delay system $H(z)$, the kepstrum f_n was defined by the relation:

$$\log H(z) = \sum_{n=0}^{\infty} f_n z^n, \ |z| < 1$$

The complex variable z represents a point within the unit circle (i.e., $|z| < 1$). However, we can consider the limit of $\log H(z)$ as the interior point z approaches a point on the circumference of the unit circle; that is, we consider the limit of $\log H(z)$ as $z \to e^{-i\omega}$. This limit is:

$$\log H(\omega) = \sum_{n=0}^{\infty} f_n e^{-in\omega}$$

Separating the above into real and imaginary parts, we obtain:

$$\log |H(\omega)| + i\theta(\omega) = \left(f_0 + \mathrm{Re}\left[\sum_{n=1}^{\infty} f_n e^{-in\omega} \right] \right) + i \left(I_m \left[\sum_{n=1}^{\infty} f_n e^{-in\omega} \right] \right) \tag{5}$$

By equating real parts in eq.5, it can be shown that:

$$f_n = 2f_n' \text{ for } n = 1, 2, 3, \ldots$$

$$f_0 = f_0' \quad \text{ for } n = 0$$

$$f_{-n}^{\bigstar} = 2f_n' \text{ for } n = -1, -2, -3, \ldots$$

where:

$$f_n' = \frac{1}{2\pi} \int_{-\pi}^{\pi} \log |H(\omega)| e^{in\omega} \, d\omega \text{ for } n = 0, \pm 1, \pm 2, \ldots \tag{6}$$

and the superscript $\bigstar$ denotes the complex conjugate, i.e., $f_n^{\bigstar}$ is the complex conjugate of f_n. Thus, the kepstrum f_n is found in terms of $\log |H(\omega)|$, thereby allowing us to obtain the minimum-delay system $H(z)$ via (2). Let us now look at the imaginary part of eq.5, namely, by equating the imaginary parts in (5) it can be shown that:

$$\theta(\omega) = \frac{-1}{2\pi} P \int_{-\pi}^{\pi} \cot\left(\frac{\omega - \omega'}{2}\right) \log |H(\omega')| \, d\omega' \tag{7}$$

where we have incorporated the result of (6) into (7); the symbol P denotes that the integral has its Cauchy principal value. The kepstrum was used as an intermediate step in deriving (7), which shows that for a minimum-delay system $H(z)$, knowledge of only the log magnitude spectrum is sufficient to determine the phase spectrum $\theta(\omega)$. In the mathematics literature, eq.7 and various versions of the Schwarz formula (1) are often referred to as *Poisson's formulas*. In the context of linear system theory, eq.7 is known as the *Hilbert transform*, where $\theta(\omega)$ and $\log |H(\omega)|$ form a Hilbert transform pair.

Bogert et al. (1963) were the first to use the kepstrum as a signal processing operation. Bogert et al. were primarily interested in measuring the time interval between the arrival of certain seismic waves, excited by either natural or man-made disturbances. They defined the term "cepstrum" as the power spectrum of the logarithm of the power spectrum of the observed time series. We see that the "cepstrum" of Bogert et al. involved no phase spectrum information, i.e., it utilized magnitude spectrum and power spectrum information only. Their "cepstrum" was effective in detecting or measuring the time delay of a *single* echo, even when other methods (such as the autocovariance method) were ineffective. However, it is important to note that Bogert et al. recognized the limitations of their technique and the difficulties encountered in the multiple echo determination problem. Although first used in geophysical applications, the cepstrum has made its way into other fields of research, for example, speech and sonar signal processing.

Schafer (1969) was concerned with the recovery of signals generated by a convolutional process as opposed to the determination of echo arrival times and called his method homomorphic deconvolution. Through a logarithmic (homomorphic) transformation of the z-transform of the observed process, convolution of two signal components is mapped into the addition of their kepstra. By appropriate filtering (sometimes called liftering) in the kepstral domain, the desired signal can be recovered. Schafer used the term "complex cepstrum" to account for the fact that both the magnitude and phase spectra of the observed signal are utilized. To date, Oppenheim and Schafer (1975), Tribolet (1977), and many others have studied the effectiveness of the "complex cepstrum" in deconvolving reflection seismograms. Our word "kepstrum" means the same as their term "complex cepstrum". Because the kepstrum of a real-time sequence is real, the use of the word "kepstrum" is less confusing than the term "complex cepstrum".

In summary, we have traced the mathematical evolution of the kepstrum, starting with the logarithmic potential problem of the 1800's to the current deconvolution process in oil exploration. In the next section, we shall review the normal-incidence reflection seismogram so as to link the physics of reflection seismology to the mathematical properties of the kepstrum.

THE ARMA MODEL USED IN REFLECTION SEISMOLOGY

It is important that we review the evolution of the well known normal-incidence convolutional model used as a basis for many deconvolution techniques. This model in its final form is simple and linear, and as a result, can be easily evaluated by conventional computational methods. However, without understanding the basic underlying assumptions and physical approximations involved in the model development, one cannot view this simplicity in the proper perspective. If we are going to use the kepstrum as a deconvolution technique, let us begin by examining its relationship to the physics of reflection seismology.

Reflection seismology is a method of mapping the subsurface structure of the earth. A source at or near the surface is set off. The source may be a dynamite explosion, a weight-dropping machine, a vibration-inducing machine, or some other means to transmit seismic energy into the earth. The seismic waves travel from the source to subsurface layers within the earth where they are reflected. The returning seismic waves are then recorded by instruments on the surface. If the entire process is assumed to be linear, then the surface source is the excitation, the reflection seismogram is the output or set of observables, and the earth is the governing linear system.

Part 1. Transmission of a deep source through the earth

Consider an inhomogeneous system (the earth) bounded by (sandwiched in) two homogeneous infinite half-spaces, the air and the basement rock. Fig.1 depicts the situation. Now in earthquake seismology, the source is considered to be embedded deep within the earth's subsurface. The classical approach to determining the response of the earth to a deep source excitation was to solve the governing partial differential equations, which described the wave propagation through the earth into the air, subject to the initial-value and boundary-

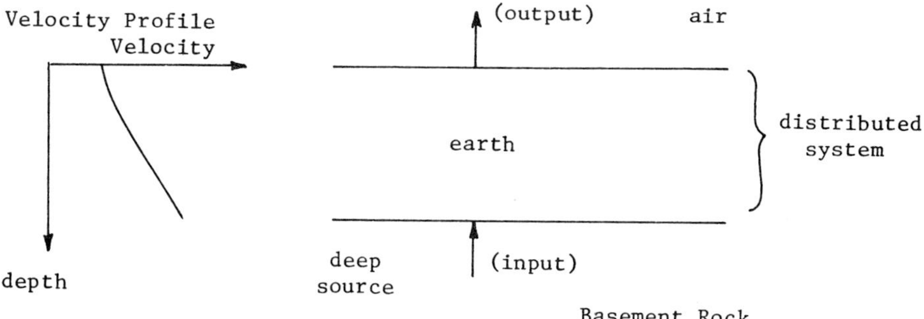

Fig.1. The inhomogeneous earth excited by a deep source and considered as a distributed parameter system. The associated velocity profile of this system is considered as a continuous function of depth.

value conditions. Hence, due to the presence of the governing partial differential equations, the earth was considered as a distributed-parameter system. Research in classical seismology is concerned with the analytic solutions of these partial differential equations, and except in the simplest cases, such solutions are mathematically extremely difficult to find.

Consider now the idea of modeling the earth by a lumped parameter system. For example, we could quantize the continuous velocity profile (distribution) shown in Fig.1. This quantization is equivalent to considering the earth as a multi-layered system, with the velocity of propagation in each layer and the depth of a layer determined by the quantization procedure. A description of the earth as a lumped-parameter system is given in Fig.2.

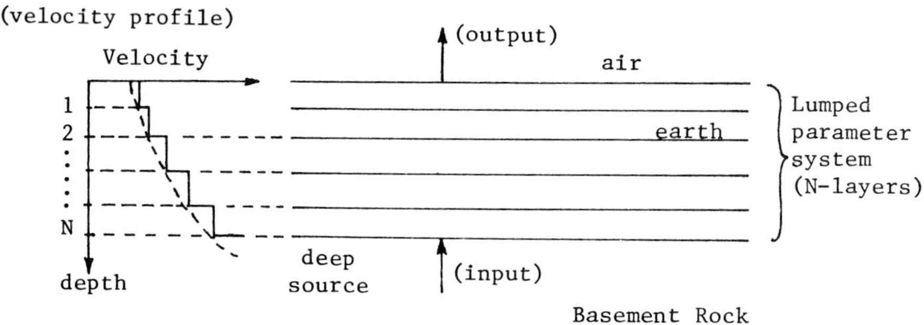

Fig.2. The multi-layered earth excited by a deep source and considered as a lumped-parameter system. The continuous velocity profile is now represented as a discontinuous profile.

Now if the time of signal propagation through a layer is short compared with the duration of the signal, then the lumped-parameter assumption is valid. By choosing the depth of each layer to be very small, i.e., considering many layers, we can satisfy the lumped-parameter conditions. In terms of transmission lines and linear network theory, Fig.1 represents a distributed circuit-element transmission line while Fig.2 represents a lumped circuit-element transmission line. Note that the model for the lumped-parameter system contains a discrete velocity profile distribution, with each discontinuity in the velocity representative of one transmission line section. Thus, the lumped-parameter system is modelled by many sections of transmission lines.

Now if we further assume that the system of Fig.2 is linear and time-invariant, we can expect that the input (ϵ_n) and output (x_n) satisfy the linear difference equation:

$$x_n + a_1 x_{n-1} + \ldots a_N x_{n-N} = \epsilon_n \tag{8}$$

where the a_n's represent the difference equation operator (i.e., constant parameters associated with our lumped-parameter system which has N degrees of freedom represented by the N layers). Our reason for considering a lumped-

parameter system with the difference eq.8 is that difference equations are
easily solved by digital computers. Thus, the difficulty of analytic solutions of
wave motion is eliminated by the numerical solution of eq.8.

On physical grounds, we know that eq.8 represents a stable system, i.e., for
every bounded input we observe a bounded output. Now given the sequences
x_n and ϵ_n, we shall define their respective z-transforms by:

$$X(z) = \sum_{n=-\infty}^{\infty} x_n z^n, \quad E(z) = \sum_{n=-\infty}^{\infty} \epsilon_n z^n \tag{9}$$

With the definition in eq.9, the z-transform of each side of eq.8 gives:

$$X(z)\,[1 + a_1 z + a_2 z^2 + \ldots + a_N z^N] = E(z)$$

or:

$$\frac{X(z)}{E(z)} = \frac{1}{1 + a_1 z + a_2 z^2 + \ldots + a_N z^N} \tag{10}$$

Eq.10 represents the transfer function of an "all-pole" linear time-invariant
system. Our physical claim that the lumped-parameter model represents a
stable system is equivalent to the mathematical condition that the transfer
function $X(z)/E(z)$ contains no poles inside the unit circle. Let us now formulate this property in terms of the difference equation operator $(1, a_1, a_2, \ldots,$
$a_N)$. The z-transform of this operator yields the denominator of eq.10. Thus,
the property that the transfer function $X(z)/E(z)$ has no poles inside the unit
circle is equivalent to the property that:

$$1 + a_1 z + a_2 z^2 + \ldots + a_N z^N \neq 0 \text{ for } |z| < 1 \tag{11}$$

We can therefore state that the operator $(1, a_1, a_2, \ldots, a_N)$ yields a stable difference equation if and only if its z-transform has no zeroes inside the unit
circle.

Now if we replace the input to our lumped-parameter system by a unit impulse, i.e., let $\epsilon_n = \delta_n$ where:

$$\delta_n = \begin{cases} 1 \,, n = 0 \\[2ex] 0 \,, n \neq 0 \end{cases}$$

then the impulse response, denoted by b_n, is a stable time sequence. This condition follows from physical reasoning. But since:

$$B(z) = \frac{1}{1 + a_1 z + a_2 z^2 + \ldots + a_N z^N} \tag{12}$$

it follows that $B(z)$ has no zeroes or poles inside the unit circle. Now b_n is the sequence which is transmitted through the earth into the air when a unit impulse (deep source) excites the earth. From actual observation of field seismic data, b_n has the bulk of its energy concentrated at its beginning and is rapidly attenuated due to successive reflections and refractions within the earth. Sequences that possess this "front-loaded" property are called *minimum-delay* sequences.

Now if $A(z) = 1 + a_1 z + a_2 z^2 + \ldots + a_N z^N$ is the z-transform of the operator $(1, a_1, a_2, \ldots, a_N)$, then from eq.12 we observe that:

$$B(z)A(z) = 1$$

or:

$$b_n \star a_n = \delta_n$$

which shows that the operator a_n and transmission impulse response b_n are mutually inverse sequences. But since a_n and b_n are mutually inverse sequences, then we can conclude that a_n is a stable difference equation operator if and only if b_n is minimum-delay. That is, from purely physical grounds, we arrive at the important result:

stable difference equation operator a_n $\Leftrightarrow$ minimum-delay response b_n

Because b_n represents all the multiple reflections and refractions resulting from an impulsive input δ_n, b_n also represents the system reverberation; b_n = transmission (impulse) response = layered system reverberation wavelet. Thus, a_n can also be viewed as the deconvolution operator which removes the layered system reverberation.

Part 2. Reflections caused by surface source (internal primary reflections)

Next, let us consider the N-layered (lumped-parameter) system of Fig.2 with the "deep" upgoing source replaced by a downgoing unit impulse (source) initiated at the surface. This situation, along with a pictorial definition of an internal primary reflection, is described in Fig.3.

Now in Part 1, we claimed that an upgoing unit impulse (source) incident on the lowest interface gave rise to the transmitted (i.e., system reverberation) waveform b_n. If we consider this situation in the context of linear time-invariant system theory, then the unit impulse δ_n is the *cause* and the system reverberation b_n is the *effect*. Now if the cause is delayed by m units of time, i.e., δ_{n-m}, then the effect is also delayed by m units of time, i.e., b_{n-m}. Also, if the cause is scaled by a constant C, i.e., $C\delta_n$, then the effect is also scaled by the constant C, i.e., Cb_n. These properties, together with the property of superposition, characterize linear time-invariant systems. Thus, instead of considering the downgoing unit impulse (source), let us replace it by a set of hypothetical (i.e., mathematical) upgoing sources at the reflecting interfaces of our N-layered system. Since we know the "effect" b_n due to the upgoing

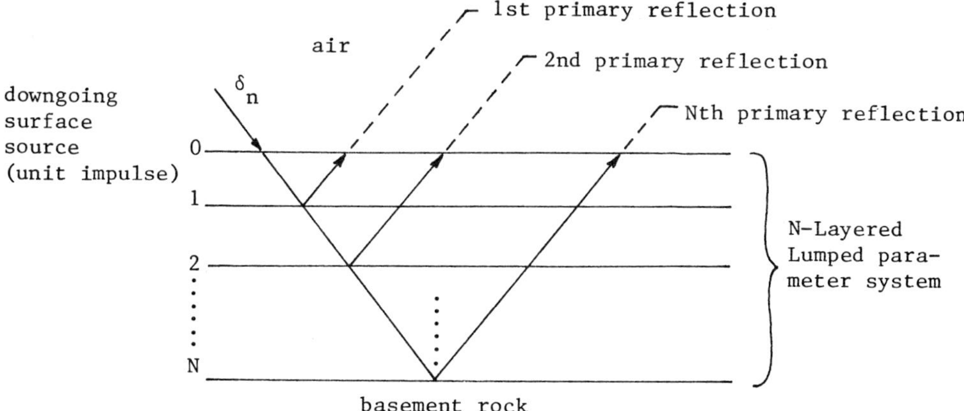

Fig.3. Internal primary reflections caused by a downgoing unit-impulse applied at the surface. For clarity, ray paths are drawn at oblique incidence but wave-motion analysis is for
normal incidence.

"cause" δ_n, we can then apply the linear time-invariant system properties to
deduce the effect due to a downgoing unit impulse at the surface. Fig.4 shows
the downgoing source impulse replaced by upgoing sources of strength $\epsilon_n = r_n$
where r_n is the reflection coefficient of interface n. From physical experience
gained in seismic prospecting, it is known that generally these reflection coef-
ficients are small in magnitude, that is, $|\epsilon_n| = |r_n| < 0.1$.

Each of these hypothetical sources gives rise to a reverberation wavelet due
to the layering. Further, to a first approximation, each reverberation wavelet
is the same shape, namely the system reverberation wavelet b_n. Silvia (1977)
examines this assumption in some detail and shows the mathematical conditions under which it is a good approximation to a more exact stratified earth
model. Taking into account the time delays associated with the source locations (see Fig.4), we reason as follows:

If an upgoing unit spike δ_n $\xrightarrow[\text{(gives rise to)}]{}$ transmission impulse-response b_n

then $\epsilon_1 \delta_{n-1}$ $\rightarrow$ $\epsilon_1 b_{n-1}$

$\epsilon_2 \delta_{n-2}$ $\rightarrow$ $\epsilon_2 b_{n-2}$

$\epsilon_N \delta_{n-N}$ $\rightarrow$ $\epsilon_N b_{n-N}$

Thus, using the superposition property of linear systems, we can add the
responses due to all the hypothetical sources to obtain the total response h_n.
Doing so, we get:

$$h_n = \epsilon_1 b_{n-1} + \epsilon_2 b_{n-2} + \ldots + \epsilon_N b_{n-N}$$

$$= \epsilon_n \star b_n \tag{13}$$

Eq.13 represents the reflection seismogram resulting from a downgoing unit impulse initiated at the surface. Further, h_n is the resultant of all the system reverberations (reverberation wavelets b_n) due to the upgoing hypothetical

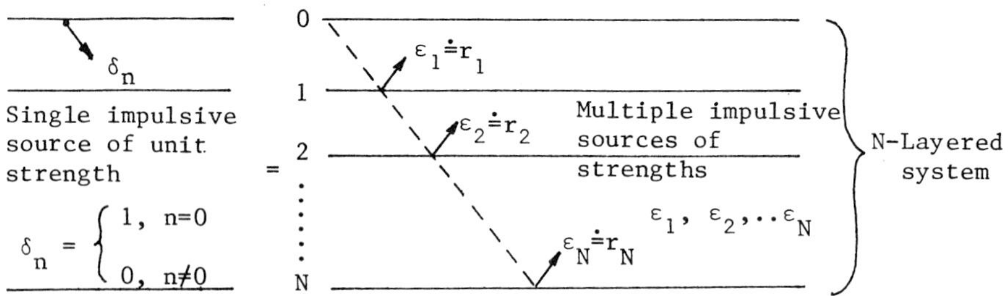

Fig.4. The downgoing unit impulse surface source δ_n replaced by hypothetical (i.e., mathematical) upgoing impulsive sources of strength $\epsilon_k \doteq r_k$.

sources ϵ_n with strength given by the reflection coefficient sequence r_n. The z-transform of eq.13 gives:

$$H(z) = E(z)B(z) \tag{14}$$

where:

$$E(z) = \epsilon_1 z + \epsilon_2 z^2 + \ldots + \epsilon_N z^N \tag{15}$$

Substitution of eqs.12 and 15 into 14 yields:

$$H(z) = \frac{\epsilon_1 z + \epsilon_2 z^2 + \ldots + \epsilon_N z^N}{1 + a_1 z + a_2 z^2 + \ldots + a_N z^N} \tag{16}$$

which gives the transfer function of a normal-incidence reflection seismogram and is the ARMA model used in reflection seismology. It is important to note that the same result can be obtained by deriving the exact expression for an N-layered earth model then making the approximation that all the reflection coefficients are small in magnitude, i.e., $|r_n| < 0.1$. Under this approximation, $\epsilon_1 \doteq r_1, \epsilon_2 \doteq r_2, \ldots, \epsilon_N \doteq r_N$; that is, the hypothetical sources are exactly the reflection coefficients (Silvia (1977)).

Since the layered system is assumed to be both linear and time-invariant, then the reflection seismogram x_n due to an arbitrary source pulse s_n is:

$$x_n = h_n \star s_n$$

$$= \epsilon_n \star b_n \star s_n = \epsilon_n \star (b_n \star s_n) \tag{17}$$

$$= \epsilon_n \star w_n$$

where $w_n = b_n \star s_n$ is defined as a composite wavelet consisting of the reverberation wavelet b_n and source pulse s_n. Thus, eq.17 describes the normal-

incidence reflection seismogram, where b_n represents the autoregressive component and $s_n{}^{\star}\epsilon_n$ the moving-average component of the reflection seismogram.
The basic deconvolution problem is to filter x_n such that we can best recover
the reflection coefficient sequence ϵ_n. Before we discuss the role of the
kepstrum in deconvolution, let us investigate some of its basic properties.

PROPERTIES OF THE KEPSTRUM

For a minimum-delay system $H_{\min}(z)$, the *kepstrum* f_n is defined as the coefficients of the Taylor-Maclaurin series:

$$\log H_{\min}(z) = F(z) = \sum_{n=0}^{\infty} f_n z^n \, , \, |z| < 1 \tag{18}$$

However, the concept of the kepstrum is not limited to minimum-delay
systems. Let us now consider a nonminimum-delay system $H(z)$. If we assume
that $H(z)$ has no zeroes or poles in an annular region including the unit circle,
then $\log H(z)$ is analytic in this annular region. As a result, $\log H(z)$ has a
Laurent expansion:

$$\log H(z) = \sum_{n=-\infty}^{\infty} f_n z^n \tag{19}$$

in this annular region. The coefficients for this Laurent series are the coefficients of a stable noncausal system. We call these coefficients the *kepstrum* of
the nonminimum-delay system $H(z)$.

Let us now return to the case of a minimum-delay system:

$$H_{\min}(z) = \sum_{n=0}^{\infty} h_n z^n \tag{20}$$

From eqs.18 and 20 we obtain:

$$H_{\min}(z) = \exp\left[F(z)\right]$$

or:

$$\sum_{n=0}^{\infty} h_n z^n = \exp\left[\sum_{n=0}^{\infty} f_n z^n\right] \tag{21}$$

A formula for obtaining the minimum-delay sequence h_n from the kepstrum
f_n can be obtained as follows. Taking the derivative of eq.21 with respect to z
gives:

$$\frac{\mathrm{d}}{\mathrm{d}z}\left(\sum_{n=0}^{\infty} h_n z^n\right) = \exp\left[\sum_{n=0}^{\infty} f_n z^n\right] \frac{\mathrm{d}}{\mathrm{d}z}\left(\sum_{n=0}^{\infty} f_n z^n\right) \tag{22}$$

Dividing eq.22 by eq.21 we obtain:

$$\frac{\dfrac{d}{dz}\left(\sum_{n=0}^{\infty} h_n z^n\right)}{\sum_{n=0}^{\infty} h_n z^n} = \frac{d}{dz}\left(\sum_{n=0}^{\infty} f_n z^n\right) \tag{23}$$

Expanding eq.23 we get:

$$\sum_{n=0}^{\infty} (n+1)h_{n+1}z^n = \left(\sum_{n=0}^{\infty} h_n z^n\right)\left(\sum_{n=0}^{\infty} (n+1)f_{n+1}z^n\right) \tag{24}$$

The inverse z-transform of eq.24 gives the recursive relation (Oppenheim and Schafer, 1975):

$$(n+1)h_{n+1} = \sum_{k=0}^{n} h_k(n+1-k)f_{n+1-k} \quad \text{for } n = 0, 1, 2, \ldots \tag{25}$$

which is useful for extracting the minimum-delay sequence h_n from the kepstrum f_n. For $n = 0, 1, 2, 3, \ldots$ eq.25 gives:

$$n = 0: \quad h_1 = h_0 f_1$$

$$n = 1: \quad h_2 = h_0 f_2 + \frac{h_1}{2}f_1 = h_0\left(f_2 + \frac{1}{2}f_1^2\right)$$

$$n = 2: \quad h_3 = h_0 f_3 + \frac{2}{3}h_1 f_2 + \frac{1}{3}h_2 f_1 = h_0\left(f_3 + f_1 f_2 + \frac{1}{6}f_1^3\right)$$

$$n = 3: \quad h_4 = h_0 f_4 + \frac{3}{4}h_1 f_3 + \frac{1}{2}h_2 f_2 + \frac{1}{4}h_2 f_1$$

$$= h_0\left(f_4 + f_3 f_1 + \frac{1}{2}f_1^2 f_2 + \frac{1}{2}f_2^2 + \frac{1}{24}f_1^3\right)$$

Let us now interpret the kepstral coefficient f_0. We have from eq.21 with $z = 0$ that:

$$h_0 = \exp(f_0) \text{ or } f_0 = \log h_0$$

Thus, it is often convenient to normalize the sequence h_n so that its leading coefficient is unity, i.e., $h_0 = 1$; thus, the leading kepstral coefficient is zero, i.e., $f_0 = 0$.

Let us now examine some of the properties of the kepstrum. At this point it is convenient to introduce the symbols a_n and b_n for the time sequences and the corresponding Greek symbols α_n and β_n for their respective kepstra. Thus, the kepstrum of a minimum-delay sequence a_n with leading coefficient $a_0 = 1$ is the causal sequence $\alpha_1, \alpha_2, \alpha_3, \ldots$ with leading term $\alpha_0 = 0$. If we let $L(a_n)$ denote the z-transform of a_n, then the kepstrum can be described by the composition of transformations L, log, and L^{-1} given by:

$$\alpha_n = L^{-1}\{\log[L(a_n)]\} \tag{26}$$

Hence, to find the kepstrum of a minimum-delay sequence h_n whose leading coefficient is unity, we compute the z-transform of the sequence, take the natural logarithm of this transform, and find the inverse z-transform of the quantity $\log H(z)$.

For convenience, we shall define the kepstral operator K given by the successive transformations L, log and L^{-1} such that:

$$K(\cdot) = L^{-1}\{\log[L(\cdot)]\} \tag{27}$$

where $K(h_n) = f_n$ and $h_0 = 1$

Let us now consider various properties of the kepstrum, i.e., convolution, reverse sequence, inverse sequence, and reverse-inverse sequence.

Convolution. Given the minimum-delay sequences a_n and b_n whose leading coefficients are unity, then:

$$K(a_n \star b_n) = K(a_n) + K(b_n) \tag{28}$$

where:

$$K(a_n) = 0, \alpha_1, \alpha_2, \alpha_3, \ldots$$

$$K(b_n) = 0, \beta_1, \beta_2, \beta_3, \ldots$$

Hence, the kepstrum maps convolution into addition. To see this, we write:

$$L(a_n \star b_n) = A(z)B(z)$$

Then:

$$\log[L(a_n \star b_n)] = \log[A(z)B(z)]$$
$$= \log A(z) + \log B(z)$$

Taking the inverse z-transform of the above result we get:

$$L^{-1}\{\log[L(a_n \star b_n)]\} = L^{-1}[\log A(z)] + L^{-1}[\log B(z)]$$

or:

$$K(a_n \star b_n) = K(a_n) + K(b_n)$$

Reverse sequence. Given the minimum-delay sequence b_n with leading coefficient $b_0 = 1$, we form the reverse sequence b_{-n} by folding b_n about the origin. The resulting sequence is called *minimum-advance*. Hence:

$$K(b_{-n}) = \beta_{-n} \tag{29}$$

We can show this result by first noting that:

$$L(b_n) = \sum_{n=0}^{\infty} b_n z^n = B(z)$$

and:

$$L(b_{-n}) = \sum_{n=0}^{\infty} b_n z^{-n} = B(z^{-1})$$

Recall that:

$$\log B(z) = \beta_1 z + \beta_2 z^2 + \beta_3 z^3 + \ldots$$

Then:

$$\log B(z^{-1}) = \beta_1 z^{-1} + \beta_2 z^{-2} + \beta_3 z^{-3} + \ldots$$

and we observe that given $K(b_n) = \beta_n$, $K(b_{-n})$ is found by folding β_n about the origin. In words:

minimum-delay sequence $\leftrightarrow$ *causal* kepstrum

minimum-advance sequence $\leftrightarrow$ *purely noncausal* kepstrum

Inverse sequence. Given the minimum-delay sequence b_n with leading coefficient $b_0 = 1$, we define the inverse sequence b_n^{-1} by the relation:

$$b_n \star b_n^{-1} = \delta_n = \begin{cases} 1, & n = 0 \\ 0, & n \neq 0 \end{cases}$$

Then:

$$K(b_n^{-1}) = -\beta_n \tag{30}$$

Since $L(b_n \star b_n^{-1}) = 1$, then $B(z) L(b_n^{-1}) = 1$

or:

$$L(b_n^{-1}) = \frac{1}{B(z)} = [B(z)]^{-1}$$

Now,

$$\log [L(b_n^{-1})] = \log [B(z)]^{-1} = -\log B(z)$$

Thus, it follows that:

$$K(b_n^{-1}) = -\beta_n$$

In words:

minimum-delay sequence $\leftrightarrow$ *causal* kepstrum

inverse of minimum-delay sequence $\leftrightarrow$ *negative* of causal kepstrum

Reverse-inverse sequence. Given the minimum-delay sequence b_n with leading coefficient $b_0 = 1$, we define the sequence b_{-n}^{-1} as the inverse of the reverse sequence. Then:

$$K(b_{-n}^{-1}) = -\beta_{-n} \tag{31}$$

To see this, we write:

$$L(b_{-n}^{-1}) = [B(z^{-1})]^{-1}$$

Then:

$$\log\,[L(b_{-n}^{-1})] = \log\,[B(z^{-1})]^{-1} = -\log B(z^{-1})$$

From the reverse-sequence property we can state:

$$K(b_{-n}^{-1}) = -\beta_{-n}$$

In words:
minimum-delay sequence $\leftrightarrow$ *causal* kepstrum
reverse of inverse of minimum-delay sequence $\leftrightarrow$ *negative* of purely noncausal
   kepstrum
   The above properties bring out the essential features and symmetries of the kepstrum. Thus, given the kepstrum β_n of a minimum-delay sequence b_n with leading coefficient $b_0 = 1$, we can find the kepstrum of the reverse sequence b_{-n}, inverse sequence b_n^{-1} and inverse-reverse sequence b_{-n}^{-1} by simple negations and reflections of the kepstrum β_n. For additional kepstral properties, see Oppenheim and Schafer (1975) and Tribolet (1977).
   In the above discussion, we developed the relation (eq.24):

$$\sum_{n=0}^{\infty} (n+1)h_{n+1}z^n = \left(\sum_{n=0}^{\infty} h_n z^n\right)\left(\sum_{n=0}^{\infty} (n+1)f_{n+1}z^n\right)$$

which can be written as:

$$\sum_{n=0}^{\infty} (n+1)f_{n+1}z^n = \frac{1}{H(z)}\sum_{n=0}^{\infty} (n+1)h_{n+1}z^n \tag{32}$$

where $H(z) = \sum_{n=0}^{\infty} h_n z^n$ and $h_0 = 1$

The inverse z-transform of eq.32 gives:

$$K(h_n) = f_{n+1} = \sum_{k=0}^{n} \left(\frac{k+1}{n+1} \right) h_{k+1} h_{n-k}^{-1} \tag{33}$$

for $n = 0, 1, 2, \ldots$ and $h_0 = 1$

which *explicitly* gives the kepstrum f_n in terms of the minimum-delay sequence h_n.

Let us now relate the mathematical properties of the kepstrum to the physics of the normal-incidence seismogram by next discussing the topic of deconvolution.

THE ROLE OF THE KEPSTRUM IN DECONVOLUTION

Recall that the reflection seismogram x_n can be approximately regarded as the output of an ARMA model with transfer function:

$$H(z) = \frac{E(z)}{A(z)} = \frac{\epsilon_1 z + \epsilon_2 z^2 + \ldots + \epsilon_N z^N}{1 + a_1 z + a_2 z^2 + \ldots + a_N z^N}$$

where ϵ_n = the reflection coefficient sequence and is the desired output of a deconvolution filter. Unfortunately, this system is not always minimum-delay, since the numerator polynomial coefficients $\epsilon_1, \epsilon_2, \ldots, \epsilon_N$ are the reflection coefficients of an N-layered system and are dependent upon a particular geographic area. Thus, we are not always sure if all the zeroes of $E(z)$ lie outside the unit circle, and in practice, the reflection coefficient sequence is generally mixed-delay, i.e., it contains zeroes both inside and outside the unit circle. However, the reverberation sequence b_n, where $B(z) = 1/A(z)$ is always minimum-delay. If we excite the earth with an arbitrary mixed-delay source pulse s_n, then the reflection seismogram x_n is the convolution:

$$x_n = s_n \star \epsilon_n \star b_n \tag{34}$$

$$= \epsilon_n \star w_n$$

where $w_n = s_n \star b_n$ is the composite mixed-delay wavelet.

The method of predictive deconvolution (Robinson (1954)) assumes that the reflection coefficient sequence is a purely random sequence with a flat power spectrum. This assumption is usually met in practice (White and O'Brien, 1974). Such an assumption allows the method of predictive deconvolution to design least-squares deconvolution (noise-whitening) filters.

From the convolution property of the last section, the kepstrum $K(x_n)$ of the reflection seismogram is:

$$K(x_n) = K(\epsilon_n \star s_n \star b_n)$$

$$= \hat{\epsilon}_n + \sigma_n + \beta_n \tag{35}$$

where $K(\epsilon_n) = \hat{\epsilon}_n$, $K(s_n) = \sigma_n$ and $K(b_n) = \beta_n$. The method of homomorphic deconvolution is concerned with processing the kepstrum $K(x_n)$ so that σ_n (source pulse kepstrum) and β_n (reverberation kepstrum) may be eliminated, leaving the desired kepstrum $\hat{\epsilon}_n$. Applying the inverse kepstral operator K^{-1} to the "filtered" kepstrum $\hat{\epsilon}_n$ would then yield the desired reflection coefficient sequence.

It appears that deconvolution via the kepstrum is less restrictive than the method of predictive deconvolution, where one assumes that the reflection coefficients are an uncorrelated sequence and that the source pulse be minimum-delay in order to achieve "optimum" results. Further, the kepstrum appears to be a more robust processor in the sense that the technique is not tied to any physical attributes of the model and that a minimum-delay source pulse assumption is not needed.

However, both the method of predictive deconvolution and the method of kepstral deconvolution require an assumption of the phase spectrum of the source pulse (Treitel and Robinson, 1977). For example, let the canonical representation of the source pulse be:

$$s_n = p_n \star s_n^{(0)} \tag{36}$$

where p_n is a causal all-pass operator, $s_n^{(0)}$ is the minimum-delay source wavelet, and s_n is the nonminimum-delay source wavelet. The reflection seismogram for a nonminimum-delay source pulse can then be expressed as:

$$x_n = \epsilon_n \star b_n \star p_n \star s_n^{(0)} \tag{37}$$

If we knew the all-pass system p_n, i.e., if we had knowledge of the source phase characteristic, then we could operate on the seismogram with the inverse all-pass p_n^{-1} to yield:

$$p_n^{-1} \star x_n = \epsilon_n \star b_n \star s_n^{(0)} \tag{38}$$

which is more suitable to the method of predictive deconvolution.

Often, the source pulse kepstrum σ_n is clustered around the kepstral origin (also called "low-time" portion of the kepstrum). See Stoffa et al. (1974) and Ulrych (1971). By a suitable window with the proper cutoff "quefrencies", the source pulse can be theoretically recovered. In essence, the selection of the cutoff quefrencies determines how well we can estimate the source pulse phase and magnitude characteristics. Thus, both deconvolution techniques require some knowledge of the phase characteristic of the source pulse.

CONCLUSIONS

The idea of the kepstrum appeared in the classical works of Poisson and Schwarz concerning the logarithmic potential problem. Szegö and Kolmogorov used the kepstrum to factor the power spectrum of a non-deterministic process.

Because the normal-incidence reflection seismogram is approximated by a convolutional process, the kepstrum offers a solution to the deconvolution of reflection seismograms. We developed additional properties of the kepstrum which bring out its essential features and symmetries. As a result, the kepstrum evolves as an important mathematical construct in the theory of signal analysis and time series, and takes the place among such established functions as the spectrum and autocorrelation.

REFERENCES

Bateman, H., 1944. Partial Differential Equations of Mathematical Physics. Dover, New York.
Båth, M., 1968. Mathematical Aspects of Seismology. Elsevier, Amsterdam.
Båth, M., 1974. Spectral Analysis in Geophysics. Elsevier, Amsterdam.
Bogert, B.P., Healy, M.J.R. and Tukey, J.W., 1963. The quefrency analysis of time series for echoes. In: M. Rosenblatt (Editor), Proc. Symp. Time Series Analysis. Wiley, New York, pp.209—243.
Kolmogorov, A.N., 1939. Sur l'interpolation et extrapolation des suites stationnaires. C.R. Acad. Sci. Paris, 208: 2043—2045.
Kulhánek, O., 1975. Introduction to Digital Filtering in Geophysics. Elsevier, Amsterdam.
Oppenheim, A.V. and Schafer, R.W., 1975. Digital Signal Processing. Prentice-Hall, Englewood Cliffs, N.J.
Poisson, S.D., 1823. Sur la distribution de la chaleur dans les corps solides. J. Ec. R. Polytech., Ser.I, 19: 1—162.
Robinson, E.A., 1954. Predictive Decomposition of Time Series with Applications to Seismic Exploration. Thesis, MIT, Cambridge, Mass. (Also in Geophysics, 32: 418—484.)
Schafer, R.W., 1969. Echo removal by discrete generalized linear filtering. Res. Lab. Electron. MIT, Tech. Rep., 466.
Schwarz, H.A., 1872. Zur Integration der partiellen Differentialgleichung. J. Reine Angewandte Math., pp.218—254.
Silvia, M.T., 1977. Deconvolution of Geophysical Time Series. Thesis, Northeastern University, Boston, Mass.
Stoffa, P.L., Buhl, P. and Bryan, G.M., 1974. The application of homomorphic deconvolution to shallow-water marine seismology — Part I. Models. Geophysics, 39: 401—416.
Szegö, G., 1915. Ein Grenzwertsatz über die Toeplitzschen Determinanten einer reellen positiven Funktion. Math. Ann., 76: 490—503.
Treitel, S. and Robinson, E.A., 1977. Deconvolution — homomorphic or predictive? IEEE Trans. Geosc. Electron., GE-15(1): 11—13.
Tribolet, J., 1977. Seismic Applications of Homomorphic Signal Processing. Thesis, Dept. of Electr. Eng. Comput. Sci., MIT, Cambridge, Mass.
Ulrych, T.J., 1971. Application of homomorphic deconvolution to seismology. Geophysics, 36: 650—660.
White, R.E. and O'Brien, P.N.S., 1974. Estimation of the primary seismic pulse. Geophys. Prosp., 22: 627—651.

CHAPTER 20

Geoexploration, 16 (1978) 1—19
© Elsevier Scientific Publishing Company, Amsterdam — Printed in The Netherlands

ITERATIVE IDENTIFICATION OF NON-INVERTIBLE AUTOREGRESSIVE MOVING-AVERAGE SYSTEMS WITH SEISMIC APPLICATIONS

ENDERS A. ROBINSON

Department of Electrical Engineering, Northeastern University, Boston, Mass. (U.S.A.)

(Received September 28, 1977)

ABSTRACT

Robinson, E.A., 1978. Iterative identification of non-invertible autoregressive moving-average systems with seismic applications. In: C.H. Chen (Editor), Computer-Aided Seismic Analysis and Discrimination. Geoexploration, 16: 1—19.

Let A be an invertible (i.e. minimum-delay) polynomial of degree m and let E be any polynomial of degree n. Then $S = E/A$ represents the transfer function of an autoregressive moving-average (ARMA) system. The numerator E may also be invertible, in which case the ARMA system S is invertible. In seismic exploration the sequence with z-transform given by S represents a model of an observable seismogram, where E is the z-transform of the reflection coefficient sequence of the unknown (and desired) deep reflecting interfaces and where A represents the (invertible) z-transform of the unknown inverse reverberation filter due to the surface layers. Because there is no physical reason for the reflection coefficient sequence to be invertible, the seismic model $S = E/A$ will in general be a non-invertible ARMA system. Given such a non-invertible sequence with z-transform S, the problem is to estimate E and A under the restriction that A is invertible. We give an iterative identification scheme based on the following algorithm. Initially we let the initial estimate $E^{(o)}$ be unity. We then apply an iterative algorithm which for the kth step has inputs S and $E^{(k-1)}$ and outputs $A^{(k)}$ and $E^{(k)}$, where $A^{(k)}$ is required to be invertible. Iteration is terminated when S and $E^{(k)}/A^{(k)}$ are approximately the same, and then $E^{(k)}$ and $A^{(k)}$ are the required estimates of E and A, respectively.

INVERTIBILITY AND MINIMUM-DELAY

In most physical applications time is a continuous parameter. However, observations are often made at discrete time intervals, and even when continuous observations are made, they are often digitized for computer processing. As a result, the mathematical principles of discrete-time processes are important. Also, certain concepts that are mathematically quite difficult in continuous time have counterparts in discrete time that are much easier to handle.

One such concept is that of invertibility. In discrete time the general principle of invertibility may be grasped with use of time sequences of very short time duration. We want to give some simple examples before presenting the general theory. We denote a time sequence as b_k where the subscript k is the

513

time parameter. Time is given by $t = k \Delta t$ where Δt is the digitization interval. When the time sequence has finite time duration, we may write it as a vector; for example, $b = (b_0, b_1, \ldots, b_n)$. A time sequence may represent the impulse response function of a filter. A filter is said to be stable provided that its impulse response function b has finite energy; that is, provided that:

$$\sum_{-\infty}^{\infty} b_k^2 < \infty$$

A filter is said to be *causal* provided that its impulse response function is one-sided; that is, provided that $b_k = 0$ for $k < 0$. The z-transform (according to the Laplace definition) of a stable causal time sequence b_k is defined as:

$$B(z) = \sum_{k=0}^{\infty} b_k z^k$$

Let us now give some introductory examples. Suppose that we have an input sequence $b = (b_0, b_1)$ of length two, a filter $f = (f_0, f_1)$ of length two, and the resulting output $c = (c_0, c_1, c_2)$ of length necessarily three. The output is determined from the input and filter by convolution:

$$c = (c_0, c_1, c_2) = (b_0 f_0, \; b_0 f_1 + b_1 f_0, \; b_1 f_1)$$

In order to lead into the concept of invertibility we first want to consider the design of spiking filters. A spiking filter is a filter designed so that the output c is as close as possible to a spike; for example, the zero-delay spike $(1,0,0)$, or the unit-delay spike $(0,0,1)$. or the two-delay spike $(0,0,1)$. We say that the two-term filter condenses or compresses the two-term input into a spike output. Actually a spiking filter is a special case of a wave-shaper, in which one chooses (f_0, f_1) so that c comes out as closely as possible to some desired output (d_0, d_1, d_2) which is prescribed. We use the least-squares criterion in making c as close as possible to d, so we minimize the sum of squares of the difference between them; namely:

$$J = (d_0 - c_0)^2 + (d_1 - c_1)^2 + (d_2 - c_2)^2$$
$$= (d_0 - f_0 b_0)^2 + (d_1 - f_0 b_1 - f_1 b_0)^2 + (d_2 - f_1 b_1)^2$$

If we set the partial derivative of J with respect to f_0 and the partial derivative with respect to f_1 each equal to zero, we get the set of simultaneous equations:

$$(b_0{}^2 + b_1{}^2)f_0 + (b_1 b_0)f_1 = b_0 d_0 + b_1 d_1$$
$$(b_1 b_0)f_0 + (b_0{}^2 + b_1{}^2)f_1 = b_0 d_1 + b_1 d_2$$

which we may solve in order to determine the filter coefficients f_0 and f_1. In the particular case when the desired output is a zero-delay spike $(d_0, d_1, d_2) = (1,0,0)$, the corresponding filter is called the zero-delay spiking filter for the input (b_0, b_1, b_2). The normal equations for the zero-delay spiking filter (f_0, f_1) are:

$$(b_0^2 + b_1^2) f_0 + b_1 b_0 f_1 = b_0$$

$$b_1 b_0 f_0 + (b_0^2 + b_1^2) f_1 = 0$$

with solution (f_0, f_1), except for a constant factor, given by:

$$(b_0^2 + b_1^2, -b_0 b_1)$$

We see that the z-transform (Laplace definition):

$$(b_0^2 + b_1^2) - b_0 b_1 z$$

has a zero $z = (b_0^2 + b_1^2)/b_0 b_1$ outside the unit circle, and has no zeros inside the unit circle. In fact, we will show that all zero-delay spiking filters have z-transforms (Laplace definition) with no zeros inside the unit circle: that is, we will show that all zero-delay spiking filters are minimum-phase.

Let us now give a numerical example. Suppose that the input sequence is $b = (1, 0.5)$. The zero-delay spiking filter is then $f = (0.95, -0.38)$. The actual output is $c = (0.95, 0.10, -0.19)$ which we compare to the desired output $d = (1, 0, 0)$. This result shows how well a two-term filter can condense the two-term input $(1, 0.5)$ into a spike.

Now we would like to find the equations to determine an $(m + 1)$-term filter that shapes an $(n + 1)$-term input into a prescribed desired output. This problem represents the case of a *wave-shaping filter*, which can be specialized to the case of a spiking filter by letting the desired output be a spike. Let the input sequence be $b = (b_0, b_1, \ldots, b_n)$. We want to construct the shaping filter $f = (f_0, f_1, \ldots, f_m)$. When the input goes into the filter the output is produced according to the convolution equation:

$$c_k = \sum_j f_j b_{k-j}$$

In this summation we can consider the limits of summation as minus infinity to plus infinity provided that we consider any terms off-the-ends of the finite sequences as zero. We now introduce the desired output sequence d which has the same number of terms $(m + n + 1)$ as the output sequence c. The shaping filter f is chosen so that c and d are alike (in the least-squares sense). That is, we minimize the sum of squared errors given by:

$$J = \sum_k (d_k - c_k)^2 = \sum_k \left(d_k - \sum_j f_j b_{k-j}\right)^2$$

Setting $\partial J/\partial f_i = 0$ for $i = 0, 1, 2, \ldots, m$ we obtain the set of equations, called the *normal equations*, given by:

$$\sum_{j=0}^{m} f_j r_{i-j} = g_i \qquad (i = 0, 1, 2, \ldots, m)$$

where:

$$r_{i-j} = \sum_k b_{k-j} b_{k-i}, \quad g_i = \sum_k d_k b_{k-i}$$

are the transient autocorrelation of b and the transient cross-correlation of d and b, respectively.

The minimum value of J (i.e., the minimum sum of squared errors) reduces to:

$$J = \sum_k (d_k - c_k)^2 = \sum d_k^2 - \sum f_j g_j$$

We will have occasion to use this formula for the minimum value of J shortly.

Let us now examine the normal equations given above. We notice that the left-hand side of the normal equations depends only upon the autocorrelation of the input b and not on b itself. If the desired output is the zero-delay spike $d = (1,0,0, \ldots ,0)$, then the right-hand side becomes:

$$g_0 = b_0, \, g_1 = 0, \, g_2 = 0, \ldots, g_m = 0$$

Thus, in the case of a zero-delay spiking filter, the wave-shape of the input sequence b itself does not enter into the normal equations, except for the initial value b_0, which affects the filter as only a scale factor. Therefore, the normal equations in the case of a zero-delay spiking filter depend upon the input sequence only through its autocorrelation. Any two input sequences with the same autocorrelation would have the same zero-delay spiking filter (except for a constant factor). Because autocorrelations contain no phase information, it is an interesting problem as to what the phase spectrum of the filter is.

Let us now show that the phase spectrum of the zero-delay spiking filter is minimum-negative-phase (i.e., minimum-phase-lag). Using the above formula for the minimum value of J, we see that the minimum value of J for the zero-delay spiking filter is:

$$J = \sum_k d_k^2 - \sum_j f_j g_j = 1 - f_0 g_0 = 1 - f_0 b_0$$

In order for $1 - f_0 b_0$ to indeed be a minimum, we see that f_0 must have the same sign as b_0. Moreover, f_0 must be as large as possible in magnitude. This requirement means that the filter must have as much of its energy concentrated in its leading coefficient as possible, or in other words, the filter must be minimum-delay. Thus, we have come to the important conclusion that a zero-delay spiking filter is minimum-delay. Because the concepts of minimum-phase and minimum-delay are identical, the zero-delay spiking filter has a minimum-phase-lag spectrum.

Now let us look at the other side of the coin. In order for $1 - f_0 b_0$ to be small when we consider all input sequences b with the same transient autocorrelation, the leading coefficient b_0 should be large in magnitude. In other words, the smallest mean-square spiking error occurs when the input sequence

b is minimum-delay. In summary, all zero-delay spiking filters f are necessarily minimum-delay, and the best case of spiking occurs when the input sequence b is also minimum-delay. In fact, for a minimum-delay input sequence $b = (b_0, b_1', \ldots, b_n)$ the mean-square spiking error $J = 1 - f_0 b_0$ tends to zero as the number of coefficients in the zero-delay spiking filter $(f_0, f_1, \ldots, f_m)$ tends to infinity ($m \to \infty$). On the other hand, for a non-minimum-delay input sequence, the mean-square spiking error does not tend to zero as the number of coefficients in the zero-delay spiking filter tends to infinity. We, therefore, conclude that perfect spiking can be achieved by a stable causal spiking filter if and only if the input sequence is minimum-delay. In other words, given the sequence b we can find a stable causal filter f such that:

$$f \star b = (1,0,0,\ldots)$$

if and only if b is minimum-delay. Any stable causal sequence b that can be converted into a zero-delay spike by a stable causal filter f is said to be invertible. We have thus established that a stable causal sequence is invertible if and only if it is minimum-delay. For that reason, we can use the terms "invertible sequence" and "minimum-delay sequence" interchangeably, even as we use the terms "minimum-phase sequence" and "minimum-delay sequence" interchangeably.

In summary, the zero-delay spiking filter for a stable causal sequence b as input may be described as the *least-squares inverse* of b. In case b is minimum-delay, then if the number of terms in the least-squares inverse is allowed to increase, the least-squares inverse tends to the exact stable causal inverse of b.

CONVERSION TO INVERTIBILITY

In many applications we know from the physics of the situation that a certain finite-length causal sequence must be invertible. However, suppose that the estimation method used does not necessarily give an invertible sequence. Thus we must have a method for converting a sequence which is not necessarily invertible into an invertible sequence. Let us describe one such method. This method consists of the following two steps:

(1) Compute the zero-delay spiking filter a^{-1} of the given filter a. As we have seen in the preceding section, the filter a^{-1} is necessarily invertible.

(2) Now take this approach one step further; namely, compute the zero-delay spiking filter of the filter a^{-1}. This second zero-delay spiking filter is necessarily invertible, and in fact it is the required invertible counterpart of the given filter a.

In summary, the invertible counterpart of the finite-length causal sequence a is the least-squares inverse of the least-squares inverse of a.

SEISMIC APPLICATIONS

Let us consider the problem of the elimination of water reverberations
from seismic records. The water reverberation problem in marine seismic
operations may be described as follows. The water-air interface is a strong
reflector, with reflection coefficient nearly -1. If the water-bottom interface is also a strong reflector, then the water layer represents a non-attenuating medium bounded by two strong reflecting interfaces and hence represents an energy trap. A seismic pulse generated in this energy trap will be
successively reflected between the two interfaces. Consequently, reflections
from deep horizons below the water layer will be obscured by the water
reverberations (Fig.1).

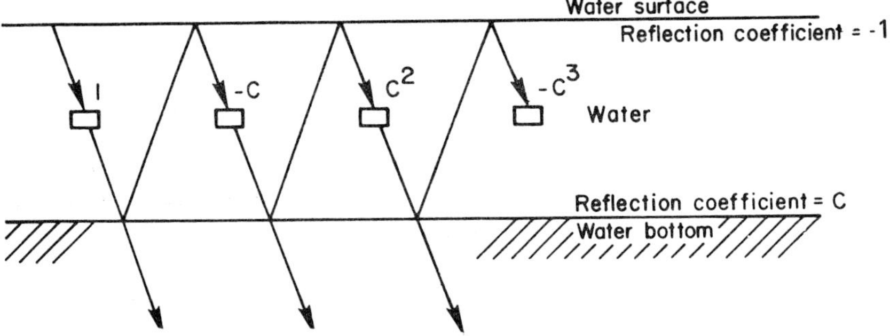

Fig.1. Water-confined reverberation. For clarity the raypaths have been drawn as slanting
lines, although in our model they are perpendicular to the two interfaces.

Let us consider a down-traveling impulsive plane wave (i.e., we assume
that the source wavelet is a unit spike) and let us suppose that we have a
transducer which measures only down-traveling motion in the water layer.
The signal received by the transducer represents the successive reflections
from the air/water interface. We let T represent the two-way traveltime in
the water; for example, if the water depth is 100 ft. and the water velocity
is 5,000 ft./sec, then the two-way traveltime would be $T = 40$ millisec.
Referring to Fig.1, we see that the coefficient 1 occurs at time 0 and
represents the initial downgoing spike. The coefficient $-c$ occurs at time T
and represents the second downgoing spike, which has suffered a reflection
at the bottom (reflection coefficient c) and a reflection at the surface (reflection coefficient -1). The coefficient c^2 occurs at discrete time $2T$ and
represents the third downgoing spike, which has suffered two reflections at
the bottom (c^2) and two reflections at the surface ($-1)^2$. The coefficient $-c^3$
occurs at discrete time $3T$ and represents the fourth downgoing spike,
which has suffered three reflections at the bottom (c^3) and three reflections
at the surface ($-1)^3$, etc. Thus the *water-confined reverberation spike-train* is
of the form:

$$1 \quad -c \quad c^2 \quad -c^3 \quad c^4 \quad \ldots \tag{1}$$

time: $0 \quad T \quad 2T \quad 3T \quad 4T \ldots$

The reflection coefficient c cannot exceed unity in magnitude; generally c will be less than unity in magnitude for otherwise the water layer would represent a perfect energy trap and no energy would penetrate to the deeper layers. Thus in the case where $|c| < 1$, the water-confined reverberation spike-train (1) represents a convergent sequence. The z-transform (Laplace definition) of the water-confined reverberation spike-train is:

$$1 - cz^T + c^2 z^{2T} - c^3 z^{3T} + \ldots$$

which is in the form of a geometric series, which can be summed to give:

$$\frac{1}{1 + cz^T}$$

The z-transform of the water-confined reverberation elimination filter would therefore have the form:

$$1 + cz^T$$

so that the impulse response function would be:

$$1 \quad c \tag{2}$$

time: $0 \quad T$

To verify that the water-confined reverberation elimination filter (2) does indeed convert the water-confined reverberation spike-train (1) into a spike (and thereby eliminates the reverberation) we may convolve (1) with (2) and thereby obtain the spike $(1,0,0,0,\ldots)$ which represents the impulsive plane wave without the water reverberation.

In our analysis so far we have only considered the water reverberation. effect itself, and have not considered the effect of the water layer on deep subsurface reflections. To do so, we must take into account that the water layer acts as a filter, and the seismic energy passes through this filter once as it goes down to the deep reflecting horizon and again as it returns to the surface. That is, the water layer acts on the deep reflection data twice, once going down to the deep strata and once again coming back to the surface. Hence we have, in effect, a situation where the deep reflection data pass through two cascaded sections of the water layer, so that the overall z-transform of the filtering introduced by the water layer is the square of the z-transform due to a single pass, as given in the foregoing paragraph (Fig. 2).

The z-transform of the reverberation spike-train resulting from a deep reflecting horizon is:

$$(1 - cz^T + c^2 z^{2T} - c^3 z^{3T} + \ldots)^2$$

or:

$$\frac{1}{(1 + cz^T)^2}$$

In the time domain, we see that the *deep-reflection reverberation spike-train* is given by:

$$
\begin{array}{cccccc}
1 & -2c & 3c^2 & -4c^3 & 5c^4 & \ldots \\
\text{time: } 0 & T & 2T & 3T & 4T & \ldots
\end{array}
\tag{4}
$$

Expression (4) may be obtained by convolving the water-confined reverberation spike-train (1) with itself. The reverberation spike-train (4) may be interpreted as being made up of the deep reflection spike at time 0, and the

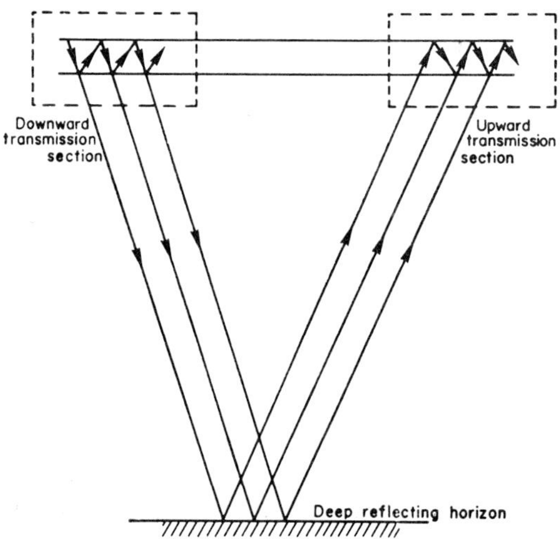

Fig.2. Deep-reflection reverberation-producing filter as two cascaded sections.

attached reverberation spikes at times T, $2T$, $3T$, $4T$, . . . To eliminate the reverberation part of spike-train (4) we must pass spike-train (4) through the inverse filter whose z-transform is given by:

$$(1 + cz^T)^2$$

or:

$$1 + 2cz^T + c^2 z^{2T}.$$

The impulse response function of this inverse filter is:

$$
\begin{array}{ccc}
1 & 2c & c^2 \\
\text{time: } 0 & T & 2T
\end{array}
\tag{5}
$$

Convolving spike-train (4) with spike-train (5) we see that the output is indeed a spike $(1,0,0,0, . . .)$ which represents the direct deep reflection spike (at time 0) with the attached reverberation spikes eliminated.

A primary reflection on the seismic record is an event resulting from the transmission of the source pulse directly down to a deep reflecting interface and then directly back up to the receiver. Neglecting everything except the reflection coefficient ϵ of the deep reflecting horizon, the primary reflection due to a unit spike source would appear as a spike ϵ at time t on the seismogram, where t is the two-way travel time from the surface to the deep reflecting interface. The z-transform of this primary event would be ϵz^t. Suppose now that there are $n + 1$ deep interfaces; the z-transform of the resulting primary events would be:

$$\epsilon_0 z^{t_0} + \epsilon_1 z^{t_1} + \ldots + \epsilon_n z^{t_n} \tag{6}$$

where $\epsilon_0, \epsilon_1, \ldots, \epsilon_n$ are the reflection coefficients and $t_0, t_1, \ldots, t_n$ are the two-way times of the interfaces, respectively.

As we have seen, the water layer acts on the deep reflection data twice, once in going down and once again in coming back up; so the overall z-transform of the filtering introduced by the water layer is given by expression (3). The z-transform of the seismogram resulting from the $n + 1$ deep interfaces is the product of expressions (3) and (6); namely, the rational function:

$$S(z) = \frac{\epsilon_0 z^{t_0} + \epsilon_1 z^{t_1} + \ldots + \epsilon_n z^{t_n}}{(1 + cz^T)^2} \tag{7}$$

Because of the physical nature of reverberations, the denominator of this rational function is necessarily invertible. However, there is no physical reason for the numerator to be invertible, except in special cases.

Eq. 7 represents the model for a seismogram in the case of a single surface layer. In the case of more complicated surface-layering, the seismogram model would take the form of:

$$S(z) = \frac{E(z)}{A(z)} \tag{8}$$

where $E(z) = \epsilon_0 z^{t_0} + \epsilon_1 z^{t_1} + \ldots + \epsilon_n z^{t_n}$

represents the (generally non-invertible) sequence $\epsilon_0, \epsilon_1, \ldots, \epsilon_n$ of reflection coefficients of the deep layers, and:

$$A(z) = a_0 + a_1 z + \ldots + a_m z^m$$

represents the (necessarily invertible) sequence $a_0, a_1, \ldots, a_m$ of the inverse reverberation filter.

Let us now briefly review ARMA systems (Box and Jenkins, 1970). The acronym ARMA stands for autoregressive moving average. The z-transform of such a system is a rational function, where the denominator polynomial (which represents the autoregressive component) must be invertible and where the numerator polynomial (which represents the moving-average component) may or may not be invertible. If the numerator polynomial is invertible, then the ARMA process is invertible. If the numerator is not invertible, then the ARMA process is not invertible.

In summary, eq. 8 represents a model of a marine seismogram as a non-invertible ARMA system. Let s_k be the values of the seismogram, i.e., s_k is the causal stable sequence with Laplace z-transform $S(z)$. We find it sometimes convenient to write eq. 8 in symbolic form as:

$$s = \epsilon/a \tag{9}$$

where s is the seismogram sequence, ϵ is the reflection coefficient sequence, and a is the inverse reverberation filter sequence.

Most treatments of ARMA systems are concerned with systems where the MA part is invertible, so the entire ARMA system is invertible. However, as we have seen, the ARMA model for a seismogram in general is not invertible. We can observe the seismogram s. We would like to determine the sequences ϵ and a from the sequence s. The purpose of this paper is to give an algorithn which carries out this identification.

NUMERICAL EXAMPLE

First, let us give a numerical example of the algorithm. In order to construct a simple numerical example, let us assume that we use the coarse time grid given by $t_0 = 0, t_1 = 1, \ldots, t_n = n$ and that the two-way travel time in the water is $T = 1$ in terms of this grid. The value $t_0 = 0$ represents the arrival time of the first deep reflection (i.e., the reflection from deep interface 0), so the time origin is the arrival time of the first deep event. The transfer function of the seismogram (for a single surface layer) in terms of this time grid is:

$$S(z) = \frac{\epsilon_0 + \epsilon_1 z + \epsilon_2 z^2 + \ldots + \epsilon_n z^n}{(1 + cz)^2}$$

which we can write as:

$$S(z) = \frac{E(z)}{A(z)}$$

where:

$$E(z) = \epsilon_0 + \epsilon_1 z + \epsilon_2 z^2 + \ldots + \epsilon_n z^n$$

and:

$$A(z) = 1 + 2cz + c^2 z^2$$

Let us now give a numerical example. Suppose there are $n + 1 = 10$ deep interfaces, and suppose that the first 19 observations of the resulting marine seismogram are (reading consecutively as words on a page)

$s = (s_0, s_1, \ldots, s_{18}) =$ (1.0000, 0.0000, 0.1875, $-$0.9063, $-$0.7148,
0.1992, $-$0.6057, $-$0.8153, $-$0.1198, 0.9911,
0.5030, 0.1896, 0.0633, 0.0198, 0.0060,
0.0017, 0.0005, 0.0001, 0.0000)

Given these observations, the problem is to find the water bottom reflection coefficient c and the deep reflection sequence $\epsilon = (\epsilon_0, \epsilon_1, \epsilon_2, \ldots, \epsilon_9)$. Because the observed seismogram is not known in absolute terms but only to within an arbitrary scale factor, the resulting reflection coefficient sequence ϵ will be equal to the actual physical reflection coefficient sequence only to within a scale factor.

The method of solution we propose is an iterative scheme based on an algorithm which, for the kth step, has inputs s and $\epsilon^{(k-1)}$ and outputs $a^{(k)}$ and $\epsilon^{(k)}$. The seismogram sequence s is the numerical sequence of 19 observations given above, and remains fixed for each step. The outputs $a^{(k)}$ and $\epsilon^{(k)}$ of the kth step are supposed to converge to the sequence a and ϵ, respectively. In order to test whether we are getting convergence, we divide the z-transform of $a^{(k)}$ into the z-transform of $\epsilon^{(k)}$ to obtain the sequence $s^{(k)}$. We write the sequence $s^{(k)}$ symbolically as:

$$s^{(k)} = \epsilon^{(k)}/a^{(k)}$$

If the sequence $s^{(k)}$ computed at the end of the kth step is close to the seismogram sequence $s = \epsilon/a$, then we say that we have convergence, and stop the iteration. The computed $\epsilon^{(k)}$ and $a^{(k)}$ then represent the required identification of ϵ and a, respectively, in the ARMA model of the seismogram.

In order to start the iteration, we set $\epsilon^{(0)}$ equal to the zero-delay spike:

$$\epsilon^{(0)} = (1,0,0,\ldots, 0)$$

The first iteration gives:
$a^{(1)} = (1, -0.32, 0.18)$
$\epsilon^{(1)} = (1, -0.32, 0.37, -0.97, -0.39, 0.27, -0.80, -0.58, 0.04, 0.88)$
The second iteration gives:
$a^{(2)} = (1, -0.37, 0.14)$
$\epsilon^{(2)} = (1, -.37, 0.33, -0.98, -0.35, 0.34, -0.78, -0.56, 0.10, 0.92)$

The third iteration gives:
$a^{(3)} = (1, -0.41, 0.12)$
$\epsilon^{(3)} = (1, -0.41, 0.30, -0.98, -0.32, 0.39, -0.77, -0.54, 0.14, 0.95)$
Continue iterating. The twentieth iteration gives:
$a^{(20)} = (1, -0.50, 0.624)$
$\epsilon^{(20)} = (1, -0.50, 0.25, -1.00, -0.25, 0.50, -0.75, -0.50, 0.25, 1.00)$
At this point the values of s and of $s^{(20)} = \epsilon^{(20)}/a^{(20)}$ are equal to within three decimal places. As a result we conclude that $\epsilon^{(20)}$ gives the deep reflection coefficients (to within a scale factor) and $a^{(20)} = (1, -0.50, 0.624) = (1, 2c, c^2)$ gives the water bottom reflection coefficient as $c = -0.25$.

DESCRIPTION OF THE ALGORITHM

The algorithm is based on the least-squares wave-form shaping filter, as derived in the first section. We recall that, in order to derive the shaping filter f, we must specify the input b and the desired output d. The shaping filter f is determined in such a way that the actual output $c = b \star f$ approximates the desired output d in a least-squares sense.

Let us now introduce a short-hand notation to describe the operation of a shaping filter. This short-hand notation has the general form

$$b \star f \doteq d$$

where we adhere to the convention that b and d are given, and f is determined in such a way that the convolution $b \star f$ approximates d in the least-squares sense. We remember the sequences on each end (i.e., b and d) are given, and the sequence in the middle (i.e., f) is determined. Subroutine SHAPE (Robinson, 1967) computes the least-squares shaping filter; in subroutine SHAPE the sequences b and d are subroutine inputs and the filter sequence f is the subroutine output.

Let us now describe the algorithm. At the beginning of step k we have the seismogram s and the approximate deep reflection coefficient sequence $\epsilon^{(k-1)}$. We apply subroutine SHAPE as:

$$s \star a^{(k)} \doteq \epsilon^{(k-1)}$$

to yield the preliminary estimate of $a^{(k)}$. However, we want to guarantee that $a^{(k)}$ is an invertible sequence, and in order to do so, we apply subroutine SHAPE as:

$$a^{(k)} \star [a^{-1}]^{(k)} \doteq (1,0,0,\ldots,0)$$

to yield the least-squares inverse $[a^{-1}]^{(k)}$. We then apply subroutine SHAPE as:

$$[a^{-1}]^{(k)} \star a^{(k)} = (1,0,0,\ldots,0)$$

to yield the second estimate of $a^{(k)}$. Because of the minimum-delay property of least-squares inverses, this second estimate $a^{(k)}$ is necessarily invertible. Next we normalize this $a^{(k)}$ by dividing each element of $a^{(k)}$ by its leading element; the result is the final estimate of $a^{(k)}$. The final estimate of $\epsilon^{(k)}$ is given as the convolution of $a^{(k)}$ with s; that is:

$$\epsilon^{(k)} = a^{(k)} \star s$$

We thus have found $a^{(k)}$ and $\epsilon^{(k)}$, and the description of the algorithm is complete. Initially, we start with $\epsilon^{(0)} = (1,0,0,\ldots,0)$ and we end when each of $\epsilon^{(k)}$ and $a^{(k)}$ has stabilized.

The algorithm is given in a diagram as follows (Fig. 3):

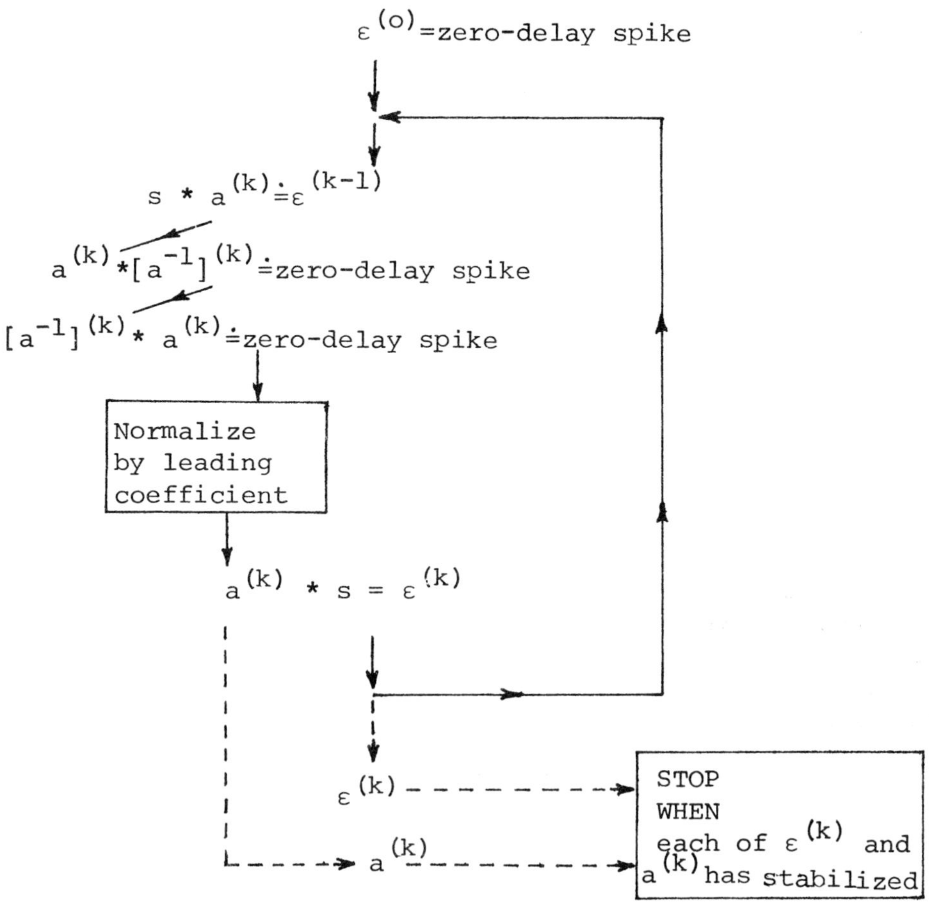

Fig. 3. Algorithm for a non-invertible system.

COMPARISON WITH DURBIN'S ALGORITHM

According to Kendall and Stuart (1966), Durbin has given an algorithm for finding the parameters ϵ and a of an invertible ARMA process $s = \epsilon/a$. Whereas in our algorithms we require only a to be invertible (which is a requirement for a stable causal ARMA system), Durbin in his algorithm requires both ϵ and a to be invertible, so s is invertible. The following diagram (Fig.4) gives a modified version of Durbin's algorithm. These modifications were made to express the essential content of such an invertible algorithm in a completely symmetrical form. We have not drawn the starting and stopping points of the algorithm, but only its internal workings.

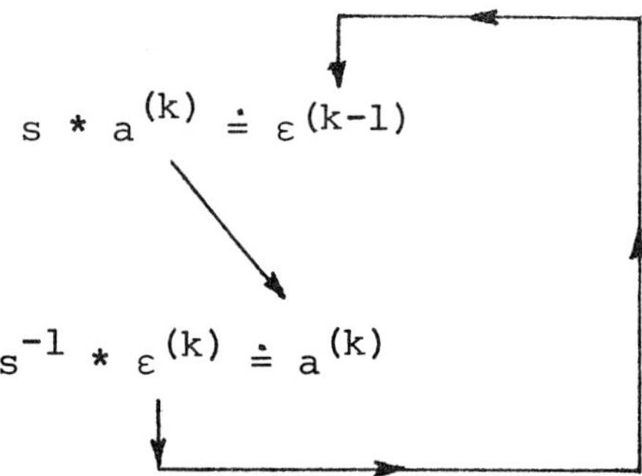

Fig. 4. Algorithm for an invertible system.

MAXIMUM ENTROPY

Let x_t (where the time index t takes on integer values) be a real stationary process. The autocorrelation function is defined as:

$$r_s = E(x_{t+s}x_t)$$

where E denotes expectation and the integer s is the time shift. For convenience, we normalize the autocorrelation function so that $r_0 = 1$. For each positive integer n we can define a prediction-error series ϵ_t and a hindsight-error series η_t associated with that value of n. The prediction-error time series is given by:

$$\epsilon_t = \sum_{s=0}^{n} a_s x_{t-s} \qquad (\text{where } a_0 = 1),$$

and where the prediction-error operator coefficients $a_0 = 1, a_1, a_2, \ldots, a_n$ are chosen to minimize $E\epsilon_t^2$. If we go through this minimization procedure, we find that the required operator coefficients $1, a_1, \ldots, a_n$ can be found from the first part of the autocorrelation $1, r_1, \ldots, r_n$ by solving the *Toeplitz normal equations:*

$$\sum_{s=0}^{n} a_s r_{t-s} = 0 \quad \text{for } t = 1, 2, \ldots, n.$$

The minimum value of $E\epsilon_t^2$ is:

$$E\epsilon_t^2 = a_0 r_0 + a_1 r_1 + \ldots + a_n r_n = v_n \ (\text{where } a_0 = 1, r_0 = 1).$$

We call the quantity v_n the (prediction-error) variance. Let us now define the hindsight-error series η_t. The hindsight-error time series for a given value of n is:

$$\eta_t = \sum_{s=0}^{n} a_s x_{t+s} \ (\text{where } a_0 = 1).$$

If we minimize $E\eta_t^2$, we find that the required hindsight-error operator coefficients $1, a_1, \ldots, a_n$ satisfy the same set of normal equations as for the prediction-error operator, and the minimum value of $E\eta_t^2$ is also equal to v_n. Thus the operator coefficients of the prediction-error operator and the hindsight-error operator are identical, and in addition:

$$E\epsilon_t^2 = E\eta_t^2 = v_n.$$

The prediction error ϵ_t is the forecasting error resulting from approximating x_t in terms of the past values $x_{t-1}, \ldots, x_{t-n}$, whereas the hindsight error η_t is the retrospection error resulting from approximating x_t in terms of the future values $x_{t+1}, \ldots, x_{t+n}$. The difference between the two operators is that the prediction-error operator acts on past values of the time series, whereas the hindsight-error operator acts on future values.

Let us now consider a given set of data points $x_{t-n}, x_{t-(n-1)}, \ldots, x_{t-1}$ corresponding to the time interval from $t-n$ to $t-1$ inclusive. Let us also consider the corresponding prediction error and hindsight error associated with this fixed set of data points. The prediction error ϵ_t is the error in forecasting x_t from the past datum points $x_{t-n}, x_{t-n-1}, \ldots, x_{t-1}$. The hindsight error η_{t-n-1} is the error in retrospectively approximating x_{t-n-1} from the future datum points $x_{t-n}, x_{t-n-1}, \ldots, x_{t-1}$. Both sets of datum points are the same, and represent all the datum points between the points x_{t-n-1} and x_t that we wish to approximate. Thus η_{t-n-1} represents the error in x_{t-n-1} and ϵ_t represents the error in x_t, after we have removed the effect of the common intervening variables $x_{t-n}, x_{t-n-1}, \ldots, x_{t-1}$. As a result, the correlation between η_{t-n-1} and ϵ_t represents the net correlation between x_{t-n-1} and x_t after removal of the effects of all intervening variables. This net correlation is called the partial autocorrelation and is given by the partial autocorrelation coefficient c_{n+1} defined as:

$$c_{n+1} = \text{partial autocorrelation coefficient} = \frac{E(\epsilon_t \eta_{t-n-1})}{\sqrt{E\epsilon_t^2}\,\sqrt{E\eta_{t-n-1}^2}}$$

In order to find an expression for the partial autocorrelation coefficient in terms of well-known quantities, let us first consider the cross-correlation between the prediction error series ϵ_t and the given time series x_t. This cross-correlation, denoted by ψ_s, is:

$$\psi_s = E\,(\epsilon_t x_{t-s}) = E\left[\left(\sum_{j=0}^{n} a_j x_{t-j}\right) x_{t-s}\right] = \sum_{j=0}^{n} a_j E(x_{t-j} x_{t-s})$$

$$= \sum_{j=0}^{n} a_j r_{s-j}$$

The cross-correlation between the hindsight-error series η_t and the given time series x_t is:

$$E(\eta_t x_{t-s}) = \sum_{j=0}^{n} a_j r_{s+j}$$

which is equal to ψ_{-s} since $r_{-s-j} = r_{s+j}$. At this point let us observe that the Toeplitz normal equations require that $\psi_s = 0$ for $s = 1,2,3,\ldots, n$; that is, the Toeplitz normal equations require that the cross-correlation ψ_s vanishes in the interval $1 \leqslant s \leqslant n$.

Let us now find an expression for the covariance term in the numerator of the partial autocorrelation. We have:

$$E(\epsilon_t \eta_{t-n-1}) = E\left(\epsilon_t \sum_{s=0}^{n} a_s x_{t-n-1+s}\right)$$

$$= \sum_{s=0}^{n} a_s E(\epsilon_t x_{t-n-1+s}) = \sum_{s=0}^{n} a_s \psi_{n+1-s}$$

$$= \psi_{n+1} + a_1 \psi_n + a_2 \psi_{n-1} + \ldots + a_n \psi_1 .$$

Because of the Toeplitz normal equations, this equation becomes:

$$E(\epsilon_t \eta_{t-n-1}) = \psi_{n+1} .$$

In a similar way, we find:

$$v_n = E\epsilon_t^2 = E\eta_t^2 = \psi_0 .$$

Thus the partial autocorrelation coefficient between x_t and x_{t-n-1} is:

$$c_{n+1} = \text{partial autocorrelation coefficient} = \frac{\psi_{n+1}}{\sqrt{\psi_0}\sqrt{\psi_0}} = \frac{\psi_{n+1}}{\psi_0} = \frac{\psi_{n+1}}{v_n}$$

$$= \frac{a_0 r_{n+1} + a_1 r_n + \ldots + a_n r_1}{a_0 r_0 + a_1 r_1 + \ldots + a_n r_n}$$

The prediction-error variance v_n is given by (Robinson, 1967, p.144):

$$v_n = r_0 (1-c_1^2)(1-c_2^2)\ldots(1-c_n^2)$$

If we let n toward infinity, we obtain the final prediction error as:

$$v_\infty = r_0 (1-c_1^2)(1-c_2^2)(1-c_3^2)\ldots$$

We now want to consider the class of all stationary processes having autocorrelations all with the same initial section $r_0, r_1, \ldots, r_n$. All members in this class have partial autocorrelations all with the same initial section $c_1, c_2, \ldots, c_n$. Thus all members of this class have final prediction errors of the form:

$$v_\infty = [r_0(1-c_1^2)(1-c_2^2)\ldots(1-c_n^2)][(1-c_{n+1}^2)(1-c_{n+2}^2)\ldots]$$

where the expression in the first set of brackets on the right is fixed. The final prediction error v_∞ is a measure of the entropy of the process, because the larger v_∞, the larger the ultimate unpredictability and hence the larger the entropy. Therefore, if we maximize v_∞, we obtain the maximum entropy member of this class. The maximization can be done by inspection; namely, the maximum is obtained by letting $c_{n+1} = c_{n+2} = \ldots = 0$, and the resulting maximum is:

$$v_\infty(\max) = r_0(1-c_1^2)(1-c_2^2) \ldots (1-c_n^2).$$

FIXED ENTROPY ESTIMATE OF THE SEISMIC ACOUSTIC LOG

Often in geophysics we know the geologic structure well enough so that we can actually experiment with fixed entropy models. As an illustration let us turn to the marine seismic model; that is, the model of a layered medium with a perfect reflector at the upper interface ($c_0 = -1$) (Robinson, 1967, o.133).

As is well known, the initial portion of a seismogram is often the most noisy and the most difficult to work with. In the ideal case, the marine seismogram due to a downgoing unit spike has a downgoing component of the form $1, r_1, r_2, \ldots$ and an upgoing component of the form $r_1, r_2, \ldots$ where the coefficients r_k make up an autocorrelation function with $r_0 = 1$. That is, the r_k constitute a positive definite function. We also know that the partial autocorrelation coefficients c_k are the reflection coefficients of the subsurface layers. As is well known, the Toeplitz recursion generates the partial autocorrelation coefficients c_k from the autocorrelation coefficients r_k in a sequential manner $k = 1, 2, 3, \ldots$ However, when we apply this Toeplitz method to an actual marine seismogram r_k, the errors due to noise in the initial portion of the seismogram are propagated sequentially in the computation of the reflection coefficients for $k = 1, 2, 3, \ldots$ As a result, the Toeplitz method results in larger and larger errors in the reflection coefficients as we go deeper and deeper into the earth, these errors being due to noise in the initial part of the seismogram.

In order to avoid this sequential generation of errors, we propose a simultaneous method of estimating the reflection coefficients from a marine seismogram, or equivalently a simultaneous method of estimating the partial autocorrelation coefficients from the autocorrelation coefficients. This method is an iterative method based on non-linear regression. The autocorrelation coefficient r_k is a non-linear function of the partial autocorrelation coefficients. We assume that all the partial autocorrelation coefficients $c_1, c_2, \ldots, c_n$ have fixed values except for a subset of m coefficients which can have arbitrary values. Usually these fixed values will be zero, corresponding to maximum entropy (i.e., no physical interface at a given depth), although in some instances certain non-zero values may be chosen on physical grounds. Let $\theta_1, \theta_2, \ldots, \theta_m$ denote the subset of m partial autocorrelation coefficients with arbitrary values. Then the autocorrelation coefficients are non-linear functions of the unknown $\theta_1, \theta_2, \ldots, \theta_m$; let these functions be:

$$r_1 = f_1(\theta_1, \ldots, \theta_m)$$
$$r_2 = f_2(\theta_1, \ldots, \theta_m)$$
$$\cdots\cdots\cdots\cdots\cdots$$
$$r_n = f_n(\theta_1, \ldots, \theta_m)$$

These functions can be computed by the computer program (see Robinson, 1967, p.137) if we are given the values of $\theta_1, \ldots, \theta_m$. However, we are given the values of $r_1, r_2, \ldots, r_n$ and we want to find $\theta_1, \theta_2, \ldots, \theta_m$.

We let $\hat{\theta}_1, \hat{\theta}_2, \ldots, \hat{\theta}_m$ be the initial guesses as to the values of the partial autocorrelation coefficients in the subset. We then expand the function $f_k(\theta_1, \ldots, \theta_m)$ in a Taylor expansion and retain the linear terms only. We thereby obtain:

$$f_k(\theta_1, \ldots, \theta_m) \doteq f_k(\hat{\theta}_1, \ldots, \hat{\theta}_m) + \sum_{i=1}^{m} b_{ki} (\theta_i - \hat{\theta}_i)$$

where:

$$b_{ki} = \left. \frac{\partial f_k(\theta_1, \ldots, \theta_m)}{\partial \theta_i} \right|_{\theta_1 = \hat{\theta}_1,\ \theta_2 = \hat{\theta}_2,\ \ldots,\ \theta_m = \hat{\theta}_m}$$

That is, the coefficient b_{ki} is the partial derivative evaluated at $\hat{\theta}_1, \hat{\theta}_2, \ldots, \hat{\theta}_m$. In practice, the b_{ki} are evaluated by a numerical approximation to the derivative instead of by an analytic expression. We want to choose $\theta_1, \theta_2, \ldots, \theta_m$ such that the mean square error:

$$\sum_{k=1}^{n} [r_k - f_k(\theta_1, \ldots, \theta_m)]^2$$

is a minimum. Using our Taylor approximation this mean square error becomes:

$$\sum_{k=1}^{n} [r_k - f_k(\hat{\theta}_1, \ldots, \hat{\theta}_m) - \sum_{i=1}^{m} b_{ki} u_i]^2$$

where the discrepancy u_i is defined as $u_i = \theta_i - \hat{\theta}_i$. If we differentiate the above expression for the mean square error with respect to the u_i and set the derivative s equal to zero, we get a set of normal equations which are linear in the unknowns u_i. We can solve these normal equations for the u_i. Using these values of the u_i we obtain the refined estimate of the partial autocorrelations given by $\theta_i = u_i + \hat{\theta}_i$. We can now use these refined estimates as our initial guesses and repeat the process. We keep iterating in this manner until successive iterations do not materially change the values of the partial autocorrelation coefficients $\theta_1, \ldots, \theta_m$.

In practice one can incorporate a source wavelet into the model as well as options that reduce or control the noise. Once the reflection coefficients c_i are determined the well-known relationship:

$$c_i = \frac{Z_i - Z_{i+1}}{Z_i + Z_{i+1}}$$

where Z_i is the acoustical impedance in layer i can be used to graph the acoustic log, i.e., the graph of acoustical impedance as a function of (two-way) travel time. The lower curve in Fig.5 shows a seismic trace as a function

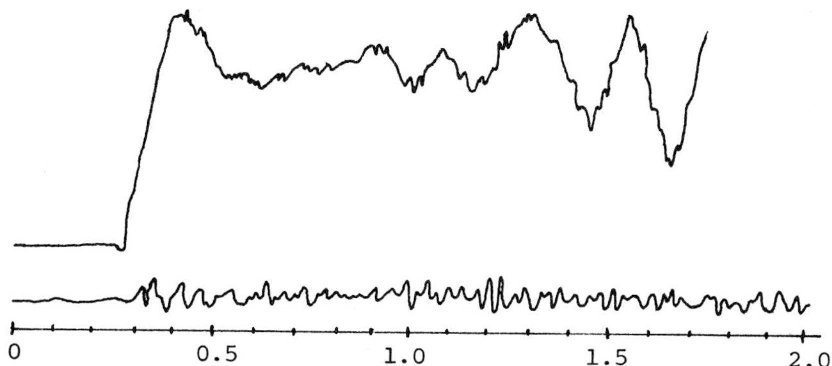

Fig. 5.A seismic trace (lower curve) and its corresponding acoustical log (upper curve).

of (two-way) travel time in seconds. The upper curve shows the corresponding
seismic acoustical log computed from the seismic trace by the method given
in this section. This seismic trace was recorded by Digicon Inc. on a marine
survey in 1969 in the Arctic Ocean north of Canada.

SUMMARY AND CONCLUSIONS

We have given an algorithm to compute the parameters of the moving-
average (MA) and the autoregressive (AR) components of a non-invertible
ARMA system from its impulse response function. We have applied this
algorithm to a marine seismogram, which can be regarded as the impulse
response of a non-invertible ARMA system, in order to determine the reflection coefficients of the deep interfaces. A diagram of the algorithm is given
and compared to the corresponding algorithm of Durbin for the case of an
invertible ARMA system. From the general principles illustrated in these dia-
grams other algorithms along similar lines may be readily devised. Finally
we have given an iterative method to compute the acoustical impedance log
from a seismic trace taking into account all primary and multiple reflections.

REFERENCES

Box, G.E.P. and Jenkins, G.M., 1970. Time Series Analysis, Forecasting and Control. Holden-
Day, San Francisco.
Robinson, E.A., 1967. Multichannel Time Series Analysis with Digital Computer Pro-
grams. Holden-Day, San Francisco.
Kendall, M.G. and Stuart, A., 1966. The Advanced Theory of Statistics, 3. Charles Griffin,
London, 484 pp.

CHAPTER 21

Iterative Least-Squares Procedure for ARMA Spectral Estimation

E. A. Robinson

With 12 Figures

In recent years an important new approach to signal processing has come into importance in science and engineering. It has become practical to represent information-bearing waveforms digitally and to do signal processing on the digital representation of the waveform. We can now manipulate physical signals by use of digital computers in ways that would have been totally impractical with continuous representations of the signals. Once the digital representation of a waveform is in computer memory, the waveform acquires a permanence which permits a wide range of processing operations, and at any point the waveform can be restored to a continuous form with a new time base. As a result, the fitting of finite-parameter models of digital signals is no longer only of theoretical importance but is a processing technique that can be used in day-to-day operations. In the previous chapter a detailed treatment of these finite-parameter models is given. In this chapter we wish to elaborate on the noninvertible autoregressive moving-average (ARMA) model.

The impulse response function of an ARMA model has a z transform given by the ratio of two polynomials. For stability, the denominator polynomial must necessarily be minimum delay. However, the numerator polynomial may or may not be minimum delay. If the numerator polynomial is minimum delay, then the ARMA model is said to be invertible. Otherwise, the ARMA model is said to be noninvertible. This chapter deals with the noninvertible model, and provides an iterative means of estimating both the numerator and denominator polynomials from the impulse response function. From these polynomials, we can then obtain an estimate of the spectral density function.

1 Basic Time Series Models

In recent years there has been increased emphasis on methods to construct finite-parameter models of a time series. Such models are based upon the idea that a time series $\{x_n\}$ (where the integer n is the time index) is generated from a series of random uncorrelated innovations $\{\varepsilon_n\}$. Specifically we assume that 1) the innovations have zero expectation, 2) the expectation of the product $\varepsilon_n\varepsilon_k$ of two different innovations (i.e. $n \neq k$) is zero, and 3) the expectation of the square ε_n^2 of any innovation a constant σ^2. Such a series of uncorrelated random variables with zero mean and constant variance is called a white-noise series.

533

The time series $\{x_n\}$ is then modeled as the output of a linear filter with a white-noise series as input. A finite-parameter model is defined as a model in which the linear filter can be represented by a finite number of coefficients.

Three basic finite-parameter models are the autoregressive model, the moving average model, and the autoregressive-moving average model. The autoregressive model is also called the autorecursive model, or AR model. The moving average is also called the move-ahead or MA model. The autoregressive-moving average model is a hybrid or mixed model with both AR and MA components, and is called the ARMA model.

The AR model is extremely useful in the representation of many physical time series. In this model the current value x_n of the time series is expressed as a finite linear combination of previous values of the time series and the current innovation ε_n. An autoregressive time series of order p may thus be represented as

$$x_n = -a_1 x_{n-1} - a_2 x_{n-2} - \ldots - a_p x_{n-p} + \varepsilon_n. \tag{4.1}$$

The reason for the name "autoregressive" is that a linear model relating a "dependent" variable x_n to a set of "independent" variables $x_{n-1}, x_{n-2}, \ldots, x_{n-p}$ plus an error term ε_n is referred to as a regression model. In this case, of course, the variable x_n is regressed on previous values of itself, and so the model is called autoregressive.

The reason for the minus signs appearing before the coefficients of the autoregressive representation above is that we want to transfer those terms to the left-hand side, and thus obtain

$$x_n + a_1 x_{n-1} + a_2 x_{n-2} + \ldots + a_p x_{n-p} = \varepsilon_n. \tag{4.2}$$

This is the standard representation of an AR model of a stationary time series. We recognize the left-hand side of this equation as the convolution of the operator $\{1, a_1, a_2, \ldots, a_p\}$ with the time series $\{x_n\}$. This operator is called the autoregressive operator. The z transform of this operator is the polynomial

$$A(z) = 1 + a_1 z^{-1} + a_2 z^{-2} + \ldots + a_p z^{-p}. \tag{4.3}$$

The symbol z^{-1} can either be interpreted as a complex variable or as a unit-delay operator. The unit-delay operator z^{-1} is defined as

$$z^{-1}[x_n] = x_{n-1}, \tag{4.4}$$

thus we see that

$$z^{-m}[x_n] = x_{n-m}. \tag{4.5}$$

With the interpretation of z^{-1} as the unit-delay operator, the fundamental autoregressive representation may be written as

$$A(z)[x_n] = \varepsilon_n, \tag{4.6}$$

if we divide each side of this equation by $A(z)$ we obtain

$$x_n = \frac{1}{A(z)}[\varepsilon_n]. \tag{4.7}$$

This equation shows that the autoregressive process x_n is the output of a finite feedback system with input ε_n. The polynomial $A(z)$ is called the feedback polynomial. If we make the expansion

$$H(z) = \frac{1}{A(z)} = 1 + h_1 z^{-1} + h_2 z^{-2} + h_3 z^{-3} + \ldots \tag{4.8}$$

then the coefficients $1, h_1, h_2, \ldots$ represent the impulse response of the AR system. In order for the AR system to be stable, and thus for the process x_n to be stationary, the feedback polynomial $A(z)$ must be strictly minimum delay, that is, $A(z)$ must not have any zeros outside or on the unit circle. We can write the autoregressive process as

$$x_n = H(z)[\varepsilon_n] \tag{4.9}$$

which is

$$x_n = \varepsilon_n + h_1 \varepsilon_{n-1} + h_2 \varepsilon_{n-2} + \ldots. \tag{4.10}$$

This equation states that x_n is the convolution of the infinitely long operator $\{1, h_1, h_2, \ldots\}$ with the innovation series $\{\varepsilon_n\}$. The operator $\{1, h_1, h_2, \ldots\}$ is the inverse of the autoregressive operator $\{1, a_1, a_2, \ldots, a_p\}$.

Another model of great physical importance is the moving average model. In this model, the time series $\{x_n\}$ is linearly dependent on a finite number of previous innovations; that is

$$x_n = \varepsilon_n + b_1 \varepsilon_{n-1} + b_2 \varepsilon_{n-2} + \ldots + b_q \varepsilon_{n-q}. \tag{4.11}$$

The name moving average is somewhat misleading because the weights $1, b_1, b_2, \ldots, b_q$ which multiply the innovations need not be positive and need not sum to one, as would ordinarily be implied by the term moving average. However the nomenclature moving average is in common use for this model. The moving average operator has z transform

$$B(z) = 1 + b_1 z^{-1} + b_2 z^{-2} + \ldots + b_q z^{-q}. \tag{4.12}$$

Interpreting z^{-1} as the unit-delay operator, the moving average model may be written economically as

$$x_n = B(z)[\varepsilon_n]. \tag{4.13}$$

Equation (4.13) shows that the moving average process $\{x_n\}$ is the output of a finite feedforward (i.e., feedfront) system with input $\{\varepsilon_n\}$. The polynomial $B(z)$ is called the feedforward polynomial, or the move-ahead polynomial.

Finally we come to the ARMA model. To achieve greater flexibility in the fitting of physical time series, it is often advantageous to include both autoregressive and moving average components in the model. This leads to the ARMA model

$$A(z)[x_n] = B(z)[\varepsilon_n] \tag{4.14}$$

or

$$x_n + a_1 x_{n-1} + \ldots + a_p x_{n-p} = \varepsilon_n + b_1 \varepsilon_{n-1} + \ldots + b_q \varepsilon_{n-q}.$$

We can write the ARMA model as

$$x_n = \frac{B(z)}{A(z)}[\varepsilon_n]. \tag{4.15}$$

This equation shows that the ARMA process $\{x_n\}$ is the output of a finite feedback-feedfront system with input $\{\varepsilon_n\}$. The polynomial $A(z)$ is the feedback polynomial and the polynomial $B(z)$ is the feedfront polynomial. If we make the expansion

$$H(z) = \frac{B(z)}{A(z)} = h_0 + h_1 z^{-1} + h_2 z^{-2} + \ldots \tag{4.16}$$

then coefficients h_0, h_1, h_2, ... represent the impulse response of the ARMA system. In order for the ARMA system to be stable, the feedback polynomial $A(z)$ must be strictly minimum delay, that is, $A(z)$ must have no zeros outside or on the unit circle. A stable ARMA system with white noise input ε_n yields a stationary output x_n.

2 Parsimony and the Physical Model

The three models which we discussed in the previous section may be described as models with both statistical and dynamic components. In each case the model expressed the time series as the output of a linear filter with a white noise input. The white noise input represents the statistical aspect of the model and the linear filter represents the dynamic aspect of the model. This approach is in keeping with the concept of a typical time series as being neither completely random nor completely predictable. The white noise series is completely random and thus has zero predictability. On the other hand, the linear filter is characterized by a set of coefficients which are fixed; these coefficients are parameters and not random variables. The resulting time series has both elements of randomness and non randomness. Such a time series can be predicted with some degree of success but not perfectly. In this respect, we think of the innovation series as being the random component of the time series, and the impulse response function of the filter as the dynamic component. The time series itself is thus the convolution of the random and dynamic components. Given a stationary time series $\{x_n\}$ there is, of course, no way to unravel its random and dynamic components without further information. This unraveling process, which takes apart the convolution of the random with the dynamic, is called deconvolution.

In the above discussion we have mentioned the need for further information. This remark is a key point in time series analysis. As a result we would like to link time series methods closely with the empirical and theoretical evidence provided by the subject matter under investigation. This approach is used in conjunction with the parsimony approach to time series analysis [4.1]. The parsimony approach is so named because the general principle of parsimony is used. As we have seen, the mathematical models which we employ contain certain coefficients or parameters whose values must be estimated from the data. The parsimony approach is the approach based upon the concept of employing in practice the smallest possible number of parameters for adequate representation. The central role played by this principle of parsimony in the estimation of parameters is evident in almost all scientific research.

With this in mind let us now return to the subject of deconvolution. Ideally in our time series model, we want to identify the random component and the dynamic component with actual physical entities or processes. The success of the deconvolution is not measured solely by the use of the least number of parameters but also by how well the estimated quantities agree with the physical world. A few coefficients that fit the time series but have nothing to do with the physical phenomenon do not represent a physical deconvolution. However, a physical deconvolution with a small number of parameters is to be preferred over one with a large number of parameters. The key point is that the number of parameters should be minimized under the condition that results can be verified in practice or in theory by direct physical measurements in the real world. In summary, the principle of parsimony must be used in conjunction with the physics of the problem.

Much is known about the physics of seismic wave propagation through the earth, and the results of a seismic time series analysis made on the basis of surface observations can be directly verified by drilling a hole into the earth (that is, an oil well). In order to illustrate these points, we will now discuss the process of seismic deconvolution and the type of spectral estimate obtained by such an approach.

3 Seismic Deconvolution and the Resulting Spectral Estimate

In order to understand seismic deconvolution, we must simplify. It is the process of simplification which brings out the essential features of the sedimentary layers when they are subjected to a seismic source of energy. A mathematical equation has little value unless we can link its meaning with the physical structure of the earth. Of course such linkage cannot usually be made in all generality, so we resort to simplification in order to gain as much understanding as we can. It is better to understand a few simple things well than to have a whole book of mathematical equations which are not connected to the physical action of the earth. The science of geophysics is indeed the study of the physics of the earth.

With this approach in mind let us look at a single horizontal interface between two sedimentary layers. Throughout we measure wave motion in terms of units which represent square root of energy. A unit spike by definition has amplitude one at time zero and amplitude zero elsewhere; thus the energy content of a unit spike is one-squared, which is one unit of energy. Similarly, a spike of amplitude 0.5 would contain 0.5 squared, or 0.25 units of energy. A wavelet with amplitudes

$$a_0 = 1, \quad a_1 = 0.5, \quad a_2 = -0.2, \quad a_t = 0 \quad \text{for} \quad t > 2$$

has total energy

$$1 + (0.5)^2 + (-0.2)^2 = 1.29.$$

Suppose now that a downgoing unit spike strikes an interface between two layers. As we know from classical physics, some of the energy is transmitted through the interface and some is reflected back from the interface. The law of conservation of energy says that the sum of the reflected and transmitted energies is equal to the incident energy. By physical reasoning we also know that we can define a reflection coefficient ε and a transmission coefficient τ. For the given downgoing unit spike the upgoing reflected spike has amplitude ε and the transmitted downgoing spike has amplitude τ. This relationship is illustrated in Fig. 4.1. In all the figures we are representing plane waves propagating at normal incidence, but ray paths are given a horizontal displacement to simulate the passage of time in the horizontal direction.

The law of conservation of energy states that

$$1^2 = \varepsilon^2 + \tau^2$$

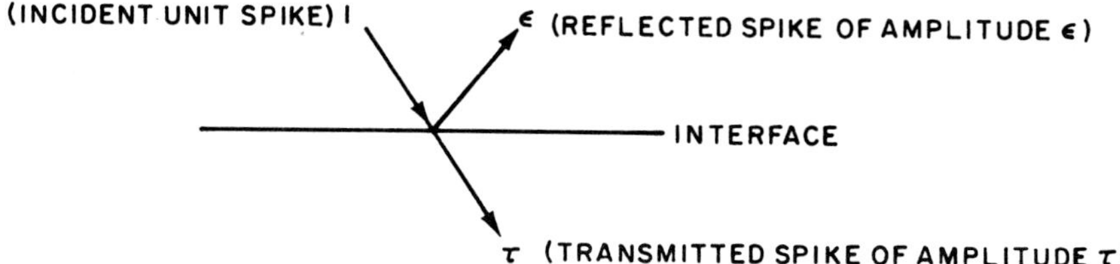

Fig. 4.1. Reflected and transmitted spikes due to a downgoing incident spike

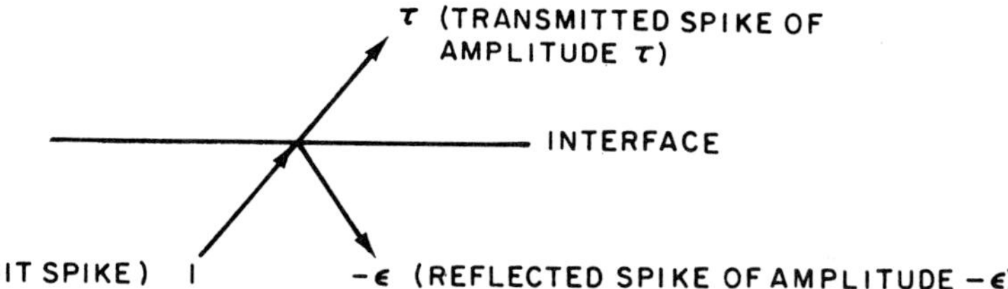

Fig. 4.2. Reflected and transmitted spikes due to an upgoing incident spike

so ε and τ are related by

$$\tau = \sqrt{1 - \varepsilon^2}. \tag{4.17}$$

From physical reasoning it also follows that a given upgoing unit spike striking the interface from below gives rise to a downgoing reflected spike of amplitude $-\varepsilon$ and an upgoing transmitted spike of amplitude τ, see Fig. 4.2.

The law of conservation of energy is

$$1^2 = (-\varepsilon)^2 + \tau^2$$

which is the same as before.

The transfer function is defined as the ratio of the z transforms of the output to the input. If we let $T(z)$ represent the transfer function of the transmitted wave as output with respect to a downgoing incident wave we have

$$T(z) = \frac{\tau}{1}$$

which in this simple case is the constant τ.

Let us now look at a single sedimentary layer with air above and granite below. We have two interfaces with reflection coefficients ε_0 and ε_1 and the corresponding transmission coefficients

$$\tau_0 = \sqrt{1 - \varepsilon_0^2} \quad \text{and} \quad \tau_1 = \sqrt{1 - \varepsilon_1^2}, \tag{4.18}$$

see Fig. 4.3.

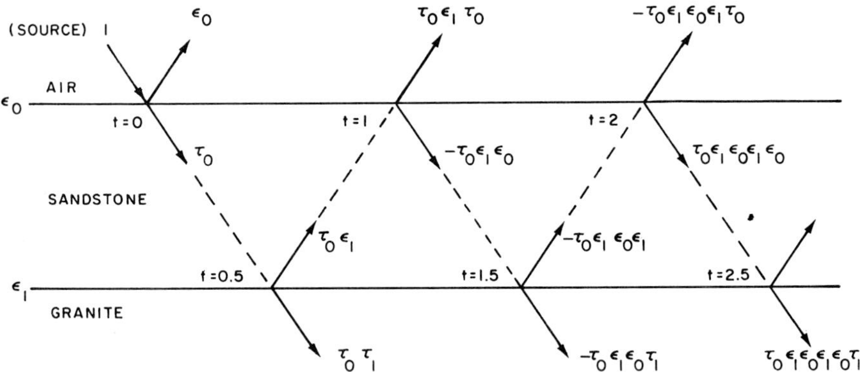

Fig. 4.3. Case of a single layer subject to an incident downgoing unit spike as source. Time t is measured horizontally to the right

As depicted in Fig. 4.3 we subject the layer to an incident downgoing spike from above. This source energy results in a sequence of reflected and transmitted waves. Let us trace this action. We suppose that it takes 0.5 time units for a spike to travel through the sedimentary layer. Thus the two-way travel time in the sandstone is one time unit. At time 0, the incident spike gives rise to a reflected spike ε_0 and a transmitted spike τ_0. The reflected spike escapes but the transmitted spike travels down to the lower interface, arrives at time 0.5, and gives rise to a transmitted spike $\tau_0\tau_1$ and a reflected spike $\tau_0\varepsilon_1$. This transmitted spike escapes but the reflected spike travels up to the upper interface, arriving at time 1, and gives rise to a transmitted spike $\tau_0\varepsilon_1\tau_0$ and a reflected spike $\tau_0\varepsilon_1(-\varepsilon_0)$. This transmitted spike escapes but the reflected spike travels down to the lower interface, arriving at time $t=1.5$, and the process is repeated again and again as the energy reverberates within the sandstone layer. At each whole unit of time (i.e., $0, 1, 2, 3, \ldots$), some energy escapes into the air, and each unit plus one-half (i.e., $0.5, 1.5, 2.5, \ldots$) of time some energy escapes into the granite, so after an infinite amount of time there is no more energy left in the sandstone.

At all times the law of conservation of energy holds. The total amount of energy is the energy of the source, namely one. This total energy must be equal at any time to the sum of the energy in the layer plus the escaped energy. For example, just after $t=0.5$ the wave motion in the layer is a spike of amplitude $\tau_0\varepsilon_1$ and the escaped wave motion is the spike of amplitude ε_0 in the air and the spike of amplitude $\tau_0\tau_1$ in the granite. We have

$$(\tau_0\varepsilon_1)^2 + \varepsilon_0^2 + (\tau_0\tau_1)^2 = \tau_0^2(\varepsilon_1^2 + \tau_1^2) + \varepsilon_0^2$$

$$= \tau_0^2 + \varepsilon_0^2 = 1$$

as it should.

The transmitted wavelet is made up of all the spikes that escape into the granite. Thus the transmitted wavelet is

$$(\tau_0\tau_1, -\tau_0\varepsilon_1\varepsilon_0\tau_1, \tau_0(\varepsilon_1\varepsilon_0)^2\tau_1, -\tau_0(\varepsilon_1\varepsilon_0)^3\tau_1, \ldots),$$

where the initial coefficient is at $t=0.5$, the next at $t=1.5$, and so on. However, for convenience we can choose a new time scale, so we may regard the initial coefficient in the transmitted wavelet as occurring at time 0, the next at time 1, and so on. In other words, the new time scale has its origin at the time break of the transmitted wavelet. The z transform of the transmitted wavelet is then

$$\begin{aligned}T(z)&=\tau_0\tau_1-\tau_0\varepsilon_1\varepsilon_0\tau_1 z^{-1}+\tau_0(\varepsilon_1\varepsilon_0)^2\tau_1 z^{-2}-\tau_0(\varepsilon_1\varepsilon_0)^3\tau_1 z^{-3}+\ldots\\&=\tau_0\tau_1[1-(\varepsilon_1\varepsilon_0 z^{-1})+(\varepsilon_1\varepsilon_0 z^{-1})^2-(\varepsilon_1\varepsilon_0 z^{-1})^3+\ldots].\end{aligned}\qquad(4.19)$$

The expression within brackets is a geometric series which may be summed. As a result we have

$$T(z)=\frac{\tau_0\tau_1}{1+\varepsilon_0\varepsilon_1 z^{-1}}.\qquad(4.20)$$

This expression is the transfer function of a single layer.

Instead of the complicated way in which we derived this transfer function, we can derive it directly as follows. The layer acts as a feedback loop. Each round trip the wave makes in the layer represents one passage through the feedback loop. Let x_t be an arbitrary source wavelet incident on the top interface and let y_t be the resulting transmitted wavelet. Let us choose the point that represents the mixer of the feedback system as a point in space just above the bottom layer, see Fig. 4.4. The input wavelet x_t must go through the top interface and then travel to the bottom layer to arrive at the mixer. Thus at the mixer the input wavelet appears as $x_t\tau_0 z^{-1/2}$, where τ_0 represents the transmission through the top interface and $z^{-1/2}$ represents the time delay of 0.5 required for the wavelet to travel through the layer. The output of the mixer is y_t/τ_1. That is, we must divide out the transmission coefficient τ_1 as the mixer is just above the bottom interface. The basic feedback loop then delays the output of the mixer by one time unit (representing the round trip passage time in the layer) and multiplies this delayed output by ε_1 representing the bounce from the bottom interface and also by $-\varepsilon_0$ representing the bounce from the top interface. That is, the basic feedback loop forms the quantity

$$(y_t/\tau_1)z^{-1}\varepsilon_1(-\varepsilon_0).$$

The mixer then adds this quantity to the input to give the output; that is

$$(y_t/\tau_1)z^{-1}\varepsilon_1(-\varepsilon_0)+x_t\tau_0 z^{-1/2}=y_t/\tau_1.$$

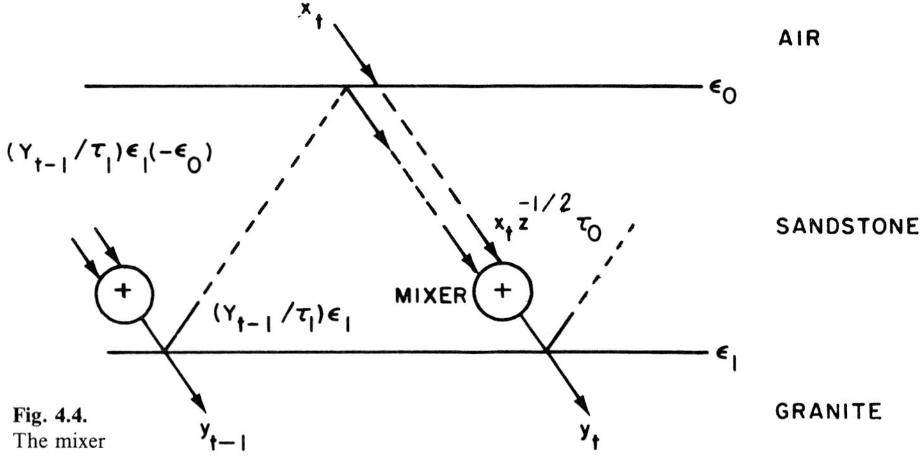

Fig. 4.4.
The mixer

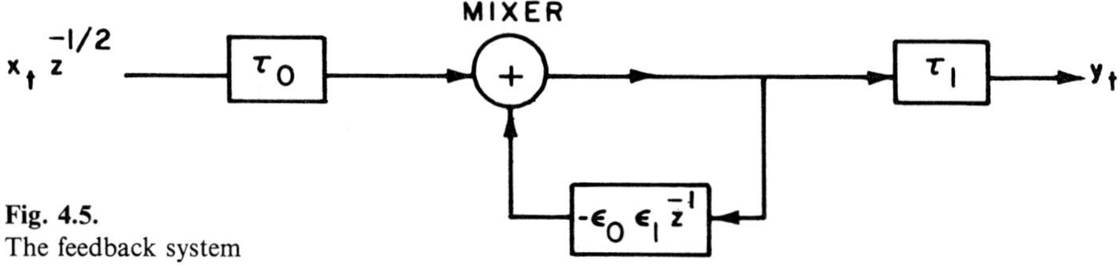

Fig. 4.5.
The feedback system

Solving this equation we have

$$(y_t/\tau_1)(1+\varepsilon_0\varepsilon_1 z^{-1}) = x_t\tau_0 z^{-1/2}$$

or

$$\frac{y_t z^{1/2}}{x_t} = \frac{\tau_0\tau_1}{1+\varepsilon_0\varepsilon_1 z^{-1}}. \tag{4.21}$$

Note that the $z^{1/2}$ on the left corresponds to the time advance of 0.5 which we introduced when we measured the transmitted wavelet at its first break. Thus we obtain the same transfer function as before (see Fig. 4.5), namely, (20).

The principle of reciprocity states that if we interchange source and receiver we still obtain the same transfer function. Figure 4.6 (left side) depicts such an interchange, where $T^1(z)$ denotes the new transfer function. However, we can flip the entire picture upside down, as shown in Fig. 4.6 (right side) to obtain the equivalent situation. Note that the order and sign of the reflection coefficients have been reversed. Thus we see that $T'(z)$ can be obtained from $T(z)$ by using $-\varepsilon_1$, $-\varepsilon_0$ instead of ε_0, ε_1. We have

$$T'(z) = \frac{\tau_1\tau_0}{1+(-\varepsilon_1)(-\varepsilon_0)z^{-1}}, \tag{4.22}$$

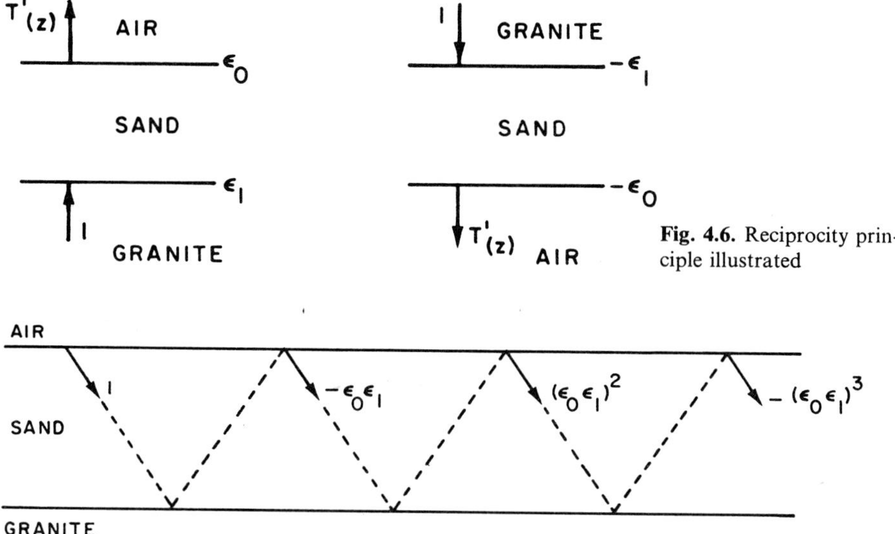

Fig. 4.6. Reciprocity principle illustrated

Fig. 4.7. The basic reverberating system

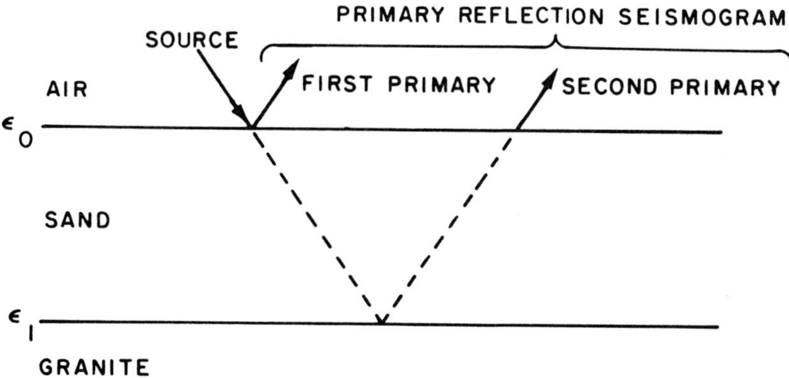

Fig. 4.8. The primary reflection seismogram

and we see that $T'(z)$ is identical to $T(z)$, as predicted by the principle of reciprocity.

Let us now look at the transfer function $T(z)$. In the expression for $T(z)$ the quantity

$$\frac{1}{1+\varepsilon_0\varepsilon_1 z^{-1}} = 1 - \varepsilon_0\varepsilon_1 z^{-1} + (\varepsilon_0\varepsilon_1)^2 z^{-2} - (\varepsilon_0\varepsilon_1)^3 z^{-3} + \ldots \tag{4.23}$$

represents the basic reverberation within the sedimentary layer, as depicted in Fig. 4.7.

Now let us construct the reflection seismogram due to a downgoing unit spike source incident on the top interface. There are two primary reflections, namely, the direct reflection off the top interface and the direct reflection of the bottom interface, as seen in Fig. 4.8.

Neglecting all other effects except the two basic reflections, the primary reflection seismogram would consist of simply the two reflection coefficients; that is, the primary reflection seismogram would be the two-term wavelet $\{\varepsilon_0, \varepsilon_1\}$ with z transform

$$E(z) = \varepsilon_0 + \varepsilon_1 z^{-1}. \tag{4.24}$$

The wavelet $\{\varepsilon_0, \varepsilon_1\}$ is the reflection coefficient series. However, as we know from physical considerations, these primary reflections make the whole sandstone layer reverberate. Thus the reflection seismogram may be considered as the output of the basic reverberation system (as shown in Fig. 4.7) subject to the input of the primary reflection seismogram (as depicted by Fig. 4.8). That is, the reflection seismogram $\{u_0, u_1, u_2, \ldots\}$ is equal to the convolution of the reflection coefficient series with the basic reverberation; i.e.,

$$\{u_0, u_1, u_2, \ldots\} = \{\varepsilon_0, \varepsilon_1\} * \{1, -\varepsilon_0\varepsilon_1, (\varepsilon_0\varepsilon_1)^2, -(\varepsilon_0\varepsilon_1)^3, \ldots\} \tag{4.25}$$

where the asterisk denotes the process of convolution. Let $U(z) = u_0 + u_1 z + u_2 z^2 + \ldots$ be the z transform of the reflection seismogram. Then the above convolutional equation becomes

$$U(z) = (\varepsilon_0 + \varepsilon_1 z^{-1})[1 - \varepsilon_0\varepsilon_1 z^{-1} + (\varepsilon_0\varepsilon_1)^2 z^{-2}$$
$$- (\varepsilon_0\varepsilon_1)^3 z^{-3} + \ldots] \tag{4.26}$$

which is

$$U(z) = \frac{\varepsilon_0 + \varepsilon_1 z^{-1}}{1 + \varepsilon_0\varepsilon_1 z^{-1}}. \tag{4.27}$$

The basic idea of deconvolution can now be stated simply as the following. Given the reflection seismogram, find the reflection coefficient series. From the above equation we can write

$$(1 + \varepsilon_0\varepsilon_1 z^{-1})(u_0 + u_1 z^{-1} + u_2 z^{-2} + \ldots) = \varepsilon_0 + \varepsilon_1 z^{-1} \tag{4.28}$$

which in the time domain is

$$\{1, \varepsilon_0\varepsilon_1\} * \{u_0, u_1, u_2, \ldots\} = \{\varepsilon_0, \varepsilon_1\}. \tag{4.29}$$

This equation states that the reflection coefficient series is the convolution of the operator $\{1, \varepsilon_0\varepsilon_1\}$ with the reflection seismogram. We say that the

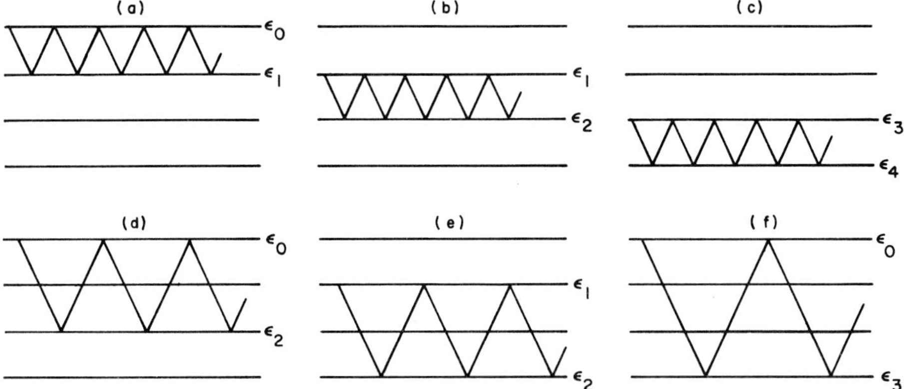

Fig. 4.9a–f. Reverberations. **(a–c)** First-order; **(d, e)** second-order linked; **(f)** third-order

operator $\{1, \varepsilon_0\varepsilon_1\}$ deconvolves the reflection seismogram to yield the reflection coefficient series. Thus the operator $\{1, \varepsilon_0\varepsilon_1\}$ is the required deconvolution operator.

Because the above discussion contains all the basic ideas, we can now write down the results for the multilayered sedimentary system which occurs in actual seismic prospecting. First we must examine the types of reverberations which can occur. We add enough hypothetical layers as necessary in our mathematical model so that the round trip travel time in each layer is one. A first-order reverberation is defined as one in which the closed loop travel time is one; a second-order reverberation is defined as one with time 2 and so on. Furthermore, we define a linked reverberation as one which involves only physically adjacent (i.e., connected) layers. These ideas will become clear if we examine a three-layer sedimentary system, see Fig. 4.9. Of course, all first-order reverberations are connected because they all involve just one layer. The feedback boxes for the three first-order reverberations shown in Fig. 4.9 are respectively

$$-\varepsilon_0\varepsilon_1 z^{-1}, \qquad -\varepsilon_1\varepsilon_2 z^{-1}, \qquad -\varepsilon_2\varepsilon_3 z^{-1},$$

which can be incorporated in one box as

$$-(\varepsilon_0\varepsilon_1 + \varepsilon_1\varepsilon_2 + \varepsilon_2\varepsilon_3)z^{-1}.$$

We recognize the expression in parenthesis as the first lag autocorrelation a_1 of the reflection coefficient series $\{\varepsilon_0, \varepsilon_1, \varepsilon_2, \varepsilon_3\}$, so this box is simply $-a_1 z^{-1}$. The feedback boxes of the two second-order linked reverberations are

$$-\varepsilon_0\varepsilon_2 z^{-2}, \qquad -\varepsilon_1\varepsilon_3 z^{-2}$$

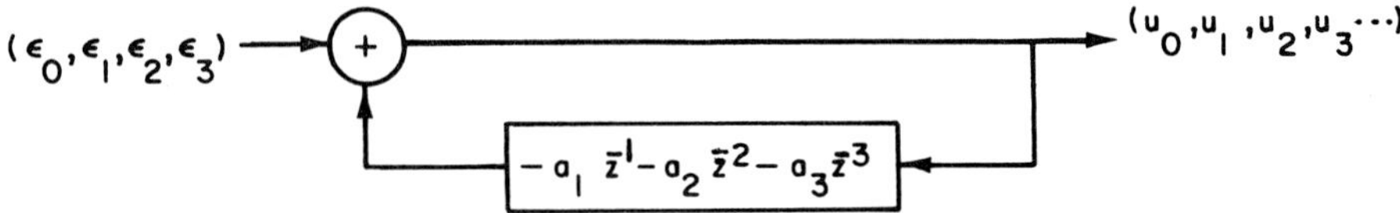

Fig. 10. Structure of reflection seismogram

which can be incorporated in one box as

$$(\varepsilon_0\varepsilon_2+\varepsilon_2\varepsilon_3)z^{-2}.$$

We recognize the expression in parenthesis as the second-lag autocorrelation a_2 of the reflection coefficient series $\{\varepsilon_0, \varepsilon_1, \varepsilon_2, \varepsilon_3\}$ so this box is simply $-a_2z^{-2}$. The third-order reverberation for a three-layer system must necessarily be linked. Its feedback box is

$$-\varepsilon_0\varepsilon_3z^{-3}$$

which may be written as $-a_3z^{-3}$. The entire feedback box for the linked reverberations is thus

$$-a_1z^{-1}-a_2z^{-2}-a_3z^{-3}.$$

The reflection coefficient series has z transform

$$\varepsilon_0+\varepsilon_1z^{-1}+\varepsilon_2z^{-2}+\varepsilon_3z^{-3}.$$

The reflection seismogram is (approximately) obtained by passing the reflection coefficient series through the feedback filter, as seen in Fig. 4.10.

The z transform of the reflection seismogram is therefore (approximately) given by [4.2]

$$U(z)=\frac{\varepsilon_0+\varepsilon_1z+\varepsilon_2z^2+\varepsilon_3z^3}{1+a_1z+a_2z^2+a_3z^3} \tag{4.30}$$

which we write as

$$U(z)=\frac{E(z)}{A(z)}. \tag{4.31}$$

The deconvolution operator is thus $\{1, a_1, a_2, a_3\}$, as given by the denominator coefficients in the above expression.

The above discussion of course is an ideal one, as it involves the entire reflection seismogram. In the case of N layers the above expression for $U(z)$

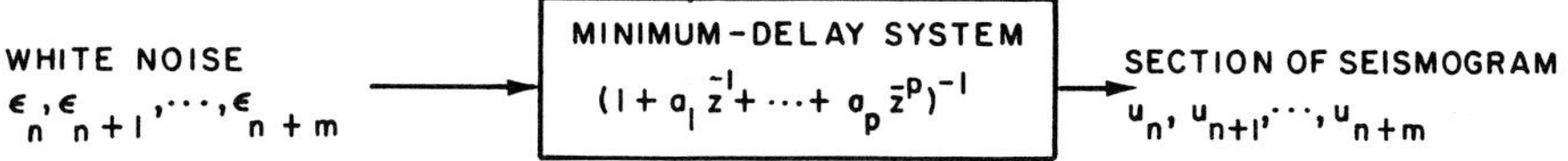

Fig. 4.11. Model of section of seismogram

becomes

$$U(z) = \frac{\varepsilon_0 + \varepsilon_1 z^{-1} + \ldots + \varepsilon_N z^{-N}}{1 + a_1 z^{-1} + \ldots + a_N z^{-N}}. \tag{4.32}$$

In actual practice we assume that the autocorrelation $a_1, a_2, \ldots, a_N$ damps out so that we need only consider the first p coefficients $a_1, a_2, \ldots, a_p$ where $p < N$. Also we only analyze a section of the seismogram at a time, say the section u_n, $u_{n+1}, u_{n+2}, \ldots, u_{n+m}$ from $t = n$ to $t = n + m$. Then the z transform of this section is approximately

$$u_n + u_{n+1} z^{-1} + \ldots + u_{n+m} z^{-m} = \frac{\varepsilon_n + \varepsilon_{n+1} z^{-1} + \ldots + \varepsilon_{n+m} z^{-m}}{1 + a_1 z^{-1} + \ldots + a_p z^{-p}} \tag{4.33}$$

which we write as

$$U_s(z) = \frac{E_s(z)}{A(z)}. \tag{4.34}$$

Now we appeal to the random reflection hypothesis, which states that the reflection coefficients $\varepsilon_n, \varepsilon_{n+1}, \ldots, \varepsilon_{n+m}$ within the given section are mutually uncorrelated (i.e., white). The random reflection hypothesis is generally true for many actual sections of sedimentary layers in the earth's crust. Also, as a physical fact, we know that a reflection seismogram damps out, so that the feedback system that generates the seismogram must be stable. This stability property means that the denominator polynomial $1 + a_1 z^{-1} + \ldots + a_p z^{-p}$ is minimum delay. Therefore, its reciprocal $(1 + a_1 z^{-1} + \ldots + a_p z^{-p})^{-1}$ is also minimum delay. Thus, we see that we have the following model for the given section shown in Fig. 4.11.

Using the above expression for $U_s(z)$ we have

$$U_s(z) U_s(z^{-1}) = \frac{E_s(z) E_s(z^{-1})}{A(z) A(z^{-1})} \tag{4.35}$$

which, if we let $z = e^{j\omega}$, becomes

$$|U_s(\omega)|^2 = \frac{|E_s(\omega)|^2}{|A(\omega)|^2}. \tag{4.36}$$

By the random reflection hypothesis the power spectrum $|E_s(\omega)|^2$ is flat, namely, a constant σ^2. The quantity $|U_s(\omega)|^2$ is the spectral density of the given section of seismic trace. If the autocorrelation coefficients of this given section are ϕ_0, $\phi_1, \dots$ (where $\phi_{-t} = \phi_t$), then we have

$$|U_s(\omega)|^2 = \sum_{t=-\infty}^{\infty} \phi_t e^{-i\omega t}. \tag{4.37}$$

Returning now to z, we have

$$\sum_{t=-\infty}^{\infty} \phi_t z^{-t} = \frac{\sigma^2}{A(z)A(z^{-1})} \tag{4.38}$$

which is

$$A(z)\sum \phi_t z^{-t} = \frac{\sigma^2}{A(z^{-1})}. \tag{4.39}$$

Because the right-hand side of (4.39) has an expansion in only nonpositive powers of z^{-1}, the left-hand side also must have an expansion in only nonpositive powers of z^{-1}. Hence, in particular the coefficients of z^{-k} for $k = 1, 2, \dots, p$ must be zero. These coefficients obtained from the left-hand side are

$$\begin{aligned} \phi_1 + a_1\phi_0 + \dots + a_p\phi_{p-1} &= 0 \\ &\cdots \\ \phi_p + a_1\phi_{p-1} + \dots + a_p\phi_0 &= 0. \end{aligned} \tag{4.40}$$

This set of equations is the set of well-known normal equations, and they can be solved for the unknown $a_1, a_2, \dots, a_p$. The constant σ^2 can then be found from the equation

$$\phi_0 + a_1\phi_1 + \dots + a_p\phi_p = \sigma^2. \tag{4.41}$$

We can now summarize the procedure. Compute the autocorrelation coefficients $\phi_0, \dots, \phi_p$ of the given section of the seismogram. Then solve the normal equations for the deconvolution operator $a_1, a_2, \dots, a_p$. Finally, deconvolve the seismogram (i.e., convolve the given section of the seismogram by the deconvolution operator $\{1, a_1, a_2, \dots, a_p\}$) to obtain the white reflection coefficient series $\{\varepsilon_n, \varepsilon_{n+1}, \dots, \varepsilon_{n+m}\}$. This series is called the deconvolved seismic trace.

This method of deconvolution is called statistical deconvolution because it is based on the random reflection hypothesis, namely, the hypothesis that the reflection coefficients within a given earth section form a white noise series. Statistical deconvolution provides an estimate of the spectral density of the given section of seismic trace. This spectral estimate is the maximum-entropy spectral estimate of *Burg* [4.3]; namely

$$|U_s(\omega)|^2 = \frac{\sigma^2}{|A(\omega)|^2},$$ (4.42)

where $A(\omega)$ is the discrete Fourier transform of the operator; that is,

$$A(\omega) = 1 + a_1 e^{-j\omega} + a_2 e^{-j\omega 2} + \ldots + a_p e^{-j\omega p}.$$ (4.43)

4 Design of Shaping and Spiking Filters

In this section we want to consider the case where the forms of the input signal and the desired output signal are known.

The design problem to be considered may be formulated as follows. Given a signal $\{x_n\}$ we want to operate on $\{x_n\}$ in some manner so as to obtain the best approximation to the signal $\{s_n\}$. That is, let $T[x_n]$ be the best approximation to $\{s_n\}$. More generally, we may let $T[x_n]$ be the best approximation to the shifted signal $\{s_{n+\alpha}\}$, where α is a time shift. At this point, the criterion of best approximation and the form of the operator are unspecified.

Let us now consider an ideal system $\{f_n\}$. An ideal system would transform the input into the desired output. If the ideal system is a linear shift–invariant system, then

$$\{s_{n+\alpha}\} = \{f_n\} * \{x_n\}.$$ (4.44)

In terms of Fourier transforms, this equation becomes

$$S(\omega)e^{j\omega\alpha} = F(\omega)X(\omega).$$ (4.45)

The solution of the above equation gives the filter characteristics of the *ideal filter* as

$$F(\omega) = \frac{S(\omega)e^{j\omega\alpha}}{X(\omega)}.$$ (4.46)

This result represents the formal solution to the problem. However, for actual computations this equation is of no help, because the impulse response f_n of the ideal system $F(\omega)$ will generally be an infinitely long two-sided operator. For practical purposes, a digital filter with a finite number of coefficients is required, so let us try to approximate the ideal filter f_n by a finite-length moving average

(MA) filter h_n where

$$h_n = 0 \quad \text{for} \quad n < 0 \quad \text{and for} \quad n > M. \tag{4.47}$$

(We recall that an MA filter is defined as a causal linear shift-invariant filter with a finite-length impulse response. Alternatively, a MA filter can be defined as a linear shift-invariant filter that has a transfer function equal to a polynomial in z^{-1}). Thus, we want to approximate the frequency response $F(\omega)$ of the ideal filter by the frequency response

$$H(\omega) = \sum_{n=0}^{M} h_n e^{-j\omega n} \tag{4.48}$$

of the required MA filter.

Such an approximation procedure means that a certain amount of information contained in the infinite-length impulse response $\{f_n\}$ must be lost in order to obtain the approximate finite-length impulse response $\{h_n\}$. Consequently, we shall need some sort of averaging process to carry out this approximation. We want the difference

$$F(\omega) - H(\omega)$$

to be small in some sense. For example, we might choose the coefficients $\{h_n\}$ so that the mean-squared difference

$$\frac{1}{2\pi} \int_{-\pi}^{\pi} |F(\omega) - H(\omega)|^2 d\omega$$

is a minimum. However, in this expression we see that the squared difference $|F(\omega) - H(\omega)|^2$ is given a uniform weighting for all frequencies. It would make more sense to weight the squared difference according to the energy spectrum $|X(\omega)|^2$ of the input signal. Thus, let us choose the MA coefficients $\{h_n\}$ so that the mean-weighted-squared difference, or mean-square error,

$$I = \frac{1}{2\pi} \int_{-\pi}^{\pi} |F(\omega) - H(\omega)|^2 |X(\omega)|^2 d\omega \tag{4.49}$$

is a minimum. This equation can be written as

$$I = \frac{1}{2\pi} \int_{-\pi}^{\pi} |F(\omega)X(\omega) - H(\omega)X(\omega)|^2 d\omega. \tag{4.50}$$

Substituting (4.46) in (4.50), we obtain

$$I = \frac{1}{2\pi} \int_{-\pi}^{\pi} |S(\omega)e^{j\omega\alpha} - H(\omega)X(\omega)|^2 d\omega. \tag{4.51}$$

Now $S(\omega)$ is the spectrum of the desired output signal $\{s_n\}$ and $H(\omega)X(\omega)$ is the spectrum of the actual output signal $\{h_n * x_n\}$. If we apply Parseval's theorem to this equation, we obtain

$$I = \sum_{n=-\infty}^{\infty} |s_{n+\alpha} - h_n * x_n|^2 . \tag{4.52}$$

Because we are dealing with real signals, the above equation becomes

$$I = \sum_{n=-\infty}^{\infty} \left(s_{n+\alpha} - \sum_{k=0}^{M} h_k x_{n-k} \right)^2 . \tag{4.53}$$

In order to minimize I, we set its partial derivatives with respect to each of h_0, $h_1, \ldots, h_M$ equal to zero. We obtain

$$\frac{\partial I}{\partial h_j} = -2 \sum_{n=-\infty}^{\infty} \left(s_{n+\alpha} - \sum_{k=0}^{M} h_k x_{n-k} \right) x_{n-j} = 0$$

for $j = 0, 1, 2, \ldots, M$. We may therefore write

$$\sum_{k=0}^{M} h_k \left(\sum_{n=-\infty}^{\infty} x_{n-k} x_{n-j} \right) = \sum_{n=-\infty}^{\infty} s_{n+\alpha} x_{n-j} \tag{4.54}$$

for $j = 0, 1, 2, \ldots, M$. We recognize the autocorrelation coefficient

$$r_{j-k} = \sum_{n=-\infty}^{\infty} x_{n-k} x_{n-j} . \tag{4.55}$$

Also, we define the cross-correlation coefficient

$$g_{j+\alpha} = \sum_{n=-\infty}^{\infty} s_{n+\alpha} x_{n-j} . \tag{4.56}$$

Then, (4.54) becomes

$$\sum_{k=0}^{M} h_k r_{j-k} = g_{j+\alpha} \tag{4.57}$$

for $j = 0, 1, 2, \ldots, M$. This set of equations are called the *Toeplitz normal equations*. The autocorrelation coefficients and the cross-correlation coefficients can be computed from the known input signal x_n and desired output signal s_n. Then the Toeplitz normal equations can be solved to find the required filter coefficients h_n. Because the output $\{h_n\} * \{x_n\}$ of this filter approximates the desired output signal $\{s_n\}$ in a least-squares sense, this filter is called the least-squares shaping filter.

Because of the special form of the simultaneous equations called the Toeplitz form, the equations may be solved by an efficient recursive procedure called the Toeplitz recursion [4.4, 5].

A special case of the least-squares shaping filter is the least-squares *spiking filter*. A spiking filter is a shaping filter for which the desired output signal is a spike. Thus, the signal $\{s_n\}$ is the unit impulse with coefficients

$$s_n = \delta_n = \begin{cases} 1, & n=0 \\ 0, & n \neq 0. \end{cases} \tag{4.58}$$

A spiking filter for time-shift $\alpha = 0$ is called a zero-delay spiking filter. A well-known theorem [4.6] states that the least-squares zero-delay spiking filter is minimum delay.

Because the concepts of minimum phase and minimum delay are identical, the least-squares zero-delay spiking filter has a minimum-phase-lag spectrum.

Let us now examine the normal equations given above. We notice that the left-hand side of the normal equations depends only upon the autocorrelation of the input $\{x_n\}$ and not on $\{x_n\}$ itself. Because the desired output for a zero-delay spiking filter is the spike $\{1, 0, 0, ..., 0\}$, the right-hand side of the normal equations are

$$g_0 = x_0, g_1 = 0, g_2 = 0, ..., g_m = 0. \tag{4.59}$$

Thus, in the case of a zero-delay spiking filter, the wave shape of the input sequence $\{x_n\}$ itself does not enter into the normal equations, except for the initial value x_0, which affects the filter as only a scale factor. Therefore, the normal equations in the case of a zero-delay spiking filter depend upon the input sequence only through its autocorrelation. Any two input sequences with the same autocorrelation would have the same zero-delay spiking filter (except for a constant factor). Because autocorrelations contain no phase information, it is an interesting consequence that the phase spectrum of the filter is the minimum-phase-lag spectrum.

For a zero-delay spiking filter the minimum value of the mean-square spiking error is

$$I = 1 - h_0 x_0. \tag{4.60}$$

In order for this expression for I to be small when we consider all input sequences with the same autocorrelation, the leading coefficient x_0 should be large in magnitude. In other words, the smallest mean-square spiking error occurs when the input sequence is minimum delay. In summary, all zero-delay spiking filters are necessarily minimum delay, and the best case of spiking occurs when the input sequence is also minimum delay. In fact, for a minimum delay input sequence the mean-square spiking error tends to zero as the number of coefficients in the zero-delay spiking filter $\{h_0, h_1, ..., h_M\}$ tend to infinity $(M \to \infty)$. On the other hand, for a nonminimum-delay input sequence, the mean-square spiking error does not tend to zero as the number of coefficients in the zero-delay spiking filter tend to infinity. We, therefore, conclude that perfect spiking can be achieved by a stable infinitely long causal spiking filter if and only if the input sequence is minimum delay. In other words, given the causal input x_n we can find a stable infinitely long causal filter $\{h_n\}$ such that

$$\{h_n\} * \{x_n\} = \{1, 0, 0, ...\} \tag{4.61}$$

if and only if $\{x_n\}$ is minimum delay. Any stable causal sequence $\{x_n\}$ that can be converted into a zero-delay spike by a stable causal filter is said to be invertible. Thus a stable causal sequence is invertible if and only if it is minimum delay. For that reason, we can use the terms "invertible sequence" and "minimum-delay sequence" interchangeably, even as we use the terms "minimum-phase sequence" and "minimum-delay sequence" interchangeably.

In summary, the least-squares finite-length zero-delay spiking filter for a stable causal sequence $\{x_n\}$ as input may be described as the least-squares inverse of $\{x_n\}$. In case $\{x_n\}$ is minimum delay, then if the number of terms in the least-squares inverse is allowed to increase, the least-squares inverse tends to the exact stable causal inverse of $\{x_n\}$.

5 Invertibility

In many applications we know from the physics of the situation that a certain finite-length causal sequence must be invertible. However, suppose that the estimation method used does not necessarily give an invertible sequence. Thus we must have a method for converting a sequence which is not necessarily invertible into an invertible sequence. In order to obtain a meaningful procedure, we must impose some sort of restriction on this conversion process. Let the restriction be the condition that the given sequence and the resulting invertible sequence have (approximately) the same autocorrelation.

Let us designate the given finite causal sequence $\{a_0, a_1, ..., a_M\}$ simply by the voctor a. Let us denote the least-squares inverse (i.e., zero-delay spiking filter) of a by a^{-1}. In turn let us denote the least-squares inverse (i.e., zero-delay spiking filter) of a^{-1} by $(a^{-1})^{-1}$.

We can now describe a method of converting a finite-length causal sequence to an invertible finite-length causal sequence with approximately the same autocorrelation. This method consists of the following two steps:

1) Compute the zero-delay spiking filter a^{-1} of the given filter a. As we have seen in the preceding section, the filter a^{-1} is necessarily invertible.

2) Now take this approach one step further; namely, compute the zero-delay spiking filter $(a^{-1})^{-1}$ of the filter a^{-1}. This second zero-delay spiking filter is necessarily invertible, and in fact it is the required invertible counterpart of the given filter a.

In summary, the invertible counterpart of the finite-length causal sequence a is the least-squares inverse of the least-squares inverse of a.

6 Spectral Estimates of the Components of a Noninvertible ARMA System

As we have seen in Sect. 3, a reflection seismogram $\{u_0, u_1, u_2, \ldots\}$ has the z transform

$$U(z) = u_0 + u_1 z^{-1} + u_2 z^{-2} + \ldots \tag{4.62}$$

which can be approximately modeled as the ratio

$$U(z) = \frac{E(z)}{A(z)}. \tag{4.63}$$

In this equation the feedfront polynomial $E(z)$ is the z transform

$$E(z) = \varepsilon_0 + \varepsilon_1 z^{-1} + \varepsilon_2 z^{-2} + \ldots + \varepsilon_N z^{-N} \tag{4.64}$$

of the reflection coefficient series $\{\varepsilon_0, \varepsilon_1, \ldots, \varepsilon_N\}$ of the earth. There is no physical reason why $E(z)$ should be minimum delay. However, because the layered earth is a stable filter, the feedback polynomial

$$A(z) = 1 + a_1 z^{-1} + a_2 z^{-2} + \ldots + a_N z^{-N} \tag{4.65}$$

must be minimum delay. The coefficients $a_1, a_2, \ldots, a_N$, respectively, are approximately equal to the autocorrelation coefficients $\gamma_1, \gamma_2, \ldots, \gamma_N$ of the reflection coefficient series provided that each reflection coefficient is much smaller than one in magnitude [4.2].

The z transform of an ARMA system is a rational function, where the denominator polynomial (which represents the autoregressive component) must be invertible and where the numerator polynomial (which represents the moving average component) may or may not be invertible. If the numerator polynomial is invertible, then the ARMA process is invertible. If the numerator is not invertible, then the ARMA process is not invertible.

The equation $U(z) = E(z)/A(z)$ represents a model of a seismogram as a noninvertible ARMA system. We find it sometimes convenient to write the seismogram in symbolic form as

$$u = \frac{\varepsilon}{a}, \tag{4.66}$$

where u is the seismogram sequence, ε is the reflection coefficient sequence, and a is the inverse reverberation filter sequence.

Most treatments of ARMA systems are concerned with systems where the MA part is invertible, so the entire ARMA system is invertible. However, as we have seen, the ARMA model for a seismogram in general is not invertible. We can observe the seismogram u. We would like to determine the sequence ε and a from the sequence u. The purpose of this chapter is to give an algorithm which carries out this identification.

The algorithm is based on the least-squares wave-form shaping filter. We recall that, in order to derive the shaping filter h, we must specify the input x and the desired output s. The shaping filter h is determined in such a way that the actual output $h * x$ approximates the desired output s in a least-squares sense. (In this discussion we let the time-shift parameter α be zero.)

Let us now introduce a short-hand notation to describe the operation of a shaping filter. This short-hand notation has the general form (here special use is made of the approximately equal symbol $\simeq$)

$$x * h \simeq s, \tag{4.67}$$

where we adhere to the convention that x and s are given, and h is determined in such a way that the convolution $x * h$ approximates s in the least-squares sense. We remember the sequences on each end (i.e., x and s) are given, and the sequence in the middle (i.e., h) is determined. Subroutine SHAPE [4.7] computes the least-squares shaping filter; in subroutine SHAPE the sequences x and s are subroutine inputs and the filter sequence h is the subroutine output.

Let us now describe the algorithm. At the beginning of step k we have the seismogram u and the approximate reflection coefficient sequence $\varepsilon^{(k-1)}$. We apply subroutine SHAPE as

$$u * a^{(k)} \simeq \varepsilon^{(k-1)} \tag{4.68}$$

to yield the preliminary estimate of $a^{(k)}$. However, we want to guarantee that $a^{(k)}$ is an invertible sequence, and in order to do so, we apply subroutine SHAPE as

$$a^{(k)} * [a^{-1}]^{(k)} \simeq \{1, 0, 0, \ldots, 0\} \tag{4.69}$$

to yield the least-squares inverse $[a^{-1}]^{(k)}$. We then apply subroutine SHAPE as

$$[a^{-1}]^{(k)} * a^{(k)} \simeq \{1, 0, 0, \ldots, 0\} \tag{4.70}$$

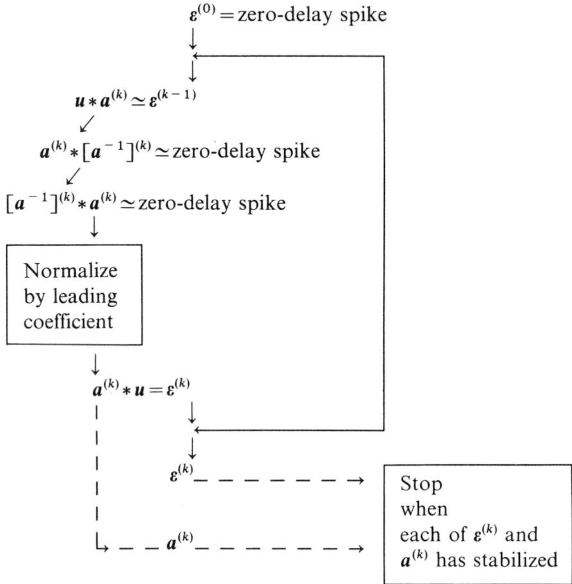

Fig. 4.12. Flow graph of the algorithm

to yield the second estimate of $a^{(k)}$. Because of the minimum-delay property of least-squares inverses, this second estimate $a^{(k)}$ is necessarily invertible. Next we normalize this $a^{(k)}$ by dividing each element of $a^{(k)}$ by its leading element; the result is the final estimate of $a^{(k)}$. The final estimate of $\varepsilon^{(k)}$ is given as the convolution of $a^{(k)}$ with u, that is,

$$\varepsilon^{(k)} = a^{(k)} * u. \tag{4.71}$$

We thus have found $a^{(k)}$ and $\varepsilon^{(k)}$, and the description of the algorithm is complete. Initially, we start with $\varepsilon^{(0)} = (1, 0, 0, \ldots, 0)$ and we end when each of $\varepsilon^{(k)}$ and $a^{(k)}$ has stabilized.

We can diagram the algorithm as shown in Fig. 4.12.

6.1 Example

In order to construct a simple numerical example, let us consider the case of a marine seismogram. Let the two-way travel time in the water layer represent one time unit, and let the water bottom reflection coefficient be c. Let the reflection coefficients of the deep interfaces be $\{\varepsilon_0, \varepsilon_1, \ldots, \varepsilon_N\}$. Then the z transform of the seismogram is approximately

$$U(z) = \frac{\varepsilon_0 + \varepsilon_1 z^{-1} + \varepsilon_2 z^{-2} + \ldots + \varepsilon_M z^{-M}}{(1 + cz^{-1})^2} \tag{4.72}$$

which we can write as

$$U(z) = \frac{E(z)}{A(z)} \qquad (4.73)$$

where

$$E(z) = \varepsilon_0 + \varepsilon_1 z^{-1} + \varepsilon_2 z^{-2} + \dots + \varepsilon_M z^{-M} \qquad (4.74)$$

and

$$A(z) = 1 + 2cz^{-1} + c^2 z^{-2}. \qquad (4.75)$$

Let us now give a numerical example. Suppose there are $M+1=10$ deep interfaces, and suppose that the first 19 observations of the resulting marine seismogram are (reading consecutively as words on a page)

$$
\begin{aligned}
U = \{ &1.0000, \quad 0.0000, \quad 0.1875, \quad -0.9063, \quad -0.7148, \\
&0.1992, \quad -0.6057, \quad -0.8153, \quad -0.1198, \quad 0.9911, \\
&0.5030, \quad 0.1896, \quad 0.0633, \quad 0.0198, \quad 0.0060, \\
&0.0017, \quad 0.0005, \quad 0.0001, \quad 0.0000 \}.
\end{aligned}
$$

Given these observations, the problem is to find the water bottom reflection coefficient c and the deep reflection coefficient sequence $\varepsilon = \{\varepsilon_0, \varepsilon_1, \varepsilon_2, \dots, \varepsilon_9\}$. Because the observed seismogram is not known in absolute terms but only to within an arbitrary scale factor, the resulting reflection coefficient sequence ε will be equal to the actual physical reflection coefficient sequence only to within a scale factor.

The method of solution is the iterative scheme based on the algorithm which, for the kth step, has inputs u and $\varepsilon^{(k-1)}$ and outputs $a^{(k)}$ and $\varepsilon^{(k)}$. The seismogram sequence u is the numerical sequence of 19 observations given above, and remains fixed for each step. The outputs $a^{(k)}$ and $\varepsilon^{(k)}$ of the kth step are supposed to converge to the sequence a and ε respectively.

In order to start the iteration, we set $\varepsilon^{(0)}$ equal to the zero-delay spike;

$$\varepsilon^{(0)} = \{1, 0, 0, \dots, 0\}.$$

The first iteration gives:

$$a^{(1)} = \{1, -0.32, 0.18\}$$

$$\varepsilon^{(1)} = \{1, -0.32, 0.37, -0.97, -0.39, 0.27, -0.80, -0.58, 0.04, 0.88\}.$$

The second iteration gives:

$$a^{(2)} = \{1, -0.37, 0.14\}$$
$$\varepsilon^{(2)} = \{1, -0.37, 0.33, -0.98, -0.35, 0.34, -0.78, -0.56, 0.10, 0.92\}.$$

The third iteration gives:

$$a^{(3)} = \{1, -0.41, 0.12\}$$
$$\varepsilon^{(3)} = \{1, -0.41, 0.30, -0.98, -0.32, 0.39, -0.77, -0.54, 0.14, 0.95\}.$$

Continue iterating. The twentieth iteration gives:

$$a^{(20)} = \{1, -0.50, 0.624\}$$
$$\varepsilon^{(20)} = \{1, -0.50, 0.25, -1.00, -0.25, 0.50, -0.75, -0.50, 0.25, 1.00\}.$$

At this point the values of the sequences have stabilized. As a result, we conclude that $\varepsilon^{(20)}$ gives the deep reflection coefficients (to within a scale factor) and

$$a^{(20)} = \{1, -0.50, 0.624\} = \{1, 2c, c^2\}$$

gives the water bottom reflection coefficient as $c = -0.25$.

In conclusion, this method allows one to decompose the impulse response of an ARMA system into estimates of its feedback and feedfront components. In this example, the spectral estimate of the feedfront component is

$$E(\omega) = 1 - 0.5e^{-j\omega} + 0.25e^{-j\omega 2} + \ldots + 0.25e^{-j\omega 8} + 1.0e^{-j\omega 9}$$

and the spectral estimate of the feedback component is

$$A(\omega) = 1 - 0.5e^{-j\omega} + 0.624e^{-j\omega 2}.$$

This feedback spectral estimate necessarily has minimum phase lag.

7 Conclusions

Three fundamental types of finite-parameter time series models are the autoregressive (AR) process which is a feedback model, the moving average (MA) process which is a feedfront model, and the hybrid (ARMA) process which is a feedback-feedfront model. The fitting of time series models involves the use of the principle of parsimony together with the linking of the model with physically observable quantities in the real world. These principles are illustrated with the model for seismic deconvolution where the parameters of the model can be identified with feedback loops within the sedimentary layers of the earth. Shaping and spiking filters are interpreted as filters which are data dependent, that is, the mean-squared error is weighted according to the energy spectrum of the input signal. The zero-delay spiking filter corresponding to a given input signal is called the least-squares inverse of that input signal. Any signal may be converted to minimum delay by simply computing the least-squares inverse of its least-squares inverse. Given the impulse response of an ARMA system, its feedback and feedfront components can be computed by an iterative scheme. The feedback component of a stable ARMA system must necessarily be minimum delay, but the feedfront component may or may not be minimum delay. If the feedfront component is minimum delay the ARMA system is invertible; if the feedfront component is not minimum delay then the ARMA system is not invertible. Each step of the iterative scheme involves first computing the feedback component and converting it to minimum delay and then estimating the feedfront component. This conversion at each step guarantees that the ARMA system is stable at each step of the iteration. The results of the iterative scheme allow us to compute the spectrum of the feedfront component which may or may not be minimum phase lag, and the spectrum of the feedback component which is necessarily minimum phase lag.

We have given an algorithm to compute the parameters of the moving average (MA) and the autoregressive (AR) components of a noninvertible ARMA system from its impulse response function. We have applied this algorithm to a reflection seismogram, which can be regarded as the impulse response of a noninvertible ARMA system, in order to determine the reflection coefficients of the deep interfaces. A diagram of the algorithm is given. From the general principles illustrated in this chapter, other algorithms along similar lines may be readily devised.

References

4.1 G.E.P.Box, G.M.Jenkins: *Time Series Analysis, Forecasting, and Control* (Holden-Day, San Francisco 1970)

4.2 E.A.Robinson, M.T.Silvia: *Digital Signal Processing and Time Series Analysis* (Holden-Day, San Francisco 1978)

4.3 J.P.Burg: "Maximum Entropy Spectral Analysis"; Ph.D. Thesis, Stanford University, Stanford, Calif. (1975)

4.4 N.Levinson: J. Math. Phys. **25**, 261–278 (1946)

4.5 R.A.Wiggins, E.A.Robinson: J. Geophys. Res. **70**, 1885–1891 (1965)

4.6 E.A.Robinson, H.Wold: "Minimum-Delay Structure of Least-Squares and eo-ipso Predicting Systems", in *Brown University Symposium on Time Series Analysis* (Wiley and Sons, New York 1963) pp. 192–196

4.7 E.A.Robinson: *Multichannel Time Series Analysis with Digital Computer Programs* (Holden-Day, San Francisco 1978)

Reprint from

Topics in Applied Physics

Volume 34: Nonlinear Methods of Spectral Analysis

Editor: S. Haykin

CHAPTER 22

COLLECTION OF FORTRAN II PROGRAMS FOR FILTERING AND SPECTRAL ANALYSIS OF SINGLE CHANNEL TIME SERIES

SUMMARY

This report gives a description of Fortran II programs for filtering and spectral analysis. All programs are tested.

CONTENTS

1. INTRODUCTION

A signal refers to a quantity that varies with time. Because time represents a unidirectional flow from past to future, the field of signal analysis has developed as a particular mathematical discipline. In turn this distinguishing characteristic of time must be carried and incorporated in digital computer programs for signal analysis.

In this paper we use the FORTRAN II computer language, and we presuppose that the reader has a knowledge of that language equivalent to that given in a standard textbook such as Elliott I. Organick: *A Fortran Primer* (Addison-Wesley Publishing Company, Reading, Massachusetts, 1962).

When we speak of "fortran programs" as in the title, we mean more specifically "fortran subroutines". In other words, everything that we present is in subroutine form; this means that our programming efforts are, as nearly as possible, independent of any particular computer, and can be used in conjunction with a main routine on practically any digital computer. The purpose of the main routine is to bring data into the machine, to call the subroutines that perform the necessary calculations, and to get the results out of the machine. As we have said, we are only interested in the subroutines; the main routine, which is to a large extent dependent on the input—output hardware available on a particular machine, is left for the user to devise.

2. The "standard" package of subroutines

The subroutines in this package are so basic that we have chosen to call them "standard". The names of the standard subroutines are:

ZERØ	(zero)
MØVE	(move)
IMPULS	(impulse)
SCALE	(scale)
DØT	(dot, or inner, product)
DØTR	(dot product reverse)
SYMDØT	(symmetrical dot product)
FØLD	(polynomial multiply, or convolve)
CØNSYM	(convolution with symmetrical filter)
CRØSS	(cross—correlate)
CRØSST	(cross—correlate)
TUKEY	(Tukey autocorrelation)
MINSN	(minimum, with sign)
MAXSN	(maximum, with sign)
NØRME	(normalize with respect to RMS energy)
NØRM 1	(normalize with respect to first element)
REVERS	(reverse vector)
POLYDV	(polynomial divide, or deconvolve)
MAINE	(symmetric matrix inverse)
MAIN 1	(fast version of MAINE)
MATRA	(matrix transpose)
PØLYEV	(polynomial evaluation, for a real value of its argument)
IPLYEV	(polynomial evaluation, for a complex value of its argument)
PSQRT	(polynomial square-root)
GENSYM	(generate a symmetric vector given one side)
MAP	(matrix print)

We frequently deal with a group of variables that form a single collection. If these variables can be related to each other by subscript notation, we then call the collection an array. The purpose of subroutine ZERØ is to put the floating-point number zero (0.0) into each storage location of an array. In other words this subroutine clears out any old numbers that might be in these storage locations, and replaces them by clean fresh zeros.

Let LX be a fortran integer variable, and X be an array of fortran floating-point variables, which more specifically may be written as

$$X(1), \ X(2), \ X(3), \ldots, \ X(LX)$$

That is, LX (called the length of X) is the number of elements in the array X. The call statement of subroutine ZERØ is

CALL ZERØ (LX, X)

The result is that each element of the array X becomes 0.0; that is

$$X(1) = 0.0, X(2) = 0.0, \ldots, X(LX) = 0.0$$

The program for subroutine ZERO is:

```
CZERO IS TO STORE ZERO IN AN ARRAY
      SUBROUTINE ZERO (LX, X)
      DIMENSION X(2)
      IF (LX) 30,30, 10
   10 DO 20 I = 1,LX
   20 X (I)=0.0
   30 RETURN
      END
```

The purpose of subroutine MØVE is to move an array to another storage location. The call statement is:

CALL MØVE (LX, X, Y)

The result is that the numbers that were in array X are now in both array X and array Y. The program is:

```
CMOVE IS MOVE ARRAY TO ANOTHER STORAGE LOCATION
      SUBROUTINE MOVE (LX, X, Y)
      DIMENSION X(2), Y(2)
      DO 10  I=1,LX
   10    Y(I)=X(I)
      RETURN
      END
```

The purpose of subroutine IMPULS is to put the Kronecker impulse function into an array. The call statement is:

CALL IMPULS (LD, D, K)

The result is that each element of array D, of length LD, is zero, except element D(K) which is unity. The program is:

```
CIMPULS IS KRONECKER IMPULSE FUNCTION
      SUBROUTINE IMPULS (LD, D, K)
      DIMENSION D(2)
      DO 10  I=1,LD
   10    D (I)=0.0
      D (K)=1.0
      RETURN
      END
```

The purpose subroutine SCALE is to multiply each element of an array by a scale factor. The call statement is:

CALL SCALE (S, LX, X)

The result is that each element of array X, of length LX, has been multiplied by the scale factor S. The program is:

```
CSCALE IS SCALE AN ARRAY BY MULTIPLYING EACH ELEMENT
C BY A SCALE FACTOR
      SUBROUTINE SCALE (S, LX, X)
      DIMENSION X(2)
      DO 10 I=1,LX
   10 X (I)=S*X (I)
      RETURN
      END
```

The *dot product* of the vector $x=(x_1, x_2, \ldots, x_n)$ with the vector $y = (y_1, y_2, \ldots, y_n)$ is defined as

$$x_1 y_1 + x_2 y_2 + \ldots + x_n y_n$$

The purpose of subroutine DØT is to form the dot product of two vectors. The call statement is:

CALL DØT (L, X, Y, ANS)

The x vector is in array X of length L $(= n)$; the y vector is in array Y also of length L $(= n)$. Here L, X, Y represent subroutine inputs; the subroutine output is ANS (standing for "answer"), which is the required dot product. The program is:

```
CDOT IS DOT PRODUCT
      SUBROUTINE DOT (L, X, Y, ANS)
      DIMENSION X(2), Y(2)
      ANS=0.0
      IF(L)  30,30,10
   10 DO 20  I=1,L
   20 ANS=ANS+X(I)*Y(I)
   30 RETURN
      END
```

The dot product reverse of the vector $x = (x_1, x_2, \ldots, x_n)$ with the vector $y = (y_1, y_2, \ldots, y_n)$ is defined as

$$x_1 y_n + x_2 y_{n-1} + \ldots + x_{n-1} y_2 + x_n y_1$$

The purpose of subroutine DØTR is to form the dot product reverse of two vectors. The call statement is:

CALL DØTR (L, X, Y, ANS)

Here L, X, Y are inputs and ANS, the required dot product reverse, is the output. The program is:

```
CDOTR IS DOT PRODUCT REVERSE
      SUBROUTINE DOTR (L, X, Y, ANS)
      DIMENSION X(2), Y(2)
      ANS=0.0
      IF (L)   30, 30, 10
   10 DO 20  I=1,L
      J=L—I+1
   20 ANS=ANS+X(I)*Y(J)
   30 RETURN
      END
```

Subroutine SYMDØT computes a symmetrical dot product. The program with the explanation given by comment cards is:

```
CSYMDOT       SYMMETRICAL DOT PRODUCT
      SUBROUTINE SYMDOT(Y,F,N,D)
C     COMPUTES,
C                         N—1
C     D = F(N)*Y(N) + SUM (F(N—I)*(Y(N—I)+Y(N+I)))
C                         I=1
C     AND DOES SO RATHER EFFICIENTLY.
C     FOR NORMAL USE THE FOLLOWING MEANINGS APPLY
C     Y=DATA    (DATA LENGTH OF 2*N—1 IS OPERATED ON)
C     F=ONE LOBE + CENTER POINT OF A SYMMETRICAL FILTER,
C                                                    F(N)=CENTER
C     N=LENGTH OF F  (ONE LOBE + CENTER)            F( 1)=END
C     D=DOT PRODUCT  (A SINGLE NUMBER)
C     PROGRAM OCCUPIES 75 REGISTERS
      DIMENSION Y(100), F(100)
      D=0.0
      M=N—1
      K=N+M
C     COMPUTES FROM THE OUTSIDE IN TO THE CENTER
      DO 10 J=1,M
      D=F(J)*(Y(J)+Y(K))+D
      K=K—1
   10 CONTINUE
      D=D+F(N)*Y(N)
      RETURN
      END
```

The *convolution* (or folding) of the vector $a = (a_0, a_1, \ldots, a_m)$ with the vector $b = (b_0, b_1 \ldots, b_n)$ is defined as the vector $c = (c_0, c_1, \ldots, c_{m+n})$ whose elements are given by the formula

$$c_k = \sum_{j=0}^{k} a_j b_{k-j} = a_0 b_k + a_1 b_{k-1} + a_2 b_{k-2} + \ldots + a_k b_0$$

(where the elements of any vector outside the given range are taken to be zero).

Alternatively we may think of the elements c_k as the coefficients of the polynomial formed by the product of the polynomials whose coefficients are the elements of the two given vectors; that is

$$c_0 + c_1 z + c_2 z^2 + \ldots + c_{m+n} z^{m+n}$$
$$= (a_0 + a_1 z + a_2 z^2 + \ldots + a_m z^m)(b_0 + b_1 z + b_2 z^2 + \ldots + b_n z^n)$$

The purpose of subroutine FØLD is to calculate the convolution of two vectors. The call statement is:

CALL FØLD (LA, A, LB, B, LC, C)

The inputs are LA, A, LB, B; the outputs are LC, C. The a vector is in array A, of length LA $(= m + 1)$; the b is in array B, of length LB $(= n + 1)$. The required c vector is found in array C, of length LC $(= m + n + 1)$. A serious and inexcusable defect of the FORTRAN language is that the integers that are used as subscripts are not allowed to have numerical values that are zero or negative. Consequently, whenever a mathematical vector starts with the subscript 0, we have no choice but to make the FORTRAN vector start with the subscript 1.

Thus in the present case we have:

$$a_0 = A(1), \quad a_1 = A(2), \ldots, \quad a_m = A(LA), \text{ where } LA = m + 1$$
$$b_1 = B(1), \quad b_1 = B(2), \ldots, \quad b_n = B(LB), \text{ when } LB = n + 1$$
$$c_0 = C(1), \quad c_1 = C(2), \ldots, \quad c_{m+n} = C(LC), \text{ when } LC = m + n + 1$$

The program for subroutine FØLD is:

```
CFOLD IS FOLDING (CONVOLUTION)
      SUBROUTINE FOLD (LA, A, LB, B, LC, C)
      DIMENSION A(2), B(2), C(2)
      LC=LA+LB—1
      CALL ZERO (LC, C)
      DO   10   I=1,LA
      DO   10   J=1,LB
      K=I+J—1
   10 C(K)=A(I)*B(J)+C(K)
      RETURN
      END
```

It is seen that this program is based on the analogy of multiplication of polynomials.

Subroutine CØNSYM performs convolution with a symmetrical filter.

The program with its explanation is:

```
CCONSYM                         CONVOLUTION WITH SYMMETRICAL FILTER
      SUBROUTINE CONSYM(N,DATA,L,FILT,OUTPUT)
C        FILTER DOES NOT GO OFF ENDS OF DATA
C        N=DATA LENGTH
C        DATA=DATA
C        L=FILTER LENGTH OF ONE LOBE + CENTER POINT   ,
C        FILT=FILTER, JUST ONE LOBE+CENTER,FILT(1)=CENTER POINT,
C                                          FILT (N)=END
C        OUTPUT=DATA CONVOLVED WITH FILTER,HAS LENGTH N—L—L+2
C        EQUIVALENCE (DATA, OUTPUT) IS ALLOWED
C        SUBPROGRAMS,SYMDOT,REVERS
C        COMPUTES
C                                    L
COUTPUT(J)=FILT(1)*DATA(J+L—1)+SUM(FILT(I)*(DATA(J+I+L—2)+
C                              I=2                  DATA(J—I+L)))
C          FOR J=1,2,3,....,N—2*L+2
C        PROGRAM OCCUPIES 69 REGISTERS
      DIMENSION DATA(1000),FILT(100), OUTPUT(1000)
      LAGS=N—L—L+2
      CALL REVERS (L,FILT)
      DO 10 I=1,LAGS
      CALL SYMDOT(DATA(I),FILT,L,OUTPUT(I))
   10 CONTINUE
      CALL REVERS(L,FILT)
      RETURN
      END
```

The *cross-product* (or cross-correlation) of the vector $(x_0, x_1, \ldots, x_m)$ with the vector $(y_0, y_1, \ldots, y_n)$ is defined as the vector $(c_0, c_1, \ldots, c_l)$ whose elements are given the formula

$$c_k = \sum_{j=0}^{m} x_j y_{j-k}$$

(where values of x_j outside the range $j = 0, 1, 2, \ldots, m$ and values of y_i outside the range $i = 0, 1, 2, \ldots, n$ are taken to be 0.0).

The number of elements in the c vector is $l + 1$; The integer l is arbitrary and is specified in advance. The purpose of subroutine CRØSS is to compute the cross-product of two vectors. The call statement is:

CALL CRØSS (LX,X,LY,Y,LC,C)

The inputs are LX, X, LY, LC; the output is C. The x vector is in array X, of length LX ($=m + 1$); the y vector is in array Y, of length LY ($= n + 1$). The required c vector is found in array C, of length LC ($= l + 1$). The program for subroutine CRØSS is:

```
CCROSS IS CROSS PRODUCT
      SUBROUTINE CROSS (LX,X,LY,Y,LC,C)
      DIMENSION X(2), Y(2), C(2)
      DO 10 I=1,LC
```

```
   10 CALL  DOT(XMINoF(LY+I—1,LX)—I+1,X(I),Y,C(I))
      RETURN
      END
```

If X and Y are the same array, CRØSS gives us the autocorrelation.

A cross-correlation subroutine based on FØLD is CRØSST. The calling statement is:

CALL CRØSST (N, X, M, Y, C)

where the subroutine inputs are

    N = length of X
    X = one vector to be cross-correlated
    M = length of Y
    Y = other vector to be cross-correlated

and the subroutine output is:

    C = cross-correlation

The program is (note: subroutine REVERS is given later in this section):

```
CCROSST

      SUBROUTINE  CROSST(N,X,M,Y,C)
      DIMENSION  X(100),Y(100),C(100)
      CALL  REVERS (M,Y)
      CALL  FOLD(N,X,M,Y,LC,C)
      CALL  REVERS (M,Y)
      RETURN
      END
```

Subroutine TUKEY computes the Tukey autocorrelation. The program with its explanation is:

```
   C   TUKEY IS TUKEY AUTOCORRELATION
       SUBROUTINE  TUKEY(LX,X,LACOR,ACOR,S)
   C       TUKEY COMPUTES THE AUTOCORRELATION OF X USING THE
   C       TUKEY APPROXIMATION.
   C       INPUTS ARE LX,X — THE TIME SERIES — AND LACOR
   C       OUTPUT IS ACOR WHICH IS TUKEY AUTOCORRELATION
   C                                             (NORMALIZED)
   C       OUTPUT IS S — UNNORMALIZED ZEROTH LAG
   C                                       AUTOCORRELATION
       DIMENSION  X(2),ACOR(2)
       DO  15  IT=1, LACOR
       ACOR(IT)=0.
       MM=LX—IT+1
       DO  10   J=1,MM
       I=J+IT—1
   10  ACOR(IT)=X(I)*X(J)+ACOR(IT)
       SCALE=1.0/FLOATF(LX—IT+1)
```

```
    15 ACOR(IT)=SCALE*ACOR(IT)
       S=ACOR(1)
       SCALE=1./ACOR(1)
       DO 20 I=1,LACOR
    20 ACOR(I)=ACOR(I)*SCALE
       RETURN
       END
```

The purpose of subroutine MINSN is to find the minimum element of an array, taking into account the algebraic signs of the elements. The calling statement is:

CALL MINSN (LX, X, XMIN, INDEX)

The inputs are LX, X; the outputs are XMIN, INDEX. The given array is of length LX and is in X. The minimum element of this array is XMIN = X(INDEX), that is, INDEX is the subscript of the minimal element XMIN. The program is:

```
       SUBROUTINE MINSN(LX,X,XMIN,INDEX)
       DIMENSION X(2)
       XMIN=X(1)
       DO 10 I=1,LX
    10 XMIN=MIN1F(XMIN,X(I))
       DO 20 J=1,LX
       INDEX=J
       IF(X(J)—XMIN)20,30,20
    20 CONTINUE
    30 RETURN
       END
```

Subroutine MAXSN is just like MINSN, except that it finds the maximum element. The program is:

```
       SUBROUTINE MAXSN(LX,X,XMAX,INDEX)
       DIMENSION X(2)
       XMAX=X(1)
       DO 10 I=1,LX
    10 XMAX=MAX1F(XMAX,X(I))
       DO 20 J=1,LX
       INDEX=J
       IF(X(J)—XMAX)20,30,20
    20 CONTINUE
    30 RETURN
       END
```

The purpose of subroutine NØRME is to normalize an array by dividing each element by the RMS energy of the array. The calling statement is:

CALL NORME (LX, X)

The inputs are LX, X where LX is the number of elements in array X; the output is X, now normalized. The program is:

```
       CNORME IS NORMALIZATION OF ARRAY BY ITS RMS ENERGY
          SUBROUTINE NORME (LX,X)
          DIMENSION X(2)
          IF(LX)   30,30,10
       10 CALL  DOT  (LX,X,X,XX)
          Z=SQRTF(XX)
          DO 20   I=1,LX
    20    X(I)=X(I)/Z
    30    RETURN
          END
```

A similar program is NØRM 1, which normalizes an array by its first element.
The program is:

```
        CNORM1 IS NORMALIZATION OF ARRAY BY ITS FIRST TERM
          SUBROUTINE NORM1(LX,X)
          DIMENSION X(2)
          IF(LX)  30,30,10
    10    DO 20   I=1,LX
    20    X(I)=X(I)/X(1)
    30    RETURN
          END
```

The purpose of subroutine REVERS is to reverse, that is, interchange,
the order of the elements of an array. The calling statement is:

```
                    CALL REVERS (N, X)
```

The inputs are N, X, where N is the number of elements in array X; the output
is X, which now has the same elements but in the reverse order. The program is:

```
          SUBROUTINE REVERS(N,X)
          DIMENSION X(100)
          NN=N/2
          DO 10 I=1,NN
          J=N—I
          TEMP=X(I)
          X(I)=X(J+1)
          X(J+1)=TEMP
    10    CONTINUE
          RETURN
          END
```

The purpose of subroutine PØLYDV is to divide one polynomial by another,
that is, to deconvolve one vector by another. The program, with the explanation given by comment cards, is:

```
          SUBROUTINE POLYDV (N,DVS,M,DVD,L,Q)
    C     PERFORM DIVISION OF POWER SERIES OF THE FORM
    C     (DVD(1)+X*DVD(2)+X*X*DVD(3)+...)/(DVS(1)+X*DVS(2)+X*X*DVS
    C                                                     (3)+...)
    C     THE RESULT IS = Q(1)+X*Q(2)+X*X**Q(3)+...
    C     N IS THE NUMBER OF COEFFICIENTS IN THE DIVISOR POLYNOMIAL.
    C     M IS THE NUMBER OF COEFFICIENTS IN THE DIVIDEND POLYNOMIAL.
```

```
C       L IS THE NUMBER OF COEFFICIENTS IN THE QUOTIENT POLYNOMIAL.
C       DVS IS THE INPUT DIVISOR POLYNOMIAL COEFFICIENTS.
C       DVD IS THE INPUT DIVIDEND POLYNOMIAL COEFFICIENTS.
C       Q IS THE OUTPUT QUOTIENT POLYNOMIAL COEFFICIENTS.
C       DVS(1) MUST BE NON-ZERO
C       THIS COULD BE REPROGRAMMED TO ALLOW EQUIVALENCE(DVD,Q).
        DIMENSION  DVS(10),  DVD(10),  Q(10)
        NM = N—1
    5   DO 8 I=1,L
    8   Q(I)  = 0.
C       MOVE THE USED PORTION OF DVD TO Q
        MML=XMINoF(M,L)
        DO 10 I=1,MML
   10   Q(I)  = DVD(I)
        DO 50 I = 1,L
        Q(I)  = Q  (I)/DVS(1)
        IF (I—L)30,20,30
   20   RETURN
   30   K = I
        ISUB = XMINoF(NM,L—I)
        DO 40  J = 1,ISUB
        K = K+1
        Q(K)=Q(K)—Q(I)*DVS(J+1)
   40   CONTINUE
   50   CONTINUE
C       PROGRAM NEVER GETS HERE
        END
```

The purpose of subroutine MAINE is to invert a symmetric matrix. The call statement is:

```
CALL MAINE (N, A, B)
```

The inputs are N, A; the output is B. The symmetric matrix has order N, and is stored fortranwise in array A. The inverse of this matrix is found stored fortranwise in array B. The program is:

```
CMAINE IS SYMMETRIC MATRIX INVERSE
        SUBROUTINE MAINE(N,A,B)
        DIMENSION A(2),B(2)
        B(1)=1.0/A(1)
        IF(N—1)60,60,2
    2   CALL ZERO(N*N—1,B(2))
        DO 50  M=2,N
        K=M—1
        MM=M+(M—1)*N
        EK=A(MM)
        DO 10  I=1,K
        DO 10  J=1,K
        MI=M+(I—1)*N
        IJ=I+(J—1)*N
        JM=J+(M—1)*N
```

```
   10  EK=EK—A(MI)*B(IJ)*A(JM)
       B(MM)=1.0/EK
       DO 30  I=1,K
       IM=I+(M—1)*N
       DO 20  J=1,K
       IJ=I+(J—1)*N
       JM=J+(M—1)*N
   20  B(IM)=B(IM)—B(IJ)*A(JM)/EK
       MI=M+(I—1)*N
   30  B(MI)=B(IM)
       DO 40  I=1,K
       IM=I+(M—1)*N
       DO 40  J=1,K
       MJ=M+(J—1)*N
       IJ=I+(J—1)*N
   40  B(IJ)=B(IJ)+B(IM)*B(MJ)*EK
   50  CONTINUE
   60  RETURN
       END
```

Subroutine MAIN1 is a speeded-up version of MAINE. The program is:

```
C       MAIN1 IS FAST VERSION OF MAINE (SYMMETRIC MATRIX INVERSE)
        SUBROUTINE MAIN1(N,A,B)
        DIMENSION  A(2),B(2)
        B(1)=1.0/A(1)
        IF(N—1)60,60,2
     2  NN=N*N
        DO 4  I=2,NN
     4  B(I)=0.0
        MM=1
        KN=0
        DO 50 M=2,N
        K=M—1
        MM=MM+N+1
        KN=KN+N
        EK=A(MM)
C       KN IS (M—1)*N
        MI=M—N
        DO 10 I=1,K
        MI=MI+N
        IJ=I—N
        JM=KN
        DO 10 J=1,K
        IJ=IJ+N
        JM=JM+1
    10  EK=EK—A(MI)*B(IJ)*A(JM)
        B(MM)=1.0/EK
        MI=M—N
        IM=KN
        DO 30 I=1,K
        IM=IM+1
        IJ=I—N
        JM=KN
        DO 20 J=1,K
```

```
      IJ=IJ+N
      JM=JM+1
   20 B(IM)=B(IM)—B(IJ)*A(JM)*B(MM)
      MI=MI+N
   30 B(MI)=B(IM)
      IM=KN
      DO 40 I=1,K
      IM=IM+1
      MJ=M—N
      IJ=I—N
      DO 40 J=1,K
      MJ=MJ+N
      IJ=IJ+N
   40 B(IJ)=B(IJ)+B(IM)*B(MJ)*EK
   50 CONTINUE
   60 RETURN
      END
```

The purpose of subroutine MATRA is to transpose an M by N matrix stored fortranwise in array A; the transposed matrix is placed fortranwise in array ATRAN. The calling is:

$$\text{CALL MATRA (A, M, N, ATRAN)}$$

One may place the ATRAN region in the A region by the call statement

$$\text{CALL MATRA (A, M, N, A)}$$

The program is:

```
      SUBROUTINE MATRA (A,M,N,ATRAN)
      DIMENSION A(12),ATRAN(12)
C           M(ROWS)  N(COLUMNS)  INPUT
      NM=N*M
C           ALL VALUES ARE NOW TO BE MADE ODD
      DO 1 I=1,NM
B   1 ATRAN(I)=A(I)+00000000001
      NMMI=NM—1
      DO 2 L=2,NMMI
B     IF(ATRAN(L)*00000000001)5,2,5
B   5 TEMPB=A(L)*77777777776
      K=L
    3 KM=(K—1)/M
      I=K—KM*M
      J=KM+1
      K=N*(I—1)+J
      TEMPA=A(K)
      ATRAN(K)=TEMPB
B     TEMPB=TEMPA*77777777776
      IF(K—L)3,2,3
    2 CONTINUE
      RETURN
      END
```

The purpose of subroutine POLYEV is to evaluate a polynomial. The program, with the explanation given by comment cards, is:

```
CPOLYEV                    EVALUATE POLYNOMIAL
       SUBROUTINE POLYEV(N,C,X,A)
C      COMPUTES THE FORMULA
C
C
C      A=C(1)+C(2)*X+C(3)*X²+C(4)*X³+...+C(N)*X^(N—1)
C
C      EXAMPLE, IF N=5, THE FORMULA IS EVALUATED IN THE
C      FOLLOWING WAY
C      A=C(1)+  X*(C(2)+X*(C(3)+X*(C(4)+X*C(5))))
C      THUS ONLY N—1 ADDS AND N—1 MULTIPLYS ARE REQUIRED
C      TIMING — APPROXIMATELY   .4N MILLISEC ON IBM 709
       DIMENSION C(100)
       A=0.
       DO 10 I=1,N
       J=N—I
    10 A=X*A+C(J+1)
       RETURN
       END
```

Subroutine POLYEV evaluates a polynomial for a real value of the argument X; a similar subroutine IPLYEV evaluates a polynomial for a complex value of its argument.

The purpose of subroutine PSQRT is to compute the square root of a polynomial. The program with explanation is:

```
CPSQRT                  COMPUTE SQUARE ROOT OF A POLYNOMIAL
       SUBROUTINE PSQRT(N,C,M,A)
C      GIVEN THE POLYNOMIAL
C
C
C      P=C(1)+C(2)*X+C(3)*X²+C(4)*X³+...+C(N)*X^(N—1)
C
C      FIND THE FIRST M COEFFICIENTS OF ITS SQUARE ROOT.
C      INPUTS—
C      N=THE NUMBER OF COEFFICIENTS IN THE GIVEN
C      POLYNOMIAL
C      C=THE VECTOR OF COEFFICIENTS OF THE GIVEN
C      POLYNOMIAL
C      M=THE DESIRED NUMBER OF COEFFICIENTS IN THE SQUARE
C      ROOT POLYNOMIAL
C      OUTPUTS
C      A=THE VECTOR OF COEFFICIENTS OF THE OUTPUT
C      POLYNOMIAL
C      TIMING—APPROXIMATELY ON IBM 709
C           3+ M +.06*M*M     MILLISECONDS
C      RESTRICTION — BOTH N AND M MUST BE GREATER THAN OR
         EQUAL TO 4, HOWEVER C(I) MAY EQUAL ZERO FOR I LESS
         THAN 4.
```

```
C              C(1) MUST NOT = ZERO.
               DIMENSION A(100), C(100)
               A(1)=SQRTF(C(1))
               TA=2.*A(1)
               A(2)=C(2)/TA
               A(3)=(C(3)—A(2)*A(2))/TA
               DO 100  I=4,M
               IF(I—N)  20,20,10
       10  PA=0.
               GO TO 30
       20  PA=C(I)
       30  CONTINUE
               PS=0.
               IH=I/2
               DO 4c J=2,IH
               K=I—J
       40  PS=PS+A(J)*A(K+1)
               PA=PA—2.*PS
               IF(2*IH—I)  50,60,50
       50  PA=PA—A(IH+1)*A(IH+1)
       60  A(I)=PA/TA
      100  CONTINUE
               RETURN
               END
```

Subroutine GENSYM generates a symmetric vector given one side. Its program is:

```
CGENSYM              GENERATE A SYMMETRIC VECTOR GIVEN ONE SIDE
               SUBROUTINE  GENSYM(N,A,B)
C              INPUTS
C              ANY SERIES A OF LENGTH N
C              OUTPUTS
C              A SERIES B DEFINED IN THE FOLLOWING WAY
C              B(I)=A(N—I+1)      FOR I=1  TO I=N
C              B(I)=A(I—N+1)      FOR I=N  TO I=N+N—1
C              THUS FOR EXAMPLE IF A IS THE CENTER PLUS ONE LOBE OF
C              AN AUTOCOR.
C              THEN B IS THE WHOLE AUTOCOR. WITH CENTER AT B(N)
C              EQUIVALENCE (A,B)      IS ALLOWED.
               DIMENSION  A(100),B(100)
               DO 10  I=1,N
               J=N+I
       10  B(J—1)=A(I)
               IN=N+N
               DO 20  I=1,N
               J=IN—I
       20  B(I)=B(J)
               RETURN
               END
```

Subroutine MAP prints out a matrix. Its program is:

```
CMAP IS MATRIX PRINT
               SUBROUTINE  MAP(N1,N2,ID1,NCL,FMT,A)
```

```
C         EXAMPLE OF USE
C         CALL MAP (N1,N2,ID1,10,9H (10F9.4),A)
C    N1 = MAXIMUM VALUE OF I IN A(I,J)
C    N2 = MAXIMUM VALUE OF J IN A(I,J)
C   ID1 = FIRST INDEX IN DIMENSION STATEMENT OF A IN CALLING
C         PROGRAM
C  NCL = NUMBER OF COLUMNS PER LINE ON PRINTED PAGE
C  FMT = FORMAT OF DATA OUTPUT
C     A = MATRIX
          DIMENSION A(2),FMT(2)
          L=NCL*ID1
          LAST=ID1*N2
          IMIN=1
          IMAX=XMIN0F(L,N2*ID1)
          J=N1
          K=1
   10     WRITE OUTPUT TAPE 6,FMT,(A(I),I=IMIN,IMAX)
          IMIN=IMIN+1
          IF(IMIN—J)      10,10,20
   20     J=J+L
          WRITE OUTPUT TAPE 6,22
   22     FORMAT(1H1)
          IMIN=L*K+1
          K=K+1
          IF(LAST—IMIN)  25,30,30
   25     IMAX=XMIN0F(IMIN+L—1,LAST)
          GO TO 10
   30     RETURN
          END
```

3. The single-channel filter package

The subroutines in this package are:

EUREKA	(Levinson least-squares iteration method)
NEXTA	(required by EUREKA)
NEXTC	(required by EUREKA)
INVTØP	(inverse Toeplitz matrix)
SHAPE	(shaping filter)
SPIKE	(spiking filter)
SIDE	(Simpson sideways iteration)
SPIKEE	(spiking filter, more efficient than SPIKE)
FILLS1	(Robinson least-squares iteration method)
AJAX1	(required by FILLS1)
ALBE1	(required by FILLS1)
FIRM1	(required by FILLS1)
WVPRED	(wavelet prediction)
BNDPAS	(band-pass filter)
BNDPS2	(band-pass filter)

TREN   (traveling energy)
TRAP   (traveling power)
TRAF   (traveling correlation filter)
REMAV   (remove arithmetic average)
RMSDEV (root-mean-square deviation)
PHASE   (constant phase-shift filter)

Subroutine EUREKA together with its auxiliary subroutines NEXTA and NEXTC solves the least-squares normal equations for the coefficients of the discrete single-channel Wiener filter by the Levinson method, as given on page 136-139 of Norbert Wiener:

Extrapolation, Interpolation and Smoothing of Stationary Time Series (John Wiley, New York, 1949). Briefly, the Levinson method makes use of an auxiliary sequence $C_k^{(M)}$, and one iteration involves finding

$$C_0^{(M+1)}, \; C_1^{(M+1)}, \; \ldots, \; C_{M+1}^{(M+1)}$$

from

$$C_0^{(M)}, \; C_1^{(M)}, \; \ldots, \; C_M^{(M)}$$

For example, suppose that we know the solution of the equations:

$$\begin{bmatrix} r_0 & r_1 & r_2 & r_3 \\ r_1 & r_0 & r_1 & r_2 \\ r_2 & r_1 & r_0 & r_1 \end{bmatrix} \begin{bmatrix} C_0^{(2)} \\ C_1^{(2)} \\ C_2^{(2)} \\ -1 \end{bmatrix} = \begin{bmatrix} 0 \\ 0 \\ 0 \\ 0 \end{bmatrix} \tag{1}$$

for $M = 2$. We then transpose these equations to:

$$\begin{bmatrix} r_1 & r_0 & r_1 & r_2 \\ r_2 & r_1 & r_0 & r_1 \\ r_3 & r_2 & r_1 & r_0 \end{bmatrix} \begin{bmatrix} -1 \\ C_2^{(2)} \\ C_1^{(2)} \\ C_0^{(2)} \end{bmatrix} = \begin{bmatrix} 0 \\ 0 \\ 0 \\ 0 \end{bmatrix} \tag{2}$$

Next we augment (1) up and to the left, thus defining δ:

$$\begin{bmatrix} r_0 & r_1 & r_2 & r_3 & r_4 \\ r_1 & r_0 & r_1 & r_2 & r_3 \\ r_2 & r_1 & r_0 & r_1 & r_2 \\ r_3 & r_2 & r_1 & r_0 & r_1 \end{bmatrix} \begin{bmatrix} 0 \\ C_0^{(2)} \\ C_1^{(2)} \\ C_2^{(2)} \\ -1 \end{bmatrix} = \begin{bmatrix} \delta \\ 0 \\ 0 \\ 0 \\ 0 \end{bmatrix} \tag{3}$$

Then we augment (2) up and to the right, thus defining σ:

$$
\begin{bmatrix}
r_0 & r_1 & r_2 & r_3 & r_4 \\
\hline
r_1 & r_0 & r_1 & r_2 & r_3 \\
r_2 & r_1 & r_0 & r_1 & r_2 \\
r_3 & r_2 & r_1 & r_0 & r_1
\end{bmatrix}
\begin{bmatrix}
-1 \\
C_2^{(2)} \\
C_1^{(2)} \\
C_0^{(2)} \\
0
\end{bmatrix}
=
\begin{bmatrix}
\sigma \\
0 \\
0 \\
0 \\
0
\end{bmatrix}
\tag{4}
$$

We then form: $(3) - C_0^{(3)}\,(4) = 0$, thus defining $C_0^{(3)}$ by $\delta - C_0^{(3)}\,\sigma = 0$

$$
\begin{bmatrix}
r_0 & r_1 & r_2 & r_3 & r_4 \\
r_1 & r_0 & r_1 & r_2 & r_3 \\
r_2 & r_1 & r_0 & r_1 & r_2 \\
r_3 & r_2 & r_1 & r_0 & r_1
\end{bmatrix}
\left\{
\begin{bmatrix}
0 \\
C_0^{(2)} \\
C_1^{(2)} \\
C_2^{(2)} \\
-1
\end{bmatrix}
- C_0^{(3)}
\begin{bmatrix}
-1 \\
C_2^{(2)} \\
C_1^{(2)} \\
C_0^{(2)} \\
0
\end{bmatrix}
\right\}
= 0
\tag{5}
$$

or

$$
\begin{bmatrix}
r_0 & r_1 & r_2 & r_3 & r_4 \\
r_1 & r_0 & r_1 & r_2 & r_3 \\
r_2 & r_1 & r_0 & r_1 & r_2 \\
r_3 & r_2 & r_1 & r_0 & r_1
\end{bmatrix}
\begin{bmatrix}
C_0^{(3)} \\
C_0^{(2)} - C_0^{(3)}C_2^{(2)} \\
C_1^{(2)} - C_0^{(3)}C_1^{(2)} \\
C_2^{(2)} - C_0^{(3)}C_0^{(2)} \\
-1
\end{bmatrix}
= 0
\tag{6}
$$

Finally we define

$$
\begin{bmatrix}
C_0^{(3)} \\
C_1^{(3)} \\
C_2^{(3)} \\
C_3^{(3)} \\
-1
\end{bmatrix}
=
\begin{bmatrix}
C_0^{(3)} = \delta/\sigma \\
C_0^{(2)} - C_0^{(3)}C_2^{(2)} \\
C_1^{(2)} - C_0^{(3)}C_1^{(2)} \\
C_2^{(2)} - C_0^{(3)}C_0^{(2)} \\
-1
\end{bmatrix}
\tag{7}
$$

Then (6) becomes

$$
\begin{bmatrix}
r_0 & r_1 & r_2 & r_3 & r_4 \\
r_1 & r_0 & r_1 & r_2 & r_3 \\
r_2 & r_1 & r_0 & r_1 & r_2 \\
r_3 & r_2 & r_1 & r_0 & r_1
\end{bmatrix}
\begin{bmatrix}
C_0^{(3)} \\
C_1^{(3)} \\
C_2^{(3)} \\
C_3^{(3)} \\
-1
\end{bmatrix}
=
\begin{bmatrix}
0 \\
0 \\
0 \\
0 \\
0
\end{bmatrix}
\tag{8}
$$

which shows that $C_0^{(3)}$, $C_1^{(3)}$, $C_2^{(3)}$, $C_3^{(3)}$ are
the desired coefficients for $M + 1 = 3$.

The programs are:

```
      CEUREKA IS FILTER (LEAST SQUARES)
            SUBROUTINE EUREKA (LR,R,G,A,C)
C           C(I)  I=1,2*LR  WORKING SPACE
            DIMENSION R(2),G(2),A(2),C(4)
            A(1)=G(1)/R(1)
            C(1)=—1.
            IF(LR—1)30,30,2
         2  DO 20 LA=2,LR
            CALL NEXTC(LA—1,R,C,C(LR+1))
            DO 10  J=1,LA
            I=LR+J
        10  C(J)=C(I)
        20  CALL NEXTA(G(LA),LA—1,A,C,R,A)
        30  RETURN
            END

      CNEXTC IS CALLED BY EUREKA
            SUBROUTINE NEXTC(LC,R,C,CP)
C           NON EQUIVALENCE (C,CP)
            DIMENSION R(2),C(2),CP(3)
            CALL DOT(LC,R(2),C,CONST1)
            CALL DOTR(LC,R,C,CONST2)
            CP(1)=CONST1/CONST2
            IF(LC—1)      30,30,10
        10  DO 20  I=2,LC
            J=LC—I+1
        20  CP(I)=C(I—1)—CP(1)*C(J)
        30  CP(LC+1)=—1.
            RETURN
            END

      CNEXTA IS CALLED BY EUREKA
            SUBROUTINE NEXTA(G,LA,A,C,R,AP)
C           MAY BE EQUIVALENCE (A,AP)
C           A(I)  I=1,LA
C           C(I)  I=1,LA+1
C           R(I)  I=1,LA+1
C           AP(I)  I=1,LA+1
            DIMENSION A(2),C(3),R(3),AP(3)
            CALL DOTR(LA,A,R(2),T1)
            CALL DOTR(LA+1,C,R,T2)
            AP(LA+1)=(T1—G)/T2
            DO 10  I=1,LA
        10  AP(I)=A(I)—AP(LA+1)*C(I)
            RETURN
            END
```

One example of the use of subroutine EUREKA is for the calculation of the inverse of a Toeplitz matrix (*cf.* Grenander and Szegö: *Toeplitz Forms and their Applications*, University of California Press). This is accomplished by subroutine INVTØP, which calls EUREKA. The calling statement is:

```
      CALL INVTØP (LR, R, Q, SPACE)
```

The inputs are LR, R; the output is Q; working space is SPACE. LR is the order of the Toeplitz matrix and R is its first row. The inverse matrix is found in Q. The program is:

```
CINVTOP IS INVERSE TOEPLITZ MATRIX
        SUBROUTINE INVTOP (LR,R,Q,SPACE)
C       INVERSE MATRIX STORED BY ROWS OR COLUMNS IN Q(I),
C       I=1,LR*LR
C       WORKING SPACE(I),I=1,3*LR
        DIMENSION R(3),Q(9),SPACE(9)
        DO 10  K=1,LR
        CALL IMPULS (LR, SPACE,K)
        J=LR*(K—1)+1
        CALL EUREKA (LR,R,SPACE,Q(J),SPACE(LR+1))
   10   CONTINUE
        RETURN
        END
```

Three numerical examples computed by subroutine INVTØP are:

Example (*a*):

LR = 5				
R = 10.00000	4.00000	—1.00000	—4.00000	—4.00000
Q = 0.14899	—0.05657	0.02727	0.03232	0.02677
—0.05657	0.16566	—0.07273	0.01010	0.03232
0.02727	—0.07273	0.16364	—0.07273	0.02727
0.03232	0.01010	—0.07273	0.16566	—0.05657
0.02677	0.03232	0.02727	—0.05657	0.14899

Example (*b*):

LR = 5				
R = 10.00000	4.00000	1.00000	4.00000	4.00000
Q = 0.15323	—0.06452	0.03763	—0.06452	—0.01344
—0.06452	0.17921	—0.08602	0.06810	—0.06452
0.03763	—0.08602	0.16129	—0.08602	0.03763
—0.06452	0.06810	—0.08602	0.17921	—0.06452
—0.01344	—0.06452	0.03763	—0.06452	0.15323

Example (*c*):

LR = 5				
R = 16.72605	0.	10.75437	0.	2.44141
Q = 0.12867	0.	—0.12045	—0.	0.05867
0.	0.10192	—0.	—0.06553	0.
—0.12045	—0.	0.21468	0.	—0.12045
—0.	—0.06553	0.	0.10192	—0.
0.05867	0.	—0.12045	—0.	0.12867

Another example of the use of subroutine EUREKA is for calculation of shaping filters (*cf.* E. A. Robinson, "Wavelet composition of time series", *Econometric Model Buildning, Essays on the Causal Chain Approach*, edited by

H. Wold, North Holland Publishing Company, Amsterdam, 1964, page 37-106). This is accomplished by subroutine SHAPE. The calling statement is:

        CALL SHAPE (LB,B,LD,D,LA,A,LC,C,ASE,SPACE)

where the subroutine inputs are:

LB = length of B
 B = wavelet that is input to shaping filter
LD = length of D
 D = wavelet that is deisred output of filter
LA = length of A

and the subroutine outputs are

  A = coefficients of shaping filter
 LC = length of C
  C = wavelet that is actual output of filter
ASE = average squared error between desired and actual outputs of filter
SPACE = working space

The program is:

```
CSHAPE IS SHAPING FILTER
        SUBROUTINE SHAPE (LB,B,LD,D,LA,A,LC,C,ASE,SPACE)
C       INPUTS ARE LB,B,LD,D,LA
C       OUTPUTS ARE A,LC,C,ASE
C       SPACE (I),I=1,4*LA WORKING SPACE
C       MAY BE EQUIVALENCE (C,D)
        DIMENSION B(2),D(1),A(4),C(5),SPACE(16)
        CALL CROSS(LB,B,LB,B,LA,SPACE)
        CALL CROSS(LD,D,LB,B,LA,SPACE(LA+1))
        CALL EUREKA(LA,SPACE,SPACE(LA+1),A,SPACE(2*LA+1))
        CALL DOT(LD,D,D,DD)
        CALL DOT(LA,A,SPACE(LA+1),AG)
        ASE=(DD—AG)/DD
        CALL FOLD(LA,A,LB,B,LC,C)
        RETURN
        END
```

Some numerical examples computed by subroutine SHAPE are:

Example (a):

```
    LB =    2
     B = —1.25000     1.00000     (minimum-delay)
    LD = 5
     D =    2.00        1.00        0.00        1.00        2.00
    LA =    2
     A = —1.08041    —1.01483
    LC =    3
     C =    1.35051     0.18813    —1.01483
   ASE =    0.71109
 SPACE =    2.56250    —1.25000    —1.50000    —1.25000
```

Example (b):

LB =	2				
B =	1.00000	—1.25000	(maximum-delay)		
LD =	5				
D =	2.00	1.00	0.00	1.00	2.00
LA = 2					
A =	0.63388	0.69945			
LC =	3				
C =	0.63388	—0.09290	—0.87432		
ASE =	0.88251				
SPACE =	2.56250	—1.25000	0.75000	1.00000	

The above two examples show that it is more difficult to shape the maximum-delay input into this particular desired output than the minimum-delay input.

Example (c):

LB =	3				
B =	1.56250	—1.76770	1.00000	(minimum-delay)	
LD =	5				
D =	2.00	1.00	0.00	1.00	2.00
LA =	2				
A =	0.90811	1.01672			
LC =	4				
C =	1.41892	—0.01663	—0.88916	1.01672	
ASE =	0.61621				
SPACE =	6.56617	—4.52973	1.35730	2.56250	

Example (d):

LB =	3				
B ==	1.00000	—1.76770	1.56250	(maximum-delay)	
LD ==	5				
D =	2.00	1.00	0.00	1.00	2.00
LA =	2				
A =	0.58120	0.79120			
C =	0.58120	—0.23618	—0.49049	1.23625	
ASE =	0.78375				
SPACE =	6.56617	—4.52973	0.23230	2.56250	

The above two examples show again that the minimum-delay input does a better job of shaping for this particular desired output.

Example (e):

LB =	5				
B =	2.00	1.00	0.00	—1.00	—2.00
LD =	5				
D =	2.00	1.00	0.00	1.00	2.00
LA =	2				
A =	—0.	0.			
LC =	6				
C =	—0.	0.	0.	0.	0.
	—0.				
ASE =	1.00000				
SPACE =	10.00000	4.00000	0.	0.	

Here it is not possible to shape the input into the desired output; the average square error is 100 percent.

Subroutine SPIKE, which computes spiking filters (*cf.* E. A. Robinson, *op. cit.*), also makes use of subroutine EUREKA through subroutine SHAPE. The calling statement is:

CALL SPIKE (LB, B, LA, A, INDEX, ASE, S)

where the subroutine inputs are

    LB = length of B
     B = wavelet that is input to spiking filter
    LA = length of A

and the subroutine outputs are

        A = coefficients of spiking filter
    INDEX = optimum spike position
      ASE = average squared error
        S = working space

The program is:

```
      CSPIKE IS SPIKING FILTER
            SUBROUTINE SPIKE(LB,B,LA,A,INDEX,ASE,S)
            DIMENSION B(2),A(2),S(14)
      C     INPUTS ARE LB,B,LA
      C     OUTPUTS ARE A,INDEX,ASE
      C     WORKING SPACE IS S(I),I=1,2*(LA+LB-1)+4*LA
            LD=LA+LB-1
            DO 10 I=1,LD
            CALL IMPULS(LD,S(LD+1),I)
            CALL SHAPE(LB,B,LD,S(LD+1),LA,A,LD,S(LD+1),S(I),S(2*LD+1))
      10    CONTINUE
            CALL MINSN(LD,S,ASE,INDEX)
            CALL IMPULS(LD,S(LD+1),INDEX)
            CALL SHAPE(LB,B,LD,S(LD+1),LA,A,LD,S(LD+1),ASE,S(2*LD+1))
            RETURN
            END
```

The values of ASE for each spike position are found in the first LD = LA + LB — 1 cells of S.

Two numerical examples computed by subroutine SPIKE are:

Example (a):

    LB = 5
     B = 2.00000   1.00000   0.   1.00000   2.00000
    LA = 8
     A = 0.10995   —0.03986   0.09315   —0.10514   —0.15317
          0.07600   —0.15306   0.35695

```
        INDEX  =  12
          ASE  =  0.28610
            S  =  0.28610      0.35022     0.35150     0.35908     —0.30458
                  0.34852      0.34852     0.30458     0.35908      0.35150
                  0.35022      0.28610
```

Example (*b*):

```
          LB  =  5
           B  =    2.00000     1.00000     0.         —1.00000    —2.00000
          LA  =    8
           A  =  —0.02714    —0.11459     0.11508   —0.11487    —0.09633
                 —0.07348     0.16259   —0.35686
        INDEX  =  12
          ASE  =    0.28627
            S  =    0.28627     0.28899     0.38070     0.38788     0.32764
                    0.32851     0.32851     0.32764     0.38788     0.38070
                    0.28899     0.28627
```

A computationally more efficient spiking filter subroutine is SPIKEE; it
makes use of Simpson sideways iteration, as performed by subroutine SIDE,
whose program is:

```
CSIDE MUST BE USED IN CONJUNCTION WITH EUREKA TO ITERATE
CSIDEWAYS
        SUBROUTINE SIDE (G,LA,A,C,R,AP)
C       MAY BE CALLED WITH EQUIVALENCE (A,AP)
        DIMENSION A(2),C(2),R(2),AP(2)
        CALL DOT(LA—1,A,R(2),T1)
        CALL DOT(LA—1,C,R(2),T2)
        CALL DOTR(LA,C,R,T3)
        ALA=A(LA)
        AP1=(—G+T1+ALA*T2)/T3
        IF(LA—1) 40,30,10
   10   DO 20 I=2,LA
        J=LA—I+2
   20   AP(J)=A(J—1)+ALA*C(J—1)—AP1*C(I—1)
   30   AP(1)=AP1
   40   RETURN
        END
```

The program for subroutine SPIKEE, with the explanation given by comment
cards, is:

```
        SUBROUTINE SPIKEE(LB,B,LA,A,LC,C,INDEX,ERRORS,SPACE)
C       INPUTS ARE
C         LB=LENGTH OF INPUT WAVELET
C          B=INPUT WAVELET
C         LA=LENGTH OF FILTER
C       OUTPUTS ARE
C          A=FILTER FOR OPTIMUM SPIKE POSITION
C         LC=LA+LB—1=LENGTH OF ACTUAL OUTPUT
C          C=ACTUAL OUTPUT FOR OPTIMUM SPIKE POSITION
C       INDEX=OPTIMUM SPIKE POSITION AS FORTRAN SUBSCRIPT
```

```
C          ERRORS=ACTUAL SUMS OF SQUARED ERRORS, ONE SUM FOR
C             EACH SPIKE POSITION FROM 1 TO LC
C       SPACE TAKES 4*LA CELLS
        DIMENSION  B(1),A(1),C(1),ERRORS(1),SPACE(1)
        LC=LA+LB—1
        CALL  CROSS(LB,B,LB,B,LA,SPACE)
        DO  4  I=1,LC
        CALL  IMPULS(LC,C,I)
        CALL  CROSS(LC,C,LB,B,LA,SPACE(LA+1))
        IF(I—1)1,1,2
      1 CALL  EUREKA(LA,SPACE,SPACE(LA+1),A,SPACE(2*LA+1))
        GO TO 3
      2 CALL  SIDE(SPACE(LA+1),LA,A,SPACE(3*LA+1),SPACE,A)
      3 CALL  DOT(LA,A,SPACE(LA+1),Q)
      4 ERRORS(I)=1.—Q
        CALL  MINSN(LC,ERRORS,EMIN,INDEX)
        CALL  IMPULS(LC,C,INDEX)
        CALL  SHAPE(LB,B,LC,C,LA,A,LC,C,EMIN,SPACE)
        RETURN
        END
```

The following two tables give a complete set of spike filters and their actual
outputs.

A subroutine that does the same job as EUREKA, but instead uses the
Robinson method of solving the normal equations for the coefficients of the
discrete single-channel Wiener filter (cf. E. A. Robinson, "Mathematical
development of discrete filters for the detection of nuclear explosions", Journal
of Geophysical Research, vol. 68, pp. 5559-5567) is subroutine FILLS1. The
calling statement is:

$$\text{CALL FILLS1 (LF, R, G, F, S)}$$

where the subroutine inputs are

        LF = length of R = length of G = length of F
         R = autocorrelation from lag zero to lag LF—1
         G = crosscorrelation

and the subroutine outputs are

         F = filter coefficients
         S = working space

Subroutine FILLS1 requires the auxiliary subroutines AJAX1, ALBE1, and
FIRM1. The programs are:

```
CFILLS1 IS FILTER (LEAST SQUARES)
        SUBROUTINE FILLS1(LF,R,G,F,S)
        DIMENSION  R(2),G(2),F(2),S(2)
```

TABLE I

—0.79174	—0.62308	—0.48556	—0.37233	—0.27771	—0.19697	—0.12609	—0.06151
0.01032	—0.77885	—0.60695	—0.46541	—0.34713	—0.24622	—0.15761	—0.07688
0.01290	0.02644	—0.75869	—0.58176	—0.43392	—0.30777	—0.19701	—0.09610
0.01612	0.03305	0.05164	—0.72720	—0.54240	—0.38471	—0.24626	—0.12013
0.02015	0.04132	0.06454	0.09100	—0.67800	—0.48089	—0.30783	—0.15016
0.02519	0.05165	0.08068	0.11375	0.15251	—0.60111	—0.38479	—0.18770
0.03149	0.06456	0.10085	0.14219	0.19063	0.24861	—0.48098	—0.23463
0.03936	0.08070	0.12606	0.17773	0.23829	0.31076	0.39877	0.29328
0.04920	0.10087	0.15758	0.22217	0.29786	0.38845	0.49846	0.63340

A complete set of digital spike filters. These filters are designed so as to transform the input wavelet $b = (— 1.25, 1.00)$ as well as possible into an output spike. Each filter has eight coefficients, i.e. $(a_0, a_1, a_2, \ldots, a_7)$, and appears as a row of the above table. The first row is the spike filter designed to produce a spike at time $k = 0$; the second row, at time $k = 1$; $\ldots$; the last row at time $k = 8$. The actual outputs of the spike filters are shown in the following table.

TABLE 2

0.98968	—0.01290	—0.01612	—0.02015	—0.02519	—0.03149	—0.03936	—0.04920	—0.06151
—0.01290	0.98388	—0.02015	—0.02519	—0.03149	—0.03936	—0.04920	—0.06151	—0.07688
—0.01612	—0.02015	—0.97481	—0.03149	—0.03936	—0.04920	—0.06151	—0.07688	—0.09610
—0.02015	—0.02519	—0.03149	0.96064	—0.04920	—0.06151	—0.07688	—0.09610	—0.12013
—0.02519	—0.03149	—0.03936	—0.04920	0.93849	—0.07688	—0.09610	—0.12013	—0.15016
—0.03149	—0.03936	—0.04920	—0.06151	—0.07688	—0.90390	—0.12013	—0.15016	—0.18770
—0.03936	—0.04920	—0.06151	—0.07688	—0.09610	—0.12013	0.84984	—0.18770	—0.23463
—0.04920	—0.06151	—0.07688	—0.09610	—0.12013	—0.15016	—0.18770	0.76537	—0.29328
—0.06151	—0.07688	—0.09610	—0.12013	—0.15016	—0.18770	—0.23463	—0.29328	0.63340

The complete set of actual outputs for the digital spike filters given in the foregoing table. Each actual output has nine coefficients, i.e. (c_0, c_1, c_2, . . . , c_8), and appears as a row in the table. The first row is the actual output of the spike filter designed to produce a spike at time $k = 0$; the second row, at time $k = 1$; . . . ; the last row, at time $k = 8$. The input is always the minimum-delay wavelet $b = (—1.25, 1.00)$; hence, the best job is done by the spike filter for time $k = 0$, as seen by the actual spike 0.98968, and the worst job is done by the spike filter for time $k = 8$, as seen by the actual spike 0.63340. The following table gives the amplitude and phase spectrum of the input wavelet $b = (1.25, 1.00)$.

```
C     WORKING SPACE S(I) I=1,7+2*LF
      DO 10 L=1,LF
      CALL AJAX1(L,R,S(8),S(LF+8),S)
      CALL ALBE1(L,R,S(8),S(LF+8),S)
      CALL FIRM1(L,R,G,F,S(LF+8),S)
   10 CONTINUE
      CALL DOT(LF,F,G,S(7))
      RETURN
      END


CAJAX1 IS CALLED BY FILLS1 TO JACK UP A AND B
      SUBROUTINE AJAX1(L,R,A,B,S)
      DIMENSION R(2),A(2),B(2),S(2)
      IF (L—1) 10,10,20
   10 A(1)=1.0
      B(1)=1.0
      GO TO 60
   20 A(L)= —S(1)*S(3)
      B(L)= —S(4)*S(6)
      IF(L—2) 60,60,40
   40 L1=L—1
      DO 50 S=2,L1
      K=L—J+1
      S(7)=A(J)
      A(J)=A(L)*B(K)+A(J)
   50 B(K)=B(L)*S(7)+B(K)
   60 RETURN
      END


CALBE1 IS CALLED BY FILLS1 TO COMPUTE ALPHA AND BETA
      SUBROUTINE ALBE1(L,R,A,B,S)
      DIMENSION R(2),A(2),B(2),S(2)
      CALL DOTR(L,A,R(2),S)
      CALL DOT(L,B,R,S(2))
      S(3)=1.0/S(2)
      CALL DOTR(L,B,R(2),S(4))
      CALL DOT(L,A,R,S(5))
      S(6)=1.0/S(5)
      RETURN
      END


CFIRM1 IS CALLED BY FILLS1 TO COMPUTE FILTER COEFFICIENTS
      SUBROUTINE FIRM1(L,R,G,F,B,S)
      DIMENSION F(2),B(2),R(2),S(2),G(2)
      CALL DOTR(L—1,F,R(2),S(7))
      F(L)=(G(L)—S(7))*S(3)
      L1=L—1
      IF(L1) 30,30,10
   10 DO 20 J=1,L1
      K=L—J+1
   20 F(J)=F(L)*B(K)+F(J)
   30 RETURN
      END
```

An example of the use of subroutine FILLS1 is given by subroutine WVPRED, which performs wavelet prediction. The calling statement is:

CALL WVPRED (LB, B, LEAD, LA, A, LC, C, S)

where the subroutine inputs are

$$LB = \text{length of } B$$
$$B = \text{wavelet to be predicted (input to prediction filter)}$$
$$LEAD = \text{prediction lead}$$
$$LA = \text{length of } A$$

and where the subroutine outputs are

$$A = \text{coefficients of prediction filter}$$
$$LC = \text{length of } C$$
$$C = \text{actual output of prediction filter}$$
$$S = \text{working space, with}$$
$$S(1) = \text{autocorrelation at lag zero}$$
$$S(2) = ASE = \text{average squared error}$$
$$S(3) = ASFE = \text{average squared future error}$$
$$S(4) = ASPE = \text{average squared past error}$$

The program is:

```
      SUBROUTINE  WVPRED(LB,B,LEAD,LA,A,LC,C,S)
C     INPUTS LB,B,LEAD,LA
C     OUTPUTS A,LC,C,S(1)=ZERO AUTOCOR,S(2)=ASE,S(3)=ASFE,S(4)=
C      ASPE
C     WORKING SPACE S I=1,LA+LEAD+2*LA
      DIMENSION  B(2),A(2),C(2),S(2)
      CALL CROSS  (LB,B,LB,B,LA+LEAD,S)
      J=LA+LEAD
      CALL  FILLS1(LA,S,S(LEAD+1),A,S(J+1))
      CALL  FOLD(LB,B,LA,A,LC,C)
      CALL  DOT(LA,A,S(LEAD+1),AG)
      S(2)=1.0—AG/S(1)
      CALL  DOT(LEAD,B,B,S(3))
      S(3)=S(3)/S(1)
      S(4)=S(2)—S(3)
      RETURN
      END
```

Subroutine BNDPAS computes a band-pass filter. The program with its explanation is:

```
CBNDPAS IS BAND PASS FILTER
      SUBROUTINE  BNDPAS(N,DT,FL,FH,FILT)
C     PRODUCES ONE LOBE OF A SYMMETRICAL FILTER WHICH
C     PASSES ONLY FREQUENCIES BETWEEN FL AND FH AND
C     PASSES THEM WITH ZERO PHASE SHIFT
```

```
C    N=THE LENGTH OF THE ONE LOBE PLUS ONE CENTER POINT
CFILT=THE OUTPUT FILTER OF LENGTH N
C   FL=THE LOWER FREQUENCY PASSED
C   FH=THE HIGHER FREQUENCY PASSED
C   DT=THE TIME SPACING BETWEEN DATA POINTS
C       IF THE USER SPECIFIES FL=FH OR
C       (FH—FL) LESS THAN 1./DT*N,
C       HE WILL BE RETURNED THE NARROWEST FILTER CONSISTENT
C       WITH THE UNCERTAINTY PRINCIPLE, CENTERED AT (FH+FL)/2
C       IF FH IS HIGHER THAN THE FOLDING FREQUENCY (FH*DT
C         GREATER THAN 0.5)
C       THE SUBROUTINE RETURNS
        DIMENSION FILT(10)
        IF(FH*DT—0.5)8,8,70
     8  FN=N
        IF ((FH—FL)—1./(DT*FN))   10,30,30
    10  FC=(FH+FL)/2.
        WL=6.2831853*(FC*DT—.5/FN)
        WH=6.2831853*(FC*DT+.5/FN)
        GO TO 40
    30  WL=FL*DT*6.2831853
        WH=FH*DT*6.2831853
    40  FILT(1)=WH—WL
        DO 50  I=2,N
        FI=I—1
        FILT(I)=(SINF(WH*FI)—SINF(WL*FI))/FI
    50  CONTINUE
        DO 60  I=1,N
        FILT(I)=FILT(I)/3.14159265
    60  CONTINUE
    70  RETURN
        END
```

A related subroutine is BNDPS2, which is:

```
CBNDPS2
        SUBROUTINE BNDPS2(N,A,B,C,D)
C       SAME AS BNDPAS EXCEPT GETS TWO SIDED FILTER CENTERED
C       AT (N+1)/2
C       BNDPAS IN TRANSFER VECTOR
        DIMENSION D(100)
        D(N)=0.
        M=(N+1)/2
        CALL BNDPAS(M,A,B,C,D(M))
        MM=M+M
        DO 10  I=1,M
        J=MM—I
    10  D(I)=D(J)
        RETURN
        END
```

Two similar subroutines are TREN (traveling energy) and TRAP (traveling power). The programs with their explanations are:

```
       CTREN IS TRAVELING ENERGY
             SUBROUTINE TREN(LY,Y,L,LE,E)
C      INPUTS ARE LY,Y AND L WHICH IS SPECIFIED LENGTH OF
C         ENERGY OPERATOR
C      OUTPUTS ARE LE,E—THE TRAVELING SUM OF SQUARES OF L
C      TERMS
C      BECAUSE OF END EFFECTS L MUST NOT EXCEED LY
             DIMENSION  Y(2),E(2)
             E(1)=0.
             DO 10   I=1,L
      10     E(1)=E(1)+Y(I)*Y(I)
             IF(LY—L) 25,25,15
      15     LE=LY—L+1
             DO 20   I=2,LE
             J=I+L—1
      20     E(I)=E(I—1)—Y(I—1)*Y(I—1)*Y(J)*Y(J)
      25     RETURN
             END

       CTRAP IS TRAVELING POWER
             SUBROUTINE TRAP(LY,Y,L,LP,P)
C      INPUTS ARE LY,Y AND L WHICH IS SPECIFIED LENGTH OF
C      POWER OPERATOR
C P =  OUTPUT TRAVELING (SUM SQS OF L TERMS)/L
C      P(I)=(SUM YSQ FROM I TO I+L)/L, L MUST NOT BE GREATER
C      THAN N
C      BECAUSE OF END EFFECTS, N—L+1=LENGTH OF P
C      BOX—CAR SPECTRAL WINDOW FOR DC POWER
             DIMENSION Y(2),P(2)
             P(1) = 0.
             DO 10   I=1,L
      10     P(1)=P(1)+Y(I)**2
             IF(LY—L) 25,25,15
             LP=LY—L+1
             DO 20   I=2,LP
             J=I+L—1
      20     P(I)=P(I—1)—Y(I—1)*Y(I—1)+Y(J)*Y(J)
      25     FL=L
             DO 30   I=1,LP
      30     P(I)=P(I)/FL
             RETURN
             END
```

A very useful subroutine is TRAF, which computes a traveling correlation filter. It is

```
       CTRAF IS TRAVELING CORRELATION FILTER
             SUBROUTINE TRAF(N,DATA,L,FILTER,ISHIFT,ISUM,OUTPUT)
C      N=DATA LENGTH
C      DATA=SERIES TO BE FILTERED
C      L=LENGTH OF FILTER
C      FILTER=THE CORRELATION FILTER
C      ISHIFT=SHIFTING INCREMENT
C      ISUM=DENSITY OF SUMMATION, = 1, EVERY POINT, 2=EVERY
C      OTHER ETC
```

```
C      OUTPUT=CROSS CORRELATION OF FILTER WITH DATA
C      OUTPUT IS DENSELY PACKED REGARDLESS OF VALUES OF
C      ISHIFT AND ISUM
C      OUTPUT IS OF LENGTH (N—L+1)/ISHIFT, ROUNDED DOWN
C      EQUIVALENCE (DATA,OUTPUT) IS ALLOWED
       DIMENSION DATA (2),FILTER(2), OUTPUT(2)
       K=1
       LS=N—L+1
       DO 20   I=1,LS,ISHIFT
       TEMP=0.0
       DO 10   J=1,L,ISUM
       II=I+J
 10    TEMP=TEMP+DATA(II—1)*FILTER(J)
       OUTPUT(K)=TEMP
 20    K=K+1
       RETURN
       END
```

The purpose of subroutine REMAV is to remove the arithmetic average
from an array Y of length LY. The average is stored in AVERAG. The program is:

```
CREMAV REMOVES THE ARITHMETIC AVERAGE
       SUBROUTINE REMAV(LY,Y,AVERAG)
       DIMENSION Y(2)
       S=0.
       DO 10   I=1,LY
 10    S=S+Y(I)
       AVERAG=S/FLOATF(LY)
       DO 20   I=1,LY
 20    Y(I)=Y(I)—AVERAG
       RETURN
       END
```

The purpose of subroutine RMSDEV is to compute the RMS (root-mean-
square) deviation of an array Y of length N. The RMS deviation is stored in
cell DEV. The program is:

```
CRMSDEV
       SUBROUTINE RMSDEV(N,Y,DEV)
C      THIS ROUTINE RETURNS SQRTF(SUM(Y—YMEAN)**2)/N)
       S=0.
       DO 10   I=1,N
 10    S=S+Y(I)
       YMEAN=S/FLOATF(N)
       SS=0.
       DO 20   I=1,N
 20    SS=SS+(Y(I)—YMEAN)**2
       DEV=SQRTF(SS/FLOATF(N))
       RETURN
       END
```

The purpose of subroutine PHASE is to compute a constant phaseshift filter. The program is:

```
CPHASE
            SUBROUTINE PHASE (RAD, N, FILT)
      C         GENERATE A CONSTANT PHASE SHIFT FILTER WHERE
      C RAD=THE DESIRED PHASE SHIFT (DELAY FOR CORRELATION, AD-
      C         VANCE FOR CONVOLUTION)
      C    N=THE LENGTH OF THE FILTER
      CFILT=THE OUTPUT FILTER
      C         N IS USUALLY AN ODD NUMBER, BUT IT ISNT NECESSARILY SO
      C         PASSES ALL FREQUENCIES WITH EQUAL AMPLITUDE
      C         THE OUTPUT OF THIS FILTER IS AT FILT(N/2+1)
      C         PUT CENTER OF FILTER AT N/2+1
      C         IF N IS EVEN, FIRST TERM OF OPERATOR WILL BE ZERO
            DIMENSION FILT(100)
            DO 5  I=1,N
        5 FILT(I)=0.
            IC=N/2+1
            FILT(IC)=COSF(RAD)
            X=SINF(RAD)/1.5707963
            ICO=IC+1
            DO 10  I=ICO,N,2
            FILT(I)=X/FLOATF(I—IC)
            J=IC+IC—I
            FILT(J)=—FILT(I)
       10 CONTINUE
            RETURN
            END
```

4. Spectral package

Here we will give some elementary spectral subroutines. These are intended only for small-scale jobs; large jobs would be done by means of ultra-fast spectral subroutines such as are found in the MIT Geophysics Program Set (cf. S. M. Simpson, *Magnetic Tape Copies of M.I.T. Geophysics Program Set I (Time Series Programs for the IBM 709, 7090)*, Scientific Report No. 4 of Contract AF 19(604) 7378, M.I.T., Cambridge, Mass, 1962).

The subroutines in our spectral package are:

COSTAB   (cosine table)
SINTAB   (sine table)
COSP   (cosine or sine spectrum)
SPLIT   (split data into even and odd parts)
FASTFT   (fast Fourier transform)
DANWT   (Daniell weighting of autocorrelation)
ASPECT   (auto-spectrum)
ARCTAN   (arctangent)
XPHAZ   (cross phase spectra)

FT   (Fourier transform)
COSTR   (cosine transform)
SMOOTH   (smooth by Tukey-Hamming formula)
DRUM   (make phase curve continuous, as around a drum)
POLAR   (polar coordinates)
CAST   (cosine and sine transform)

Two basic subroutines are COSTAB, which generates a cosine table, and SIN-TAB, which generates a sine table. The programs are:

```
      CCOSTAB GENERATES FULL WAVELENGTH COS TABLE
    C     TABLE LENGTH=2*M—1 AND TABLE(1)=TABLE(2*M—1)=1.0.
          SUBROUTINE COSTAB(M,TABLE)
          DIMENSION TABLE(2)
          FM=M+M—2
          MM=M+M—1
          DO 10  I=1,MM
       10 TABLE(I)=COSF(FLOATF(I—1)*6.2831853/FM)
          RETURN
          END
```

```
      CSINTAB GENERATES FULL WAVELENGTH SINE TABLE
    C     TABLE LENGTH=2*M—1 AND TABLE(1)=TABLE(2*M—1)  =  0.0
          SUBROUTINE SINTAB(M,TABLE)
          DIMENSION TABLE(2)
          FM=M+M—2
          MM=M+M—1
          DO 10  I=1,MM
       10 TABLE(I)=SINF(FLOATF(I—1)*6.2831853/FM)
          RETURN
          END
```

The purpose of subroutine COSP is to compute the kth value of either a cosine transform or a sine transform. The calling statement is:

$$\text{CALL COSP (N, DATA, TABLE, M, K, C)}$$

The subroutine inputs are:

$$N = n + 1$$
$$DATA = x_0, x_1, x_2, \ldots, x_n$$

TABLE = either a full wavelength cosine table of length

$$2*M—1 \text{ with } TABLE(1) = TABLE(2*M—1) = 1.0,$$
or a full wavelength sine table of length
$$2*M—1 \text{ with } TABLE(1) = TABLE(2*M—1) = 0.0$$

$$M = m+1$$
$$K = k+1, \text{ where } 0 \leq k \leq m$$

The subroutine output is either

$$C = c_k \equiv \sum_{i=0}^{n} x_i \cos \left(\frac{\pi\, ik}{m} \right)$$

if TABLE is the cosine table, or

$$C = s_k = \sum_{i=0}^{n} x_i \sin \left(\frac{\pi\, ki}{m} \right)$$

if TABLE is the sine table. TABLE is conveniently generated by subroutine CØSTAB in the case of the cosine table, or by subroutine SINTAB in the case of the sine table. The program for subroutine CØSP is:

```
CCOSP COMPUTES (FOR A GIVEN K BETWEEN 1 AND M INCLUSIVE)
C          N
C     C=SUM(COSF(PI*(I—1)*(K—1)/(M—1))*DATA(I)).
C          I=1
C     INPUT IS A FULL WAVELENGTH COSINE TABLE OF LENGTH
C     2*M—1 FOR WHICH
C     TABLE(1)=TABLE(2*M—1)=1.0,
C     OR ALTERNATIVELY BY THE CORRESPONDING FULL WAVE-
C     LENGTH SIN TABLE
C     (THE ABOVE FORMULA THEN IS COMPUTED WITH SINE
C     REPLACING COSINE).
      SUBROUTINE COSP(N,DATA,TABLE,M,K,C)
      DIMENSION DATA(2), TABLE(2)
      J=1
      C=0.
      KK=K—1
      MM=M+M—1
      MMM=MM—1
      DO 20  I=1,N
      C=C+DATA(I)*TABLE(J)
      J=J+KK
      IF(J—MM)   20,20,10
   10 J=J—MMM
   20 CONTINUE
      RETURN
      END
```

The basic loop requires 1 floating multiplication, 1 floating addition, 1 fixed addition, and 1 fixed subtraction; this takes about 200 μ sec in FORTRAN on a 7090-1604 type machine.

The purpose of subroutine SPLIT is to split data into its even and odd parts. The program is:

```
C GIVEN DATA(I) FOR  I=1,N
CSPLIT FINDS FOR  I=1,N
C      EVEN(I)=.5*(DATA(N—1+I)+DATA(N+1—I))
C      ODD(I)=.5*(DATA(N—1+I)—DATA(N+1—I))
       SUBROUTINE SPLIT(N,DATA,EVEN,ODD)
       DIMENSION DATA(2),EVEN(2),ODD(2)
       DO 10  I=1,N
       K=N—1+I
       L=N+1—I
       EVEN(I)=.5*(DATA(K)+DATA(L))
       ODD(I)=.5 (DATA(K)—DATA(L))
   10  CONTINUE
       RETURN
       END
```

Subroutine FASTFT performs a fast Fourier transform. The program with
its explanation is:

```
CFASTFT COMPUTES FAST FOURIER TRANSFORM OF AN ODD NUMBER OF
C      POINTS WHERE
C      THE TIME ORIGIN IS AT THE MIDDLE POINT. FASTFT IS EXACT IN
C      THAT
C      CALLING IT TWICE YIELDS THE ORIGINAL DATA WITHIN A SCALE
C      FACTOR.
C      THE SUBROUTINE INPUTS ARE N AND
C      DATA(I)  I=1,2*N—1   (THE INPUT DATA)
C      THE SUBROUTINE OUTPUTS ARE
C      C(I)       I=1,N       (THE COSINE TRANSFORM ABOUT THE MIDDLE)
C      S(I)       I=1,N       (THE SINE TRANSFORM ABOUT THE MIDDLE,
C                             WHERE S(1) = S(N) = 0.0)
C      SP(I)      I=1,4*N—1 (WORKING SPACE)
       SUBROUTINE FASTFT(N,DATA,C,S,SP)
       DIMENSION DATA(2),C(2),S(2),SP(2)
       CALL SPLIT(N,DATA,SP(2*N),SP(3*N))
       CALL COSTAB(N,SP)
       DO 10  I=1,N
   10  CALL COSP(N,SP(2*N),SP,N,I,C(I))
       C(1)=C(1)/2.
       C(N)=C(N)/2.
       CALL SINTAB(N,SP)
       DO 20  I=1,N
   20  CALL COSP(N,SP(3*N),SP,N,I,S(I))
       RETURN
       END
```

The purpose of subroutine DANWT is to prepare an autocorrelation function
for subroutine COSP to compute a Daniell spectral estimate. The program
with its explanation is:

```
CDANWT TAKES AUTOCORRELATION OF HALF LENGTH N, DIVIDES AUTO(1)
C      BY 2.,
C      AND THEN WEIGHTS IT BY SINF(X)/X WHERE X=.5*(I—1)*PI/
C      (M—1),
C      AND THEN COLLAPSES IT LIKE AN ACCORDIAN TO LENGTH M.
```

```
C       AUTO IS THUS PREPARED FOR SR COSP TO COMPUTE DANIELL
C       ESTIMATE
        SUBROUTINE DANWT(N,AUTO,M)
        DIMENSION AUTO(2)
        AUTO(1)=AUTO(1)/2.
        FM=M—1
        SC=1.57079632/FM
        DO 10  I=2,N
        FI=I—1
   10   AUTO(I)=AUTO(I)*SINF(SC*FI)/(SC*FI)
        IF(M—N)  20,60,60
   20   L=—1
        K=M
        MP=M+1
        DO 50  I=MP,N
        K=K+L
        AUTO(K)=AUTO(K)+AUTO(I)
        IF(K—M)  30,40,40
   30   IF(K—1)   40,40,50
   40   L=—L
   50   CONTINUE
   60   RETURN
        END
```

The purpose of subroutine ASPECT is to compute the Daniell spectral estimate at M equi-spaced frequencies between zero and the folding (i.e. Nyquist) frequency inclusive. The program with its explanation is:

```
CASPECT COMPUTES  THE  DANIELL  ESTIMATE  AT  M  EQUISPACED
        FREQUENCIES
C       BETWEEN ZERO AND FOLDING FREQUENCY INCLUSIVE. THE
C       SPECTRAL
C       WINDOWS ARE SIDE BY SIDE (NO 50 PER SENT OVERLAP). INPUTS
C       ARE N,M, AND
C       DATA(I),I=1,N (SAMPLE FUNCTION). OUTPUT IS SPECT(I),I=1,M
C       (DANIEL
C       ESTIMATE). WORKING SPACE IS SN(I),I=1,N AND SM(I),I=1,2*M
        SUBROUTINE ASPECT(N,DATA,M,SPECT,SN,SM)
        DIMENSION DATA(2),SPECT(2),SN(2),SM(2)
        CALL CROSS(N,DATA,N,DATA,N,SN)
        CALL DANWT(N,SN,M)
        CALL COSTAB(M,SM)
        K=XMINoF(M,N)
        DO 10  I=1,M
   10   CALL COSP(K,SN,SM,M,I,SPECT(I))
        RETURN
        END
```

Subroutine ARCTAN computes the arctangent THETA of the ratio Y/X; the result lies in the correct quadrant between $-\pi$ and π. The program is:

```
        SUBROUTINE ARCTAN (X, Y, THETA)
        IF(X)  5,1,5
    1   IF(Y)  4,3,2
```

```
2       THETA  =  3.14159265/2.
        RETURN
3       THETA  =  0.
        RETURN
4       THETA  =  —3.14159265/2.
        RETURN
5       THETA  =  ATANF (ABSF(Y/X))
        IF(X)   9,6,6
6       IF(Y)   7,8,8
7       THETA  =  —THETA
8       RETURN
9       IF(Y)   11,10,10
10      THETA  =  3.14159265  —  THETA
        RETURN
11      THETA  =  —3.14159265  +  THETA
        RETURN
        END
```

Subroutine **XPHAZ** computes the cross phase spectra of two data series. The program with explanation is:

```
CXPHAZ COMPUTES CROSS PHASE SPECTRA OF TWO SERIES
C       INPUTS ARE  X(I),Y(I),I=1,N
C       OUTPUT IS PHASE(I),I=1,N,PHASE SPECTRA OF CROSS CORRELA-
C       TION OF X,Y
C       SPACE(I),I=1,8*N IS WORKING SPACE
        SUBROUTINE  XPHAZ(N,X,Y,PHASE,SPACE)
        DIMENSION  X(2),Y(2),PHASE(2),SPACE(2)
        CALL REVERS(N,X)
        CALL  FOLD(N,X,N,Y,LC,SPACE)
        CALL FASTFT(N,SPACE,SPACE(2*N),SPACE(3*N),SPACE(4*N))
        ICOS=2*N—1
        ISIN=3*N—1
        DO 10  I=1,N
        J=ICOS+1
        K=ISIN+1
10      CALL ARCTAN(SPACE(J),SPACE(K);PHASE(I))
        CALL REVERS(N,X)
        RETURN
        END
```

Subroutine **FT** computes one value of the Fourier transform by the sum of angles formulas for sine and cosine. The subroutine inputs are:

$$N = \text{length of DATA}$$

$$\text{DATA} = \text{series whose Fourier transform is desired}$$

$$W = \text{angular frequency at which the Fourier transform is evaluated}$$

and the subroutine outputs are:

$$S = \text{sine transform at angular frequency } W$$

$$C = \text{cosine transform at angular frequency } W$$

The program is:

```
CFT  COMPUTES ONE VALUE OF FOURIER TRANSFORM OF DATA AT ANG
C       FREQ W BY
C       MEANS OF SUM OF ANGLES FORMULAS FOR SINE AND COSINE.
C       OUTPUTS ARE
C         N—1                                    N—1
C       S=SUM(DATA(K+1)*SIN(W*K))AND C=SUM(DATA(K+1)*COS(W*K))
C         K=0                                    K=0
C       TIMING IS ABOUT (5.+1.2*N)/6. MILLISEC ON 7090
        SUBROUTINE  FT(N,DATA,W,S,C)
        DIMENSION  DATA(2)
        COSNW=1.
        SINNW=0.
        SINW=SINF(W)
        COSW=COSF(W)
        S=0.
        C=0.
        DO 10   I=1,N
        C=C+COSNW*DATA(I)
        S=S+SINNW*DATA(I)
        T=COSW*COSNW—SINW*SINNW
        SINNW=COSW*SINNW*SINW*COSNW
        COSNW=T
   10   CONTINUE
        RETURN
        END
```

Subroutine CØSTR computes the cosine transform of the autocorrelation;
it is also based on the sum of angles formula. The program with its explanation is:

```
CCOSTR COSINE TRANSFORM OF AUTOCOR(SHORT PROGRAM)
        SUBROUTINE COSTR(N,R,W,S)
CCOSTR          COMPUTES
C                     N—1
C       S = R(1)/2. + SUM(R(K+1)*COS(K*W)  )
C                     K=1
        DIMENSION  R(500)
        COSNW=1.
        SINNW=0.
        COSW=COSF(W)
        SINW=SINF(W)
        S=.5*R(1)
        DO 10   I=2,N
        T=COSW*COSNW—SINW*SINNW
        SINNW=  COSW*SINNW + SINW*COSNW
        COSNW=  T
   10   S=  S + R(I)*COSNW
        RETURN
        END
```

Subroutine SMOOTH computes the smoothed spectrum from the cosine
transform by the Tukey-Hamming formula.

```
      SUBROUTINE SMOOTH (SPECT, LS)
C     SMOOTH COMPUTES SMOOTHED SPECTRUM FROM THE
C     COSINE TRANSFORM USING TUKEY+HAMMING FORMULA.
C     THE SUBROUTINE INPUTS ARE
C        SPECT = UNSMOOTHED SPECTRUM
C        LS = ITS LENGTH
C     THE SUBROUTINE OUTPUT IS
C        SPECT = SMOOTHED SPECTRUM
      DIMENSION SPECT(2)
      MM=LS—1
      A=.54*SPECT(1)+.46*SPECT(2)
      B=.54*SPECT(LS)+.46*SPECT(MM)
      SJ=SPECT(1)
      SK=SPECT(2)
      DO 10   J=2,MM
      SI=SJ
      SJ=SK
      SK=SPECT(J+1)
   10 SPECT(J)=.54*SJ+.23*(SI+SK)
      SPECT(1)=A
      SPECT(LS)=B
      RETURN
      END
```

Subroutine DRUM makes a phase curve continuous; it is used by subroutine CAST. The program for DRUM is:

```
CDRUM MAKES PHASE CURVE CONTINUOUS
      SUBROUTINE DRUM(LPHZ,PHZ)
      DIMENSION PHZ(2)
      PJ=0.
      DO 40   I=2,LPHZ
      IF(ABSF(PHZ(I)+PJ—PHZ(I—1))—3.14159265)   40,40,10
   10 IF(PHZ(I)+PJ—PHZ(I—1))   20,40,30
   20 PJ=PJ+3.14159265*2
      GO TO 40
   30 PJ=PJ—3.14159265*2.
   40 PHZ(I)=PHZ(I)+PJ
      RETURN
      END
```

Subroutine PØLAR computes polar coordinates; it is also used by subroutine CAST. The program is:

```
CPOLAR COMPUTES POLAR COORDINATES
      SUBROUTINE POLAR(L,RE,XIM,AMP,PHZ)
      DIMENSION RE(2),XIM(2),AMP(2),PHZ(2)
      PI=3.14159265
      DO 110   I=1,L
      AMP(I)=SQRTF(RE(I)**2+XIM(I)**2)
      IF(XIM(I))   10,20,30
   10 IF(RE(I))   40,50,60
   20 IF(RE(I))   70,80,60
```

```
30    IF(RE(I))  90,100,60
40    PHZ(I)=ATANF(XI(I)/RE(I))—PI
      GO TO 110
50    PHZ(I)=—PI/2.0
      GO TO 110
60    PHZ(I)=ATANF(XIM(I)/RE(I))
      GO TO 110
70    PHZ(I)=—PI
      GO TO 110
80    PHZ(I)=0.0
      GO TO 110
90    PHZ(I)=ATANF(XIM(I)/RE(I))+PI
      GO TO 110
100   PHZ(I)=PI/2.0
110   CONTINUE
      RETURN
      END
```

Subroutine CAST computes the cosine and sine transform, and also the amplitude and phase spectra. The subroutine inputs are:

$$LW = \text{length of } W$$
$$W = \text{data whose transform is desired}$$
$$LT = \text{length of TCØS, TSIN, AMP, or PHZ}$$

The subroutine outputs are:

TCØS = cosine transform at equally spaced frequencies from zero to Nyqvist

TSIN = sine transform at equally spaced frequencies from zero to Nyqvist

AMP = amplitude spectrum at equally spaced frequencies from zero to Nyqvist

PHZ = phase spectrum at equally spaced frequencies from zero to Nyqvist

The program is:

```
CCAST IS COSINE AND SINE TRANSFORM
      SUBROUTINE  CAST(LW,W,LT,TCOS,TSIN,AMP,PHZ)
      DIMENSION  W(2),TCOS(3),TSIN(3),AMP(3),PHZ(3)
      DANG=3.14159265/FLOATF(LT—1)
      ANG=0.0
      DO 10  I=1,LT
      C=COSF(ANG)
      S=SINF(ANG)
      CALL  IPLYEV(LW,W,C,S,TCOS(I),TSIN(I))
10    ANG=ANG+DANG
      CALL  POLAR(LT,TCOS,TSIN,AMP,PHZ)
      CALL  DRUM(LT,PHZ)
      RETURN
      END
```

We conclude this report with Tables 3 through 10, which (except for Table 8) give some computed spectra. Table 8 gives spike filters whose spectra are given in Tables 9 and 10.

TABLE 3

Frequency in degrees	Amplitude spectrum	Phase-lag in degrees
0	0.25000	180
5	0.26835	161
10	0.31699	147
15	0.38430	138
20	0.46181	132
25	0.54473	129
30	0.63043	128
35	0.71737	127
40	0.80460	127
45	0.89148	128
50	0.97751	128
55	1.06234	130
60	1.14564	131
65	1.22717	132
70	1.30669	134
75	1.38400	136
80	1.45890	138
85	1.53121	139
90	1.60078	141
95	1.66745	143
100	1.73107	145
105	1.79152	147
110	1.84866	149
115	1.90238	152
120	1.95256	154
125	1.99911	156
130	2.04193	158
135	2.08093	160
140	2.11604	162
145	2.14718	165
150	2.17430	167
155	2.19733	169
160	2.21624	171
165	2.23099	173
170	2.24154	176
175	2.24788	178
180 (Nyqvist freq.)	2.25000	180

The amplitude and phase spectrum of the minimum-delay $b = (-1.25, 1.00)$. A minimum-delay wavelet has minimum phase-lag characteristic; the total phase-lag displacement for this wavelet from frequency 0 to 180 degrees is 180—180 = 0 degrees. The amplitude and phase spectrum of the reverse of this wavelet, namely $(1.00, -1.25)$ is given in the following table.

Table 4

Frequency in degrees	Amplitude spectrum	Phase-lag in degrees
0	0.25000	—180
5	0.26835	—156
10	0.31699	—137
15	0.38430	—123
20	0.46181	—112
25	0.54473	—104
30	0.63043	—98
35	0.71737	—92
40	0.80460	—87
45	0.89148	—83
50	0.97751	—78
55	1.06234	—75
60	1.14564	—71
65	1.22717	—67
70	1.30669	—64
75	1.38400	—61
80	1.45890	—58
85	1.53121	—54
90	1.60078	—51
95	1.66745	—48
100	1.73107	—45
105	1.79152	—42
110	1.84866	—39
115	1.90238	—37
120	1.95256	—34
125	1.99911	—31
130	2.04193	—28
135	2.08093	—25
140	2.11604	—22
145	2.14718	—20
150	2.17430	—17
155	2.19733	—14
160	2.21624	—11
165	2.23099	—8
170	2.24154	—6
175	2.24788	—3
180 (Nyqvist freq.)	2.25000	—0

The amplitude and phase spectrum of the maximum-delay wavelet $b = (1.00, -1.25)$. A maximum-delay wavelet has maximum phase-lag characteristic; the total phase-lag displacement for this wavelet from 0 to 180 degrees is $0 - (-180) = 180$ degrees. Note that the amplitude spectrum is the same as the amplitude spectrum of the wavelet $(-1.25, 1.00)$.

TABLE 5

Frequency	Amplitude spectrum	Phase-lag
0.000	3.00	0.00
0.025	2.99	0.05
0.050	2.96	0.10
0.075	2.92	0.15
0.100	2.86	0.20
0.125	2.79	0.25
0.150	2.71	0.30
0.175	2.61	0.34
0.200	2.49	0.39
0.225	2.37	0.42
0.250	2.23	0.46
0.275	2.09	0.49
0.300	1.94	0.51
0.325	1.78	0.52
0.350	1.62	0.52
0.375	1.47	0.50
0.400	1.32	0.45
0.425	1.19	0.38
0.450	1.09	0.28
0.475	1.02	0.15
0.500 (Nyqvist)	1.00	0.00

Amplitude-and-phase spectrum of minimum-delay wavelet (2, 1).

TABLE 6

Frequency F	Amplitude spectrum	Phase-lag	Sum of phase-lag	First differences
0.000	3.00	0.00	0.00	
0.025	2.99	0.10	0.15	0.15
0.050	2.96	0.20	0.30	0.15
0.075	2.92	0.31	0.46	0.16
0.100	2.86	0.42	0.62	0.16
0.125	2.79	0.52	0.77	0.15
0.150	2.71	0.63	0.93	0.16
0.175	2.61	0.75	1.09	0.16
0.200	2.49	0.86	1.25	0.16
0.225	2.37	0.98	1.40	0.15
0.250	2.23	1.10	1.56	0.16
0.275	2.09	1.23	1.72	0.16
0.300	1.94	1.37	1.88	0.16
0.325	1.78	1.51	2.03	0.15
0.350	1.62	1.67	2.19	0.16
0.375	1.47	1.85	2.35	0.16
0.400	1.32	2.05	2.50	0.15
0.425	1.19	2.28	2.66	0.16
0.450	1.09	2.54	2.82	0.16
0.475	1.02	2.83	2.98	0.16
0.500 (Nyqvist)	1.00	3.14	3.14	0.16

The first three columns give the amplitude-and-phase spectrum of the maximum-delay wavelet $(1, 2)$. The amplitude spectrum is the same as that of the maximum-delay wavelet $(2, 1)$ of the foregoing table. The fourth column is the sum of the phase-lags (at a given frequency) of the minimum-delay and maximum-delay wavelets. The fifth column gives the first differences of fourth column. It is seen that the sum of phaselags is equal to $2\pi f$.

TABLE 7

Frequency F	Amplitude spectrum	Phase-lag for min-delay wavelet	Phase-lag for max-delay wavelet	Sum of phase-lags	First difference
0.000	1.25	0.00	0.00	0.00	
0.025	1.24	0.06	0.25	0.31	0.31
0.050	1.21	0.12	0.50	0.62	0.31
0.075	1.16	0.17	0.76	0.93	0.31
0.100	1.10	0.21	1.03	1.24	0.31
0.125	1.03	0.24	1.32	1.56	0.32
0.150	0.95	0.25	1.63	1.88	0.32
0.175	0.87	0.23	1.96	2.19	0.31
0.200	0.81	0.18	2.33	2.51	0.32
0.225	0.76	0.10	2.72	2.82	0.31
0.250	0.75	0.00	3.14	3.14	0.32
0.275	0.76	—0.10	3.55	3.45	0.31
0.300	0.81	—0.18	3.95	3.77	0.32
0.325	0.87	—0.23	4.31	4.08	0.31
0.350	0.95	—0.25	4.65	4.40	0.32
0.375	1.03	—0.24	4.95	4.71	0.31
0.400	1.10	—0.21	5.24	5.03	0.32
0.425	1.16	—0.17	5.51	5.34	0.31
0.450	1.21	—0.12	5.77	5.65	0.31
0.475	1.24	—0.06	6.03	5.97	0.32
0.500 (Nyqvist)	1.25	—0.00	6.28	6.28	0.31

The second column gives the common amplitude spectrum of the wavelets (1.00, 0.00, 0.25) and (0.25, 0.00, 1.00). The third column gives the phase-lag of the minimum-delay wavelet (1.00, 0.00, 0.25) and the fourth column gives the phase-lag of the maximum-delay wavelet (0.25, 0.00, 1.00). The fifth column gives the sum of the phase-lags, which is equal to $4\pi f$.

TABLE 8

Spike at	Spike filter coefficients		
	a_0	a_1	a_2
$k = 0$	—0.70	—0.45	—0.22
$k = 1$	0.11	—0.56	—0.27
$k = 2$	0.14	0.29	—0.34
$k = 3$	0.17	0.36	0.56

Spike at	Actual output			
	c_0	c_1	c_2	c_3
$k = 0$	0.88	—0.14	—0.17	—0.22
$k = 1$	—0.14	0.82	—0.22	—0.27
$k = 2$	—0.17	—0.22	0.72	—0.34
$k = 3$	—0.22	—0.27	—0.34	0.56

Spike at	Desired output			
	d_0	d_1	d_2	d_3
$k = 0$	1.00	0.00	0.00	0.00
$k = 1$	0.00	1.00	0.00	0.00
$k = 2$	0.00	0.00	1.00	0.00
$k = 3$	0.00	0.00	0.00	1.00

The complete set of spike filters and their actual outputs for the minimum-delay input wavelet $(b_0, b_1) = (—1.25, 1.00)$.

TABLE 9

Frequency F   Phase-lag for spike filters for spike at

Frequency F	$k = 0$	$k = 1$	$k = 2$	$k = 3$
0.000	—3.14	—3.14	0.00	0.00
0.025	—3.04	—2.90	—0.55	0.21
0.050	—2.93	—2.66	—0.68	0.42
0.075	—2.84	—2.42	—0.64	0.64
0.100	—2.74	—2.19	—0.53	0.86
0.125	—2.65	—1.97	—0.38	1.08
0.150	—2.57	—1.75	—0.22	1.31
0.175	—2.50	—1.54	—0.04	1.55
0.200	—2.44	—1.34	0.14	1.81
0.225	—2.40	—1.15	0.33	2.09
0.250	—2.39	—0.96	0.53	2.39
0.275	—2.41	—0.79	0.74	2.72
0.300	—2.48	—0.63	0.96	3.11
0.325	—2.59	—0.48	1.19	3.54
0.350	—2.74	—0.35	1.43	4.00
0.375	—2.89	—0.23	1.68	4.46
0.400	—3.00	—0.14	1.95	4.89
0.425	—3.07	—0.07	2.23	5.27
0.450	—3.11	—0.02	2.52	5.63
0.475	—3.13	—0.00	2.83	5.96
0.500 (Nyqvist)	—3.14	—0.00	3.14	6.28

The phase-lag spectra for the spike filters of the foregoing table. Note how the phase-lag increases monotonically as the spike position moves from $k = 0$ to $k = 3$. The spike filter for $k = 0$ is minimum-delay; for $k = 1$ and $k = 2$, mixed-delay; and for $k = 3$, maximum-delay.

Table 10

Frequency F   Amplitude spectra for spike filters for spike at

	$k = 0$	$k = 1$	$k = 2$	$k = 3$
0.000	1.38	0.73	0.08	1.10
0.025	1.37	0.73	0.11	1.10
0.050	1.34	0.73	0.17	1.07
0.075	1.30	0.73	0.24	1.04
0.100	1.24	0.73	0.31	0.99
0.125	1.16	0.73	0.37	0.93
0.150	1.07	0.73	0.42	0.86
0.175	0.97	0.72	0.47	0.78
0.200	0.87	0.72	0.51	0.69
0.225	0.76	0.70	0.54	0.61
0.250	0.66	0.68	0.56	0.53
0.275	0.57	0.66	0.57	0.45
0.300	0.49	0.63	0.58	0.39
0.325	0.43	0.60	0.57	0.34
0.350	0.40	0.56	0.56	0.32
0.375	0.40	0.52	0.55	0.32
0.400	0.41	0.49	0.53	0.33
0.425	0.43	0.45	0.52	0.34
0.450	0.45	0.42	0.50	0.36
0.475	0.47	0.41	0.49	0.37
0.500 (Nyqvist)	0.47	0.40	0.49	0.38

The amplitude spectra for the spike filters of the foregoing two tables.

ACKNOWLEDGEMENT

I would like to thank Dr. Ulf Ericsson and Dr. Bo Jansson of the Research Institute of National Defence, Stockholm, Professor Markus Båth and Dr. Jon Claerbout of the Seismological Institute, Uppsala University, Professor Herman Wold of the Statistical Institute, Uppsala University, Dr. Sven Treitel of the Pan American Petroleum Corporation, Tulsa, and Professor Stephen M. Simpson of the Massachusetts Institute of Technology for their help and support.

CHAPTER 23

GIVEN THE POWER SPECTRUM $\Phi(\omega)$ OF THE TIME SERIES x_t $(-\infty < t < \infty)$ WITH $E[x_t] = 0$, WE WISH TO FIND THE LINEAR OPERATOR WITH OPERATOR COEFFICIENTS a_0, a_1, a_2, a_3, a_4 SUCH THAT THE PREDICTION ERROR VARIANCE $E\left[\xi_t^2\right]$ IS A MINIMUM WHERE THE PREDICTION ERROR IS GIVEN BY $\xi_t = a_0 x_t + a_1 x_{t-1} + a_2 x_{t-2} + a_3 x_{t-3} + a_4 x_{t-4}$

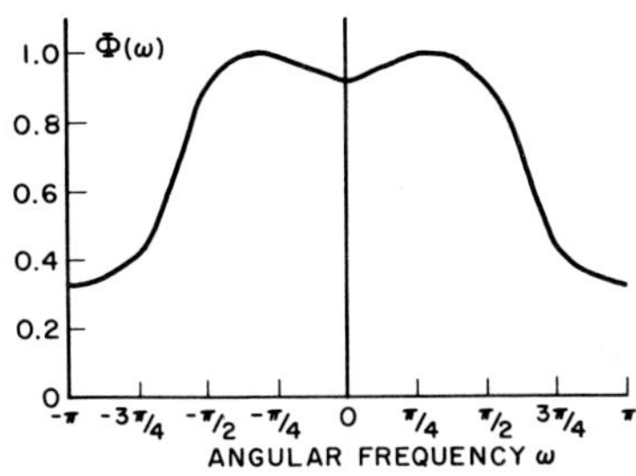

POWER SPECTRUM $\Phi(\omega)$

$$\Phi(\omega) = \frac{1}{r_0 + 2\sum_{\tau=1}^{4} r_\tau \cos \omega\tau}$$

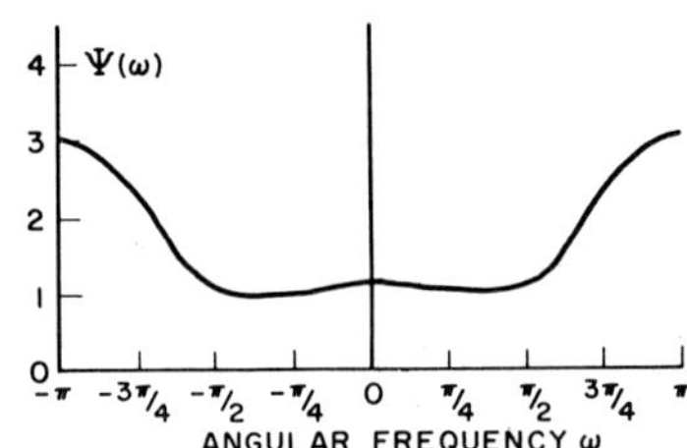

POWER TRANSFER FUNCTION $\Psi(\omega) = \dfrac{1}{\Phi(\omega)}$

$$\Psi(\omega) = |A(\omega)|^2 = r_0 + 2\sum_{\tau=1}^{4} r_\tau \cos \omega\tau = \sum_{\tau=-4}^{4} r_\tau e^{-i\omega\tau}$$

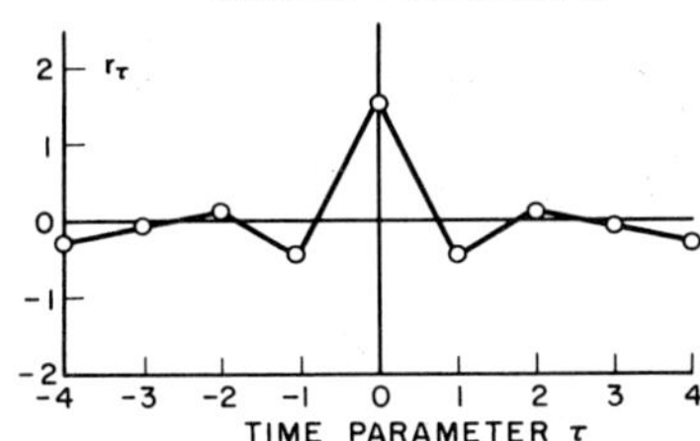

FOURIER TRANSFORM OF POWER TRANSFER FUNCTION

$$r_\tau = \frac{1}{2\pi} \int_{-\pi}^{\pi} \Psi(\omega) e^{i\omega\tau} d\omega$$

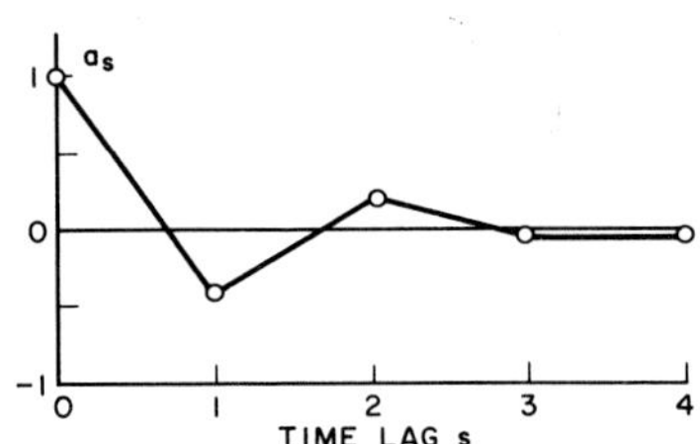

MINIMUM PHASE
STABLE OPERATOR COEFFICIENTS a_s

DETERMINED FROM SOLUTION OF THE SIMULTANEOUS EQUATIONS

$$r_0 = a_0^2 + a_1^2 + a_2^2 + a_3^2 + a_4^2$$

$$r_1 = a_0 a_1 + a_1 a_2 + a_2 a_3 + a_3 a_4$$

$$r_2 = a_0 a_2 + a_1 a_3 + a_2 a_4$$

$$r_3 = a_0 a_3 + a_1 a_4$$

$$r_4 = a_0 a_4$$

SUBJECT TO THE CONDITION THAT THE TRANSFER FUNCTION $A(\omega) = \sum_{t=0}^{4} a_t e^{-i\omega t}$ HAVE NO SINGULARITIES OR ZEROS IN THE LOWER HALF λ PLANE, WHERE $\lambda = \omega + i\sigma$

FIGURE 5   THE LEAST SQUARES LINEAR OPERATOR

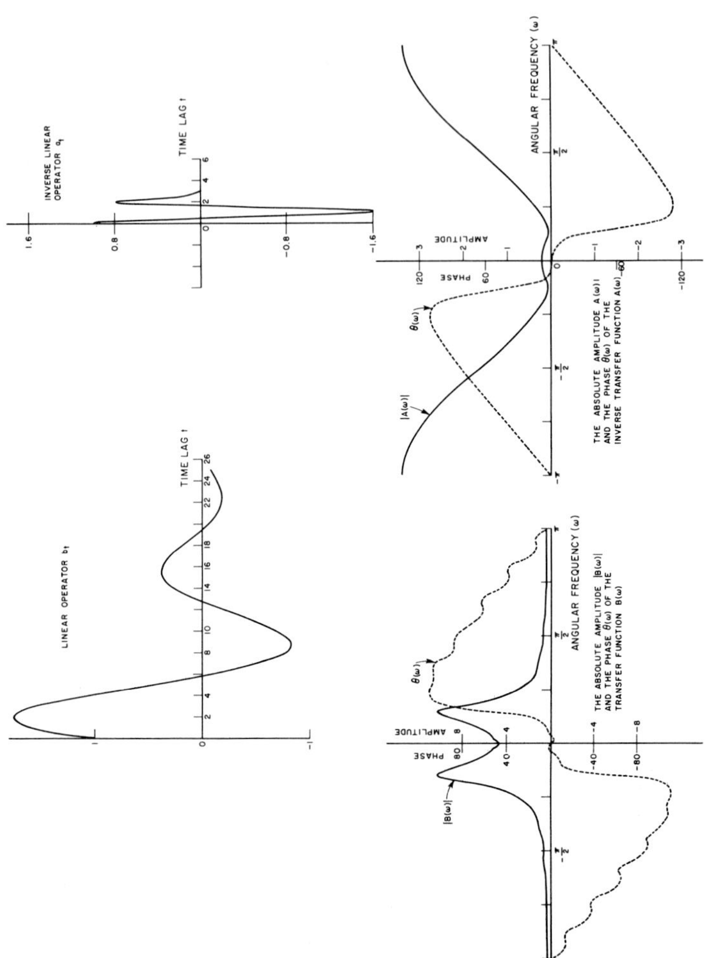

FIGURE 6   THE AUTOREGRESSIVE PROCESS $a_0 x_t + a_1 x_{t-1} + a_2 x_{t-2} = \xi_t$

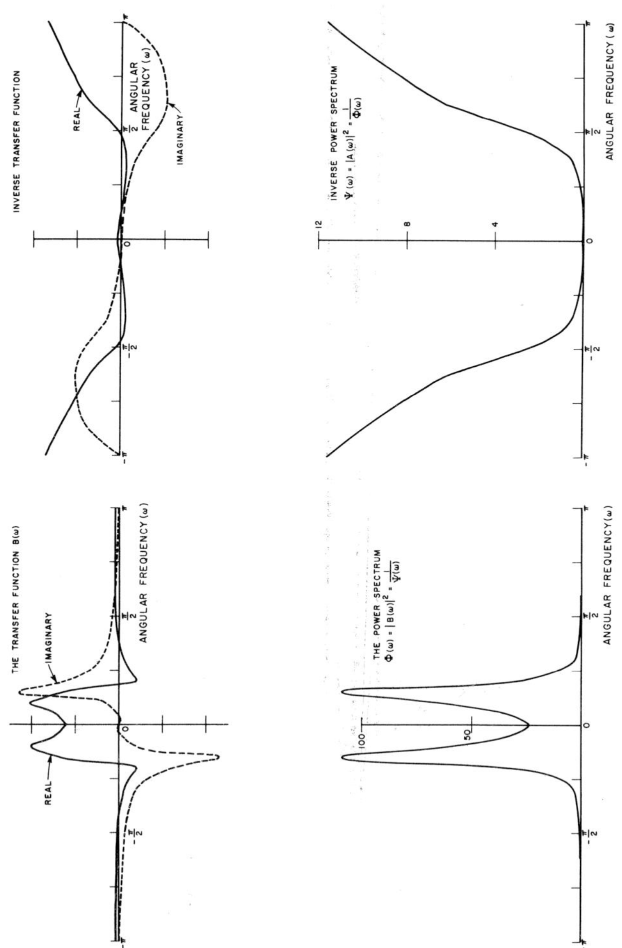

FIGURE 7 THE AUTOREGRESSIVE PROCESS $a_0 x_t + a_1 x_{t-1} + a_2 x_{t-2} = \xi_t$

ENDERS A. ROBINSON

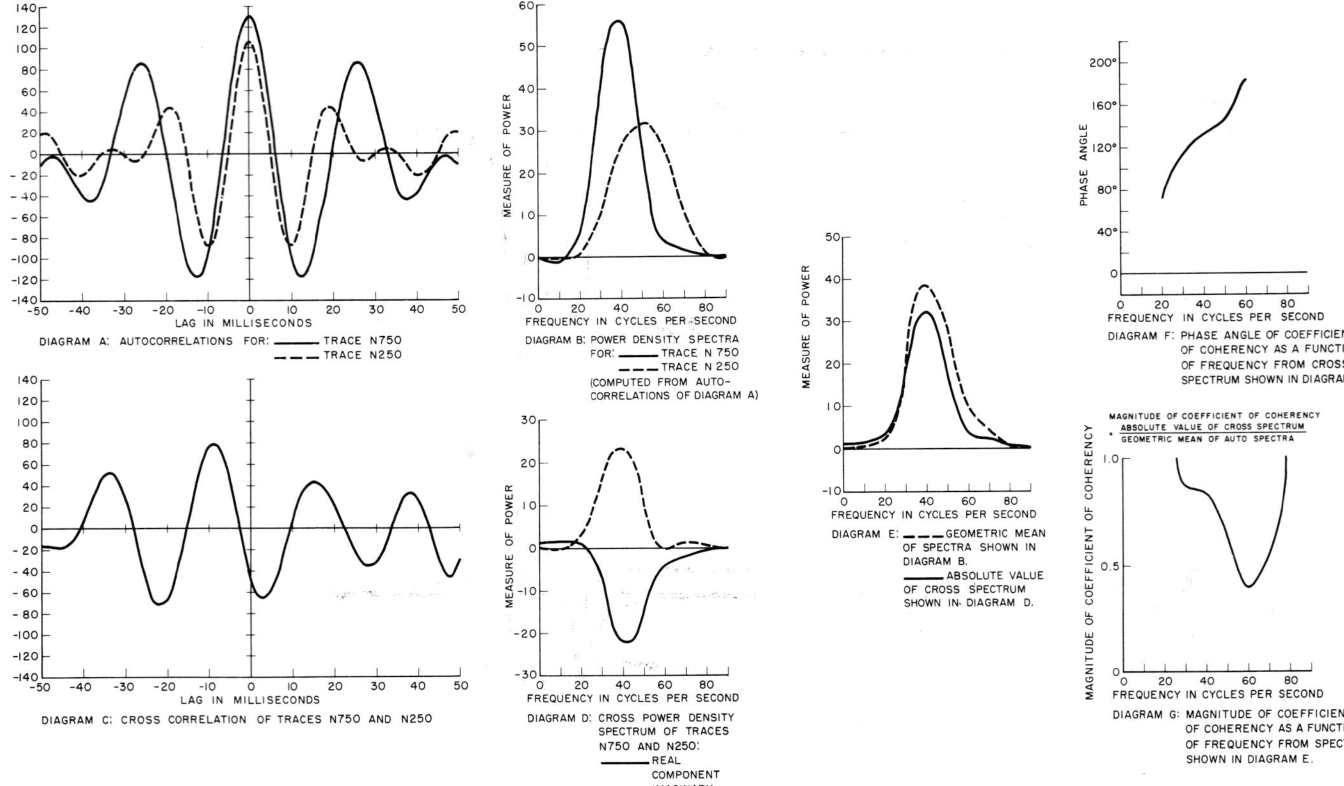

FIGURE 8 CORRELATION FUNCTIONS AND SPECTRA, ON M.I.T. RECORD NO. I FROM TIME EQUAL TO 1.05 SECONDS TO 1.225 SECONDS

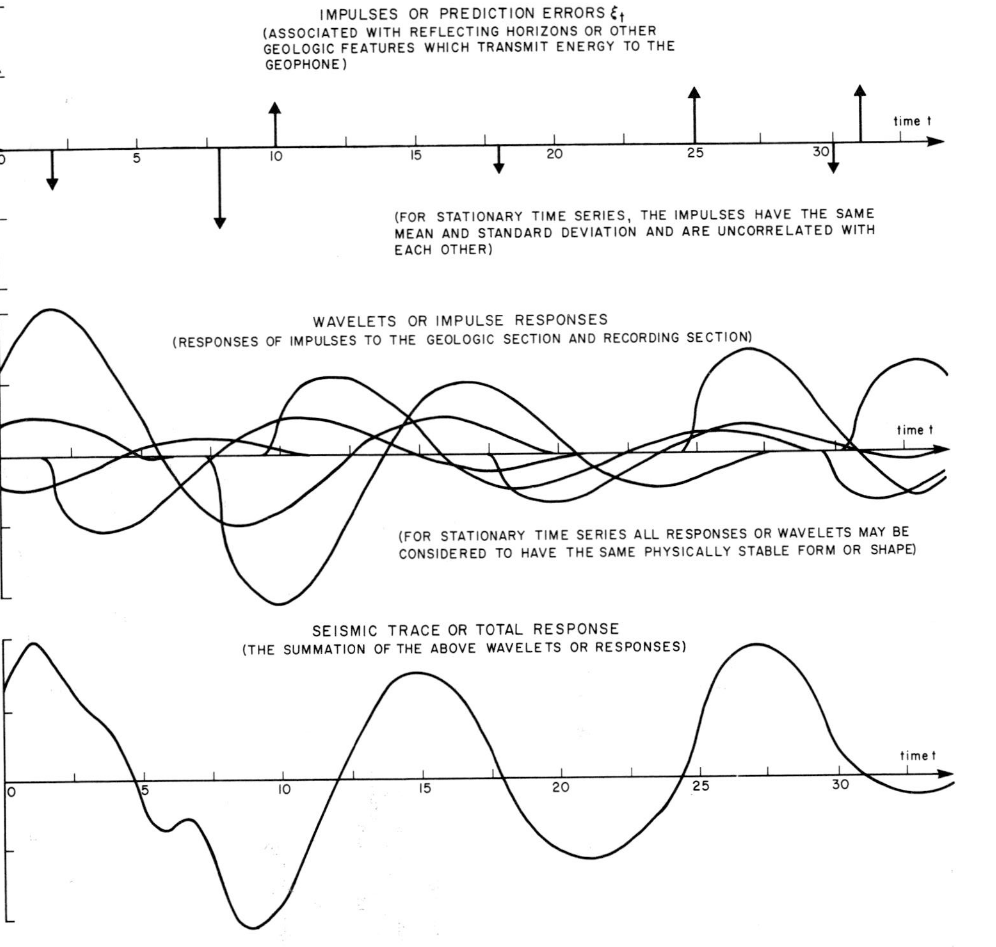

FIGURE 12   THE PREDICTIVE DECOMPOSITION OF A SEISMIC TRACE

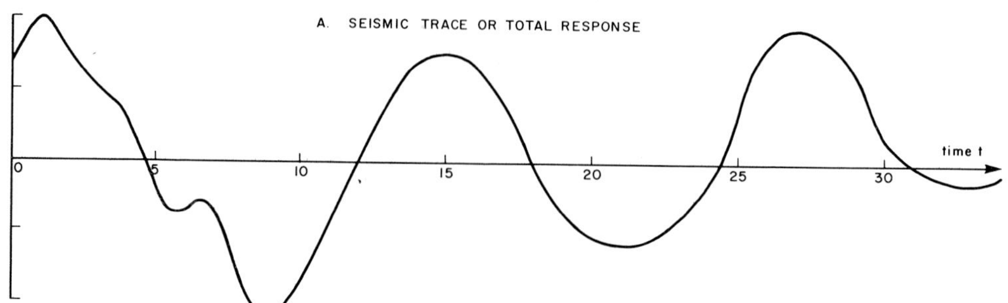

B. AUTOCORRELATION FUNCTION OF TRACE

(UNIQUELY DETERMINED FROM INFINITE TIME SERIES, BUT MAY BE ESTIMATED FROM FINITE SECTION OF TRACE)

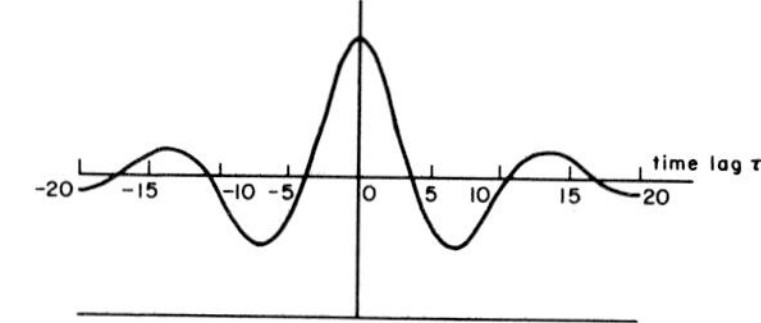

C. AUTOCORRELATION FUNCTION OF INDIVIDUAL WAVELET

(SAME AS AUTOCORRELATION FUNCTION OF TRACE)

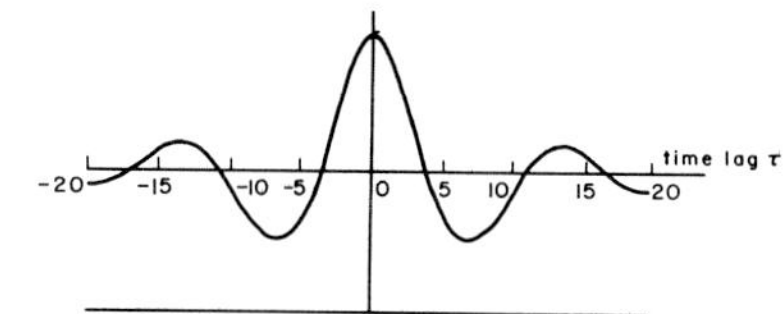

D. PHYSICALLY STABLE FORM OR SHAPE OF INDIVIDUAL WAVELET

(UNIQUELY DETERMINED FROM WAVELET AUTOCORRELATION BY METHOD OF THE FACTORIZATION OF THE SPECTRUM)

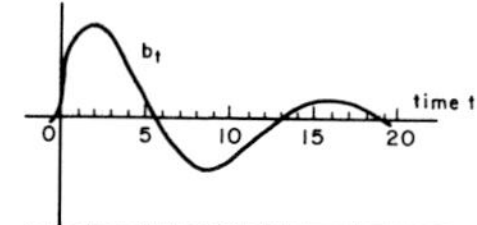

E. STABLE PREDICTION OPERATOR WITH MINIMUM PHASE CHARACTERISTIC

(UNIQUELY DETERMINED FROM WAVELET FORM)

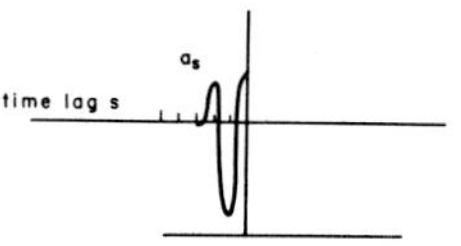

F. IMPULSE OR PREDICTION ERROR

YIELDED BY PREDICTION OPERATOR ACTING ON WAVELET. THIS IMPULSE OCCURS AT THE ARRIVAL TIME OF WAVELET

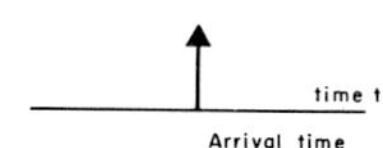

G. SERIES OF IMPULSES OR PREDICTION ERRORS

YIELDED BY PREDICTION OPERATOR ACTING ON SEISMIC TRACE (WHICH IS A SUM OF WAVELETS, WEIGHTED BY THE IMPULSE STRENGTHS). THE PREDICTION ERRORS OR IMPULSES ξ_t OCCUR AT THE ARRIVAL TIMES OF THE WAVELETS

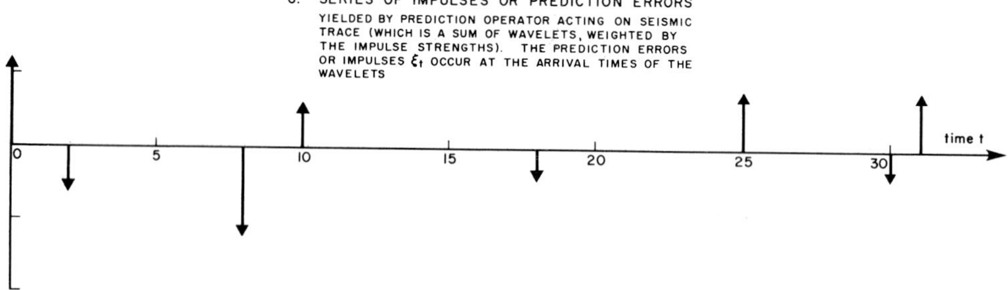

FIGURE 13   ANALYSIS OF A SEISMIC TRACE BY A PREDICTION OPERATOR

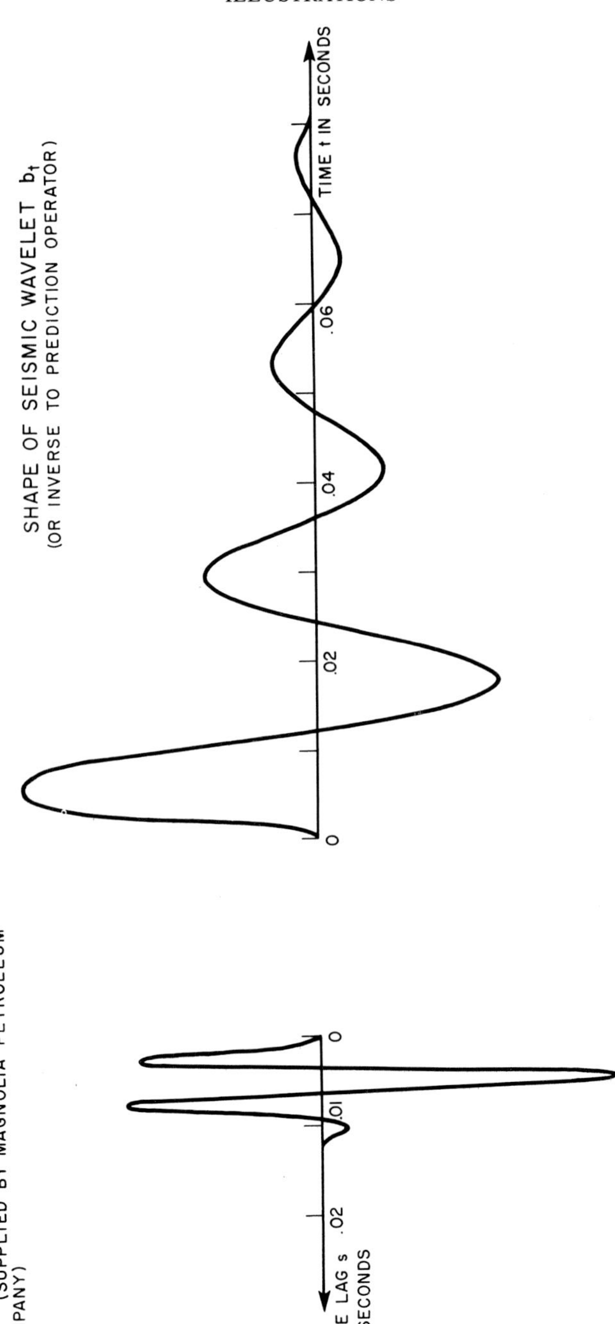

FIGURE 16   COMPUTATION OF WAVELET FROM SECTION OF TRACE ON M.I.T.
RECORD NO.1 (SUPPLIED BY THE MAGNOLIA PETROLEUM COMPANY)

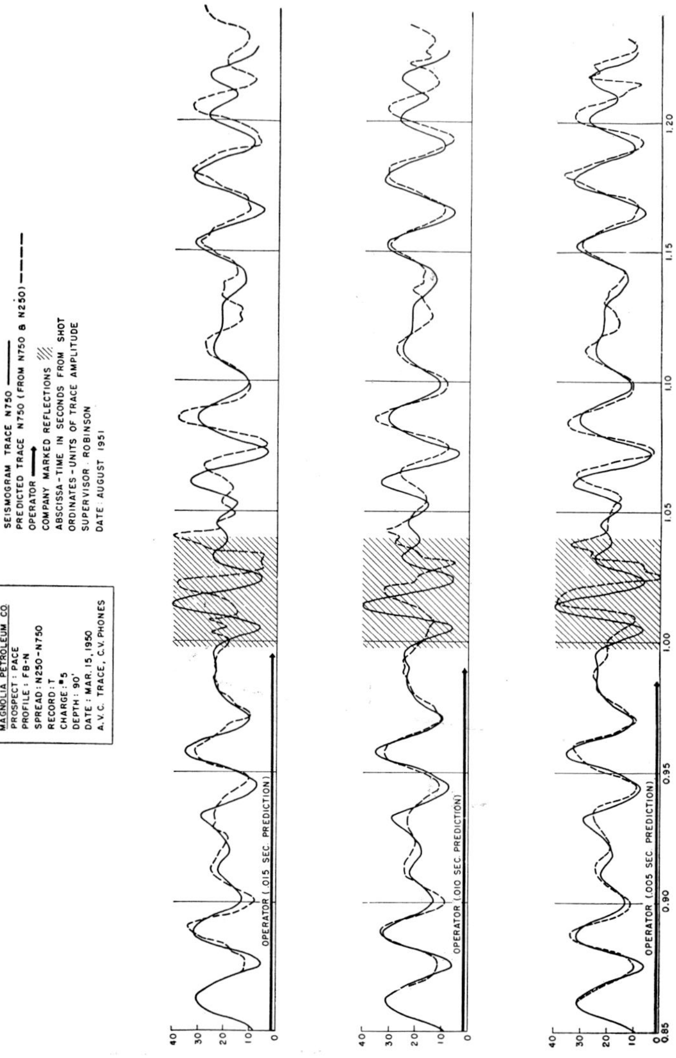

Fig. 5. Actual and predicted values of trace N750, MIT Record No. 1, for operators with various prediction distances.

SOURCES

1. Entretien de Monoco, Session 1964, Monoco, pp 124–162
2. Econometrica, Vol. 27, 1959, pp 679–684
3. Biostatistics in Pharmacology, Vol. 3 (A. L. Delaunois, ed.), Pergamon, 1979, pp 1229–1266
4. Econometric Model Building, North-Holland, Amsterdam, 1964, pp 37–106
5. Econometric Model Building, North-Holland, Amsterdam, 1964, pp 111–168
6. Institute Gulbenkian de Ciencia, Lisbon, 1967, pp 129–230
7. An Introduction to Infinitely Many Variates, Griffin, High Wycombe, 1959, pp 81–110
8. Mathematics Research Center, Madison, Wisconsin, no. 126, 1959, pp 1–15
9. Mathematics Research Center, Madison, Wisconsin, no. 76, 1959, pp 1–6
10. Proceedings of the American Mathematical Society, Vol. 11, 1960, pp 77–79
11. Time Series Analysis (M. Rosenblatt, ed.), Wiley, N.Y., 1963, pp 170–196
12. Arkiv for Matematik, Vol. 4, no. 28, 1961, pp 379–384
13. Journal of Mathematical Analysis and Applications, Vol. 6, 1963, pp 75–85
14. Teoriya Veroyatnostei i ee Primememiya, Moscow, Vol. 7(2), 1963, pp 201–211
15. Journal of Geophysical Research, Vol. 68, 1963, pp 5559–5567
16. Journal of Geophysical Research, Vol. 70, 1965, pp 1885–1891
17. Applications of Statistics (P. R. Krishnaiah, ed.), North-Holland, 1977, pp 447–460
18. Applied Time Series Analysis (D. F. Findley, ed.), Academic Press, 1978, pp 287–323
19. Geoexploration, Vol. 16, 1978, pp 55–73
20. Geoexploration, Vol. 16, 1978, pp 1–19
21. Topics in Applied Physics, Vol. 34 (S. Haykin, ed.), Springer-Verlag, 1979, pp 127–153
22. Geophysical Prospecting, Vol. 15, Supplement no. 1, 1966, pp 1–52
23. Predictive Decomposition of Time Series with Applications to Seismic Exploration, MIT Geophysical Analysis Group, no. 7, 1954

BOOKS BY ENDERS A. ROBINSON

1954 Predictive Decomposition of Time Series with Applications to Seismic Exploration
1959 An Introduction to Infinitely Many Variates
1962 Random Wavelets and Cybernetic Systems
1967 Statistical Communication and Detection with special reference to Digital Data Processing of Radar and Seismic Signals
1967 Multichannel Time Series Analysis with Digital Computer Programs
1967 Forecasting on a Scientific Basis (with H. Wold, G. Orcutt, D. Suits, P. de Wolf)
1969 The Robinson-Treitel Reader (3 editions) (with S. Treitel)
1978 Digital Signal Processing and Time Series Analysis (with M. T. Silvia)
1979 Deconvolution of Geophysical Time Series in the Exploration for Oil and Natural Gas (with M. T. Silvia)
1979 Digital Foundations of Time Series Analysis
    Volume 1. The Box-Jenkins Approach (with M. T. Silvia)
1980 Geophysical Signal Analysis (with S. Treitel)
1980 Physical Applications of Stationary Time Series
1980 University Course in Digital Seismic Methods used in Petroleum Exploration
1981 Digital Foundations of Time Series Analysis
    Volume 2. Wave-Equation Space-Time Processing (with M. T. Silvia)
1981 Time Series Analysis with Applications
1981 Least Squares Regression Analysis in Terms of Linear Algebra